Food Science Text Series

The Food Science Text Series provides faculty with the leading teaching tools. The Editorial Board has outlined the most appropriate and complete content for each food science course in a typical food science program and has identified textbooks of the highest quality, written by the leading food science educators.

For further volumes:
http://www.springer.com/series/5999

Vickie A. Vaclavik • Elizabeth W. Christian

Essentials of Food Science, 4th Edition

Vickie A. Vaclavik (Retired)
The University of Texas
Southwestern Medical Center
Dallas, Texas
USA

Elizabeth W. Christian
Department of Nutrition &
Food Science
Texas Women's University
Denton, Texas
USA

ISSN 1572-0330
ISBN 978-1-4614-9137-8 ISBN 978-1-4614-9138-5 (eBook)
DOI 10.1007/978-1-4614-9138-5
Springer New York Heidelberg Dordrecht London

Library of Congress Control Number: 2013953293

Printed on acid-free paper

Springer is part of Springer Science+Business Media (www.springer.com)

Preface

Hello. It is with great pleasure that we introduce *Essentials of Food Science* Fourth Edition!

The student of Food Science, Nutrition, Dietetics, Hospitality, and Culinary Arts enrolled in an introductory Food Science course may each benefit from working with *Essentials of Food Science*! This new edition continues to be designed to present principles of food science at an introductory level, with the non-major in mind. Appropriate chapters each include relevant parts in Nutritive Value as well as Food Safety of the commodity being discussed.

Bold, italicized words appearing in the text of each chapter are defined in a glossary at the completion of each respective chapter.

What is new: There are updates in each chapter—some significant. This better provides for internal consistency, clarification, and current thought in the field of food science. New to this edition of the book includes:

- Chapters covering Food Preservation, Food Additives, and Food Packaging are now part of a text part entitled Food Processing.
- The USDA Food Guide now becomes ChooseMyPlate.gov.
- After 100 years, *The American Dietetic Association* (ADA) has gotten a new name: *The Academy of Nutrition and Dietetics* (AND) (http://www.eatright.org).
- Also see: IFT's new campaign on food science, food facts, K-12 educational materials http://www.worldwithoutfoodscience.org.
- With the intent to further enrich student learning, there is, at the close of each chapter, a space to enter any additional "**Notes**:"as well as a "**CULINARY ALERT**!"

Thanks

Thank you to each textbook user for your feedback to the authors! We would like to express our appreciation for the review and input of Andres Ardisson Korat, M.S. Food Science, M.A. Gastronomy, a practicing Food Scientist for many years, currently redirecting his time in order to work on his Ph.D. in Nutrition. We are appreciative of those professionals who provided materials used throughout *Essentials of Food Science* to offer better explanations of the text material.

Thank you to the Lord for giving these authors great interest in food science and also the grace to meet each challenge in the process of writing!

For More Information

More information is available in texts relating to topics such as Food Chemistry, Food Engineering, Food Packaging, Food Preparation, Food Safety, Food Technology, Nutrition and Quantity Foods, Product Evaluation, and in references cited at the end of each chapter.

Enjoy!

Dallas, TX Vickie A. Vaclavik
Denton, TX Elizabeth W. Christian

About the Authors

V. A. Vaclavik, Ph.D., R.D., Retired. Dr. Vaclavik has taught for over 25 years at the college level in Dallas, TX. Included among her students are *nutrition* students at the Dallas County Community College District; *food science* and *management* students at The University of Texas Southwestern Medical Center at Dallas, Nutrition Department; and *culinary* students at the International Culinary School at the Art Institute of Dallas. She is a graduate of Cornell University, human nutrition and food; Purdue University, restaurant, hotel, institution management; and Texas Woman's University, institution management and food science.

She has been the lead author with Marjorie M. Devine Ph.D., Professor Emeritus, and Marcia H. Pimentel M.S., of *Dimensions of Food* since its third edition, having been a college student worker for the original edition. Her newest culinary text is *The Art of Nutritional Cuisine*, written with Amy C. Haynes, R.D. This book, *Essentials of Food Science*, written with Elizabeth W. Christian, is now in its fourth edition with two foreign translations.

Personally, she really likes passing on what she knows and enjoys. Prior to teaching and writing, Dr. Vaclavik worked in various foodservice operations—including hotel restaurants, Meals-on-Wheels, and more. Two of her three sons are married with children of their own!

Elizabeth Christian, Ph.D. Elizabeth Christian has been an adjunct faculty member at Texas Woman's University in Denton for 22 years, teaching both face-to-face and online classes in the Nutrition and Food Science department. Food Science has been her passion since she was a freshman in high school. She obtained her B.S. and her Ph.D. in Food Science from the Leeds University, England. After working for 5 years as a research scientist at the Hannah Dairy Research Institute in Scotland, she married an American and moved to the United States. Elizabeth and her husband currently live in Longview, TX, with their two daughters, who are in college.

Best wishes and God bless!

Contents

Part VI Baked Products

Part VII Aspects of Food Processing

Part I

Introduction to Food Components

Evaluation of Food Quality

1

Introduction

Food quality is an important concept because the food people choose depends largely on quality. Consumer preference is important to the food manufacturer, who wants to gain as wide a share of the market for the product as possible. Quality is difficult to define precisely, though it refers to the degree of excellence of a food and includes all the characteristics of a food that are significant and that make the food acceptable.

Whereas certain attributes of a food, such as nutritional quality, can be measured by chemical analysis, food acceptability is not easy to measure as it is very subjective. In fact, consumers make subjective judgments using one or more of the five senses every time they select or eat any food. For example, potato chips, celery, and some cereals have a crunchy sound when they are eaten; the taste and smell of foods can be highly appealing or unacceptable; and the appearance and feel of a food are also important in determining its acceptability.

Food quality must be monitored on a regular day-to-day basis to ensure that a uniform product is produced and that it meets the required quality control standards. Companies must also monitor the quality of their products during storage while changing ingredients and developing new lines. Objective tests using laboratory equipment are useful for routine quality control, yet they cannot measure consumer preference. The only sure way to determine what a population thinks about any food is to ask them! This is done using sensory testing and asking panelists to taste a food and give their opinion on it. Both sensory and objective tests are important in evaluating food quality, and ideally, they should correlate with or complement each other.

However, some consumer attributes can be correlated to laboratory measurements. For instance, LAB color charts can serve to produce foods within the range of acceptability for consumers. The same is true for texture, which can be correlated to viscosity for flowable products or breaking strength for hard products (more details in the "Texture" section). Flavor is the hardest attribute to predict via analytical measurements.

Aspects of Food Quality

Food quality has both subjective and nonsubjective aspects. Appearance, texture, and flavor are largely subjective attributes, whereas nutritional and bacterial quality are not. The last two qualities can be measured objectively, either by chemical analysis, by measuring bacterial counts, or using other specific tests (Sahin and Sumnu 2006). They will only be mentioned briefly in this chapter, and the subjective qualities will be discussed in detail.

Appearance

The appearance of a food includes its size, shape, color, structure, transparency or turbidity,

V.A. Vaclavik and E.W. Christian, *Essentials of Food Science, 4th Edition*, Food Science Text Series,
DOI 10.1007/978-1-4614-9138-5_1, © Springer Science+Business Media New York 2014

dullness or gloss, and degree of wholeness or damage. While selecting a food and judging its quality, a consumer takes these factors into account, as they are indeed an index of quality. For instance, the *color* of a fruit indicates how ripe it is, and color is also an indication of strength (as in tea or coffee), degree of cooking, freshness, or spoilage. Consumers expect foods to be of a certain color, and if they are not, it is judged to be a quality defect. The same is true for *size*, and one may choose large eggs over small ones, or large peaches over small ones, for example.

Structure is important in baked goods. For example, bread should have many small holes uniformly spread throughout, and not one large hole close to the top. *Turbidity* is important in beverages; for example, orange juice is supposed to be cloudy because it contains pulp, while white grape juice should be clear and without any sediment, which would indicate a quality defect.

Texture

Texture refers to those qualities of a food that can be felt with the fingers, tongue, palate, or teeth. Foods have different textures, such as crisp crackers or potato chips, crunchy celery, hard candy, tender steaks, chewy chocolate chip cookies, and creamy ice cream, to name but a few.

Texture is also an index of quality. The texture of a food can change as it is stored, for various reasons. If fruits or vegetables lose water during storage, they wilt or lose their turgor pressure, and a crisp apple becomes unacceptable and leathery on the outside. Bread can become hard and stale on storage. Products like ice cream can become gritty due to precipitation of lactose and growth of ice crystals if the freezer temperature is allowed to fluctuate, allowing thawing and refreezing.

Evaluation of texture involves measuring the response of a food when it is subjected to forces such as cutting, shearing, chewing, compressing, or stretching. Food texture depends on the rheological properties of the food (Bourne 1982). ***Rheology*** is defined as the science of deformation and flow of matter or, in other words, reaction of a food when a force is applied to it. Does it flow, bend, stretch, or break? From a sensory perspective, the texture of a food is evaluated when it is chewed. The teeth, tongue, and jaw exert a force on the food, and how easily it breaks or flows in the mouth determines whether it is perceived as hard, brittle, thick, runny, and so on. The term ***mouthfeel*** is a general term used to describe the textural properties of a food as perceived in the mouth.

Subjective measurement of texture gives an indirect evaluation of the *rheological* properties of a food. For example, a sensory panel might evaluate *viscosity* as the consistency or mouthfeel of a food. However, viscosity can be measured directly using a viscometer. Rheological properties are therefore discussed in more detail in section "Objective Evaluation" of this chapter.

> "Formulators typically turn to hydrocolloids first when trying to manipulate texture. It is important to remember that hydrocolloids vary significantly in their performance, price, ease of use and even impact on clean labeling" (Berry 2012)

Collectively known as *texturants*, some carbohydrate and proteins can impact the texture and mouthfeel of foods, while most likely contributing in a minor manner in terms of calories or flavor.

Research and Development (R&D) scientists now may use a blend of texturants in order to achieve the texture and mouthfeel desired in a specific food. "Our approach is to get ample information on the ideal end product. In many cases it is important to retain information on other stabilizing systems in the product, as well as any ingredients that may have a synergistic reaction with the texturant." (Berry 2012)

Flavor

Flavor is a combination of taste and smell and is largely subjective. If a person has a cold, food usually seems to be tasteless. Though it is not the taste buds that are affected, rather it is the sense of smell. Taste is detected by the taste buds at the tip, sides, and back of the tongue, whereas aromas are detected by the olfactory epithelium in the upper part of the nasal cavity. For any food to have an aroma, it must be volatile. These volatile substances can be detected in very small amounts (vanillin can be detected at a concentration of 2×10^{-10} mg/l of air).

Aroma is a valuable index of quality. A food will often smell bad before it looks bad, and old meat can be easily detected by its smell. (However, foods that are contaminated with pathogens may have no off-odor, so the absence of bad smell is not a guarantee that the food, such as meat, is safe to eat.)

The taste of a food is a combination of five major tastes—salt, sweet, sour, bitter, and ***umami***. It is complex and hard to describe completely. *Sweet* and *salt* tastes are detected primarily on the tip of the tongue, and so they are detected quickly, whereas *bitter* tastes are detected mainly by taste buds at the back of the tongue. It takes longer to perceive a bitter taste, and it lingers in the mouth; thus bitter foods are often described as having an aftertaste. *Sour* tastes are mainly detected by the taste buds along the side of the tongue.

Sugars, alcohols, aldehydes, and certain amino acids taste sweet to varying degrees. Acids (such as vinegar, lemon juice, and the many organic acids present in fruits) contribute the sour taste, saltiness is due to salts, including sodium chloride, and bitter tastes are due to alkaloids such as caffeine, theobromine, quinine, and other bitter compounds.

Umami is a taste that in recent times been added to the other four. It is a savory taste given by ingredients such as monosodium glutamate (MSG) and other flavor enhancers. The umami taste is significant in Japanese foods and in snack foods such as taco-flavored chips. "In the early 1900's ... merges the Japanese words for 'delicious and 'taste'." (Koetke 2013)

Taste Sensitivity

People vary in their sensitivity to different tastes. Sensitivity depends on the length of time allowed to taste a substance. Taste sensitivity varies greatly from person to person and sometimes even day to day for the same person, which makes it very difficult to measure flavor objectively. Sweet and salt tastes are detected quickly (in less than a second), because they are detected by taste buds on the tip of the tongue; in addition, they are usually very soluble compounds. Bitter compounds, on the other hand, may take a full second to be detected because they are detected at the back of the tongue. The taste may linger, producing a bitter aftertaste.

Sensitivity to a particular taste also depends on the concentration of the substance responsible for the taste. The ***threshold*** concentration is defined as the concentration required for identification of a particular substance. The threshold concentration may vary from person to person; some people are more sensitive to a particular taste than others and are, therefore, able to detect it at a lower concentration. Below the threshold concentration, a substance would not be identified yet may affect the perception of another taste. For example, *subthreshold* salt levels *increase* perceived sweetness and *decrease* perceived acidity, whereas subthreshold sugar concentrations make a food taste *less* salty than it actually is. Although it is not clear why, flavor enhancers such as MSG also affect taste sensitivity by intensifying a particular taste in a food.

Temperature of a food also affects its flavor. Warm foods generally taste stronger and sweeter than cold foods. For example, melted ice cream tastes much sweeter than frozen ice cream. There are two reasons for the effects of temperature on flavor. The volatility of substances is increased at higher temperatures, and so they smell stronger. Taste bud receptivity is also an important factor. Taste buds are most receptive in the region between 68 and 86 °F (20 and 30 °C), and so tastes will be more intense in this temperature range. For example, coffee that has cooled to room temperature tends to taste stronger and more bitter than very hot coffee.

CULINARY ALERT! Food tastes best when served at its optimum temperature. If chilled or frozen food is warmed to room temperature, or inversely, if hot food is cooled to room temperature, the flavor may be negatively impacted.

Psychological factors also affect taste sensitivity and perception. Judgments about flavor are often influenced by preconceived ideas based on the appearance of the food or on previous experience with a similar food. For example, strawberry-flavored foods would be expected to be red. However, if colored green, because of the association of green foods with flavors such as lime, it would be difficult to identify the flavor as strawberry unless it was very strong. Color intensity also affects flavor perception. A stronger color may cause perception of a stronger flavor in a product, even if the stronger color is simply due to the addition of more food coloring!

Texture can also be misleading. A thicker product may be perceived as tasting richer or stronger simply because it is thicker, and not because the thickening agent affects the flavor of the food. Other psychological factors that may come into play when making judgments about the flavor of foods include time of day (for example, certain tastes are preferred at breakfast time), general sense of well-being, health, and previous reactions to a particular food or taste.

Sensory Evaluation

Sensory evaluation has been defined as a scientific method used to evoke, measure, analyze, and interpret those responses to products as perceived through the senses of sight, smell, touch, taste, and hearing (Stone et al. 2012). This definition has been accepted and endorsed by sensory evaluation committees within both the Institute of Food Technologists and the American Society for Testing and Materials. For more detailed information on sensory evaluation, the reader is referred to the books by Lawless and Heymann (2010), and by Stone et al. (2012).

Sensory testing utilizes one or more of the five senses to evaluate foods. *Taste panels*, comprising groups of people, taste specific food samples under controlled conditions and evaluate them in different ways depending on the particular sensory test being conducted. This is the *only* type of testing that can measure consumer preference and acceptability. When it comes to public opinion of a product, there is no substitute for tasting by individual consumers.

In addition to a taste-panel evaluation, objective tests can be established that correlate with sensory testing, which give an indication of consumer acceptability, although this may not always be sufficient. In the development of *new* foods or when changing an *existing* product, it is necessary to determine consumer acceptance directly, and objective testing is not sufficient, even though it may be a reliable, objective indication of food quality.

Sensory methods may be used to determine:

1. Whether foods differ in taste, odor, juiciness, tenderness, texture, and so on
2. To what extent foods differ
3. To ascertain consumer preferences and to determine whether a certain food is acceptable to a specific consumer group

Three types of sensory testing are commonly used, each with a different goal. *Discrimination or difference tests* are designed to determine whether there is a difference between products; *descriptive tests* determine the extent of difference in specific sensory characteristics; and *affective or acceptance/preference tests* determine how well the products are liked, or which products are preferred. There are important differences between these three types of tests. It is important to select the appropriate type of test so that the results obtained are able to answer the questions being asked about the products and are useful to the manufacturer or product developer.

The appropriate tests must be used under suitable conditions, in order for results to be interpreted correctly. All testing must be carried out under controlled conditions, with controlled lighting, sound (no noise), and temperature to minimize distractions and other adverse psychological factors.

Sensory Testing Procedure

Sensory testing is carried out by members of a taste panel, preferably in individual testing booths under controlled conditions. All distractions, bias, and adverse psychological factors must be minimized so that the evaluation is truly an evaluation of the sample being tested, and not a reaction to adverse circumstances, cultural prejudice, or the opinions of other testers. The noise level must be controlled to avoid distractions, temperature and humidity should be within an acceptable range, and lighting within the booth must also be monitored. In addition, there should be no extraneous smells, which may distract people from making judgments about the product under test.

Since color has a significant effect on subjective evaluation of a product, color differences may need to be masked. This is achieved by using red lights in the booths when necessary. It is important that people rate samples that may have different color intensities on *flavor* and not simply on the fact that they *look* different. For example, two brands of cheese puffs may look different because one is a deeper shade of orange than the other, and so one could tell the difference between them simply because of their color. However, there may not be a difference in the taste. If the color difference is masked by conducting the tests under red light, any differences detected could then be attributed to flavor differences, and not to color differences.

The samples are usually placed on a tray and passed to each panelist through a hatch in front of the testing booth. The tray should contain a ***ballot*** that gives specific instructions on how to evaluate the samples and a place for the panelist's response. A cracker and water are provided, in order to cleanse the palate before tasting the samples. It is important that tasters have not eaten spicy or highly flavored food before tasting food samples, or their judgment may be impaired. Preferably, panelists should not have eaten anything immediately prior to carrying out a taste test.

Additionally, it is important that panelists cannot identify the products they are tasting and that they do not know which sample is the same as their neighbor's sample, so that there is no room for bias in the results. This is accomplished by assigning three-digit random numbers to each product. For example, if two products are being tested (denoted product 1 and 2), each product is given at least two different random numbers. Panelists sitting next to each other will not be given samples with the same number, so that they cannot compare notes and agree with each other and introduce bias into the results that way.

If two products are being tested, 50 % of the panelists must test product 1 first, and the rest must test product 2 first; the order of testing must be randomized. This eliminates bias due to sample order, and also due to any changes in experimental conditions that may occur from the beginning to the end of the test. The specific product order and random numbers seen by each panelist are detailed on a ***master sheet*** to ensure that the test design is carried out correctly.

Sensory Tests

Discrimination or difference tests are used to determine if there is a perceivable difference between products. Such tests would be used if a company was changing the source of one of its ingredients or substituting one ingredient for another. Difference tests can also be used to see if the quality of a product changes over time or to compare the shelf life of a particular product packaged in different packaging materials. For example, a difference test could be used to determine if juices keep their flavor better when stored in glass bottles rather than in plastic ones.

A *small* group of panelists may be used to conduct such tests and they may be trained to recognize and describe the differences likely to occur in the products being tested. For example, if trained panelists are testing different tea blends or flavor bases, they have more experience than an average consumer in recognizing particular flavors associated with such products, and they are more sensitive to differences, and are able to describe them better. This is partly because they have been trained to identify such flavors.

However, they are likely to be experienced tea drinkers (or tea connoisseurs) with a liking for different teas before they are trained for taste-panel work. Such people may be employees of the company doing the testing or members of a university research group. They would be expected to detect small differences in the product flavor that would go unnoticed by most of the general population. Thus, their evaluation would be important in trying to keep a tea blend constant or in determining if there is a significant flavor difference when the source of an ingredient is changed.

It may also be important to know if small differences in a product can be detected by untrained consumers, who simply like the product and buy it on a regular basis. For this reason, difference tests are often conducted using larger panels of untrained panelists.

Two of the most frequently used difference tests are the triangle test and the duo–trio test. Typical ballots for these tests are given in Figs. 1.1 and 1.2. These ballots and the one shown in Fig. 1.3 were developed at the sensory evaluation laboratory at Texas Woman's University, Denton, Texas by Dr. Clay King, in conjunction with Coca-Cola® Foods. The ballots have been used for consumer testing of beverages and other foods at the university sensory facility.

In the ***triangle test***, each panelist is given three samples, two of which are alike, and is asked to indicate the *odd* sample. The panelists are asked to taste the samples from left to right, cleansing their palate before each sample by taking a bite of cracker and a sip of water. Then they circle the number on the ballot sheet that corresponds to the sample they believe to be different. If they cannot tell, they must guess. Statistics are applied to the results to see if there is a significant difference among the products being tested.

Since the panelists have to guess which is the odd one if they cannot detect a difference, one third of them would pick the correct sample as being odd just by guessing. Therefore, more than one third of the panelists must choose the correct answer for there to be a significant difference among the products.

For example, if there are 60 members on a taste panel, 27 would need to choose the correct answer for the results to be significant at a probability level of 0.05, and 30 correct answers would be needed for significance at a probability level of 0.01. A probability level (or ***p value***) of 0.05 means that out of 100 trials, the same result would be obtained 95 times, indicating 95 % confidence that the result is valid. A probability of 0.01 is equivalent to 99 % confidence in the significance of the results, because the same result would be expected in 99 out of 100 trials. Statistical tables are available to determine the number of correct answers required for significance at different probability levels.

In the ***duo–trio test***, each panelist is given a reference and two samples. He or she is asked to taste the reference first, and then each sample, working from left to right, and circle on the ballot the sample that is the same as the reference. Again, if a panelist cannot tell which sample is the same as the reference, he or she must guess, and statistics must be applied to the results to determine whether there is a significant difference among the products. If everyone guessed, 50 % of the panelists would get the correct answer, and so for the results to be significant, more than 50 % must choose the correct answer. For a panel of 60 people, 40 must give the correct answer for the results to be significant at the 0.01 probability level. Again, tables are available to determine if results are statistically significant (Roessler et al. 1978).

Affective, ***acceptance***, or ***preference*** tests are used to determine whether a specific consumer group likes or prefers a particular product. This is necessary for the development and marketing of new products, as no laboratory test can tell whether the public will accept a new product or not. A *large* number of panelists, representing the general public, must be used; thus, consumer testing is expensive and time-consuming. A relevant segment of the population needs to test the

Fig. 1.1 Ballot for triangle sensory test (obtained from Dr. Clay King at the Sensory Testing Laboratory at Texas Woman's University, Denton, Texas)

TEST#_____ **Panelist#_____**

TRIANGLE DIFFERENCE TEST

PRODUCT_____

INSTRUCTIONS: Proceed when you are ready. (Quietly so as not to distract others.)

FOR EACH SAMPLE:

1) Take a bite of the cracker and a sip of water to rinse your mouth.

2) Two of the samples are the same and one is different. **CIRCLE** the **ODD** sample. If you can not tell, guess.

_____ _____ _____

3) Describe the reason why the ODD sample is DIFFERENT. (Please be specific.)

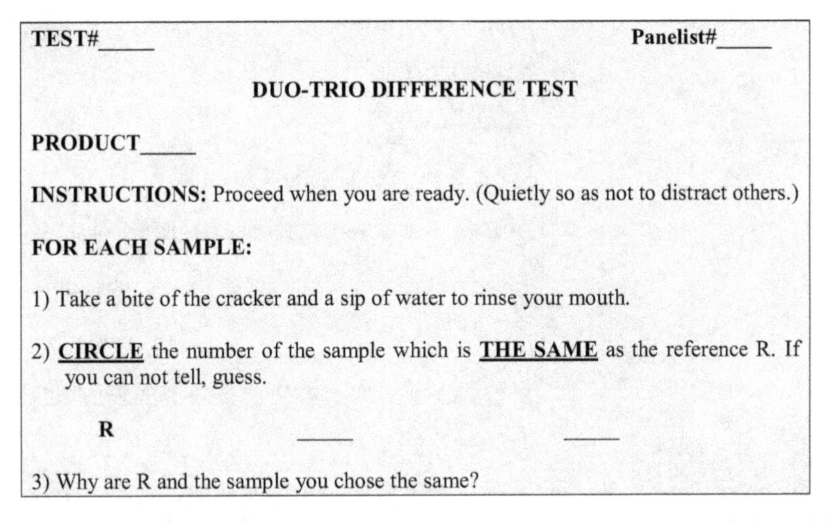

TEST#_____ **Panelist#_____**

DUO-TRIO DIFFERENCE TEST

PRODUCT_____

INSTRUCTIONS: Proceed when you are ready. (Quietly so as not to distract others.)

FOR EACH SAMPLE:

1) Take a bite of the cracker and a sip of water to rinse your mouth.

2) **CIRCLE** the number of the sample which is **THE SAME** as the reference R. If you can not tell, guess.

R _____ _____

3) Why are R and the sample you chose the same?

Fig. 1.2 Ballot for duo–trio sensory test (obtained from Dr. Clay King at the Sensory Testing Laboratory at Texas Woman's University, Denton, Texas)

product. For example, if it is being aimed at over-50s, senior citizens must make up the taste panel, and not mothers with young children. The opposite would apply if the product was aimed at young children. (Products aimed at children would have to be acceptable to mothers as well, because they would be the ones to buy it.) Ethnic products must be tested either by the group for which they are aimed, or by a wide cross section of the public if the aim is to introduce the products to a broader market than is currently interested.

Panelists are not trained for this type of sensory testing. All that is required from them is that they give their opinion of the sample(s). However, they are normally screened to make sure that they are users or potential users of the product to be tested. Typically, they are asked to fill out a *screening sheet* and answer questions about how much they like the product (or similar products), and how often they consume it. Anyone who does not like the product is asked not to take the test. The screening sheet may also ask for demographic information, such as gender and age range of the panelists. The specific questions for each screening sheet are determined by whoever sets up the test, based on the consumer group they aim to target with their product.

The simplest preference tests are ***ranking tests***, where panelists are given two or more samples and asked to rank them in order of preference. In the *paired preference* test, panelists

Fig. 1.3 Ballot for likeability and paired preference sensory tests (obtained from Dr. Clay King at the Sensory Testing Laboratory at Texas Woman's University, Denton, Texas)

TEST#____ **Panelist#____**

LIKEABILITY RATING AND PAIRED PREFERENCE TEST

PRODUCT____

INSTRUCTIONS: Proceed when you are ready. (Quietly so as not to distract others.) Evaluate one sample at a time, working from top to bottom.

FOR EACH SAMPLE:

1) Take a bite of the cracker and a sip of water to rinse your mouth.

2) Taste the sample then **CIRCLE** the number which best expresses your opinion of the sample.

SAMPLE CODE:____

Likeability 1 2 3 4 5 6 7 8 9

Scale Dislike Extremely Like Extremely

SAMPLE CODE:____

Likeability 1 2 3 4 5 6 7 8 9

Scale Dislike Extremely Like Extremely

Describe the DIFFERENCES between the two samples. (Please be specific.)

Taste the samples again, then circle the one you prefer.

____ ____

Describe the reasons why you prefer the one you chose.

are given two samples and asked to circle the one they prefer. Often, the panelists are asked to taste a sample and score it on a 9-point hedonic scale from "dislike extremely" to "like extremely". This type of test is called a ***likeability test***.

Sometimes panelists are asked to test more than one sample, to score each on the 9-point likeability scale, and then to describe the differences between the samples. This would not be a difference test, as differences in this case are usually obvious, and the point of the test is to see which product is preferred. In fact, the differences may be considerable. An example might be comparison of a chewy brand of chocolate chip cookies with a crunchy variety. The difference is obvious, although consumer preference is not obvious and would not be known without carrying out preference tests on the two products. A paired preference or ranking test may be included on a same ballot and carried out along with a likeability test. An example of a typical ballot is given in Fig. 1.3.

Descriptive tests are usually carried out by a small group of highly trained panelists. They are specialized difference tests, where the panelists are not simply asked whether they can determine differences between the two products, but rather, are asked to *rate* particular aspects of the flavor of a particular product on a scale. Flavor aspects vary depending on the type of product being studied. For example, flavor notes in tea may be bitter, smoky, and tangy, whereas flavor notes in yogurt may be acid, chalky, smooth, and sweet. A descriptive "flavor map" or profile of a product is thus developed. Any detectable changes in the product would result in changes in the flavor map.

The training required to be able to detect, describe, and quantify subtle changes in specific flavor notes is extensive. Therefore, establishment of such panels is costly. When trained, the panelists function as analytical instruments, and their evaluation of a product is not related to their like or dislike of it. The descriptive taste panel work is useful to research and development scientists, because it gives detailed information on the types of flavor differences between products.

Objective Evaluation

Objective evaluation of foods involves *instrumentation* and use of physical and chemical techniques to evaluate food quality. ***Objective testing*** uses equipment to evaluate food products instead of variable human sensory organs. Such tests of food quality are essential in the food industry, especially for routine quality control of food products.

An objective test measures one particular attribute of a food rather than the overall quality of the product. Therefore, it is important to choose an objective test for food quality that measures a key attribute of the product being tested. For example, orange juice is both acidic and sweet; thus, suitable objective tests for this product would be measurement of pH and measurement of sugar content. These tests would be of no value in determining the quality of a chocolate chip cookie. A suitable test for cookie quality might include moisture content or the force required to break the cookie.

There are various objective tests available for monitoring food quality. Fruits and vegetables may be graded for size by passing them through apertures of a specific size. Eggs are also graded in this manner, and consumers may choose among six sizes, including small, large, or jumbo-sized eggs. Flour is graded according to particle size, which is required to pass through sieves of specific mesh size.

Color may be measured objectively by several methods, ranging from simply matching the product to colored tiles to using the **Hunterlab color and color difference meter**. The color meter measures the intensity, chroma, and hue of the sample and generates three numbers for the sample under test. Thus, small changes in color can be detected. This method of color analysis is appropriate for all foods. For liquid products, such as apple juice, a **spectrophotometer** can be used to measure color. A sample is placed in the machine and a reading is obtained, which is proportional to the color and/or the clarity of the juice.

Food Rheology

Many objective methods for measurement of food quality involve measurement of a specific aspect of texture, such as hardness, crispness, or consistency. As mentioned already, texture is related to the rheological properties of food, which determine how it responds when subjected to forces such as cutting, shearing, or pulling.

Rheological properties can be divided into three main categories. A food may exhibit *elastic* properties, *viscous* properties, or *plastic* properties, or a combination. In reality, rheological properties of most foods are extremely complex and they do not fit easily into one category.

Elasticity is a property of a solid and could be illustrated by a rubber band or a coiled spring.

If a force or *stress* is applied, the material will deform (stretch or be compressed) in proportion to the amount of force applied, and when the force is removed, it will immediately return to its original position. If sufficient force is applied to a solid, it will eventually break. The force required to break the material is known as the *fracture stress*.

Various solids are more elastic than others; examples of very elastic solids are springs and rubber bands. Bread dough also has elastic properties, although its rheology is complex, and includes viscous and plastic components as well. All solid foods exhibit elastic properties to some degree.

Viscosity is a property of a liquid and could be illustrated by a piston and cylinder (or a dashpot), or by a syringe.

Viscosity is a measure of the resistance to flow of a liquid when subjected to a shearing force. The thicker the liquid, the greater is its viscosity or resistance to flow. For example,

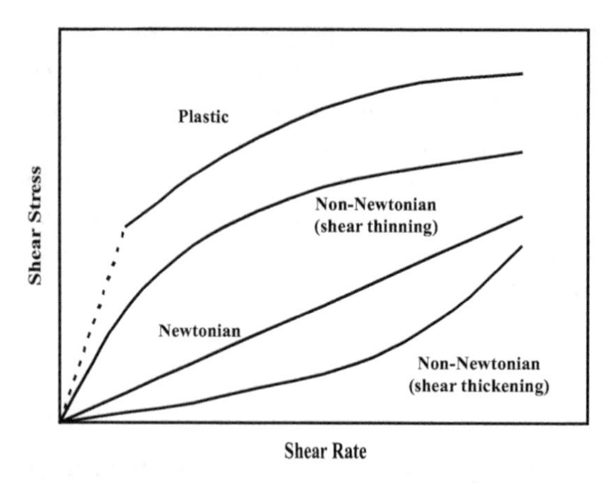

Fig. 1.4 Schematic representation of flow behavior of Newtonian and non-Newtonian liquids [modified from Bowers, 1992]

water has a low viscosity and flows readily, whereas catsup is considered "thick," has a higher viscosity, and flows relatively slowly.

Liquids can be separated into ***Newtonian*** and ***non-Newtonian*** fluids. In the case of a Newtonian liquid, the shear stress applied to the fluid is proportional to the shear rate or shear velocity of the flowing liquid. This means that the viscosity is independent of the shear rate. Therefore, viscosity will be the same, even if the viscometer used to measure it is operated at different speeds. A graph of shear stress against shear rate would give a straight line, and the viscosity could be calculated from the gradient of the line (see Fig. 1.4). The steeper the line, the greater the resistance to flow, and the greater the viscosity of the liquid.

Examples of Newtonian liquids/liquids include water, sugar syrups, and wine. However, most liquid/fluid foods are non-Newtonian, in which case the consistency or apparent viscosity depends on the amount of shear stress applied. This can be seen with catsup, which appears fairly solid, and is hard to get out of the bottle if it has been standing for a while. However, after shaking (applying a shear stress), the catsup becomes almost runny and will flow out of the bottle much more easily. If the bottle is again left to stand, the consistency of the undisturbed catsup will be regained after a short period of time. Shaking the bottle causes the molecules to align so that they flow over each other more easily, and the apparent viscosity decreases. A graph of shear stress against shear rate would not give a straight line for catsup, since the apparent viscosity is not constant. (Strictly speaking, the term "apparent viscosity" should be used for non-Newtonian liquids, whereas the term "viscosity" should be reserved for Newtonian liquids.)

A number of non-Newtonian liquids appear to get thicker when a shear stress is applied. In this case, the particles in the liquid tend to aggregate and trap pockets of liquid, thus making it harder for the molecules to flow over each other. Examples of such liquids include starch slurries and dilute solutions of a few gums, such as alginates, carboxymethylcellulose, and guar gum.

The viscosity of both Newtonian and non-Newtonian liquids is affected by temperature. Higher temperatures cause liquids to flow more readily, thus decreasing viscosity, whereas lower temperatures cause an increase in viscosity. For this reason, it is important to make measurements of viscosity at a constant temperature and to specify that temperature.

A ***plastic*** substance can be molded, because it contains a liquid, although only after a certain minimum force (the *yield stress*) is applied. At forces below the yield stress, it behaves as an elastic solid, yet above the yield stress, it behaves as a liquid. Examples of plastic substances include modeling clay, and foods such as warm chocolate, and hydrogenated vegetable shortenings that can be creamed easily.

Certain foods exhibit both elastic and viscous properties at the same time. They are termed *viscoelastic*. Bread dough is a good example of a viscoelastic material. When a force is applied, the material first deforms like an elastic solid, then it starts to flow. When the force is removed, it only partly regains its original shape.

The rheological properties of a food affect its texture and sensory properties. For example, brittleness, shortness, and hardness are related to the fracture stress of a solid food, whereas thickness and creaminess are related to the consistency or apparent viscosity of a liquid food. The rheological properties of many foods can be modified by adding stabilizers such as gums. These are added to increase viscosity, which in turn restricts movement of everything in the system and may delay undesirable changes, such as precipitation of solids or separation of emulsions.

Objective Measurement of Texture

Many objective methods for measurement of food quality involve measurement of some aspect of texture. For example, viscometers are used to measure viscosity or consistency of foods ranging from thin liquids such as oil to thick sauces such as catsup. The sophistication of these instruments also varies widely. The **Bostwick consistometer** is a simple device that involves filling a reservoir with the sample to be tested. A stopwatch is started, the gate holding the product in the reservoir is lifted, and the product is timed to flow a certain distance along the consistometer trough. At the other end of the scale, **Brookfield viscometers** are sophisticated instruments that may be used to measure viscosity under controlled temperature and when the sample is subjected to shearing forces of different magnitudes.

The **Instron Universal Testing Machine** has various attachments that allow it to measure different aspects of texture, including compressibility of bread and the force required to break a cookie or to shear a piece of meat.

The **Brabender amylograph** (Chap. 4) is an instrument that was developed to measure the viscosity of starch mixtures as they are heated in water. Another instrument with a very specific use is the mixograph, which is used to measure the ease of mixing of bread doughs.

Sophisticated equipment, such as the **mass spectrophotometer**, **gas chromatography**, and **high-performance liquid chromatography** equipment, are available in research and analytical laboratories for analysis of specific products or components.

The list of equipment used in the food industry for evaluating food quality would fill a complete textbook! Certain principles must be emphasized when considering objective tests to evaluate the quality of a food product:

- The objective test must be appropriate for the food product being tested. In other words, it must measure an attribute of the food that has a major effect on quality.
- Ideally, the objective test results should correlate with sensory testing of similar food products to make sure that the test is a reliable index of quality of the food.
- Most objective tests used to assess food quality are empirical; that is, they do not measure an absolute property of the food. However, the results are still meaningful, as long as instruments are calibrated with materials that have similar properties to the foods under test.
- Objective tests include all types of instrumental analysis, including laboratory tests to determine chemical composition, nutrient composition, and bacterial composition.
- Objective tests are repeatable and are not subject to human variation. If the equipment is properly maintained and is used correctly, it should give reliable results from day to day.

Objective tests are necessary to identify contaminants in foods and to reveal faulty processing methods as well as testing for deterioration such as rancidity in fats and oils. Objective tests are essential for routine quality control of foods and food products. However, they must correlate with sensory testing, because no single objective test can measure the overall acceptability of a specific food or food product.

An in-depth study of analysis of foods by objective methods is beyond the scope of this book. For more information, the reader is referred to Food Analysis by Nielsen (2010) and to the many other textbooks available on the subject.

Comparison of Subjective and Objective Evaluation

Both *sensory evaluation* and *objective evaluation* of food quality are essential in the food industry in order to routinely monitor food quality and to ensure that the foods being produced are acceptable to consumers. The two methods of evaluation complement each other.

Sensory testing is expensive and time-consuming, because many panelists are required to test a single product in order for the results to be meaningful. On the other hand, objective testing is efficient and, after the initial purchase of the necessary equipment, relatively inexpensive. One person can usually perform an objective test on many samples in a day, whereas it may take a day to perform a complete sensory test on one or two samples. Objective tests give repeatable results, whereas sensory tests may give variable results due to variation of human responses and opinions.

While sensory evaluation gives a judgment of the overall acceptability of a product, an objective method of evaluation is only able to measure one aspect of the food, and this may not always be sufficient to determine whether the quality of the product is acceptable. The only true judge of acceptability of a food product is a consumer! Therefore, objective tests must correlate with sensory tests to give a reliable index of food quality.

Objective tests are essential for routine quality control of food products. However, sensory evaluation is essential for product research and development. Only consumers can tell whether there is a perceivable difference in a product when the formulation or packaging is changed, and only consumers can determine whether a new product is acceptable or preferred over another brand.

Subjective vs. objective analysis—overview

Subjective/sensory analysis	Objective analysis
Uses individuals	Uses equipment
Involves human sensory organs	Uses physical and chemical techniques
Results may vary	Results are repeatable
Determines human sensitivity to changes in ingredients, processing, or packaging	Need to find a technique appropriate for the food being tested
Determines consumer acceptance	Cannot determine consumer acceptance unless correlated with sensory testing
Time-consuming and expensive	Generally faster, cheaper, and more efficient than sensory testing
Essential for product development and for marketing of new products	Essential for routine quality control

Conclusion

Food quality can be defined as the degree of excellence of a food and includes factors such as taste, appearance, and nutritional quality, as well as its bacteriological or keeping quality. Food quality goes hand in hand with food acceptability, and it is important that quality is monitored, both from a food safety standpoint and to ensure that the public likes a particular product and will continue to select it. Both sensory and objective methods are important in evaluation of food quality and the two methods complement one another. Sensory analysis is essential for development of new products, because only consumers can tell whether they like a product or not. However, objective testing is also important, especially for routine quality control of food products.

Notes

CULINARY ALERT!

Glossary

Affective or acceptance/preference tests Used to determine whether a specific consumer group likes or prefers a particular product.

Ballot Sheet of paper on which the panelist receives pertinent sample information and instructions, and on which observations are recorded during a sensory test.

Descriptive tests Specialized difference tests used to describe specific flavor attributes of a product, or to describe degree of difference between products.

Discrimination or difference tests Used to determine if there is a perceivable difference between samples.

Duo–trio test Samples include a reference food and two samples, one of which is the same as the reference.

Elasticity Ability of a material to stretch when a force is applied and to return to its original position when the force is removed.

Likeability test Panelists rate a sample on a hedonic scale from "dislike extremely" to "like extremely."

Master sheet Details the specific three-digit product numbers and positions for every panelist in a sensory test. Used to ensure that each product is seen an equal number of times in each position, so that bias is avoided.

Mouthfeel Textural qualities of a food as perceived in the mouth.

Newtonian liquid The viscosity is independent of the shear rate. Stirring or shaking does not make the liquid runnier or thicker. Examples are water, sugar syrups, and wine.

Non-Newtonian liquid/fluid Apparent viscosity depends on the shear rate. Catsup gets thinner with increasing shear rate, whereas some gums thicken with increasing shear rate.

Objective evaluation Involves use of physical and chemical techniques to evaluate food quality, instead of variable human sensory organs.

Plasticity Material flows when subjected to a certain minimum force; material can be molded.

***p*-Value** Statistical probability that a result is significant. A p value of 0.01 indicates 99 % confidence that a result is significant. In other words, out of 100 trials, the same result would be expected 99 times. The probability of the opposite result occurring is only 1 in 100 trials.

Ranking test Panelists rank two or more samples in order of preference or intensity for a particular attribute.

Rheology Science of the deformation and flow of matter, how a food reacts when force is applied; includes elasticity, viscosity, and plasticity.

Sensory testing Use of senses to evaluate products; involves consumer opinion.

Threshold Concentration required for identification of a particular substance.

Triangle test Three samples, two of which are alike, one is odd.

Umami Savory taste, given by substances such as monosodium glutamate.

Viscosity Resistance to flow of a liquid when a shear force is applied. Liquids with a low viscosity flow readily, whereas liquids with a high viscosity flow slowly.

References

Berry D (2012) Targeting texture. Food Product Design, pp 22–31

Bourne ML (1982) Food texture and rheology. Academic, New York

Bowers J (1992) Characteristics of food dispersions. In: Bowers J (ed) Food theory and applications, 2nd edn. pp 30, MacMillan, New York

Koetke C (2013) Umami's mysteries explained. Food Product Design, pp 62–68

Lawless HT, Heymann H (2010) Sensory evaluation of food. Principles and practices, 2nd edn. Springer, New York

Neilsen SS (2010) Food analysis, 4th edn. Springer, New York

Roessler EB, Pangborn RM, Sidel JL, Stone H (1978) Expanded statistical tables for estimating significance in paired-preference, paired-difference, duo–trio and triangle tests. J Food Sci 43:940–942

Sahin S, Sumnu SG (2006) Physical properties of foods: what they are and their relation to other food properties. In: Peleg M, Bagley EB (eds) Physical properties of foods. Springer, New York

Stone H, Bleibaum R, Thomas H (2012) Sensory evaluation practices, 4th edn. Academic, San Diego

Water

2

Introduction

Water is abundant in all living things and, consequently, is in almost all foods, unless steps have been taken to remove it. It is essential for life, even though it contributes no calories to the diet. Water also greatly affects the texture of foods, as can be seen when comparing grapes and raisins (dried grapes), or fresh and wilted lettuce. It gives crisp texture or turgor to fruits and vegetables, and it also affects perception of the tenderness of meat. For some food products, such as potato chips, salt, or sugar, lack of water is an important aspect of their quality, and keeping water *out* of such foods is important to maintain quality.

Almost all food processing techniques involve the use of water or modification of water in some form: freezing, drying, emulsification (trapping water in droplets or trapping oil in a water phase to give salad dressings their characteristic mouthfeel), breadmaking, thickening of starch, and making pectin gels are a few examples. Further, because bacteria cannot grow without water, the water content has a significant effect on maintaining quality of the food. This explains why freezing, dehydration, or concentration of foods increases shelf life and inhibits bacterial growth.

Water is important as a solvent or dispersing medium, dissolving small molecules to form true solutions, and dispersing larger molecules to form colloidal solutions. Acids and bases ionize in water; water is also necessary for many enzyme catalyzed and chemical reactions to occur, including hydrolysis of compounds such as sugars. It is also important as a heating and cooling medium and as a cleansing agent.

Since water has so many functions that are important to a food scientist, it is important to be familiar with some of its unique properties. When modifying the water content of a food, it is necessary to understand these functions in order to predict the changes that are likely to occur during processing of such foods.

Drinking water is available to the consumer in convenient bottled and aseptic containers in addition to the tap.

Chemistry of Water

The chemical formula for water is H_2O. Water contains strong ***covalent bonds*** that hold the two hydrogen atoms and one oxygen atom together. The oxygen can be regarded to be at the center of a tetrahedron, with a bond angle of 105° between the two hydrogen atoms in *liquid water* and a larger angle of 109° 6′ between the hydrogens in *ice* (Fig. 2.1).

The bonds between oxygen and each hydrogen atom are polar bonds, having a 40 % partial ionic character. This means that the outer-shell electrons are unequally shared between the

V.A. Vaclavik and E.W. Christian, *Essentials of Food Science, 4th Edition*, Food Science Text Series,
DOI 10.1007/978-1-4614-9138-5_2, © Springer Science+Business Media New York 2014

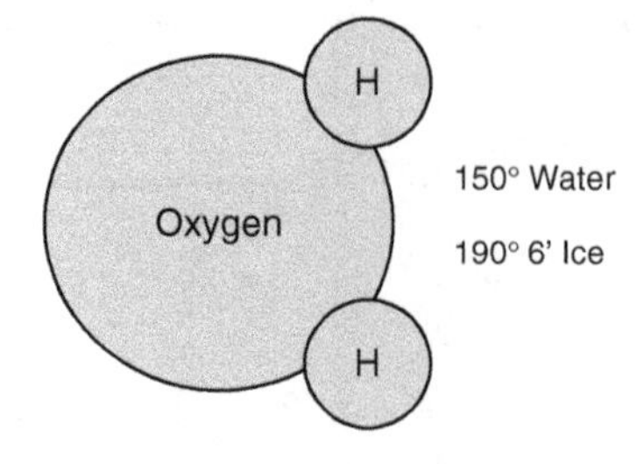

Fig. 2.1 Bond angle of water and ice

oxygen and hydrogen atoms, the oxygen atom attracting them more strongly than each hydrogen atom. As a result, each hydrogen atom is slightly positively charged and each oxygen atom is slightly negatively charged. Therefore they are able to form ***hydrogen bonds***.

A hydrogen bond is a *weak* bond between polar compounds where a hydrogen atom of one molecule is attracted to an electronegative atom of another molecule (Fig. 2.2). It is a weak bond relative to other types of chemical bonds such as covalent or ionic bonds, but it is very important because it usually occurs in large numbers and, therefore, has a significant cumulative effect on the properties of the substance in which it is found. Water can form up to four hydrogen bonds (oxygen can hydrogen bond with two hydrogen atoms).

Water would be expected to be gas at room temperature if compared with similar compounds in terms of their positions in the periodic table, yet due to the many hydrogen bonds it contains, it is liquid. Hydrogen bonds between hydrogen and oxygen are common, not just between water molecules, although between many other types of molecules that are important in foods, such as sugars, starches, pectins, and proteins.

Due to its V-shape, each molecule of water can form up to four hydrogen bonds with its nearest neighbors. Each hydrogen atom can form one hydrogen bond, and the oxygen atom can form two, which results in a three-dimensional lattice in ice. The structure of ice—frozen water, is dynamic, and hydrogen bonds are continually breaking and reforming between different water molecules. Liquid water also contains hydrogen bonds and, therefore, has a variety of ordered structures that are continually changing as hydrogen bonds break

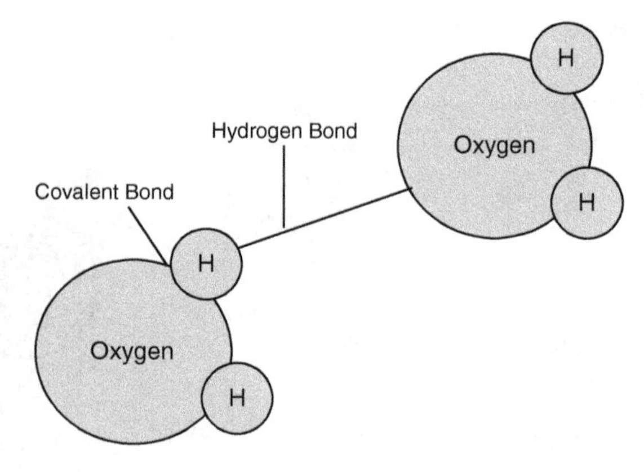

Fig. 2.2 Hydrogen and covalent bonds in water molecules

and re-form. In liquid water, it is estimated that about 80 % of water molecules are involved in hydrogen bonding at any one time at 212 °F (100 °C), whereas 90 % are involved in liquid water at 32 °F (0 °C).

For the reason that *liquid* water has a smaller bond angle than ice, the molecules can be packed together more tightly, and so the *coordination number* or, in other words, the average number of nearest neighbors is higher for water than for ice. The average distance between water molecules is also affected by temperature and increases with temperature as the molecules have more kinetic energy and can move around faster and further at higher temperatures. Both of these affect the density of water, although the coordination number has a much more dramatic effect. Ice is less dense than water because the molecules have a smaller coordination number and cannot be packed together as tightly as water. Therefore, ice floats.

As water *freezes*, its density decreases and its *volume increases* by about 9 %. This is very significant when freezing foods with high water content. Containers and equipment must be designed to accommodate the volume increase when the product freezes, for example, molds for popsicles must allow room for expansion. This volume increase also contributes to the damage to the structure of soft fruits on freezing. This is discussed in Chap. 7. As water is heated above 39 °F (4 °C), the increase in the average distance between molecules causes a slight decrease in density.

Specific Heat and Latent Heat of Water

When ice is heated, the temperature increases in proportion to the amount of heat applied. The ***specific heat*** of water is the energy (in calories or in joules) required to raise the temperature of 1 g of water by 1 °C, and is the same whether heating water or ice. It is relatively high compared to other substances due to the hydrogen bonds. The specific heat of water is 1 cal/g/°C. This means that it takes 100 cal to raise the temperature of 1 g of water from 0 to 100 °C.

Once ice has reached 0 °C, energy needs to be put in to break the hydrogen bonds and enable ice to change to the liquid form. Until the ice has been converted to liquid, there is no further change in temperature until liquid water created.

The ***latent heat of fusion*** is the energy required to convert 1 g of *ice to water* at 0 °C and is 80 cal; that is, 1 g of ice at the freezing point absorbs approximately 80 cal as it changes to the liquid state.

The ***latent heat of vaporization*** is the energy required to convert 1 g of *water into vapor* at 100 °C and is 540 cal; that is, 1 g of water at the boiling point absorbs approximately 540 cal as it becomes steam.

Both the *specific heat* and *latent heat* for water are fairly high compared with most substances, and this is an important consideration when water is used as a medium of heat transfer. It takes considerable energy to heat water, and that energy is then available to be transferred to the food. Foods *heated* in water are slow to heat. Water also must take up considerable heat to evaporate. It takes heat from its surroundings, thus, it is a good *cooling* agent.

When ice is subjected to vacuum and then heated, it is converted into vapor *without* going through the liquid phase. This phenomenon is known as ***sublimation***, and is the basis for the food processing method known as ***freeze drying***. Coffee is an example of a food product that is freeze-dried. The process is expensive and is only used for foods that can be sold at a high price, such as coffee. The coffee beans are frozen and then subjected to a high vacuum, after which radiant heat is applied until almost all of water is removed by sublimation. Freezer burn is also the result of sublimation.

Vapor Pressure and Boiling Point

Vapor Pressure

If a puddle of water is left on the ground for a day or two, it will dry up because the liquid evaporates. The water does not boil, yet individual water molecules gain enough energy to escape from the liquid as vapor. Over a period, an *open*, small pool of water will dry up in this way. If the liquid is in a *closed* container, at equilibrium, some molecules are always evaporating and vapor molecules are condensing, so there is no overall change in the system. The *vapor* (gaseous) molecules that have escaped from the *liquid* state exert a pressure on the surface of the liquid known as the ***vapor pressure***.

When the vapor pressure is *high*, the liquid evaporates (is vaporized) easily and *many* molecules exist in the vapor state; the boiling point is low. Conversely, a *low* vapor pressure indicates that the liquid does not vaporize easily and that there are few molecules existing in the vapor state. The boiling point for these liquids is higher. The liquid boils when the vapor pressure reaches the external pressure.

The vapor pressure *increases* with increasing temperature. At higher temperatures, the molecules have more energy and it is easier for them to overcome the forces holding them within the liquid and to vaporize, and so there are more molecules in the vapor state.

The vapor pressure *decreases* with addition of solutes, such as salt or sugars. In effect, the solutes dilute the water; therefore, there are less water molecules (in the same volume) available for vaporization and, thus, there will be fewer molecules in the vapor state, and the vapor pressure will be lower. Attraction to the solute also limits evaporation.

Boiling Point

Anything that lowers the vapor pressure (pressure by gas above the liquid) increases the boiling point. This is due to the fact that as the vapor pressure is lowered at a particular temperature, more energy must be put in; in other words, the temperature must be raised to increase the vapor pressure again. The external pressure does not change if salts or sugars are added, although it becomes harder for the molecules to vaporize and so the temperature at which the vapor pressure is the same as the external pressure (boiling point) will be higher. One mole of sucrose elevates the boiling point by 0.52 °C, and 1 mol of salt elevates the boiling point by 1.04 °C. Salt has double the effect of sucrose because it is ionized, and for every mole of salt, there is 1 mol of sodium ions and 1 mol of chloride ions. Salts and sugars depress the freezing point of water in a similar fashion.

If the external pressure is increased by heating in a pressure cooker or retort (commercial pressure cooker), the boiling point increases and a shorter time than normal is required to cook a particular food (the basis of preserving foods by canning). For example, food may be heated in cans in retorts, and the steam pressure is increased to give a boiling point in the range 239–250 °F (115–121 °C). Conversely, if the external pressure is decreased, for example, at high altitude, water boils at a lower temperature and so food may require a longer time to cook.

CULINARY ALERT! Even when water comes to a rapid boil in high altitude locations, its temperature is not as high as rapidly boiling water at sea level!

Water as a Dispersing Medium

Substances are either *dissolved, dispersed, or suspended* in water depending on their particle size and solubility. Each is described below. Water is the usual dispersion medium.

Solution

Water *dissolves* small molecules such as salts, sugars, or water-soluble vitamins to form a true ***solution***, which may be either **ionic** or **molecular**. (A discussion of unsaturated, saturated, and supersaturated solutions appears in Chap. 14.)

An ionic solution is formed by dissolving substances that ionize in water, such as salts, acids, or bases. Taking sodium chloride as an example, the solid contains sodium (Na^+) and chloride (Cl^-) ions held together by ionic bonds. When placed in water, the water molecules reduce the attractive forces between the oppositely charged ions, the ionic bonds are broken, and the individual ions become surrounded by water molecules, or **hydrated**. Each ion is usually surrounded by six water molecules; the ions move independently of each other.

Polar molecules, such as sugars, which are associated by hydrogen bonding, dissolve to form **molecular solutions**. When a sugar crystal is dissolved, hydrogen-bond interchange takes place and the hydrogen bonds between the polar hydroxyl groups on the sugar molecules are broken and replaced by hydrogen bonds between water and the sugar molecules. Thus, the sugar crystal is gradually hydrated; each sugar molecule being surrounded by water molecules.

Water molecules bind to polar groups on the sugar molecules by hydrogen bonds. The sugar molecules are removed from the sugar crystal and hydrated as water molecules surround them and bind to them by hydrogen bonds.

When a hydrogen-bond interchange is involved, solubility increases with increasing temperature. Heating disrupts hydrogen bonds and reduces water–water and sucrose–sucrose attraction, thus facilitating formation of hydrogen bonds between water and sucrose, and hydration of sucrose molecules. Therefore, sucrose is much more soluble in hot water than in cold water. Solutes increase the boiling point of water, and the dramatic increase in sucrose solubility with temperature, particularly at temperatures above 100 °C (the boiling point of pure water), makes it possible to determine the sucrose concentration by measuring the boiling point of sucrose solution (Chap. 13). This is important when making candies or pectin jellies.

Colloidal Dispersion

Molecules that are *too big* to form true solutions may be dispersed in water. Those with a particle size range *1–100 nm* are dispersed to form a ***colloidal dispersion*** or ***sol***. Examples of such molecules include cellulose, *cooked* starch, pectic substances, gums, and some food proteins. Colloidal dispersions are often unstable; thus, food scientists must take care to stabilize them where necessary if they occur in food products. They are particularly unstable to factors such as heating, freezing, or pH change. Changing the conditions in a stable dispersion can cause precipitation or gelation; this is desirable in some cases, for example, when making pectin jellies.

(The reader is referred to Chap. 4 for a discussion of sols and gels; *sol* is a colloid that pours—a two-phase system with a solid dispersed phase in a liquid continuous phase, for example, a hot sauce. A ***gel*** is also a two-phase system, containing an elastic solid with a liquid dispersed phase in a solid continuous phase.)

Colloid science is important to food scientists as many convenient or packaged foods have colloidal dimensions, and their stability and sensitivity to certain types of reactions can only be understood with knowledge of colloid science.

Suspension

Particles that are *larger than 100 nm* are too large to form a colloidal dispersion. These form a ***suspension*** when mixed with water. The particles in a suspension separate out over a period, whereas no such separation is observed with colloidal dispersions. An example of a suspension would be *uncooked* starch grains in water. It may be temporarily suspended and then easily settle out, no longer "suspended," but rather, falling to the bottom of the container/pan.

CULINARY ALERT! Starches remain suspended throughout the liquid by stirring. If left undisturbed, they settle downward, and a sediment is observed at the bottom of the container. Starches do not "dissolve."

Free, Bound, and Entrapped Water

Water is abundant in all living things and, consequently, in almost all foods, unless steps have been taken to remove it. Most natural foods contain water up to 70 % of their weight or greater unless they are dehydrated, and fruits and vegetables contain water up to 95 % or greater. Water that can be extracted easily from foods by squeezing or cutting or pressing is known as ***free water***, whereas water that cannot be extracted easily is termed as ***bound water***.

Bound water is usually defined in terms of the ways it is measured; different methods of measurement give different values for bound water in a particular food. Many food constituents can bind or hold onto water molecules, such that they cannot be easily removed and they do not behave like liquid water. Several characteristics of bound water include the following:

- It is not free to act as a solvent for salts and sugars.
- It can be frozen only at very low temperatures (below freezing point of water).
- It exhibits essentially no vapor pressure.
- Its density is greater than that of free water.

Bound water has more structural bonding than liquid or free water, thus it is unable to act as a solvent. As the vapor pressure is negligible, the molecules cannot escape as vapor; and the molecules in bound water are more closely packed than in the liquid state, so the density is greater. An example of bound water is the water present in cacti or pine tree needles—the water cannot be squeezed or pressed out; extreme desert heat or a winter freeze does not negatively affect bound water and the vegetation remains alive. Even upon dehydration, food contains bound water.

Water molecules bind to polar groups or ionic sites on molecules such as starches, pectins, and proteins. Water closest to these molecules is held most firmly, and the subsequent water layers are held less firmly and are less ordered, until finally the structure of free water prevails. A more detailed discussion of bound water is given in books such as Fennema's Food Chemistry (Reid and Fennema 2007).

Water may also be **entrapped** in foods such as pectin gels, fruits, and vegetables. Entrapped water is immobilized in capillaries or cells, yet if released during cutting or damage, it flows freely. Entrapped water has properties of free water and no properties of bound water.

CULINARY ALERT! Freshness of any produce is evaluated in part by the presence of water. Food items appear wilted when free water is increasingly lost through dehydration.

Water Activity (A_w)

Water activity, or A_w, is a ratio of the vapor pressure of water in a solution (P_s) to the vapor pressure of pure water at a given temperature (P_w):

$$A_w = P_s/P_w$$

A_w must be high as living tissues require sufficient level of water to maintain turgor. However, microorganisms such as bacteria, mold, and yeast multiply at a high A_w. Because their growth must be controlled, preservation techniques against spoilage due to these microorganisms take into account the water activity of the food. Less bacterial growth occurs if the water level is lowered to less than 0.85 (FDA Model Food Code). Microbial growth (especially molds) can still occur at $A_w < 0.8$. Of course, there are other factors in addition to the water that must be present for bacterial growth to occur (food, optimum pH, etc.).

Jams, jellies, and preserves are prepared using high concentrations of sugar and brines, which contain high concentrations of salt that are used to preserve hams. Sugar and salt are both effective preservatives, as they decrease A_w. Salt decreases A_w even more effectively than sugar due to its chemical structure that ionizes and attracts water.

Role of Water in Food Preservation and Shelf Life of Food

Drying and *freezing* are common food preservation techniques. Foods are dehydrated or frozen to reduce the available water and extend shelf life.

The control of water level in foods is an important aspect of food quality as water content affects the shelf life and bacterial quality of food. For example, foods may be more desirable either crispy or dry. Freezing and drying are common food preservation processes that are used to extend the shelf life of foods because they render water unavailable for pathogenic or spoilage bacteria. If the water in foods is frozen quickly, there is less damage to the food at the cellular level. Preservatives may be added to a formulation to prevent mold or yeast growth. Humectants, which have an affinity for water, are added to retain moisture in foods.

> Water content influences a food's structure, appearance taste and even susceptibility to degradation. Depending upon the foodstuff, water may function as a free-flowing liquid or be a component of a larger matrix, visibly (pudding) or invisibly (granola bar).
>
> Gums and starches can act together providing a moisture-management system, and can "help prevent staling, which results from the retrogradation of starch in baked goods. Retrogradation releases moisture over time, leading to staling... Because gums do not undergo the retrogradation process,

they can slow the staling process by holding onto moisture." (Berry 2012)

Water Hardness and Treatments

The hardness of water is measured in parts per million or in "grains," with one grain equivalent to 0.064 g of calcium carbonate. ***Soft water*** contains 1–4 grains per gallon some organic matter, and has no mineral salts. ***Hard water*** contains 11–20 grains per gallon. Water may exhibit *temporary* hardness due to iron or calcium and magnesium bicarbonate ions [$(Ca(HCO)_3)_2$ and Mg $(HCO_3)_2$]. The water may be softened by boiling (soluble bicarbonates precipitate when boiled and leave deposits or scales) and insoluble carbonates may be removed from the water.

Permanently hard water *cannot* be softened by boiling as it contains either calcium or magnesium sulfates ($CaSO_4$ or $MgSO_4$) as well as other salts that are not precipitated by boiling. Permanent hard water may only be softened by the use of chemical softeners. Hard water exhibits less cleaning effectiveness than soft water due to the formation of insoluble calcium and magnesium salts with soap, which could be prevented by the use of detergents.

Water has a pH of 7 or neutral; tap water displays a variance on either side of neutral. It may be slightly alkaline or slightly acidic depending on the source and so forth. Hard water has a pH of up to 8.5. Chlorinated water is that which has had chlorine added to kill or inhibit the growth of microorganisms. Manufacturing or processing plants may require chemically pure water to prevent turbidity, off-color, and off-flavor. Tap water may not be sufficiently pure for use in food products.

Beverage Consumption Ranking

Drinking water was ranked as the preferred beverage to fulfill daily water needs and was followed in decreasing value by tea and coffee, low-fat (1.5 % or 1 %) and skim (nonfat) milk and soy beverages, noncalorically sweetened beverages, beverages with some nutritional benefits (fruit and vegetable juices, whole milk, alcohol, and sports drinks), and calorically sweetened, nutrient-poor beverages. (Popkin et al. 2006)

Conclusion

Water is essential for life and makes up the major part of living tissue. The nature of hydrogen bonds allows water to bond with other water molecules as well as with sugar, starches, pectins, and proteins. Water absorbs energy as it changes from frozen to liquid to vapor state, and is an effective cooling medium. If water is easily extracted from foods by squeezing, or pressing, it is known as free water. Inversely, water that is not easily removed from foods and that is not free to act as a solvent is known as bound water; water in foods imparts freshness. A measure of water activity is the ratio of the vapor pressure of water in a solution to the vapor pressure of pure water. If water is unavailable for pathogenic or spoilage-causing bacteria to multiply, food is better preserved and has a longer shelf life.

Notes

CULINARY ALERT!

Glossary

Bound water Water that cannot be extracted easily; it is bound to polar and ionic groups in the food.

Colloidal dispersion Molecules, larger than those in solution, dispersed in the surrounding medium.

Covalent bonds Strong bonds that hold the two hydrogen atoms and one oxygen atom together in a water molecule.

Free water Water that can be extracted easily from foods by squeezing, cutting, or pressing.

Freeze drying A food processing method that converts ice to vapor without going through the liquid phase (sublimation).

Gel Elastic solid; a two-phase system that contains a solid continuous phase and a liquid dispersed phase.

Hard water Contains 11–20 grains per gallon. Hardness is due to calcium and magnesium bicarbonates or sulfates, which results in less effective cleaning.

Hydrogen bonds Weak bonds between polar compounds where a hydrogen atom of one molecule is attracted to an electronegative atom of another molecule.

Latent heat of fusion The energy required to convert 1 g of ice to water at 0 °C—requires 80 cal.

Latent heat of vaporization The energy required to convert 1 g of water to vapor at 100 °C—requires 540 cal.

Sol A two-phase system with a solid dispersed in a liquid continuous phase; pourable.

Soft water Contains one to four grains per gallon, no mineral salts, some organic matter.

Solution (Ionic or molecular) Small molecules dissolved in water.

Sublimation When ice is subjected to vacuum and then heated, it gets converted to vapor without going through the liquid phase; basis for freeze drying; occurs in freezer burn.

Specific heat The energy required to raise the temperature of 1 g of water by 1 °C whether heating water or ice; requires 1 cal/g/°C.

Suspension Molecules larger than those in a solution or dispersion that are mixed with the surrounding medium. A *temporary* suspension settles upon standing.

Vapor pressure The pressure vapor molecules exert on the liquid.

Water activity (A_w) The ratio of the vapor pressure of water in a solution to the vapor pressure of pure water.

References

Berry D (2012) Managing moisture. Food Prod Des June:34–42

Reid, DS and Fennema, OR (2007) Water and ice. In: Fennema OR, Damadoran S, Parkin KL (ed) Food chemistry, 4th edn. Marcel Dekker, New York

Popkin BM, Armstrong LE, Bray GM, Caballero B, Frei B, and Willett WC (2006) A new proposed guidance system for beverage consumption in the United States. Am J Clin Nutr 83:529–542. http://www.ajcn.org/cgi/content/full/83/3/529

Bibliography

Rockland LB, Stewart GF (1981) Water activity: influences on food quality. Academic, New York

Simatros D, Karel M (1988) Characterization of the condition of water in foods: physiochemical aspects. In: Seow CC (ed) Food preservation by moisture control. Elsevier Applied Science Publishers, London

Water Activity Books

Part II

Carbohydrates in Food

3 Carbohydrates in Food: An Introduction

Introduction

Carbohydrates are organic compounds containing carbon, hydrogen, and oxygen, and they may be simple or complex molecules. Historically, the term "carbohydrate" has been used to classify all compounds with the general formula $C_n(H_2O)_n$ as the hydrates of carbon. Important food carbohydrates include simple sugars, dextrins, starches, celluloses, hemicelluloses, pectins, and gums. They are an important source of energy or fiber in the diet, and they are also important constituents of foods because of their functional properties. Carbohydrates may be used as sweeteners, thickeners, stabilizers, gelling agents, and fat replacers.

The simplest carbohydrates are known as *mono*saccharides or sugars, and they have the general formula $C_nH_{2n}O_n$. The most common ones contain six carbon atoms. *Di*saccharides contain two sugar units, *tri*saccharides contain three, *oligo*saccharides contain several units, and *poly*saccharides are complex polymers containing as many as several thousand units linked together to form a molecule. These carbohydrates are discussed in this chapter.

Monosaccharides

Monosaccharides are simple carbohydrates containing between three and eight carbon atoms, yet only those with five or six carbon atoms are common. Two of the most important ones in foods are the six-carbon sugars glucose and fructose. These have the general formula $C_6H_{12}O_6$.

Examples of Monosaccharides

Glucose. Glucose is known as an ***aldose sugar*** because it contains an aldehyde group (CHO) located on the first carbon atom of the chain:

Glucose and an aldehyde group:

It is conventional to number the carbon atoms along the chain so that the carbon atom with the highest number is farthest away from the aldehyde (or functional) group. The aldehyde group is therefore located on carbon one in glucose (and in all other aldose sugars). The numbering of the carbon atoms in glucose is shown in Fig. 3.1.

Two isomers of glucose exist, which are mirror images of each other, D-glucose and L-glucose. D-Glucose is the isomer that occurs naturally.

In fact, there are two series of aldose sugars, known as the D-series and the L-series, each formed by adding CHOH groups to build the carbon chain, starting from the smallest aldose sugar, which is D- or L-glyceraldehyde (see Fig. 3.2).

Each H—C—OH group within the chain is asymmetrical (since the H and OH groups are different). The highest-numbered asymmetric carbon atom of each D-series sugar has the same configuration as D-glyceraldehyde, rather than its L-isomer. In glucose, the highest-numbered

For use with subsequent Carbohydrate food chapters

V.A. Vaclavik and E.W. Christian, *Essentials of Food Science, 4th Edition*, Food Science Text Series,
DOI 10.1007/978-1-4614-9138-5_3,

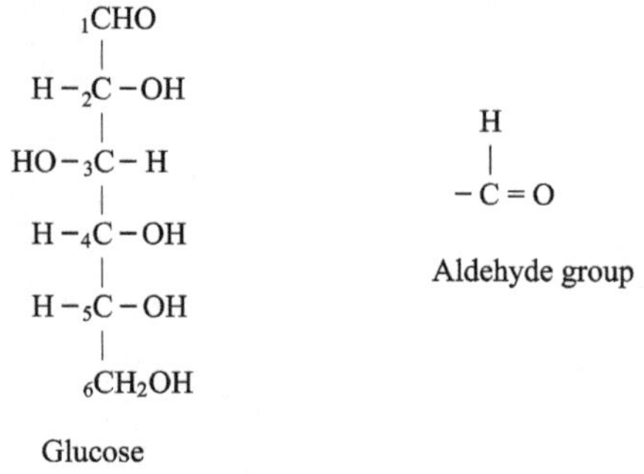

```
   1CHO
    |
H - 2C - OH
    |
HO - 3C - H            H
    |                  |
H - 4C - OH         - C = O
    |
H - 5C - OH        Aldehyde group
    |
   6CH2OH

  Glucose
```

Fig. 3.1 Glucose and an aldehyde group

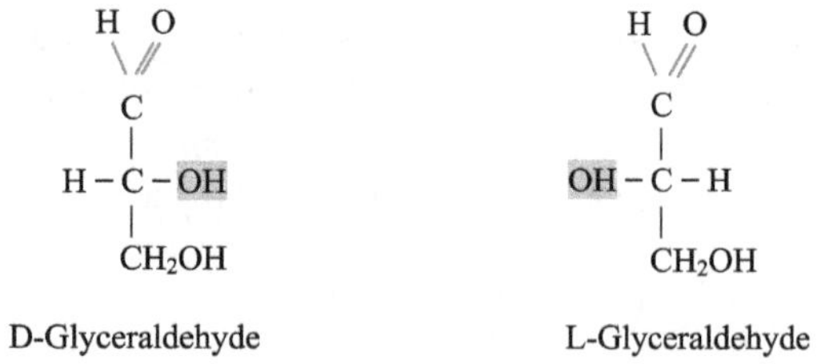

```
   H  O                      H  O
    \//                       \//
     C                         C
     |                         |
 H - C - OH               OH - C - H
     |                         |
    CH2OH                     CH2OH

D-Glyceraldehyde          L-Glyceraldehyde
```

Fig. 3.2 Mirror images of glyceraldehyde

asymmetric carbon atom is carbon-5. This is termed the ***reference carbon atom***, because its configuration determines whether the sugar belongs to the D-series or to the L-series. The hydroxyl group attached to it is called the ***reference hydroxyl group***. This group is always on the right side in a D-series sugar.

The straight-chain configuration of glucose (and of other monosaccharides) does not account for all the properties of the molecule. In reality, the straight-chain form exists in equilibrium with several possible ring configurations. In other words, the different configurations exist together in solution in a delicate balance. Glucose can exist in four ring structures: two pyranose or six-membered ring forms, and two furanose or five-membered ring forms. These exist along with the straight-chain form, as shown in Fig. 3.3.

The most common configurations for glucose are the ***pyranose*** structures, drawn according to the Haworth convention in Fig. 3.4. These are ***anomers*** and are designated ***alpha*** (α) and ***beta*** (β). They are formed when the hydroxyl group on the fifth carbon reacts with the carbonyl group (located on the first carbon, designated as Cl). As the ring closes, a new hydroxyl group is formed on Cl. This is termed the ***anomeric hydroxyl group***, and the carbon atom to which it is attached is termed the ***anomeric carbon atom***. For glucose and the other aldoses, the anomeric carbon atom is always the first carbon atom of the chain.

The anomeric hydroxyl group can project towards either side of the ring, as shown in Fig. 3.4. Hence, there are two possible pyranose structures.

For glucose and all the hexoses, the α-anomer has the anomeric hydroxyl group on the *opposite* face of the ring to carbon-6 (i.e., pointing in the opposite direction to carbon-6), when drawn according to the Haworth convention, whereas the β-anomer has the anomeric hydroxyl group on the *same* face of the ring as carbon-6 (i.e., pointing in the same direction as carbon-6). For the D-series sugars, when the ring closes, carbon-6 is always located above the plane of the ring. Therefore, in the case of the α-anomer, the anomeric hydroxyl group points *down*, or *below* the plane of the ring, whereas in the case of the β-anomer, the anomeric hydroxyl group points *up*, or *above* the plane of the ring.

> ***Alpha-anomer***—anomeric hydroxyl group is on the **opposite** face of the ring to carbon-6 **D-series sugars**—anomeric hydroxyl group points **down**
>
> ***Beta-anomer***—anomeric hydroxyl group is on the **same** face of the ring as carbon-6 **D-series sugars**—anomeric hydroxyl group points **up**

[For the chemists who prefer to define the alpha- and beta-configurations according to the reference carbon, when the anomeric hydroxyl group is formed on the same side of the ring as the reference hydroxyl group (as seen in the Fischer projection formula), the anomer is denoted alpha, whereas, when it is formed on the opposite side, it is denoted beta.]

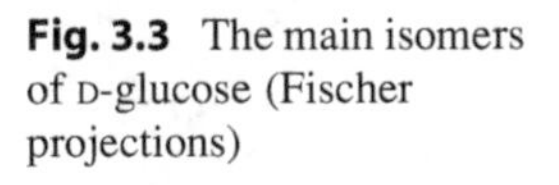

Fig. 3.3 The main isomers of D-glucose (Fischer projections)

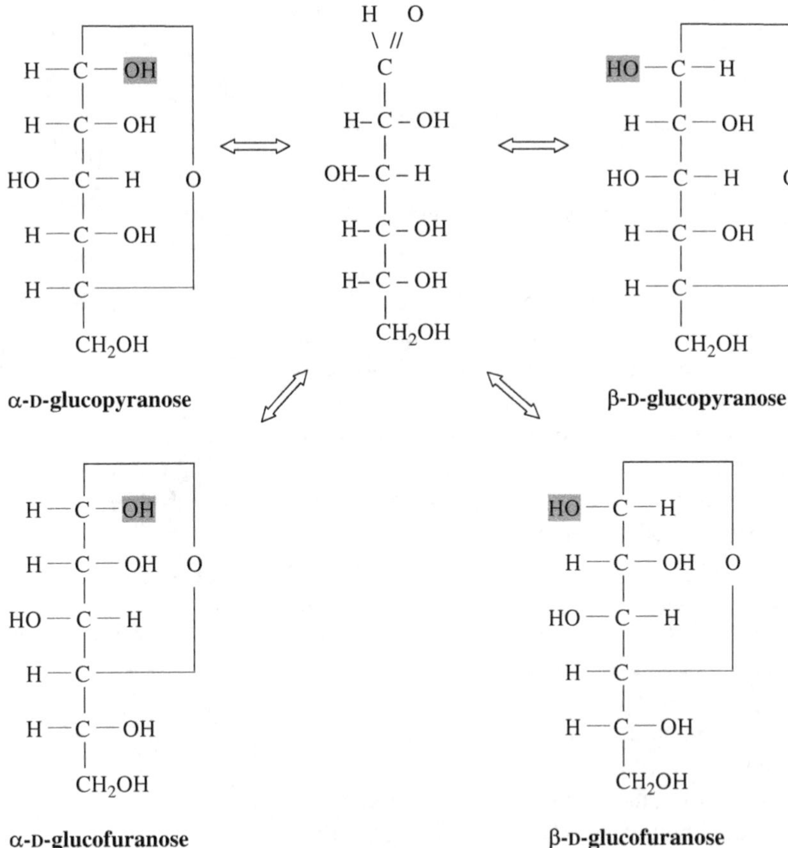

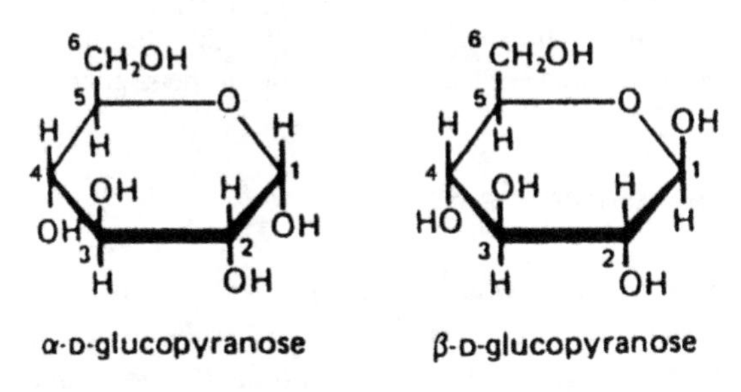

Fig. 3.4 The D-glucopyranose anomers, drawn according to the Haworth convention

In solution, the alpha- and beta-forms are in equilibrium, yet the configuration can be fixed if the molecule reacts to form a disaccharide. It is important to know whether the configuration is fixed as the alpha- or beta-configuration, because this affects properties of the molecule, including digestibility. For example, starch contains α-D-glucose molecules, and so can be digested, although cellulose contains β-D-glucose molecules and is indigestible.

Although the ring structures are drawn with flat faces in the Haworth formulae, in reality they are not planar rings, yet, rather, they are bent, and could be visualized more as a boat or a chair configuration, as shown in Fig. 3.5.

The different configurations of glucose and the relationships between them are complex, and are beyond the scope of this book. For a more in-depth treatment, interested readers are referred to books such as Food Chemistry, edited by Owen Fennema, or to basic biochemistry textbooks.

Glucose is the most important aldose sugar. Two other aldose sugars important in foods include galactose and mannose. Galactose is important as a constituent of milk sugar (lactose),

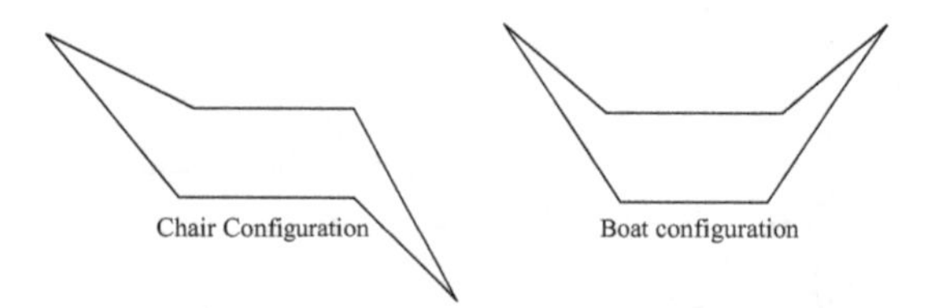

Fig. 3.5 Chair and boat configurations of monosaccharides

and mannose is used to make the sugar alcohol mannitol, which is used as an alternative sweetener in chewing gum and other food products. These are both D-series sugars. In fact, almost all naturally occurring monosaccharides belong to the D-series.

Fructose. Fructose is a six-carbon sugar, like glucose, although despite this, it is a **ketose sugar**, not an aldose, because it contains a ketone group, and not an aldehyde group (see Fig. 3.6):

Similarly to the aldose sugars, there is a D-series and an L-series of ketose sugars, but D-fructose is the only ketose of importance in foods. All ***ketose sugars*** contain a ketone group, not an aldehyde group.

In fructose, the ketone group is located on the second carbon of the chain. The second carbon atom is therefore the anomeric carbon in fructose. Fructose occurs mainly in the α- and β-***furanose***, or five-membered ring configurations, as shown in Fig. 3.7.

Both the ketone groups of a ketose sugar and the aldehyde group of an aldose sugar may be called a ***carbonyl group***. A carbonyl group contains a carbon atom double-bonded to an oxygen atom, but the other atoms are not specified. Hence, an aldehyde group is a specific type of carbonyl group, with both a hydrogen atom and an oxygen atom attached to the carbon atom. A ketone group is also a carbonyl group, because it contains an oxygen atom double-bonded to a carbon atom located within a hydrocarbon chain.

Disaccharides

Disaccharides contain two monosaccharides joined together with a special linkage, called a **glycosidic**

$$
\begin{array}{c}
{}_1CH_2OH \\
| \\
{}_2C=O \\
| \\
HO-{}_3C-H \\
| \\
H-{}_4C-OH \\
| \\
H-{}_5C-OH \\
| \\
{}_6CH_2OH
\end{array}
\qquad
\begin{array}{c}
| \\
C=O \\
|
\end{array}
$$

Fructose

Ketone group

Fig. 3.6 Fructose and a ketone group

α-D-fructofuranose

β-D-fructofuranose

Fig. 3.7 The main configurations of D-fructose

bond. Several disaccharides are important in foods—sucrose or table sugar is the most common and contains glucose and fructose. There are other important disaccharides, such as maltose, containing two glucose units, and lactose, which contains glucose and galactose. Lactose is also known as milk sugar because it is found in milk. It is the least sweet and least soluble of the sugars.

Glycosidic Bonds

A ***glycosidic bond*** is formed when the carbonyl group of one monosaccharide reacts with a hydroxyl group of another molecule and water is eliminated (see Fig. 3.8).

Formation of a glycosidic link fixes the configuration of the monosaccharide containing the involved carbonyl group in either the α- or β-position. Therefore, it is necessary to specify whether the link is an α- or a β-link. The position of the bond must also be specified. For example, when two glucose molecules are joined to make maltose, the glycosidic link occurs between carbon-1 of the first glucose molecule and carbon-4 of the second, and the configuration of the first glucose molecule is fixed in the α-position. Maltose therefore contains

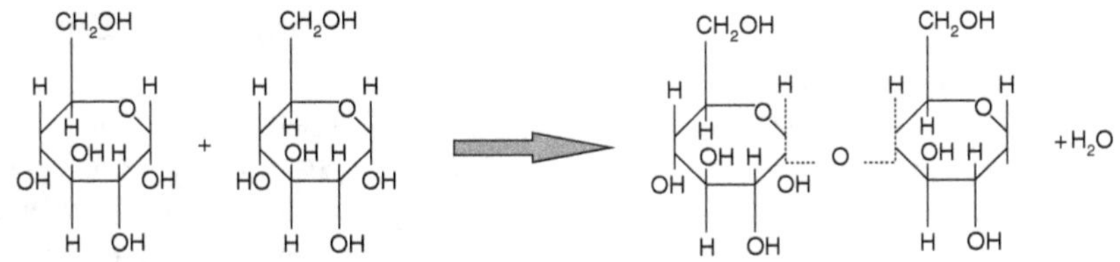

Fig. 3.8 A glycosidic bond between the carbonyl and hydroxyl groups of monosaccharides

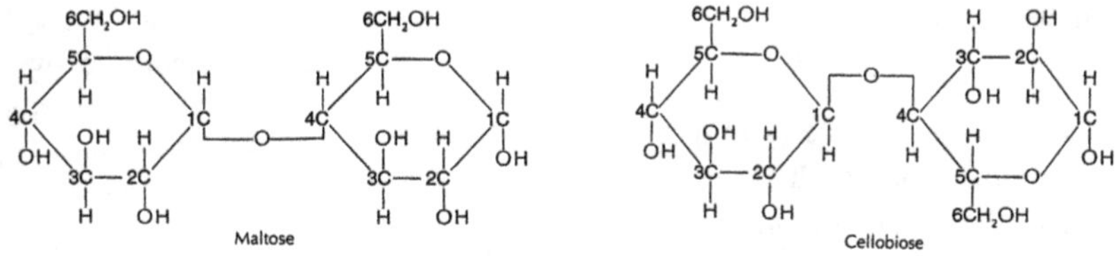

Fig. 3.9 Maltose and cellobiose

two glucose units linked by an α-1,4-glycosidic bond. The anomeric hydroxyl group that is not involved in the glycosidic bond (i.e., the one on the second glucose molecule) remains free to assume either the α- or β-configuration. Therefore, there are two forms of the disaccharide in equilibrium with each other.

Glycosidic bonds are stable under normal conditions yet can be hydrolyzed by acid and heat, or by enzymes such as sucrase, invertase, or amylases.

Glycosidic Bond

- Formed between the free carbonyl group of one monosaccharide and a hydroxyl group of another monosaccharide
- Fixes the configuration of the monosaccharide containing the involved carbonyl group in either the α- or β-position
- It is necessary to specify
  - The **configuration** of the link—whether it is an α-link or a β-link
  - The **position** of the link—it is numbered according to the respective positions of the two carbon atoms it links together. For example, an α-1,4 glycosidic link would occur between carbon-1 of the first monosaccharide

(continued)

and carbon-4 of the second monosaccharide, as occurs in maltose
- Readily hydrolyzed by
  - Heat and acid
  - Certain enzymes, such as sucrase, invertase, and amylases

Examples of Disaccharides

Maltose and Cellobiose. As has already been mentioned, maltose contains two glucose units linked by an α-1,4-glycosidic bond. When two glucose molecules are joined together and the configuration of the first glucose molecule is fixed in the β-position, cellobiose is formed. Cellobiose contains a β-1,4-glycosidic bond. The chemical formulas for maltose and cellobiose are shown in Fig. 3.9.

Maltose is the building block for *starch*, which contains α-1,4-glycosidic bonds. Alpha links can be broken down by the body, so starch is readily digested. Cellobiose is the building block for *cellulose*, which contains β-1,4-glycosidic bonds. Cellulose cannot be digested in the human body because the β-linkages cannot be broken down by the digestive enzymes. Therefore, cellulose is known as dietary fiber. (The glycosidic bonds in cellulose cross the plane of the monosaccharide

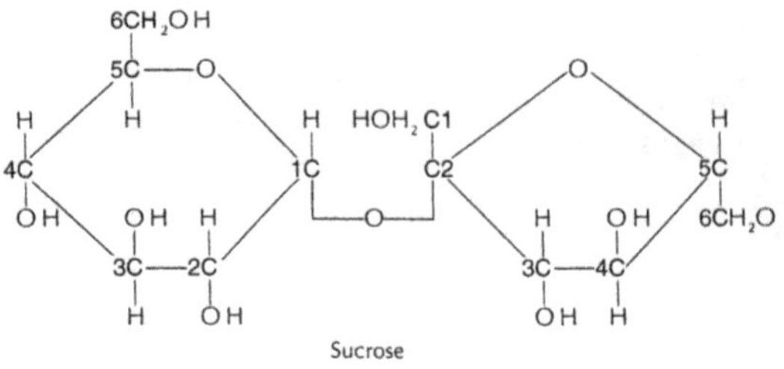

Fig. 3.10 Sucrose

units they join together, and so they may be termed ***cross-planar bonds***. It is because they are cross-planar that they are not digestible. In reality, because of the orientation of the bonds, the monosaccharide units tend to twist or flip over, as drawn in Fig. 3.9, which results in a twisted ribbon effect for the polymer chain.)

Sucrose. Sucrose is the most common disaccharide, and it contains glucose and fructose joined together by an α-1,2-glycosidic link (see Fig. 3.10). The carbonyl groups of both the glucose and the fructose molecule are involved in the glycosidic bond; thus, the configuration of each monosaccharide becomes fixed. Glucose is fixed in the α-configuration, whereas fructose is fixed in the β-configuration. Sucrose can be hydrolyzed to glucose and fructose by heat and acid, or by the enzymes invertase or sucrase. The equimolar mixture of glucose and fructose produced in this way is called ***invert sugar***. Production of invert sugar is important during the formation of candies and jellies, as invert sugar prevents unwanted or excessive crystallization of sucrose. (For further discussion of crystallization of sucrose, see Chap. 14.)

Various Properties of Sugars

Sweetness

The most obvious sensory property of sugars such as glucose, fructose, and sucrose is their **sweetness**, which varies depending on the specific sugar. Lactose (milk sugar) is the least sweet, whereas fructose is the sweetest sugar. Sugars are used as sweeteners in candies and many other food products.

Formation of Solutions and Syrups

Sugars are soluble in water and readily form syrups. If water is evaporated, crystals are formed. Sugars form **molecular solutions** due to hydrogen-bond interchange. When sugar is placed in water, the water molecules form hydrogen bonds with the sugar molecules, thus hydrating them and removing them from the sugar crystals. Solubility increases with temperature; thus, a hot sucrose solution may contain more solute than a cold one. (For a discussion of molecular solutions, see Chap. 2.)

If a hot saturated sucrose solution is cooled without disturbance, it will supercool, and a supersaturated solution will be obtained. A ***supersaturated solution*** contains more solute than could normally be dissolved at that temperature. It is unstable, and if stirred or disturbed, the extra solute will rapidly crystallize out of solution. Supersaturated solutions are necessary in candy-making. For more detail on sugar crystallization and candies, see Chap. 14.

Body and Mouthfeel

Sugars contribute body and "mouthfeel" to foods. In other words, the addition of sugar makes a food more viscous or gives it a less runny consistency. If sugar is replaced by a non-nutritive or high-intensity sweetener such as aspartame or saccharin, the consistency of the food will be watery and thin. To prevent this, another substance has to be added to give the expected body or mouthfeel to the food. Modified starches or gums are usually added to such food products to give the desired consistency without addition of sugar.

Fermentation

Sugars are readily digested and metabolized by the human body and supply energy (4 cal/g). They are also metabolized by microorganisms. This property is important in breadmaking, where sugar is fermented by yeast cells. The yeast feeds on the sugar, producing carbon dioxide, which is the leavening agent and causes bread dough to rise before and during baking.

Preservatives

At high concentrations, sugars prevent growth of microorganisms, because they reduce the water activity of food to a level below which bacterial growth cannot be supported. Sugars can, therefore, be used as preservatives. Examples of foods preserved in this manner include jams and jellies.

Reducing Sugars

Sugars that contain a free carbonyl group are known as ***reducing sugars***. All monosaccharides are reducing sugars. Disaccharides are reducing sugars only if they contain a free carbonyl group. Sucrose is not a reducing sugar because it does not contain a free carbonyl group. The carbonyl groups of glucose and fructose are both involved in the glycosidic bond and are, therefore, not free to take part in other reactions. Maltose, on the other hand, has one carbonyl group involved in the glycosidic bond, and the other carbonyl group is free; thus, maltose is a reducing sugar.

Reducing sugars give brown colors to baked goods when they combine with free amino acid groups of proteins in a browning reaction called the ***Maillard reaction*** (this reaction is discussed further in Chap. 8).

Caramelization

Sugars ***caramelize*** on heating, giving a brown color. Caramelization is caused by the decomposition of the sugars and occurs at extremely high temperatures. A variety of compounds are formed as a result, including organic acids, aldehydes, and ketones. The reaction does not involve proteins and should not be confused with the Maillard browning reaction.

Sugar Alcohols

Reduction of the carbonyl group to a hydroxyl group gives ***sugar alcohols*** such as xylitol, sorbitol, and mannitol. These compounds are sweet, although not as sweet as sucrose. However, they are not fermented as readily as sugar by microorganisms in the mouth, and so they are noncariogenic. (In other words, they do not cause tooth decay.) Therefore, they are used in chewing gum, breath mints, and other products that may be kept in the mouth for a while. Although products containing sugar alcohols may be labeled as "sugar-free," it is important to realize that sugar alcohols are not free of calories. They are not metabolized as efficiently as sugars and have a lower caloric value (between 1 and 3 kcal/g).

Sugar alcohols may be used as a low-energy bulk ingredient (in place of sugar) in many food products. Since sorbitol is mostly transformed to fructose in the body rather than glucose, it is tolerated by diabetics. Hence, it can be used to replace sugar in diabetic foods.

Oligosaccharides

Oligosaccharides contain a few (3–10) monosaccharide residues linked together by glycosidic bonds. Common ones include raffinose and stachyose. Raffinose is a ***trisaccharide*** and contains galactose, glucose, and fructose. Stachyose contains glucose, fructose, and two galactose units. Both occur in legumes such as dry beans and peas. They are not hydrolyzed or digested by the human digestive system, and become food for bacteria in the large intestine. The bacteria metabolize the carbohydrates and produce gas, causing varying degrees of discomfort.

Polysaccharides

The most important food ***polysaccharides*** are the starches, pectins, and gums. All are complex carbohydrate polymers with different properties, which depend on the sugar units that make up the molecule, the type of glycosidic linkages, and the degree of branching of the molecules. Starches are discussed in Chap. 4 and pectins and other polysaccharides are covered in Chap. 5.

Dextrins and Dextrans

Dextrins are intermediate-chain length glucose polymers formed when starch is broken down or hydrolyzed. They are larger than oligosaccharides, considerably shorter than starch molecules. Dextrins contain glucose molecules joined by α-1,4-glycosidic bonds, and they are linear polymers. They are found in corn syrups, produced by hydrolysis of starch.

Dextrans are also intermediate-chain length glucose polymers, but they contain α-1,6-glycosidic bonds. They are produced by some bacteria and yeasts.

Starch

Starch is a glucose polymer that contains two types of molecules, known as amylose and amylopectin. These are shown in Figs. 3.11 and 3.12, respectively. Both are long chains of glucose molecules joined by α-1,4-glycosidic bonds; however, amylose is a linear chain, whereas amylopectin contains branches. For every 15–30 glucose residues there is a branch, joined to the main chain by an α-1, 6-glycosidic link. The branches make amylopectin less soluble in water than amylose. Usually, the two types of starch occur together, although starches may contain only amylose or only amylopectin. They have different properties, which are discussed in Chap. 4.

Starches can also be modified to give specific functional properties in food products, so knowledge of the properties of different starches is important in the food industry. Chapter 4 gives detailed information on characteristics of different starches and their uses in foods.

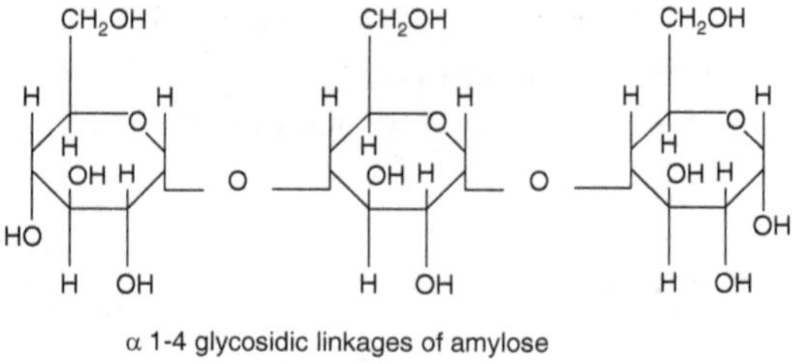

α 1-4 glycosidic linkages of amylose

Fig. 3.11 Amylose

Pectins and Other Polysaccharides

Pectins, gums, and seaweed polysaccharides are also important carbohydrates used in food products. They are discussed further in Chap. 5. Pectins occur naturally in plant food products, yet gums and seaweed polysaccharides do not come from edible plant sources. They are extracted and purified and then added to food products.

Pectins are used mainly as gelling agents in jellies, jams, and other products. They are also used as stabilizers and thickeners. They are found in fruits and vegetables, and they help to hold the plant cells together. Structurally, they are long-chain polymers of α-D-galacturonic acid, which is an acid derived from the simple sugar galactose. They are soluble in water, and, under appropriate conditions, they form gels. Their structure and properties are discussed in Chap. 5.

Gums are mainly plant extracts and include gum tragacanth and guar gum. They are highly branched polysaccharides that form very viscous solutions, trapping large amounts of water within their branches. Most do not form gels because of the high level of branching. They are useful as thickeners and stabilizers, particularly in reduced-fat salad dressings and in other convenience foods.

Seaweed polysaccharides include the agars, alginates, and carrageenans. They are classified as gums, although they are able to form gels, unlike most gums. They are useful as gelling agents, thickeners, and stabilizers in foods.

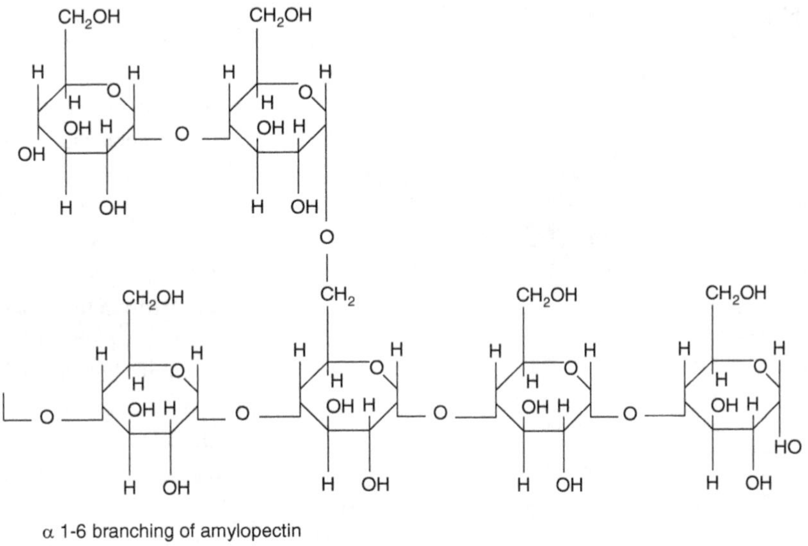

Fig. 3.12 Amylopectin

α 1-6 branching of amylopectin

Cellulose and *hemicellulose* are structural polysaccharides that provide support in plant tissues. They are not digested in the body, so they do not supply energy. However, they provide insoluble dietary fiber, which is an important part of a healthy, balanced diet.

Regarding fiber, food items may make the claim "good source of fiber" if 2.5–4.9 g of fiber per serving are present. When 5 g per serving, or more, are present, a food item may be labeled "high fiber." It is recommended by health agencies and Dietary Guidelines for Americans that men should consume 38 g of fiber per day, and that women should consume 25 g per day. There is soluble and insoluble fiber and they are structurally different (see Chap. 4).

Inulin is a polysaccharide with the general formula $(C_6H_{10}O_5)_n$. It is found in tubers and the roots of various plants and, when hydrolyzed, yields fructose.

Conclusion

Carbohydrates come in various shapes and sizes, from small sugar molecules to complex polymers containing thousands of simple sugar units. The digestible carbohydrates provide energy (4 cal/g), whereas the indigestible ones are an important source of dietary fiber. In addition to their nutritional value, carbohydrates are important as thickeners, stabilizers, and gelling agents. They are used in a wide spectrum of convenience foods, and, without them, the range of food products relished today would be greatly diminished.

Notes

CULINARY ALERT!

Glossary

Aldose Sugar containing an aldehyde group monosaccharide—single sugar unit.

Alpha-anomer The anomeric hydroxyl group is on the opposite face of the ring from carbon-6 (i.e., the two groups point in opposite directions).

Anomeric carbon atom The carbon atom that is part of the free carbonyl group in the straight-chain form of a sugar.

Anomers Isomers that differ only in the orientation of the hydroxyl group on the anomeric carbon atom; there are two forms—alpha (α) and beta (β).

Beta-anomer The anomeric hydroxyl group is on the same face of the ring as carbon-6 (i.e., the two groups point in the same direction).

Carbonyl group Contains an oxygen atom double-bonded to a carbon atom. The aldehyde group and the ketone group can both be described as a carbonyl group.

Caramelization Decomposition of sugars at very high temperatures resulting in brown color.

Cross-planar bond Formed when the hydroxyl groups on the carbon atoms involved in the formation of a glycosidic bond are oriented on opposite faces of the sugar rings. Cross-planar bonds occur in cellobiose and in cellulose. They also occur in pectin. They are not digested in the human digestive system.

Dextrans Glucose polymers joined by α-1,6-glycosidic bonds. Produced by some bacteria and yeasts.

Dextrins Glucose polymers joined by α-1,4-glycosidic bonds. Product of starch hydrolysis. Found in corn syrups.

Disaccharide Two sugar units joined together by a glycosidic bond.

Furanose Five-membered ring.

Glycosidic bond Bond that links two sugar units together; it is formed between the free carbonyl group of one sugar and a hydroxyl group of another sugar; the orientation (α or β) and position (e.g., 1,4) of the link must be specified.

Hydroxyl group The —OH group on the carbon atom.

Invert sugar An equimolar mixture of glucose and fructose, formed by hydrolysis of sucrose, either by acid and heat, or by enzymes such as invertase or sucrase.

Ketose Sugar containing a ketone group.

Maillard reaction (Maillard browning reaction) Nonenzymatic browning reaction involving a reducing sugar and a free amino acid group on a protein.

Monosaccharide Single sugar unit.

Oligosaccharide Several (3–10) sugar units joined together by a glycosidic bond.

Polysaccharide Many (hundreds or thousands of) sugar units joined together.

Pyranose Six-membered ring.

Reducing sugar Sugar that contains a free carbonyl group.

Reference carbon atom The highest-numbered asymmetric carbon atom; C5 in glucose and fructose.

Reference hydroxyl group The hydroxyl group attached to the reference carbon atom.

Sugar alcohol The result of reduction of carbonyl group to a hydroxyl group.

Supersaturated solution Solution that contains more solute than could normally be dissolved at a particular temperature.

Trisaccharide Three sugar units joined together by a glycosidic bond.

Bibliography

BeMiller JN, Huber KL (2007) Carbohydrates. In: Damodaran S, Parkin K, Fennema O (eds) Fennema's food chemistry, 4th edn. CRC Press, Boca Raton, FL

Charley H, Weaver C (1998) Foods. A scientific approach, 3rd edn. Merrill/Prentice-Hall, New York

Garrett RH, Grisham CM (2013) Biochemistry, 5th edn. Brooks/Cole/Cengage Learning, Belmont, CA

Hazen C (2012) Fiber files. Food Product Design. September:102–112

McWilliams M (2012) Foods: experimental perspectives, 7th edn. Prentice-Hall, Upper Saddle River, NJ

Penfield MP, Campbell AM (1990) Experimental food science, 3rd edn. Academic, San Diego, CA

Potter N, Hotchkiss J (1999) Food science, 5th edn. Springer, New York

Vieira ER (1999) Elementary food science, 4th edn. Springer, New York

4 Starches in Food

Introduction

Starch is a plant polysaccharide stored in roots and seeds of plants, and in the endosperm of a grain kernel. It provides humans with energy (4 cal/g), and is hydrolyzed into glucose, supplying the glucose that is necessary for brain and central nervous system functioning.

Starch grains, or ***granules***, contain long-chain glucose polymers and are *in*soluble in water. Unlike the small molecules of salt and sugar, the larger starch polymers do *not* form a true solution. Instead, starch granules form a *temporary suspension* when stirred in water. As *un*cooked granules, each may swell slightly when it absorbs water. However, once starch is *cooked*, the swelling is *ir*reversible and the starch leaches out. This characteristic of starch granules enables starch to be used as a thickener.

Overall, the characteristics of a finished starch food product are determined by several factors: the source of starch, concentration of starch used in a formulation, the temperature and time of heating, and other components used with the starch, such as acid and sugar. There are many types of starch and modified starches. These thicken, prevent curdling, and stabilize cooked salad dressings, dips, gravies, desserts, and more.

Intermediate, shorter chain products from starch breakdown, known as dextrins, may be used to simulate fat in salad dressings and frozen desserts. For example, wheat, potato, and tapioca maltodextrins may be used as fat replacers. These provide the viscosity and mouthfeel of fat in a food product, yet, with reduced calories compared to fat.

Starch Sources in the Diet

Starch sources are numerous, with common ones derived from cereal grains such as wheat, corn, or rice. Wheat yields a cloudy, thick mixture, while cornstarch produces more clear mixtures such as gravies or sauces. Vegetables, roots and tubers, including the root of cassava, and potatoes, are frequently used in the preparation of gluten-free foods, where persons with wheat allergies or intolerances do not use any wheat as a thickener. Specialty starches are available commercially and some may be available to the consumer, perhaps purchased through specialty food stores.

Another source of starch is legumes such as soybeans or garbanzo beans. As well, *sago* is a powdery starch obtained from the stems and trunks of the sago palm in tropical Asia. Sago may be used as a food thickener as well as a fabric stiffener. Fruits such as bananas may also be sources of starch.

Thus it may be seen that starch may come from a variety of sources. Depending on the source, starches may also have different crystalline structures.

V.A. Vaclavik and E.W. Christian, *Essentials of Food Science, 4th Edition*, Food Science Text Series,
DOI 10.1007/978-1-4614-9138-5_4,

Starch Structure and Composition

The starch *granules* from various grains differ in *size*, ranging in size from 2 to 150 μm. The *shape* of starches may also vary—being round or polygonal, as seen in the photomicrographs of corn, wheat, and waxy maize in Figs. 4.1, 4.2, and 4.3.

Starch is made up of two molecules, *amylose* and *amylopectin*, whose parts are connected by glycosidic linkages (see Chap. 3). ***Amylose*** molecules typically make up approximately one-quarter of starch. Amylose is a long linear chain composed of thousands of glucose units with attachment of the carbon 1 and carbon 4 of glucose units, and therefore contains α-1,4 glycosidic linkages. It forms a three-dimensional network when molecules associate upon cooling, and is responsible for the gelation of cooked, cooled starch pastes.

While those starches with a *high* amylose content are able to gel, or hold their shape when molded, starches *without* amylose thicken, although do *not* gel. Examples of the amylose content of various starch sources include:

Cereal grains—26–28 % amylose
Roots and tubers—17–23 % amylose
Waxy varieties of starch—0 % amylose

Amylopectin molecules (Chap. 3) constitute approximately three-quarters of the polymers in a starch granule. The glucose chain of amylopectin contains α-1,4 linkages, similar to amylose, however, with α-1,6 branching at every 15–30 glucose units of the chain. There is a linkage between the carbon 1 of the glucose and carbon 6 of the branch in amylopectin. The chains are highly branched and bushy (however less branched and less bushy than the *animal* storage form of carbohydrate, which is glycogen, *not* starch).

Starches with a high percentage of amylopectin will *thicken* a mixture, although can*not* form a *gel* because, unlike amylose, amylopectin molecules do not associate and form chemical linkages. The greater the amylopectin content, the more viscous the starch paste (not a gel), while the greater the amount of amylase, the stronger the *gel*.

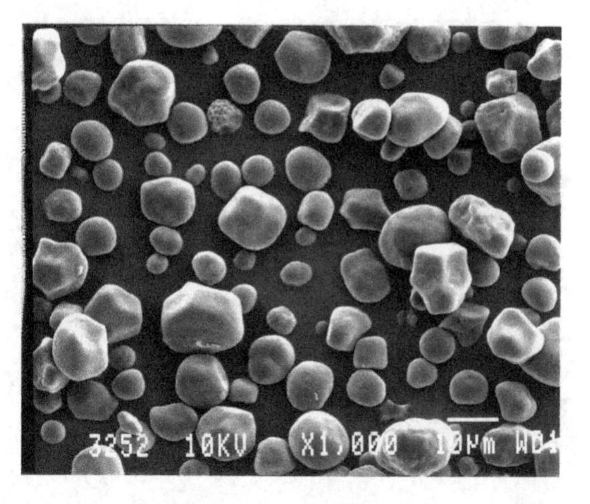

Fig. 4.1 Scanning electron micrograph of common corn cereal grains magnified 2,000 times (*Source*: Purdue University—Whistler Center for Carbohydrate Research)

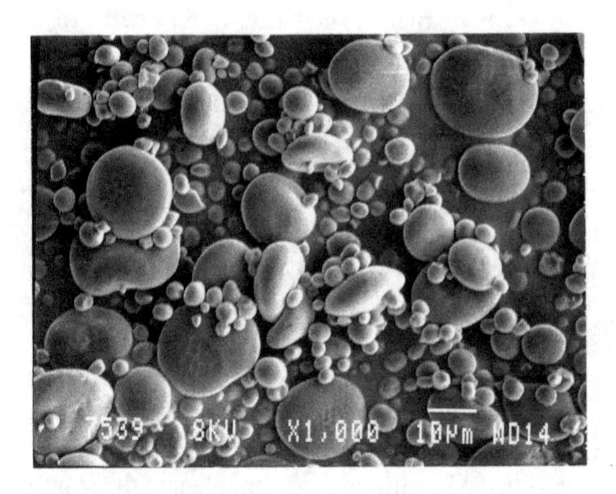

Fig. 4.2 Scanning electron micrograph of wheat magnified 600 times (*Source*: Purdue University—Whistler Center for Carbohydrate Research)

Gelatinization Process in Cooking

Steps in the process of becoming gelatinized will be enumerated in the following text. Starch in its *uncooked* stage is *in*soluble in water. Thus it cannot be referred to as "going into solution," or "dissolving." It forms a temporary ***suspension*** of large granules/particles, which are *un*dissolved in the surrounding medium, and these particles will settle to the bottom of a container of liquid unless it is continuously stirred or otherwise agitated.

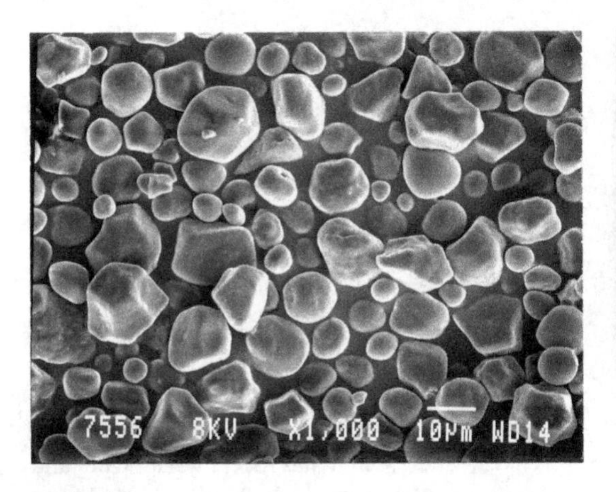

Fig. 4.3 Scanning electron micrograph of waxy maize magnified 1,000 times (*Source*: Purdue University—Whistler Center for Carbohydrate Research)

In a suspension, the starch particles may imbibe a small amount of water; however, generally, a suspension offers minimal change to the starch. Any uptake of water by the starch granule is reversible if starch is dried while still in the *un*cooked state.

Another feature of the *un*cooked starch molecule is that it exhibits a Maltese cross formation, or ***birefringence*** on the granule when it is viewed under polarized light with an electron microscope. This is due to the fact that it is a highly ordered crystalline structure, and light is refracted in two directions (Fig. 4.4).

Once cooking has begun when the starch is heated in surrounding water there occurs ***imbibition***, or the taking of water into the granule. This first occurs in *less*-dense areas, and subsequently in the *more* crystalline regions of the starch molecule. At this *initial* point this is still a *reversible* step in the gelatinization process. However, as heating continues, starch granules take up more water *ir*reversibly and swell; some short chains of amylose come out of the granules. This process, known as ***gelatinization***, is responsible for the thickening of food systems. The gelatinized starch mixtures are opaque and fragile, and the ordered crystalline structure of starch is lost.

As starch leaches out of swollen granules in the gelatinization process, the water–starch mixture becomes a ***sol***. A sol is a colloidal two-phase system containing a liquid continuous phase and a solid dispersed phase. This *solid-in-a-liquid* is pourable and has a low ***viscosity*** or resistance to flow.

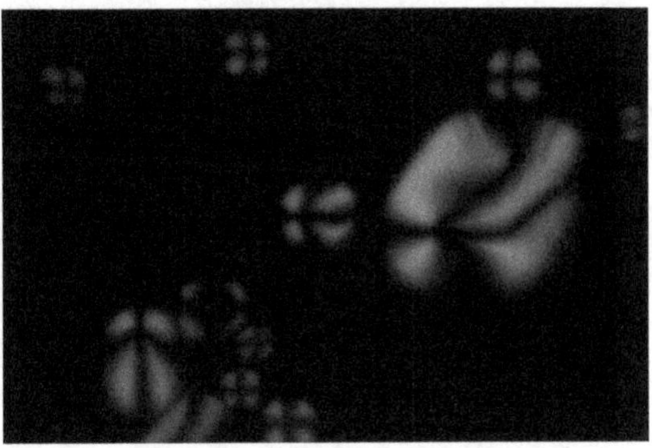

Fig. 4.4 Graph of the thickening of various starches (*Source*: Schoch TJ. Starches in foods. In: *Carbohydrates Their Roles*, Schultz HW, Cain RF, Wrolstad RW, eds. Westport, CT: AVI Publishing Company, 1969. With permission)

$$\text{SUSPENSION} \xrightarrow[\text{heat}]{} \text{SOL}$$

Gelatinization may be synonymous with *pasting*, although the two terms may also be reported as sequential occurrences (Freeland-Graves and Peckham 1996). Whether a separate process or the continuation of gelatinization, pasting occurs with the continued heating of already gelatinized starch grains. The process involves a *loss* of the ordered crystalline structure in starch, which is observed as the *disappearance* of the Maltese cross formation when starch is viewed under polarized light with an electron microscope.

The temperature at which a starch loses its ordered crystalline structure, and gelatinizes, may actually be a *range* of temperatures specific to a starch. The granules within a starch will swell and thicken mixtures at slightly different temperatures, with the larger granules swelling earlier than smaller granules.

The **steps** in the gelatinization process are as follows:

- The gelatinization temperature is *reached*—approximately 140–160 °F (60–71 °C), depending upon the starch type, and is *completed* at 190–194 °F (88–90 °C), or higher.
- The kinetic energy of the hot water molecules breaks the hydrogen bonds between the starch

molecules. Hydrogen bond interchange occurs as starch forms hydrogen bonds with water molecules instead of other starch molecules. As hydrogen bonds are formed, water is able to penetrate further into the starch granule and swelling takes place. Sufficient water must be present to enter and enlarge the starch granule.

- Diffusion of some amylose chains occurs as they leach out of the starch granules.
- Birefringence and the ordered crystalline structure of the uncooked granule is lost. Increased translucency is apparent because the refractive index of the expanded granule is close to that of water.
- Granule swelling increases as the temperature increases. The larger starch granules are the first to swell.
- Swollen granules take up more space and the mixture thickens as the enlarged granules leach amylose and possibly amylopectin.
- The starch paste continues to become thicker, more viscous, and resistant to flow as it gelatinizes.
- The final step in the gelatinization involves the necessity of cooking the gelatinized starch mix—gravy, pie filling, and so forth—for 5 min or longer to develop flavor. Unnecessary overstirring thins the cooked starch mixture because the swollen starch granules implode, rupture, and lose some of the liquid held inside the enlarged granule.

Factors Requiring Control in Gelatinization

It is important to note that starches must first be *thoroughly* gelatinized in order to produce viscous pastes or strong gels. Several factors must be controlled during gelatinization in order to produce a high quality gelatinized starch mixture. (Starches that are not *thoroughly* gelatinized *cannot* produce viscous pastes or strong gels.)

These factors include the following:

Agitation: Agitation, or stirring both initially and throughout the gelatinization process, enables granules of starch to swell independently of one another and creates a more uniform mixture, without lumps. Even so, as previously mentioned, excessive agitation after gelatinization is complete may rupture granules, and consequently thin starch mixtures.

Acid: Acid hydrolysis during cooking of starch granules results in fragmentation and the formation of ***dextrins*** or short chain polymers. Hydrolysis of the starch molecule results in less water absorption by the starch granule, thus a thinner *hot* paste and less firm *cooled* product. Therefore, the late addition of acid to a starch mixture is best, after starch has been gelatinized and begun to thicken. Acid is frequently added to starch sauces in the form of vinegar, tomatoes, fruit, or citrus juice

Enzymes: Starch may be hydrolyzed by the starch-splitting enzymes α-amylase, β-amylase, and beta-glucoamylase.

Endoenzymes such as α-amylase act anywhere on the starch chain and undamaged starch grains to degrade starch. The hydrolysis products of β-amylase are glucose, maltose, and dextrins, depending on the extent of hydrolysis that takes place, and this may be desirable in commercial breadmaking.

The *exoenzyme* β-amylase acts on α-1,4 glycosidic linkages from the nonreducing end, and on damaged amylose or amylopectin chains. This further hydrolyzes starch two glucose units at a time, thus producing maltose.

The β-amylase cannot hydrolyze starch beyond the branch points of amylopectin. The enzyme β-glucoamylase hydrolyses the α-1,4 link, producing glucose, and slowly hydrolyzes α-1,6 linkages in starch.

Fat and proteins: The presence of fat and protein (such as in meat drippings used to produce a meat gravy) initially coats or ***adsorbs*** to the surface of the starch granules causing a delay in hydration and viscosity. Fat "waterproofs" the starch granules so that water does not easily penetrate during the gelatinization process. Thus, with the presence of fat there is less granular swelling and less amylose exiting from the granule, resulting in a decreased viscosity of the starch paste and decreased gel strength.

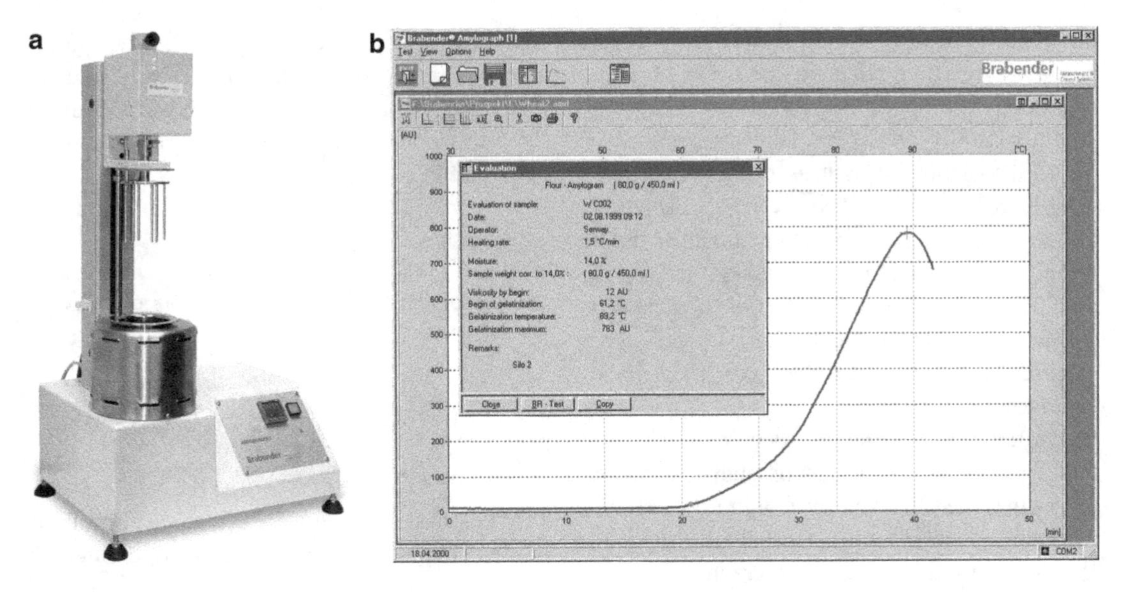

Fig. 4.5 Brabender amylograph and Brabender amylogram (*Source*: C.W. Brabender Instruments, Inc.)

Sugar: The addition of just moderate amounts of sugar, especially the disaccharides sucrose and lactose from milk, decreases starch paste viscosity, the firmness of the cooked and cooled starch product. Sugar competes with the starch for water and thus delays the absorption of water by starch granules. This prevents a speedy or complete swelling of the starch granule. Sugar also elevates the temperature required for gelatinization to occur.

As with acid addition, timing of sugar addition is significant. For a thicker mixture and gel, it is advised that a *partial* addition of sugar before the starch thickens and the remainder added later is best. Thus there is less sugar to compete with granule water absorption than if all of the sugar is added at the beginning of cooking.

If *both* acid and sugar are added to a starch mixture, there is less swelling due to the presence of sugar that competes with starch for water and there is less hydrolysis from acid, on the granule.

Salt: raises the temperature at which a starch mixture thickens.

Temperature: as previously mentioned, there is a range of temperatures, 190–194 °F (88–90 °C), that need to be met for gelatinization to occur. The *completion* is up to 203 °F (95 °C), although starches vary in their gelatinization temperature.

Length of heating: Enough time is required for all the granules to swell (especially the large ones) but as the heating is lengthened, the finished mixture may be thinner due to possible overstirring and rupturing of enlarged granules. Alternatively, cooking for a long time in an uncovered double boiler may evaporate the water that would otherwise thin the mixture.

Type of heat: *Moist heat* is necessary for gelatinization to occur. *Dry heat* causes the starch to hydrolyze, forming shorter chain dextrins. Dry heat creates "browned" flour that imparts a slightly toasted flavor and brown color to a food mixture. This browning effect may be desired in many recipes.

Rate of heating: In general, the faster a starch–water dispersion is heated, the thicker it will be at an identical endpoint temperature.

It can be seen that many factors—many "if's" must be controlled in the gelatinization process. For example, a desired three-dimensional starch structure forms IF gelatinization has occurred correctly, IF the starch is the correct type, IF the concentration of starch is sufficient, IF the heat is correctly applied, IF inclusion of added substances is properly timed, and so forth!

The viscosity of a starch and water mixture is recorded on a moving graph as the mixture is tested and stirred (Fig. 4.4). The recording instrument portrays the thickness of starch mixtures during heating, gelatinization, and cooling. It may be used in order to show the effects of α-amylase on starch mixtures, or the thickness of various starches at different times and temperatures. Evidence of dextrinization may be seen.

Further discussion of specific times and temperatures of when various starches thicken or gel may be observed by reading data from a recording instrument or recording viscometer (Fig. 4.5). As illustrated in the graph in Fig. 4.4, *root* starches such as potato and tapioca, and *waxy cereal* starches thicken earlier and at lower temperatures than cereal starches.

Gelation or Setting of Gelatinized Starch Pastes During Cooling

Amylose

Further changes in the amylose component of starch pastes occur beyond the previously discussed process of gelatinization. During *cooling*, *for* example, the amylose "sets" and forms a ***gel***—a process referred to as ***gelation***. A gel forms a colloidal, elastic and solid, three-dimensional, two-phase system with a *solid* continuous phase of amylose polymers holding a *liquid* dispersed phase. This is in contrast to the ***sol*** it was beforehand.

The reason that the gel network forms is due to the fact that as the mixture cooled, energy is reduced. Subsequently, intermittent hydrogen cross-bonds formed among *amylose*, reassociating at random intervals of the amylose molecule, forming a *gel*.

Amylopectin

On the other hand, the highly branched *amylopectin* molecules do *not* readily form bonds or a gel. Rather, amylopectin exhibits *less* tendency to re-associate or revert to a more crystalline structure than amylose. It remain a sol; however, it forms a *thick* sol upon cooling as shown below. This may be desirable in food items such as pie fillings.

$$\text{SUSPENSION} \xrightarrow[\text{heat}]{\text{imbibition}}$$

$$\overset{\text{maximum gelatinization}}{\text{SOL}} \xrightarrow[\text{cool}]{\text{gelation}} \text{GEL (or a thick sol)}$$

Gels

Starches may be selected for use based on their gelling potential as identified in the following list:

Forms gel	Does not form gel
Cornstarch	Waxy cereals
Wheat starch	Tapioca
Wheat flour	

If gels are formed, mixtures are non-pourable and of high viscosity. It is significant to know that if an appropriate flour is utilized to yield a gel, gelling requires a quantity of two times the amount of flour as cornstarch because an equal amount of flour contains additional non-starch components, such as protein that will not gel.

Retrogradation

Retrogradation refers to the occurrence where starch reverts or retrogrades to a more crystalline structure upon cooling. Both amylose and amylopectin may participate in a textural change that makes them somewhat more "gritty" with time. Retrogradation is more likely to occur in a *high amylose* starch. This occurrence is noted in baked products that become "*stale*," no longer "*fresh*" tasting or "*fresh*" handling (a "fresh" baked good indicates that the starch is still in existence as a gel form). It is also observed in leftover, long-grain rice. Due to its high amylose content, left-over rice gets hard.

Retrogradation Facts

Likely if gel has been formed improperly, the resulting amylose structures are fragile, readily losing entrapped water.

- The amylose retrogrades and recrystallizes.
- Retrogradation is likely when gel is exposed to the effects of freeze–thaw cycles, as the water is frozen and thawed. Water created from melted ice crystals is not able to reassociate with the starch, and water loss becomes apparent.

Included below are a few baking examples of retrogradation from the literature. They are used to better clarify the term.

"In baking, the starch present in bread dough and batters becomes gelatinized. During this process the starch goes from an ordered, crystalline state to a disordered, amorphous state. Upon cooling the disordered starch state begins to re-order (or retrograde), returning the starch back to its more rigid crystalline state, resulting in the firming of crumb texture in baked goods. Starch retrogradation is a time and temperature dependent process." Available from: Tessier, J. Increasing Shelf-Life without Preservatives (Bakers' Journal: July 2001)

"When the starch stays as a gel, a product is softer, and we say it is "fresh". When the starch regains its crystalline form, the product becomes firmer, and we say it is "stale". The technical term for this is starch retrogradation." Available from: Ingredients - Starch and Modified Starch http://sci-toys.com/ingredients/starch.html

"Staling as a result of changes in the starch component (i.e., a change in the amylose and amylopectin starch molecules) of the bakery product is called starch retrogradation. Starch retrogradation begins as soon as baking is complete and the product begins to cool. Amylose retrogradation is mostly complete by the time the product has cooled to room temperature. Amylopectin retrogradation requires more time than amylose retrogradation and as a result, is the primary factor resulting in staling. During the staling period, the amylopectin molecules revert back to their original firm state as rigid crystalline granules. As a result, the baked product loses moisture in the crumb, becoming firmer and less elastic." Available from: Nadia Brunello-Rimando. Bakers' Journal—Voice of the Canadian Baking Industry. May 2004. (Product and Process Development, *Consulting & Technical Services*) http://www.gftc.ca/about-us

Syneresis

Syneresis or "weeping" is water freed from a cooked, cooled starch gel. The process is a change following *gelatinization* and is caused by *gelation*. As a cooked, cooled starch gel stands, the gel ages, then further association of amylose occurs and the gel contracts, causing both water loss and shrinkage to become apparent. This is caused by retrogradation, and is the separation of a liquid from a gel, upon standing. The process is a change following *gelatinization* and is caused by *gelation*.

If cooled undisturbed, the gels remain strong, yet reassociation may be accompanied by the unacceptable water loss or *syneresis*. To control syneresis, modified starches (see "Modified Starches" section) or starches containing only non-gelling amylopectin are used in commercial products.

"Research has well established that the cooling conditions will impact the strength of the gel. Generally, if cooled too fast, the amylose will not have time to form the vital micelles necessary for the three dimensional structure. If cooled too slowly, the amylose fractions will have a chance to align too much and become too close together and the liquid portion will not be trapped in the micelles. In both instances there will be weeping and syneresis." (Oregon State University) Available from: http://food.oregonstate.edu/learn/starch.html

Separating Agents and Lump Formation

Separating agents are used in food preparation in order to prevent lumps in a starch-thickened food item. A problem in the preparation of starch-thickened mixtures is the undesirable formation of lumps. Lumps are due to the unequal swelling or "clumping" together of individual starch granules. The granules must be allowed to swell independently; thus, it becomes important to "separate" the granules with a separating agent.

For product success, one of the three ***separating agents*** such as fat, cold water, and sugar must be used. They should be added to *just* the starch/flour ingredient in order to physically *separate* the grains *prior* to its addition to a recipe. The correct use of any of these agents produces a desirable smooth-textured mixture as opposed to a lumpy mixture.

Fat. *Fat* is a separating agent. When stirred into the flour, fat forms a film around the individual starch granule allowing each granule to swell independently of other granules. Thus, a *lump-free* sauce or gravy is obtained when liquid is added and cooking occurs. Oftentimes a *roux* is made—flour is browned and then separated by agitation with liquid fat during heating.

A roux may range in color from light brown to almost black (Cajun cooking). As a starch is heated and becomes darker, the starch progressively loses its thickening ability as it undergoes dextrinization from heating. An added benefit of adding flour to *hot* meat fat drippings is that α-amylase (which thins) is destroyed.

Cold water. *Cold water* may be used to physically separate starch granules. When mixed with insoluble starch, water puts starch granules in a suspension known as a "*slurry*." The cold water–starch suspension is then slowly mixed into the hot liquid for thickening.

Cold water as a separating agent may be desirable if the product is to remain fat-free or sugar-free. *Hot* water is not a successful separating agent as hot water partially gelatinizes the starch.

Sugar. *Sugar* is a common separating agent used for a sweetened mixture. It is mixed with starch, prior to incorporation into the liquid, so that starch granules remain physically separate to allow individual swelling without lump formation.

Once starch is separated, so that the granules do not "clump together," forming lumps, the separated starch mixture is added to the other recipe ingredients. Sauces must be heated slowly and/or **stirred constantly** in order to be free of lumps. Extensive or harsh stirring after maximum gelatinization has occurred will rupture starch granules, causing the mixture to be thin.

CULINARY ALERT! Many cookbook recipes do not specify the use/proper use of a separating agent and the result is a mixture with lumps! The choice of which separating agent to use is dependent upon the desired end product—e.g., sweetened, fat-free.

Modified Starches

Natural starches may be ***modified*** chemically to produce physical changes that contribute to shelf stability, appearance, convenience, and performance in food preparation. Some "natural" starches are *not* modified chemically which may be a "plus" for concerned consumers and processors. Various examples of modified starches used in food manufacturing are described in the following text.

Pregelatinized starch is an instant starch that has been gelatinized and then dried. It subsequently swells in liquid without the application of heat. Pregelatinized starch appears in many foods, including instant pudding mixes.

Some properties of a pregelatinized starch include the following:

- Dispersible in cold water; it can thicken without heat being applied
- Can be cooked and dried, yet is able to reabsorb a lot of water in preparation without cooking the food (instant pudding)

- Undergoes irreversible change and can*not* return to its original ungelatinized condition after treatment
- A greater weight of starch is required to thicken a liquid because some rupturing and loss of starch granule contents occurred during gelatinization and drying

Cold water-swelling (CWS) starch is an instant starch that remains as an *intact granule*. It offers convenience, stability, clarity, and texture. CWS starches may be gelling or nongelling. They may be used in no-cook or cold-process salad dressings providing the thick, creamy mouthfeel in no-fat salad dressings.

Cross-linked starches are those that undergo a molecular reaction at selected hydroxyl (−OH) groups of two adjoining, intact, starch molecules. The purpose of cross-linking is to enable the starch to withstand such conditions as low pH, high shear, or high temperatures. The cross-linked starch becomes less fragile and more resistant to rupture than the original unmodified starch.Although it is more tolerant of *high* temperatures, it is *not* more tolerant of *cold* temperatures. These starches are used in many foods, especially acid food products such as pizza sauce or barbecue sauce because the modified starch is *more acid-resistant* than an unmodified starch.As a result of cross-linking, a starch swells less and is less thick.

Stabilized (substituted) starches are used in frozen foods and other foods stored at cold temperature in order to prevent gelling and subsequent syneresis. The main types of substitutions include hydroxypropylated, hydroxyethylated, and so forth.These starches prevent molecular associations and cause ionic repulsion. The stabilized starch produces pastes able to withstand several freeze–thaw cycles before syneresis occurs. This is value to the frozen food industry, and also to foods such as sauces and gravies stored at cold temperatures.Stabilized starches are *not* appropriate for foods that require prolonged *heating*. However, starches may be modified by a combination of both cross-linking and stabilization treatments. Such modification ensures that the starches are acid-, heat-, and freeze–thaw-stable. Stabilized starches have a wide range of uses in food products.

Acid-modified starch is starch that is subject to treatment in an acid slurry. A raw starch and dilute acid are heated to temperatures less than the gelatinization temperature. Once the starch is mixed into a food product, it appears less viscous in *hot* form, although it forms a strong gel upon *cooling*.More about Modified Starches as follows: http://food.oregonstate.edu/learn/starch.html

Non-food uses for Modified Starches include glue in cardboard manufacture and glue on postage stamps

Waxy Starches

Waxy starches are derived from some natural strains of barley, corn, rice, and sorghum. They do *not* contain amylose, begin to thicken at *lower* temperatures, become *less* thick, and undergo *less* retrogradation than non-waxy varieties. *Waxy* cornstarch, for example, does not have the same gel forming properties as regular cornstarch. It contains *no* gel producing *amylose*, and only *amylopectin*.

- Waxy cornstarch—contains NO amylose, is all amylopectin, and does NOT gel
- Ordinary *cornstarch*—contains 27 % amylose and forms a gel
- *High amylose cornstarch*—contains 55 % amylose and forms a gel

Waxy varieties of starch are commonly used in the preparation of pie fillings to *thicken*, however, *not gel*. They may also be cross-linked for better function.

Starch Uses in Food Systems

Starches have many uses in food preparation and are very versatile and oftentimes inexpensive. They may be introduced into foods primarily because of their *thickening* ability. For example, pureed, cooked, or instant potatoes, or pureed cooked rice may be undetectable, however, useful as thickeners. A white sauce may be added during the preparation of a tomato and milk-based soup, in order to thicken and stabilize. It aids in the control of milk protein precipitation caused from the addition of tomato acid. Starch may also be useful as a water binder and gelling agent.

Another use of starch is as a *fat replacer* in food systems. Molecularly, the amylose chains form helical or spherical shapes, holding water and providing bulk. This confers the satisfying "mouthfeel" attributes on starch. Intermediate length polymers of D-glucose, called ***malto-dextrins***, are formed from the hydrolysis of starches such as tapioca, potato, and wheat. Maltodextrins simulate the viscosity and mouth-feel of fats/oils and are used to reduce the fat content of some foods.

With the use of ordinary cross-breeding procedures, new starches are being discovered that have various applications in food systems. Baking, microwave cakes, frozen sauces, fat replacers, breadings, snacks, and gelled candies are some of the uses of starch (American Maize-Products Company, Hammond, IN). For example, **pea starch** may offer an alternative to other modified starches used in the food industry as it provides a very high viscosity *immediately* upon agitation. It is available in pregelatinized form for use in cold processed products such as dessert creams, dressings, instant soups, and sauces (Feinkost Ingredient Company, Lodi, OH). Pea starch may also be environmentally friendly as a biodegradable food packing material introduced in landfill sites.

New food starches and their uses are continually being developed. Food starches are commercially manufactured and available for use in products such as baked food, beverages, canned, frozen, and glassed foods, confections, dairy products, dry goods, meat products, and snack foods (National Starch and Chemical Company—Food Products Division, Bridgewater, NJ).

CULINARY ALERT! A starch chosen for use in food systems may involve a choice by habit, or convenience. Consumers may actually use less than the best because "it's what mom always used, so I'll use it too," or "it's here in the kitchen, so I'll use it!"

Cooking with Starch

Several of the applications of cooking with starch appear below. Cooking with the appropriate starch, in the proper concentration, timing of addition, and so forth as previously discussed are factors crucial to the success of any starch-thickened product.

Appearance

The appearance of a cooked, cooled starch mixture is influenced by the choice of starch. For example, *cereal* starches in general produce *cloudy*, thickened mixture upon cooling. Within the group of cereals, *flour* produces a *more* cloudy thickened mixture than corn*starch* because the wheat flour contains additional non-starch ingredients not present in cornstarch. A *clear gel* is produced using cornstarch.

A *clear, thickened mixture* is also produced by other non-gelling starch sources such as Arrowroot. Non-gelling may be a desirable feature of pie fillings.

Use of a Double Boiler

Cooking *over boiling water* (such as with a double boiler for household preparation) promotes temperature control and a more even gelatinization than would occur with *direct heat* cooking. A disadvantage of this cooking method is that it requires cooking for a longer time period to reach the thickening stage than a direct heat cooking method.

Tempering

Tempering involves the technique of *slowly* adding small amounts of hot starch to eggs in a recipe, in order to gradually raise the temperature thus slowly exposing eggs to heat without the danger of coagulation. In this manner, the eggs do not curdle and produce an unacceptable consistency. To achieve the desired consistency and texture of a recipe containing hot starch *and* raw eggs (in sauces, cream puffs, etc.), the process of *tempering* is used.

White Sauce

White sauces have widespread applications in cooking. The *concentration* of starch used in a formulation varies. For example, a white sauce of flour, fat, and milk may be thickened to various consistencies, for croquettes, sauces, and so forth. The concentration of flour may be as follows:

White Sauces
Thin—1 tablespoon of flour/cup of liquid
Medium—2 tablespoons of flour/cup of liquid
Thick—3 tablespoons of flour/cup of liquid

Liquid

The use of liquid type varies in starch mixtures. *Water or fruit juice* is incorporated into some foods dictated by need for clarity or flavor. Milk is usually used in a starch-thickened sauce such as white sauce. Since milk easily curdles at high temperatures, it may be made less likely to curdle if it is first thickened with flour prior to recipe addition.

CULINARY ALERT! Flourless sauces are thickened by reduction of the stock/liquid. Portions of the starchy ingredients of a soup recipe may be saved out and pureed, and then added back to the soup in order to thicken and flavor it.

Nutritive Value of Starch

Nutritive value is provided by starches. Starches are a complex carbohydrate containing 4 cal/g, and traces of protein and fat. Short chain maltodextrins derived from the hydrolysis of starch may be used in foods to partially replace fat. Maltodextrins simulate the taste of fat, and offer less calories per gram than the 9 cal/g in fat.

Not all starch is capable of carrying calories or being digested. A "*resistant starch*" is dietary fiber, with an example being whole cooked beans. Resistant starches offer benefits to the colon, namely, "roughage." Also, intestinal bacterial flora use fiber, producing vitamins such as vitamin K.

Whole grains that are *ground* to make flour are different than the whole grain from which they came. For example, they have a higher *glycemic index* than *un*ground grains. This is due to the ease of absorbing the starch into the blood as sugars.

Special nutritional needs may require a dietary restriction of wheat that may lead to use of non-wheat starches for those individuals following gluten-free diets. Assorted alternatives to wheat are corn, potato, or rice starch. Packages of potato "flour" indicate on the finer print of the label that the contents are solely potato *starch* (Ener-G Foods, Inc., Seattle, WA). A gluten-free addition of starch or even fiber may be utilized in product development (Hazen 2012).

Safety of Starches

Starches are one of many white powders used in food handling and production operations. Proper storage, including separation from other dangerous chemicals, is crucial. If used in bulk quantities, labeling of both the container and its lid (if removable) better assures safety of starches in the workplace.

Conclusion

Starch is a plant polysaccharide that is the storage form of carbohydrate in roots, seeds, and tubers. It may be derived from cereals such as corn, wheat, rice, or oats, or legumes such as soybeans, or from vegetable roots and tubers such as potatoes or arrowroot. In its uncooked stage, starch is insoluble in water. As it is heated and undergoes gelatinization, factors such as acid, agitation, use of enzymes, fat, proteins, sugar, and temperature require control. A separating agent prevents lumps in a starch mixture.

The source of starch and its concentration determine the thickening, gelling, retrogradation, and clarity of the finished product. Flour and cornstarch may be used to form gels; waxy varieties of starch do not gel. Syneresis may occur as the cooked, cooled starch mixture ages. Modification of starch granules allows starches to be used successfully in a variety of food applications. Starch may be added to foods in order to provide thickening, or product stability, or potentially, to carry flavors.

Notes

CULINARY ALERT!

Glossary

Adsorb Surface adherence of gas, liquids, or solids onto a solid.

Amylose Long, linear chain composed of thousands of glucose molecules joined by an α-1,4-gycosidic linkage.

Amylopectin Branched chains of glucose units joined by α-1,4 linkages, with α-1,6 branching occurring every 15–30 units.

Birefringence A Maltese cross appearance on each uncooked crystalline starch granule when viewed under a polarizing microscope due to light refraction in two directions.

Dextrin Glucose polymers; a product of the early stages of starch hydrolysis.

Gel Elastic solid formed upon cooling of a gelatinized starch paste; a two-phase system that contains a solid continuous phase and a liquid dispersed phase.

Gelatinization Starch granules take up water and swell irreversibly upon heating, and the organized granular pattern is disrupted.

Gelation Formation of a gel upon cooling of a gelatinized starch paste.

Granule Starch grain of long-chain glucose polymers in an organized pattern; granule shape is particular to each starch type.

Imbibition Starch granules taking up water and swelling as it is exposed to moist heat.

Maltodextrin Starch hydrolysis derivative that may be used to simulate fat in formulations.

Modified starch Specific chemical modification of natural starches to physically create properties that contribute to shelf stability, appearance, convenience, and performance in food preparation.

Retrogradation Reverting back, or reassociation of amylose as the gelatinized starch once again forms a more crystalline structure upon cooling.

Separating agent Prevents lump formation in a starch mixture. Physically separates starch grains and allows their individual swelling.

Sol A two-phase system with a solid dispersed in a liquid continuous phase.

Spherical aggregate Open, porous starch granules with spaces that can be filled and used to transport materials such as flavor, essences, and other compounds.

Starch Carbohydrate made up of two molecules—amylose and amylopectin.

Suspension Large particles undissolved in the surrounding medium. Particles are too large to form a solution or a sol upon heating.

Syneresis "Weeping" or water loss from a cooked, cooled gel due to excessive retrogradation or improper gel formation.

Viscosity Resistance to flow of a liquid when force is applied. A measure of how easily a liquid will flow. Thin liquids have a low viscosity. Thick liquids or gels have a high viscosity and flow slowly.

References

Freeland-Graves JH, Peckham GC (1996) Foundations of food preparation, 6th edn. Macmillan, New York

Hazen C (2012). Fiber files. Food Product Design September: pp 102–112

Bibliography

A. E. Staley Manufacturing Co. Decatur, IL

Bennion M (1980) The science of food. Harper and Row, New York

Bennion M, Scheule B (2009) Introductory foods, 13th edn. Prentice Hall, Upper Saddle River, NJ

5 Pectins and Gums

Introduction

Pectins and gums are important polysaccharides in foods because of their functional properties. They are widely used as gelling agents, thickeners, and stabilizers. They are constituents of plant tissue and are large, complex molecules whose exact nature is not certain. However, enough is known to understand some of their properties and to make use of their functional properties to produce convenience and special texture foods.

To identify a few, pectic acid is found in overripe fruit. Some recognizable gums are seed gums such as guar gum and locust bean gum, and common seaweed polysaccharides including carrageenan and agar.

Pectic Substances

Pectic substances including protopectin, pectinic acid, and pectic acid are an important constituent of plant tissue and are found mainly in the primary cell wall. They also occur between cell walls, where they act as intercellular cement. Although their exact nature is not clear, they can be considered as linear polymers of D-galacturonic acid joined by α-1,4-glycosidic linkages, as shown in Fig. 5.1. Some of the acid or carboxyl (COOH) groups along the chain are esterified with methanol (CH_3OH) as shown.

Each glycosidic linkage is a ***cross-planar*** bond, because it is formed by reaction of one hydroxyl group located above the plane of the first ring with another hydroxyl group located below the plane of the second ring. The configuration of these bonds causes twisting of the molecule, and the resulting polymer can be likened to a twisted ribbon. Cross-planar bonds are not readily digested in the human digestive tract, and so pectins are classified as soluble fiber.

Pectic substances may be grouped into one of the three categories depending on the number of methyl ester groups attached to the polymer. ***Protopectin*** is found in immature fruits, and is a high-molecular weight methylated galacturonic acid polymer. It is insoluble in water yet can be converted to water-dispersible pectin by heating in boiling water. It cannot form gels.

Pectinic acid is a methylated form of galacturonic acid that is formed by enzymatic hydrolysis of protopectin as a fruit ripens. High-molecular weight pectinic acids are known as pectins. Pectinic acids are dispersible in water and can form gels. ***Pectic acid*** is a shorter-chain derivative of pectinic acid that is formed as fruit overripens. Enzymes, such as polygalacturonase and pectinesterase, cause depolymerization and demethylation of the pectinic acid, respectively. Complete demethylation yields pectic acid, which is incapable of gel formation.

V.A. Vaclavik and E.W. Christian, *Essentials of Food Science, 4th Edition*, Food Science Text Series,
DOI 10.1007/978-1-4614-9138-5_5,

Fig. 5.1 Basic structure of pectic substances

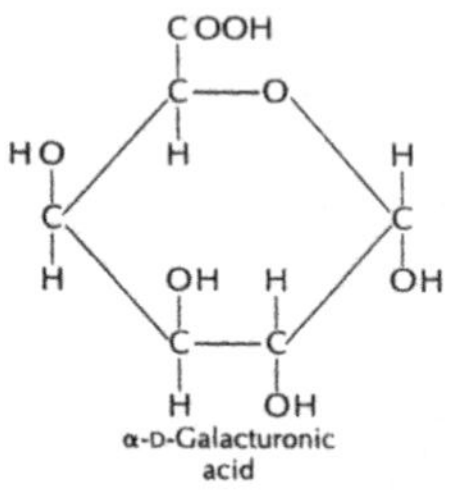

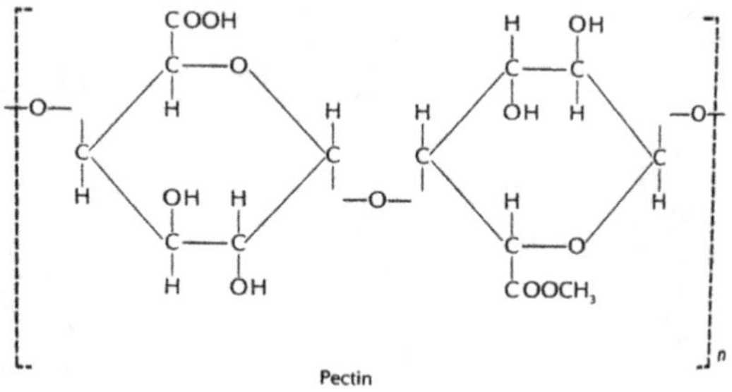

Pectic Substances
Protopectin—methylated galacturonic acid polymer found in immature fruits.
Pectinic acid—methylated galacturonic acid polymer; includes pectins.
Pectic acid—short-chain demethylated derivative of pectinic acid found in over-ripe fruits.

Pectins

Pectins are high-molecular weight pectinic acids and are dispersible in water. Some of the carboxyl groups along the galacturonic acid chain are esterified with methanol. The degree of esterification in unmodified pectins ranges from about 60 % in apple pulp to about 10 % in strawberries. (Pectins can be deliberately deesterified during extraction or processing.) According to the degree of esterification, pectins are classified as *high-methoxyl* or *low-methoxyl* pectins. The two groups have different properties and gel under different conditions.

Low-methoxyl pectins. Low-methoxyl pectins contain mostly free carboxyl groups. In fact, only 20–40 % of the carboxyl groups are esterified. Therefore, most of them are available to form cross-links with divalent ions such as calcium, as shown in Fig. 5.2.

If sufficient cross-links are formed, a three-dimensional network can be obtained that traps liquid, forming a gel. Low-methoxyl pectins can, thus, form gels in the presence of divalent ions without the need for sugar or acid.

High-methoxyl pectins. High-methoxyl pectins contain a high proportion (usually 50–58 %) of esterified carboxyl groups. Most of the acid groups are, therefore, not available to form cross-links with divalent ions, so these pectins do not form gels. However, they can be made to gel with the addition of sugar and acid. It is the high-methoxyl pectins that are commonly used to form pectin jellies.

Pectin Gel Formation

A pectin gel consists mainly of water held in a three-dimensional network of pectin molecules. Pectin is dispersible in water and forms a ***sol*** (solid dispersed in liquid continuous phase), although under the right conditions, it can be converted into a ***gel*** (liquid dispersed in solid continuous phase). This occurs when the pectin molecules interact with each other at specific points. It is not easy to form pectin gels; it requires a delicate balance of pectin, water, sugar, and acid.

Pectin is hydrophilic (water loving) due to the large number of polar hydroxyl groups and charged carboxyl groups on the molecule. When pectin is dispersed in *water*, some of the acid groups ionize, and water binds to both the charged and polar groups on the molecules. The negative charge on the pectin molecules, coupled with their attraction for water, keeps them apart so that they form a stable sol.

To form a gel, the forces keeping the pectin molecules apart must be reduced so that they can interact with each other at specific points, trapping water within the resulting three-dimensional network. In other words, the attraction of the pectin molecules for water must be *decreased* and the attraction of the pectin molecules for

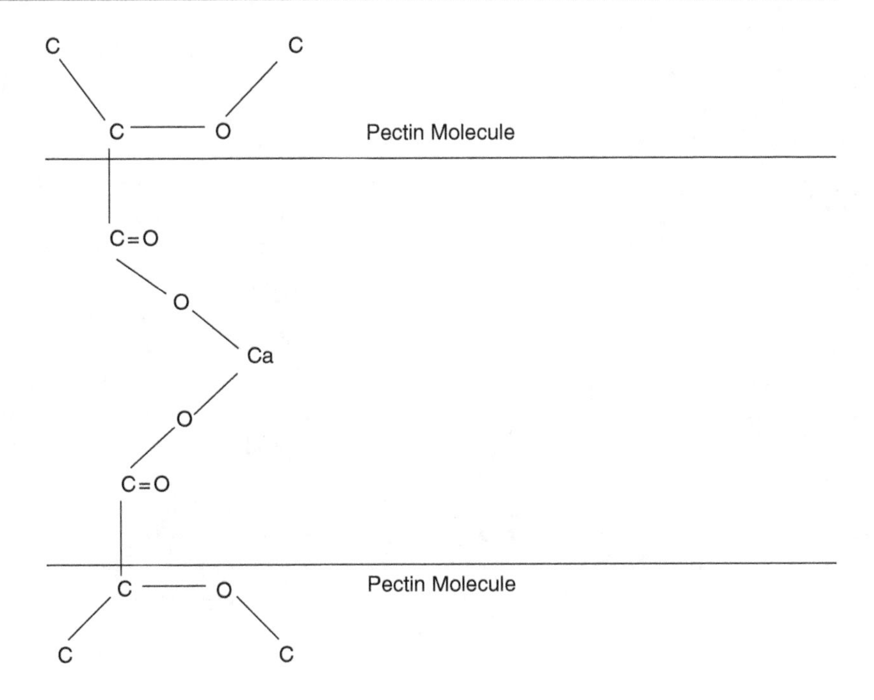

Fig. 5.2 Cross-links in low-methoxyl pectin

each other must be *increased*. This can be achieved by addition of sugar and acid.

Sugar competes for water, thus making less water available to associate with the pectin molecules. This reduces the attractive forces between the pectin and water molecules.

Acid adds hydrogen ions, reducing the pH. (The pH must be below 3.5 for a gel to form.) Carboxylic acids contain a ***carboxyl group*** (COOH), are weak acids, and are not fully ionized in solution; the un-ionized form of the acid exists in equilibrium with the ionized form.

$$-\mathrm{COOH} + \mathrm{H_2O} \leftrightharpoons -\mathrm{COO^-} + \mathrm{H_3O^+}$$

When hydrogen ions are added, they react with some of the ionized carboxyl groups to form undissociated acid groups. In other words, the equilibrium is shifted to the left, and more of the carboxylic acid is present in the un-ionized form. Thus, when hydrogen ions are added to pectin, the ionization of the acid groups is depressed and the charge on the pectin molecules is reduced. As a result, the pectin molecules no longer repel each other.

In fact, there is an attractive force between the molecules and they align and interact at specific regions along each polymer chain to form a three-dimensional network. These regions of interaction are called ***junction zones***, shown diagrammatically in Fig. 5.3. However, there are also regions of the pectin chains that are not involved in junction zones because they are unable to interact with each other. These regions form pockets or spaces between the junction zones that are able to entrap water. Hence, a gel is formed, with water trapped in the pockets of the three-dimensional pectin network.

Exactly how the junction zones form is not certain, although hydrogen bonds are thought to play an important role. The ***steric fit*** of the molecules (in other words, their ability to fit together in space) is also important. Pectin molecules contain minor components such as rhamnose and other neutral sugars that are bound to the main galacturonic acid chain by 1-2-glycosidic links. These sugars cause branches or kinks in the molecules and make it difficult for them to align and interact to form junction zones. However, there are regions of the pectin chains that do not contain these neutral sugars and it is

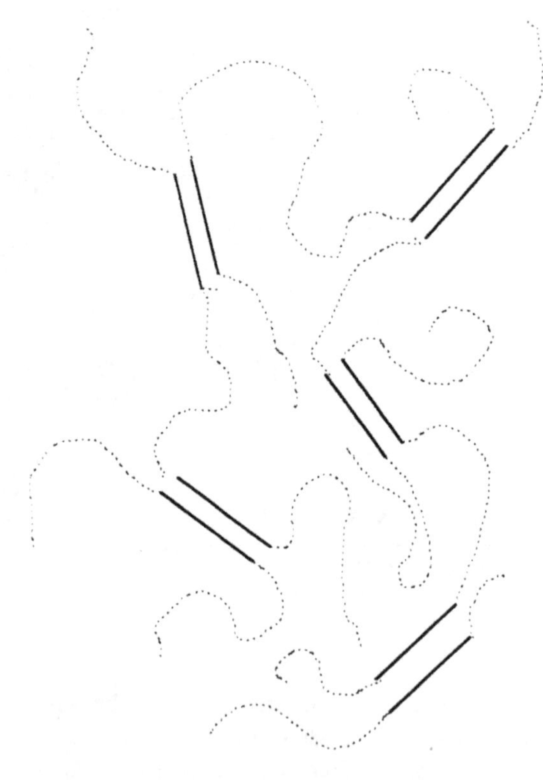

Fig. 5.3 Junction zones in a pectin gel. Generalized two-dimensional view. Regions of the polymer chain involved in junction zones are shown as—. The other regions of the chain are shown as—. Water is entrapped in the spaces between the chains. Adapted from Coultate (2009)

these regions that are thought to form the junction zones.

High-methoxyl pectins form gels in this way. Low-methoxyl pectins require divalent ions to gel, and intermediate pectins require sugar, acid, and divalent ions to gel.

Pectin Sources

Pectins with a high-molecular weight and a high proportion of methyl ester groups have the best jelly-forming ability. The pectin content of fruits is variable and depends not only on the type of fruit but also on its maturity or ripeness. If jellies or jams are made at home, it is best to add commercial pectin to ensure that there is sufficient pectin to form a gel. Purified pectin is made from apple cores and skins (apple pomace) and from the white inner skin (albedo) of citrus fruits. It is available in either liquid or granular form. The granular products have a longer shelf life than the liquids. Low-methoxyl pectin can be obtained by demethylating pectin with enzymes, acid, or alkali until it is 20–40 % esterified (Glicksman 1982). Since these pectins gel with divalent ions and need no sugar, they can be used commercially for the production of low-calorie jams, jellies, or desserts. They have also been introduced to the retail market so that such low-calorie products may be made at home.

Pectin Gel Formation

In a pectin ***sol***

- **Water** binds to ionic and polar groups on pectin.
- **Pectin** molecules are negatively charged and hydrated; therefore, they do not interact with each other.

To form a pectin ***gel***

- Attraction of pectin molecules for water must be decreased.
- Attraction of pectin molecules for each other must be increased.

This is achieved with

- **Sugar**
  - Competes for water.
  - Decreases pectin–water attraction.
- **Acid**
  - Adds hydrogen ions.
  - Depresses ionization of pectin.
  - Reduces the charge on the pectin molecules.
  - Increases pectin–pectin attraction.
- **Pectin**
  - Interacts at junction zones forming a three-dimensional network.
  - Pectin becomes the continuous phase.
- **Water**
  - Is trapped in pockets within the gel network.
  - Water becomes the disperse phase.

Some Principles of Making Jelly

This book does not attempt to describe the practical aspect of making jellies. For such information, the reader is referred to consumer information publications, or to books by authors such as Charley (Charley and Weaver 1998) or Penfield and Campbell (Coultate 2009). The intention is to highlight some of the more important scientific principles of jelly-making.

To make jelly, fruit juice (which is a source of water and acid), pectin, and sucrose are combined in a suitable pan and heated until the mixture boils. The temperature and boiling time are monitored, and boiling is continued until the desired temperature is reached.

As boiling continues, water evaporates, the concentration of sucrose increases, and the boiling point of the jelly also increases. Therefore, the boiling point can be used as an index of sucrose concentration. By measuring the temperature of the boiling jelly, it can be determined when sufficient water has been removed to give the desired sucrose concentration in the final jelly.

However, all solutes increase the boiling point of water, so it is important to allow for the effect of any additional ingredients on the boiling point; a pure sucrose solution may boil at a lower temperature than a jelly mix containing the same concentration of sucrose. In other words, a boiling jelly may contain less sucrose than expected if the effects of the additional ingredients on the boiling point are not taken into account. This could result in a runny or weak gel.

It is important to control the boiling time, not just the temperature of the boiling jelly, because chemical reactions occur in the presence of heat and acid that need to be controlled to maintain gel quality. Glycosidic links are hydrolyzed in the presence of heat and acid. Therefore, depolymerization of pectin will occur if the boiling time is too long. This will result in loss of gelling power and the gel may not set.

During the boiling process, sucrose is converted to invert sugar, and the presence of invert sugar in the jelly prevents crystallization of sucrose on storage over a long period. A short boiling time may not allow formation of sufficient invert sugar to inhibit sucrose crystallization over time, especially if the jelly is stored at refrigeration temperatures.

Boiling Time

- If too long—depolymerization of pectin occurs, and the gel may not set.
- If too short—insufficient invert sugar may be formed, and crystallization of sucrose may occur.

It has already been mentioned that a commercial pectin should be used in addition to the fruit, because the quality of pectin in the fruit varies. An overripe fruit is deficient in pectin, because demethylation and depolymerization occur as the fruit ages. Hydrolysis of only a few glycosidic bonds causes a marked drop in viscosity and gelling power and will produce a weak gel.

Gums

Gums are a group of complex hydrophilic carbohydrates containing thousands of monosaccharide units. Galactose is the most common monosaccharide found in gums; glucose is usually absent. Gums are often referred to as ***hydrocolloids***, because of their affinity for water and their size; when added to water, they form stable aqueous colloidal dispersions or sols. The molecules are highly branched, and as a result most gums are unable to form gels. However, they are able to trap or bind large amounts of water within their branches. Aqueous dispersions therefore tend to be very viscous, because it is difficult for the molecules to move around freely without becoming entangled with each other.

Main Characteristics of Gums

- Larger highly branched hydrophilic polymers
- Rich in galactose
- Hydrocolloids
- Form viscous solutions
- Most do not gel if used alone
- Seaweed polysaccharides form gels

Gums are classified as soluble fiber because they undergo little digestion and absorption in the body. Therefore, they supply relatively few calories to the diet, as compared with digestible carbohydrates such as starch.

Gums are common in a wide range of food products, including salad dressings, sauces, soups, yogurt, canned evaporated milk, ice cream and other dairy products, baked goods, meat products, and fried foods. They are used as thickening agents in food products, replacing starch. They are also used to assist in the stabilization of emulsions and to maintain the smooth texture of ice cream and other frozen desserts. They are common in reduced fat products, because they are able to increase viscosity and help to replace the texture and mouthfeel that was contributed by the fat.

Gums are obtained from plants, and can be separated into five categories: seed gums, plant exudates, microbial exudates, seaweed extracts, and synthetic gums derived from cellulose.

Seed Gums

The seed gums include guar and locust bean gums. These gums are branched polymers containing only mannose and galactose. Guar gum contains a mannose/galactose ratio of 2:1, whereas the ratio is 4:1 in locust bean gum. Guar gum is soluble in cold water, whereas locust bean gum must be dispersed in hot water. Neither gum forms a gel when used alone. However, they may be used synergistically with other gums to form gels.

Guar gum forms gels with carrageenan and guar gum. It is used to stabilize ice cream, and it is also found in sauces, soups, and salad dressings.

The presence of guar gum in the intestine seems to retard the digestion and absorption of carbohydrates and slow absorption of glucose into the bloodstream. Use of guar gum in foods may, therefore, be useful in treating mild cases of diabetes (Penfield and Campbell 1990).

Locust bean gum is typically used as a stabilizer in dairy and processed meat products. It may also be used synergistically with xanthan gum to form gels.

Plant Exudates

The plant exudates include gum arabic, which comes from the acacia tree, and gum tragacanth. These are complex, highly branched polysaccharides. Gum arabic is highly soluble in cold water, and is used to stabilize emulsions and to control crystal size in ices and glazes. Gum tragacanth forms very viscous sols, and is used to impart a creamy texture to food products. It is also used to suspend particles, and acts as a stabilizer in products such as salad dressings, ice cream, and confections.

Categories of Gums

- **Seed gums:** guar gum, locust bean gum
- **Plant exudates:** gum arabic, gum tragacanth
- **Microbial exudates:** xanthan, gellan, dextran
- **Seaweed polysaccharides:** alginates, carrageenan, agar
- **Synthetic gums:** microcrystalline cellulose, carboxymethyl cellulose, methyl cellulose

Microbial Exudates

Xanthan gum, gellan gum, dextran, and curdlan are all gums produced using fermentation by microorganisms. Of these, *xanthan* is the most common. Xanthan forms viscous sols that are stable over a wide range of pH and temperature. It does not form a gel, except when used in combination with locust bean gum. It is used in a wide range of products as a thickener and stabilizer and suspending agent. For example, most salad dressings contain xanthan gum. Xanthan gum is extremely versatile and relatively inexpensive, which makes its presence almost ubiquitous in thickened food products.

Seaweed Polysaccharides

The ***seaweed polysaccharides*** include the agars, alginates, and carrageenans. Unlike most other gums, they are able to form gels under certain conditions.

Carrageenan is obtained from red seaweeds, and especially from Irish moss. It occurs as three main fractions, known as kappa-, iota-, and lambda-carrageenan. Each is a galactose polymer containing varying amounts of negatively charged sulfate esters. Kappa-carrageenan contains the smallest number of sulfate esters, and is therefore the least negatively charged. It is able to form strong gels with potassium ions. Lambda-carrageenan contains the largest number of sulfate groups, and is too highly charged to form a gel. Iota-carrageenan forms gels with calcium ions.

The carrageenan fractions are generally used in combination. Several different formulations are available, containing different amounts of the individual fractions, and food processors are able to choose formulations that best fit their needs.

The carrageenans are used to stabilize milk products such as ice cream, processed cheese, canned evaporated milk, and chocolate milk, because of their ability to interact with proteins. The carrageenans may also be used with other gums, because of their ability to cross-link with them (see more in Food Additives chapter).

Agar is also obtained from red seaweeds. It is noted for its strong, transparent, heat-reversible gels; that is, agar gels melt on heating and re-form when cooled again. Agar contains two fractions—agarose and agaropectin, both of which are polymers of β-D- and α-L-galactose. Agaropectin also contains sulfate esters.

The **alginates** are obtained from brown seaweeds. They contain mainly D-mannuronic acid and L-guluronic acid, and they form gels in the presence of calcium ions. Calcium alginate gels do not melt below the boiling point of water; thus, they can be used to make specialized food products. Fruit purees can be mixed with sodium alginate and then treated with a calcium-containing solution to make reconstituted fruit. For example, if large drops of cherry/alginate puree are added to a calcium solution, convincing synthetic cherries are formed. Reconstituted apple and apricot pieces for pie fillings can also be made by rapidly mixing the sodium alginate/fruit puree with a calcium solution and molding the gel into suitable shapes.

Functional Roles of Gums

Gums may be used to perform one or more of the roles in food peoducts.

- **Thickeners**—salad dressings, sauces, soups, beverages
- **Stabilizers**—ice creams, icings, emulsified products
- **Control crystal size**—candies
- **Suspending agents**—salad dressings
- **Gelling agents**—fruit pieces, cheese analogs, vegan jellies
- **Coating agents**—batters for deep-fried foods
- **Fat replacers**—low-fat salad dressings, ice creams, desserts
- **Starch replacers**—baked goods, soups, sauces
- **Bulking agents**—low-fat foods
- **Source of fiber**—beverages, soups, baked goods

CULINARY ALERT! Vegan jellies can be made using products that contain agar, carrageenan, and/or other gums such as locust bean gum and xanthan gum in place of gelatin.

(One such product is *Lieber's Unflavored Jel*, which contains both carrageenan and locust bean gum, and is available from http://www.VeganEssentials.com.)

Synthetic Gums

Cellulose is an essential component of all plant cell walls. It is insoluble in water and cannot be digested by man, and so it is not a source of energy for the body. It is classified as insoluble fiber.

The polymer contains at least 3,000 glucose molecules joined by β-1,4-glycosidic linkages. Long cellulose chains may be held together in bundles forming fibers, as in the stringy parts of celery.

Synthetic derivatives of cellulose are used in foods as *nonmetabolizable* bulking agents, binders, and thickeners. *Microcrystalline cellulose*, known commercially as Avicel (FMC Corp.), is used as a bulking agent in low-calorie foods. It is produced by hydrolysis of cellulose with acid. ***Carboxymethyl cellulose*** (CMC) and *methyl cellulose* (MC) are alkali-modified forms of cellulose. The former is the most common, and it is often called simply *cellulose gum*. It functions mainly to increase the viscosity of foods. It is used as a binder and thickener in pie fillings and puddings; it also retards ice crystal growth in ice cream and the growth of sugar crystals in confections and syrups.

In dietetic foods, it can be used to provide the bulk, body, and mouthfeel that would normally be supplied by sucrose. *Methyl cellulose* forms gels when cold dispersions are heated. It is used to coat foods prior to deep fat frying, in order to limit absorption of fat. Two other forms of modified cellulose include *hydroxypropyl cellulose* and *hydroxypropylmethyl cellulose*. These are also used as batters for coating fried foods.

As well, there may be a synergistic effect of using various gums together. This is most likely the case when more than one gum name appears on a food label!

Conclusion

Pectins and seaweed polysaccharides are useful for various food products because of their gelling ability. In general, gums are important because they form very viscous solutions, but most do not gel. All these carbohydrates are important to the food industry because of their functional properties and their ability to produce foods with special textures. Used in a wide range of food products, as gelling agents, thickeners, and stabilizers, their availability has increased the choice and quality of many convenience foods. Synthetic derivatives of cellulose are important as nonmetabolizable bulking agents, thickeners, and stabilizers in a wide range of calorie-reduced foods.

Notes

CULINARY ALERT!

Glossary

Carboxyl group COOH group; weak acid group that is partially ionized in solution.

Carboxymethyl cellulose (CMC) Synthetic derivative of cellulose used as a bulking agent in foods. Also known as cellulose gum.

Cellulose Glucose polymer joined by β-1,4-glycosidic linkages; cannot be digested by humans, and so provides dietary fiber.

Cross-planar bond Formed when the hydroxyl groups on the carbon atoms involved in the formation of a glycosidic bond are oriented on opposite faces of the sugar rings. Cross-planar bonds occur in pectin and cellulose. They are not digested in the human digestive system.

Gel Two-phase system with a solid continuous phase and a liquid dispersed phase.

Gums Complex, hydrophilic carbohydrates that are highly branched and form very viscous solutions; most gums do not gel.

High-methoxyl pectin Pectin with 50–58 % of the carboxyl groups esterified with methanol.

Hydrocolloid Large molecule with a high affinity for water that forms a stable aqueous colloidal dispersion or sol. Starches, pectins, and gums are all hydrocolloids.

Junction zone Specific region where two molecules such as pectin align and interact, probably by hydrogen bonds; important in gel formation.

Low-methoxyl pectin Pectin with 20–40 % of the carboxyl groups esterified with methanol.

Pectic acid Shorter-chain derivative of pectinic acid found in overripe fruits; demethylated; incapable of forming a gel.

Pectic substances Include protopectin, pectinic acids, and pectic acids.

Pectin High-molecular-weight pectinic acid; methylated α-D-galacturonic acid polymer.

Pectinic acid Methylated α-D-galacturonic acid polymer; includes pectins; can form a gel.

Protopectin Insoluble material found in immature fruits; high-molecular weight methylated galacturonic acid polymer; cannot form a gel.

Seaweed polysaccharides Complex polysaccharides that are capable of forming gels; examples include alginates, carrageenan, and agar; used as thickeners and stabilizers in food.

Sol or dispersion Two-phase system with a solid dispersed phase and a liquid continuous phase.

Steric fit Ability of molecules to come close enough to each other in space to interact (or fit together).

References

Glicksman J (1982) Food applications of gums. In: Lineback DR, Inglett GE (eds) Food carbohydrates. AVI, Westport, CT

Charley H, Weaver C (1998) Foods. A scientific approach, 3rd edn. Merrill/Prentice-Hall, New York

Coultate TP (2009) Food. The chemistry of its components, 5th edn. Royal Society of Chemistry, Cambridge

Penfield MP, Campbell AM (1990) Experimental food science, 3rd edn. Academic, San Diego, CA

Bibliography

BeMiller JN, Huber KL (2007) Carbohydrates. In: Damodaran S, Parkin K, Fennema OR (eds) Fennema's food chemistry, 4th edn. CRC Press, Boca Raton, FL

McWilliams M (2012) Foods: experimental perspectives, 7th edn. Prentice-Hall, Upper Saddle River, NJ

Potter N, Hotchkiss J (1999) Food science, 5th edn. Springer, New York

Vieira ER (1996) Elementary food science, 4th edn. Springer, New York

6 Grains

Introduction

Throughout the world, there is a great variety of types and amounts of grain products that are selected to be consumed by individuals. The World Health Organization (WHO) and many countries including the United States stress the *nutritional* importance of grains as a foundation of a good diet.

From a *culinary* point of view, consumers see a great variety of grains included in menu offerings—from soups and salads to desserts. Grain consumption has risen substantially in popularity due in part to a committed number of Americans making more nutritious food selections. Specific grain choices may be based on food intolerances or allergies.

In this chapter the physical and chemical properties of grains are addressed. The variety of cereals, milling, type of flours used in bread making, pasta products, safety, and nutritional value are presented. Further discussion of quick breads, yeast breads, the functions of various added ingredients, and details of gluten appear in the chapter on Baked Products.

Cereals Definition

Cereal is a cultivated *grass*, such as wheat, corn, rice, and oats, which produces an edible *seed* (grain or fruit). By definition, ***cereal*** comprises all the cereal products prepared from grain, not merely cold, sweetened, boxed breakfast cereal! Depending on the composition, the cereal crops may be *processed into various items* such as the following:

- **Bread**, using flour or meal from various grains (Chap. 15)
- **Cereal**, ready-to-eat, or cooked breakfast cereal varieties; such as oatmeal
- **Oil**, from germ processing (Chap. 12)
- **Pasta**, a dried paste of various flours (and perhaps legumes, herbs, and spices)
- **Starch**, from the starchy component of endosperm (Chap. 4)

When stored *properly*, and thus protected from adverse environmental impact, insect, and animal pests, grains are extremely resistant to deterioration during storage, especially when compared to the perishable dairy, eggs, meats, or fruit and vegetable crops. Grains are utilized extensively in developing and *less* affluent countries where animal products are either not available or not used. In *more* affluent countries, many varieties of grains and whole grains, processed ready-to-eat (r-t-e) breakfast cereals, cereal bars, and so forth are routinely consumed along with animal products.

V.A. Vaclavik and E.W. Christian, *Essentials of Food Science, 4th Edition*, Food Science Text Series,
DOI 10.1007/978-1-4614-9138-5_6,

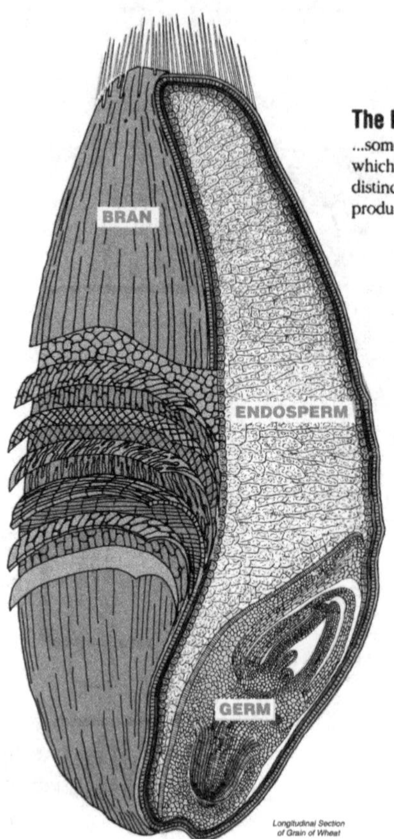

Fig. 6.1 Structure of a wheat kernel (*Source*: Wheat Foods Council)

Structure of Cereal Grains

The structure is similar in all grains. Each kernel of grain is composed of three parts: the *germ*, *endosperm*, and *bran*, and if *all* are present in a grain, it is a *whole grain*, such as whole wheat. When the bran and/or germ of the seed are removed or separated from the kernel in milling, a product is no longer "whole grain," however *refined* (Fig. 6.1). Most likely, these two terms are familiar to the reader. It is recommended by the USDA to "make half your grains whole." That advice also appears on many grain-based food products available to the consumer, such as whole grain crackers and cereals. Actual whole grain content is made available on Nutrition Facts and the Ingredients labels.

The ***germ***, or embryo, is the *inner* portion of the kernel—located on the lower end. It composes approximately 2.5 % of the seed and is where sprouting begins as the new plant grows. The germ is the kernel component with the highest percent lipid, containing 6–10 % lipid. Rancidity may result from either the lipoxidase *enzyme*, or *non-enzymatic* oxidative rancidity.

Due to this possibility of rancidity, a whole grain product may either undergo germ removal or include antioxidants such as BHA or BHT (see Additives, Chap. 18). The germ contains approximately 8 % of the kernel's protein and most of the thiamin.

Another structural part of the kernel is the ***endosperm***. It represents the greatest percentage of the kernel and is primarily starch, held as part of a protein matrix, with an exact composition that differs among grain types and varieties.

Regardless of the grain type, wheat, corn, or another grain, the endosperm is the seed component *lowest* in fat, containing *less* than the *germ*, with up to only 1.5 % of the lipid of the seed. It is also *lower* in fiber than the *bran*. The endosperm makes up approximately 83 % of the seed and has approximately 70–75 % of the protein of the kernel

Specifically regarding wheat—certain varieties or types of wheat may be carefully specified for use in various food products in order to ensure success in baking or cooking. Wheat bakes and functions differently depending on the type of wheat that is utilized. For example, wheat may be a *soft* or *hard* type, with the soft wheat variety containing more starch and less protein than hard wheat. The composition makes a difference as shown later in this chapter.

The *third* major component of a grain in addition to the germ and endosperm is the ***bran***. It is the layered, outer coat of a kernel and consists of an *outside* pericarp layer, offering protection to the seed, and an *inside* layer that includes the seed coat. The bran is often removed by abrasion or polishing in the milling process and may be used in many foods or animal feed. It is approximately 14.5 % of the seed and contains 19 % of the protein, 3–5 % lipid, and minerals such as iron.

Bran provides *cellulose* and *hemicellulose* that are both fiber or "roughage" in the diet. Yet, functionally, the individual bran may differ among grain types and varieties. For example, *wheat bran* includes an *insoluble fiber* that functions chiefly as a stool softener. *Oat bran* is a soluble fiber that functions among other ways, to reduce serum cholesterol.

If wheat is devoid of the bran, and germ, only the endosperm remains, and that is the component used in making white bread.

Composition of Cereal Grains

In *structural* composition, the various grains each contain three parts and thus the grains are similar; however, they vary in their *nutrient* composition, with each grain containing different amounts of carbohydrate, fat, protein, water, vitamins, and minerals (Tables 6.1 and 6.2). The main nutrient component of cereal grains is **carbohydrate**, which makes up 79–83 % of the dry matter of grain. It exists predominantly as starch, with fiber especially cellulose and hemicellulose, composing approximately 6 % of the grain.

Lipid (fats and oil) makes up approximately 1–7 % of a kernel, depending on the grain. For example, wheat, rice, corn, rye, and barley contain 1–2 % lipid; oats contain 4–7 %. The lipid is 72–85 % unsaturated fatty acids—primarily, oleic acid and linoleic acid.

Protein composes 7–14 % of the grain, depending on the grain. Cereals are low in the amino acids tryptophan and methionine, and although potential breeding may produce cereals higher in the amino acid *lysine*, it remains the *limiting* amino acid in cereals.

Grain consumption provides half of the protein consumed worldwide. However, in comparison to animal foods such as milk, meats, or eggs, grains from plants do *not* include all the essential amino acids contained in animal protein. Grains are not complete proteins. In fact, the protein is of *low biological value*, and therefore, less efficient in supporting body needs.

Combining the various food sources of protein is common in cultures throughout the world. For example, the preparation of traditional dishes combines the lower biological value grains with *legumes* or *nuts* and *seeds* to provide the needed amino acids to yield a *complete* dietary protein. In particular, combining beans with rice, or beans with cornbread; combining tofu and vegetables, or tofu and cashews, or eating chickpeas and sesame seed paste (tahini) known as hummus, or peanut butter on whole wheat bread, and so forth are put together (eaten in combinations) creating complete proteins. (Botanically, each of these grains, legumes, nuts, and seeds are *fruits* of a plant.)

CULINARY ALERT! All "flour" used in a recipe is not created equal. High protein or "hard" flour absorbs more water than low protein "soft" flour. Therefore, finished products using assorted "flour" will differ. The recipe

Table 6.1 Typical percent composition of common cereal grains (100 g)

Grain	Carbohydrate	Fat	Protein	Fiber	Water
Wheat flour	71.0	2.0	13.3	2.3	12.0
Rice	80.4	0.4	6.7	0.3	12.0
Corn meal	78.4	1.2	7.9	0.6	12.0
Oats, rolled	68.2	7.4	14.2	1.2	8.3
Rye flour	74.8	1.7	11.4	1.0	11.1
Barley	78.9	Trace	10.4	0.4	10.0
Non-cereal flours					
Buckwheat flour	72.1	2.5	11.8	1.4	12.1
Soybean flour, defatted	38.1	0.9	47.0	2.3	8.0

Source: Wheat Flour Institute

Table 6.2 Vitamin, mineral, and fiber content of wheat flours (100 g)

Flour	Thiamin B_1 (mg)	Riboflavin B_2 (mg)	Niacin B_3 (mg)	Iron (mg)	Fiber (g)
Whole wheat flour (whole grain)	0.66	0.14	5.2	4.3	2.8
Enriched flour (enriched)	0.67	0.43	5.9	3.6	0.3
White flour (refined)	0.07	0.06	1.0	0.9	0.3

Source: Wheat Flour Institute

must specify flour type and users must plan usage accordingly in order to ensure product success.

Significant proteins in some grains such as wheat, rye, and barley are *gliadin, secalin, and hoirdein*, respectively. To the extent that these proteins are present, flour has ***"gluten-forming potential."*** Then, with subsequent and sufficient hydration and manipulation these proteins form a gummy, elastic ***gluten*** structure (Chap. 15). *Wheat* contains both gliadin and glutenin proteins that contribute desirable strength and extensibility to the yeast dough, in bread making. Other flours without these two proteins cannot rise sufficiently, even with the use of yeast because there are no gluten stands to trap the yeasts' air and gasses.

CULINARY ALERT! Knowing that gluten may be an allergen, some individuals must follow a gluten-free diet.

CULINARY ALERT! Gluten-forming flour is high protein. Yeast is a good leaven to slowly fill the gluten structure as it readily stretches. Non-gluten-forming flour contains less protein. Baking powder and baking soda that bubble up immediately are good leavens for non-gluten-forming flour.

An additional protein, the enzyme α-amylase, is naturally present in grains and promotes dextrinization of starch molecules to shorter-chain polymers, as well as the sugars maltose and glucose. The action of α-amylase may thin starch mixtures or be detrimental to the bread-making industry, yet it is often added in the form of malt so that there is sugar to feed yeast.

In this section on proteins in grains, we have seen that worldwide, grain consumption is common, as is one grain used in combination with other grains.

For a number of nations, just *having* grains is an important issue. For other more affluent and mobile nations, a Baker may *have* sufficient grain and have concerns instead about the *baking* properties of the grain. In this latter case, for example, the chef might be concerned, not about shortages, though rather about functionality of the flour. The fact is that wheat flours *high* in protein absorb a lot of water, while *low* protein flours do not absorb much. This can mean a dry or a soupy mixture, and perhaps unsatisfactory finished foods. Armed with a knowledge of flour differences, an experienced baker knows that recipes that work for them in one region of the country may not work in another, and that all "flour" is not created equal!

Vitamins present in cereals are predominantly the B vitamins—thiamin (B_1), riboflavin (B_2), and niacin (B_3). These vitamins may be lost in the milling process andso are added back through the process of ***enrichment***. Today, there is less prevalence of the once deadly diseases beriberi and pellagra, due to cereal enrichment with thiamin and niacin, respectively (Table 6.2). Whole grain products contain some fat-soluble vitamins in the germ.

Minerals are *naturally* present at higher levels in *whole* grains than in *refined* grains. ***Fortification*** of refined flour with added iron (Table 6.2) is common. Zinc, calcium as well as vitamins may also be added at *levels beyond/not present in the original grain.*

Water is present in cereal grains at levels of 10–14 % of the grain. Of course soaking and cooking add water to cereal grains, and the grain size expands as additional water is absorbed. If a flour is *high* in protein content, it absorbs *a lot* of added water compared to *low* protein flour.

Fiber content is measured by *crude fiber* (CF) and *total dietary fiber* (TDF). These two measurements are *not* correlated. CF is composed of *cellulose* and the non-carbohydrate *lignin*. TDF includes cellulose and lignin, *plus hemicellulose, pectic substances, gums,* and *mucilages.*

Common Cereal Grains and Their Uses

Common cereal grains are noted below. While there is a great variety of cereal grains and their uses throughout the world, the most important and largest cereal grain consumed by man in the United States diet is *wheat.* That will be discussed first. Some wheat is also used for animal feed.

Wheat

Wheat has widespread uses. It may be cracked (bulgur, couscous), made into flour, and breads, cereals, and pasta and is the basis of numerous products that are recognized in diets throughout the world (more later). Some individuals exhibit an intolerance or even an allergy to wheat and its protein (see Gluten Intolerance).

As noted, the wheat kernel (wheat berry) is the most common cereal milled into flour in the United States (see Fig. 6.2). There are over 30,000 varieties of wheat grown in the United States, grouped into the following *major* classifications: hard red winter, hard red spring, soft red winter, hard white wheat, soft white wheat, and durum wheat.

This large number of wheat varieties is named according to several factors—*season planted, texture, and color.*

- Season—wheat is classified as winter wheat or spring wheat. *Winter wheat* is planted in cold seasons such as Fall and Winter and is harvested in June or July. *Spring wheat* is the spring planting and is harvested in late summer or fall seasons.
  (Spring wheat may have a continuous growth cycle with no inactive period. In areas where winters are severe, such as northern North America, wheat is planted in the spring after there is no risk of frost. In areas with very mild winters, such as India or Australia, spring wheat is sown in the autumn and grows through the winter.)
- Texture—wheat is classified as either *hard* or *soft. Hard* wheat kernels contain strong protein–starch bonds, the kernel is tightly packed, and there are minimal air spaces. *Hard* wheat flour forms elastic dough due to its *high* gluten-forming protein content and is the best flour to use for bread making. Hard *spring* wheat is 12–18 % protein and hard *winter* wheat is 10–15 % protein.
  Conversely, *soft* wheat is *lower* in protein and is desirable for cakes and pastries.
  Starch-protein bonds in the kernel break down more easily in soft wheat, than hard. (Yet inherent differences in the starch or protein components of hard and soft wheat alone do *not* sufficiently explain the differences in hardness.) Hard and soft wheat may be blended to create all-purpose flour that contains about 10.5 % protein. In the absence

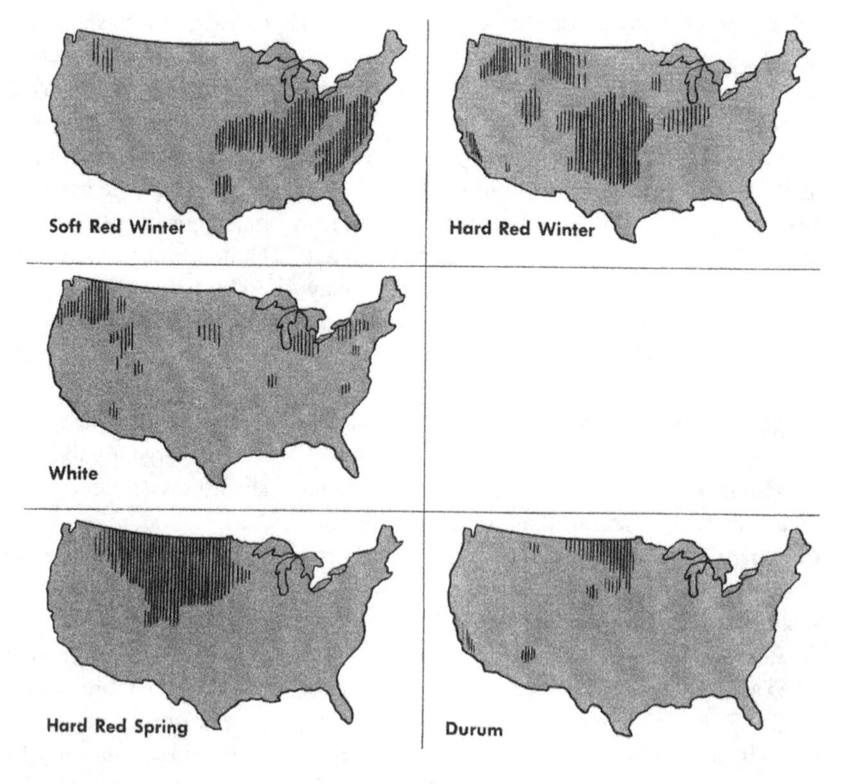

Fig. 6.2 Map showing wheat growth in the United States (*Source*: Wheat Flour Institute)

of pastry flour, "instant" (see below) flour and all-purpose flour may be combined.

- Color—*red, white,* and *amber*. The color of the grain depends on the presence of pigment, such as carotenoid. Durum wheat, for example, is hard wheat and highly pigmented. Its endosperm is milled into ***semolina*** for pasta, and couscous (most spaghetti is made from wheat).

Milling process of wheat. Specific milling tolerances of the ground wheat kernel or 'berry' must meet the Food and Drug Administration (FDA) grades to call the product "flour." When milled, each 100 lb of wheat yields approximately 72 lb of white flour and 28 lb of other product, including animal feed.

The conventional milling process (Fig. 6.3) of wheat first involves washing to remove foreign substances such as dirt or rocks. Conditioning or tempering by adjustment to water level (the addition or removal of water) of the kernel follows in order to obtain the appropriate water content and to facilitate the easy separation of the kernel components. Next, wheat is subject to coarse breaking of the kernels into *middlings*. The breaking process separates most of the kernels' outside (bran) and inside core (germ) from the endosperm. Once the endosperm is separated, it is subsequently ground multiple times in reduction rolls to become finer and finer for flour. As the bran and germ are removed, the refined flour contains streams that contain less vitamins and minerals.

If flour streams of the endosperm are blended during the milling process, various flours are created. *Straight grade* flour is a combination of *all* of the mill streams. Typically, home and bakery operations use ***patent flours*** that are 85 % straight grade flour and the combination of various highly refined mill streams.

Patent flour is the highest grade of flour; hence, the highest in value. *Short-patent* flour, such as cake flour, contains *more starch* in the starch-protein matrix and is produced by combining fewer streams than the *higher-protein, long-patent* flour. The remainder of flour, not incorporated into patent flours, is *clear* flour. It is used when color is not of importance, as it is slightly gray.

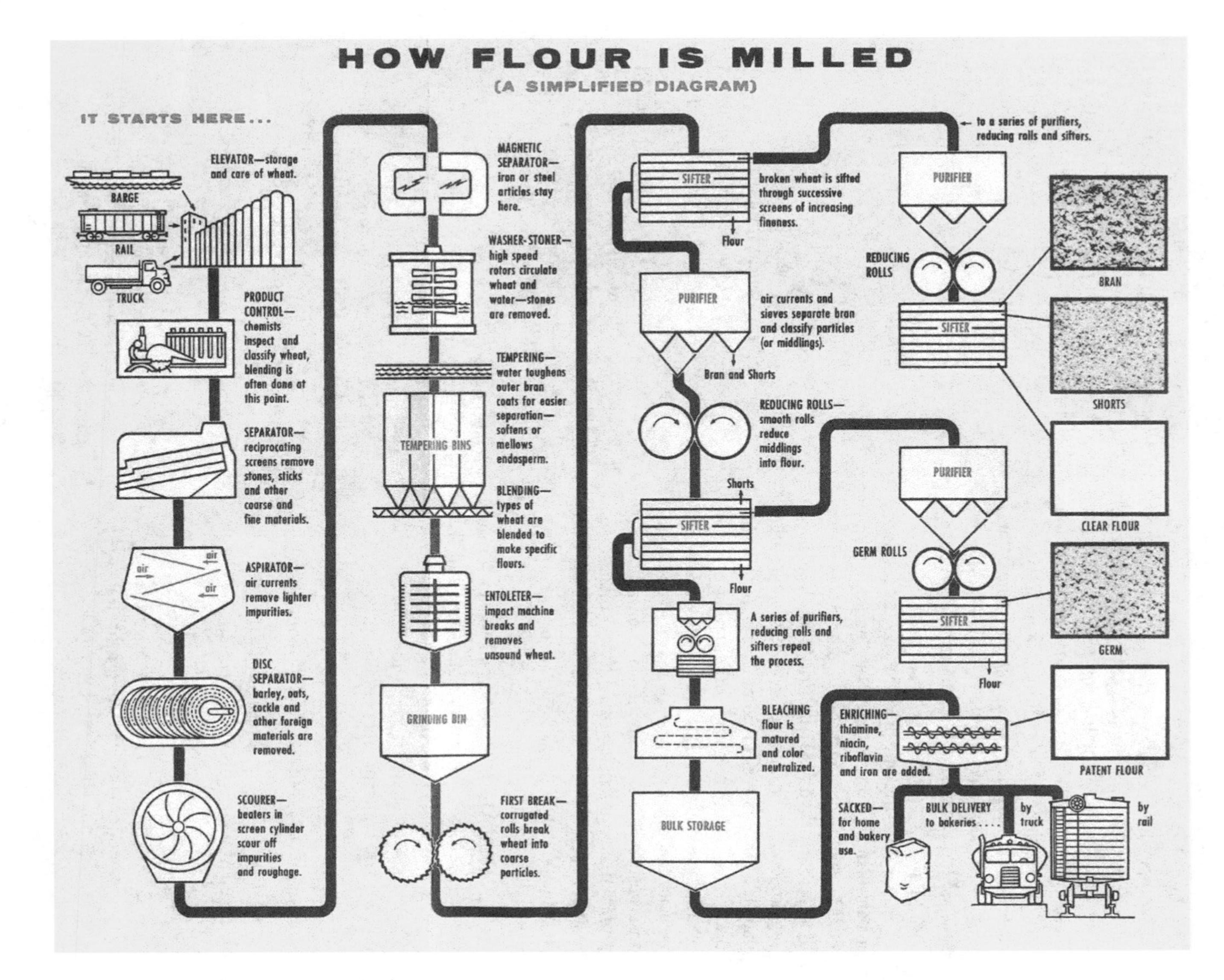

Fig. 6.3 The milling process (*Source*: Wheat Flour Institute)

It is the rule that flours from the *same* mill *vary* in composition from 1 year to the next. Also, the *various* flour production mills may produce slightly *different* flours, depending on such factors as geographic location of the crop, rainfall, soil, and temperature. It follows that this variance of crop year, mill, geographic location, and so forth may produce different baking results. For that reason, food manufacturers (and their Research and Development laboratory) constantly test flour so that variance is minimal or nonexistent. Otherwise flour may produce slightly different finished products. Of course, using *different flours* may produce *disastrous* results!

Additionally, milling produces the less common instant-blending, instantized, or "agglomerated" flour. Instant-blending flour is all-purpose flour that has been hydrated and dried, forming large "agglomerated" or clustered particles, larger than that the FDA approves for commercial white wheat flour. It has a more uniform particle size range than white wheat flour and does *not* readily pack down. Instant-blending flour is easily dispersible in water and is used when dispersibility of flour in liquid is preferred or required. It *mixes* into a formulation or recipe *better than* ordinary flour and is free-flowing, pouring like salt or sugar.

CULINARY ALERT! When a product formulation specifies a particular flour type, and that type is unavailable, the baker may combine various blends of flour to yield the correct flour product and better product results.

Milling (see Fig. 6.3) of various textures of wheat produces some of the following flours.

Hard Wheat: 10–18 % Protein

Bread flour

- It is typically made of hard red spring wheat kernels, with a high protein-to-starch ratio.
- It is capable of holding a lot of water (2 cups flour holds 1 cup water).
- It has a high gluten-forming potential forming a very strong and elastic structure, which can hold the air and gasses of yeast.
- It is not finely milled. Recall that hard spring wheat has a greater protein content than hard winter wheat. ("Gluten flour," milled from spring wheat, may contain 40–45 % protein.)

Hard and Soft Wheat Blend: 10.5 % Protein

All-purpose flour

- Combines the desirable qualities of both hard and soft wheat flour.
- It does not contain bran or germ and is known as white wheat flour, or simply "flour."
- It forms a less strong and elastic dough than bread flour.
- It may be enriched or bleached.

Soft Wheat: 7–9 % Protein

Cake flour

- Contains less protein, and more starch than all-purpose flour, and holds less water (2–3/4 cup holds 1 cup water)
- It is *low* in gluten-forming potential, is highly bleached, and finely milled (7/8 cup all-purpose flour + 2 tablespoons cornstarch = 1 cup cake flour).

Pastry flour

- It maintains intermediate characteristics of all-purpose and cake flour.
  It contains less starch than cake flour, and less protein than all-purpose flour.

Additional flour treatments involve the following:

- **Self-rising flour (phosphated flour)** contains 1–1/2 teaspoons of baking powder and ½ teaspoon of salt per cup of flour (and provides convenience!)
- ***Bleached flour*** is created when the yellowish (mainly xanthophyll) pigment is bleached by oxygen to a white color. Bleaching is achieved (1) naturally by exposure to oxygen in the air (2 or 3 months), or (2) by the chemical addition of either chlorine dioxide gas or benzoyl peroxide, *bleaching agents* which later evaporate. (Yes, even unbleached flour is bleached, naturally!) Bleaching results in finer grain and a higher volume.
- ***Matured flour*** also comes (1) naturally with age or (2) by the addition of *maturing agents*. If matured, gluten elasticity and baking properties of dough are improved because the unwanted effects of excess sulfhydryl groups are controlled. There is less polymerization of gluten protein molecules, and therefore, a less gummy dough (Chap. 20). Not all bleaching agents are maturing agents, yet chlorine dioxide (above) serves as both types of agent.
- ***Organic (chemical-free) flour*** uses grains that are grown without the application of synthetic herbicides and pesticides.

Wheat foods also include bulgur (Fig. 6.4), cracked wheat, and couscous as discussed below.

Bulgur is the whole kernel, i.e., parboiled, dried, and treated to remove a small percentage of the bran. It is then cracked and used as breakfast cereal or pilaf. Bulgur is similar in taste to wild rice.

Cracked wheat is similar to bulgur—the whole kernel broken into small pieces, yet not subject to parboiling. Whole grains should be stored in an airtight container, in a cool, dark place. *Farina* is the pulverized wheat middlings of endosperm used predominantly as a cooked cereal. It is similar in appearance to grits (corn).

Fig. 6.4 Bulgur wheat (*Source*: Wheat Foods Council)

Fig. 6.5 Couscous (*Source*: Wheat Foods Council)

Couscous is a processed form of semolina wheat (Fig. 6.5). It is popular throughout the world, especially in Northern Africa and Latin America. It is often served as a pilaf or as tabouli.

In addition to wheat, other common grains are highlighted in the text that follows.

Rice

Rice is a major cereal grain whose varieties are used as *staple foods* by people throughout the world. Thus, it may be the *major* aspect of a diet, or as well, incorporated to a *lesser degree* into the main dish, side dish, or dessert. It is commonly used in the preparation of r-t-e breakfast cereals. Rice, and rice flour, is especially important to persons with wheat allergies, or gluten intolerance, and rice is commonly eaten as a 'first food' by infants, as it is food that offers the least cereal allergy.

Rice may be eaten as the *whole grain*, or *polished*, which involves shedding the outer coat of bran. Brown rice contains the bran. Generally, rice is polished during milling in order to remove the brown hull; however, it *al*so removes some of the protein, vitamins, and minerals. When left unpolished, whole rice is *more* subject to rancidity and favors deterioration, as well as insect infestation compared to polished, white rice.

Today, most white rice is enriched with vitamins and minerals, to add back nutrients lost in milling. (Recall, the once-prevalent deadly disease, beriberi resulted from eating polished rice as a staple food. Thiamin removed in the milling process.)

Enrichment (Table 6.3) of rice is common and may be achieved by two primary methods. *One method* is to coat the grain with a powder of thiamin and niacin, waterproof it, dry it, and then coat the grain with iron before it is dried again. *Another method* of enrichment involves parboiling or "converting" rice. This process allows water-soluble bran and germ nutrients to travel to the endosperm by boiling or a pressure steam treatment. As a result nutrients are retained when the outside hull is removed. Following the steaming process, rice is subsequently dried and polished. Optional enrichment may include vitamins such as riboflavin, and vitamin D, and the mineral calcium.

Rice is grown in a variety of sizes. *Long* grain rice (with three times the length as width) is *high in amylose* content. *Medium* and *short* grain rice contain *less* amylose. Rice remains *soft* in hot form; however, leftover rice is hard because the high amylose crystallizes, or hardens as it cools.

It is recommended that rice puddings prepared with leftover rice use *medium or short grain* varieties in the original cooking process since they will contain less amylose and will not be texturally, as hard. The same medium or short grain rice is recommended for use in menu items such as sushi, where the food *should* remain soft and "stick" together.

Table 6.3 Primary nutrients for the enrichment of rice

Nutrient	mg/lb
Thiamin	2–4
Riboflavin	1.2–2.4
Niacin	16–32
Vitamin D	250–1,000
Iron	13–26
Calcium	500–1,000

Amylose content of rice

Size variety	% Amylose
Short grain	15–20 % (less amylase, more sticky)
Medium grain	18–26 %
Long grain	23–26 % (high amylase, less sticky)

CULINARY ALERT! Short grain rice is *low* in amylose. It is sticky and holds ingredients together. Therefore, in a product such as sushi, short grain, sticky varieties of rice are preferable over long grain rice.

Rice may be modified to allow flavor and aroma variety, very detectable by some palates. "Rice" may even be made from *pasta* such as when macaroni is shaped to resemble rice in products such as RiceARoni®. It may be processed into flours, starches, cereals, cooking wine, or the Japanese wine, sake. Rice "milk′ is commonly available and used. Rice flour is successfully made into items such as low-fat tortillas or noodles. Wild "rice" is actually *not* rice and, however, is derived from seeds of another reed-like water plant.

Numerous research studies have focused on shelf-stable cooked rice, ready-to-eat cereal, confectionery applications, rice oils, and flavored rice. Defatted rice bran extracts, aromatic rice, pregelatinized rice flours, starches, and rice syrups are chosen as food ingredients, depending upon the application. Rice use in a wide variety of foods continues to be common (Pszczola 2001).

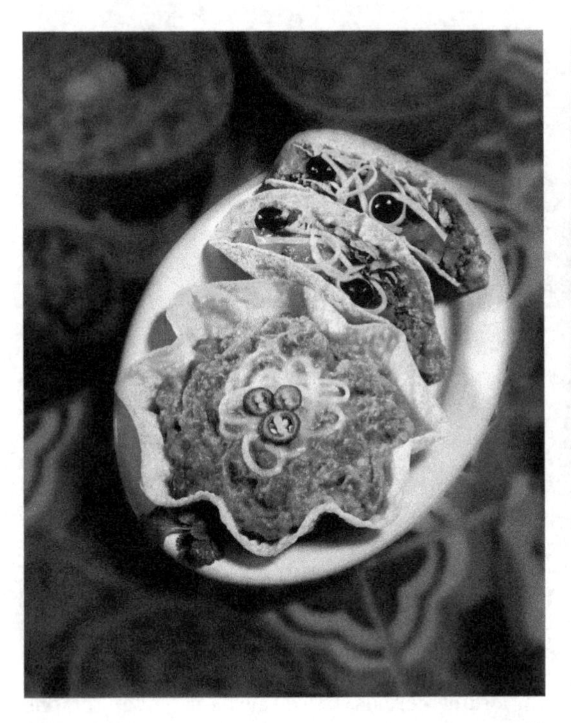

Fig. 6.6 Corn taco shells (*Courtesy of* SYSCO® Incorporated)

Corn

Corn is a staple cereal food of many people and nations, although the majority of corn is used for animal feed. It is lacking in the two essential amino acids, tryptophan and lysine, yet research continues to explore the addition of a protein trait to corn DNA

Sweet corn is actually a cereal; however, it is commonly eaten as a vegetable. *Field corn* has non-vegetable uses, including starch that is of value to growers and consumers alike. The whole kernels of special breeds of corn containing 11–16 % moisture are desirable for popcorn where the kernel increases in volume as the water escapes as steam.

- The *whole or partial kernels* may be coarsely ground (perhaps stone-ground) and used to create cornmeal or masa. Cornmeal is popular in cornbread and tamales, corn tortillas, snack foods, and items such as taco shells (Fig. 6.6). It may be soaked in alkali, such as lime (calcium hydroxide) for 20–30 min, for a better amino acid balance and greater protein availability. This soaking process may sacrifice some niacin (vitamin B_3), however it adds calcium.
- The *endosperm* of corn may be made into hominy, ground into grits, or used in r-t-e breakfast cereals or cornstarch. It may be hydrolyzed in hydrochloric acid or treated with enzymes, to produce corn syrup, or high-fructose corn syrup (HFCS) (Chap. 14).
- The *germ* yields corn oil
- Corn and its finished products (corn syrup, etc.) may be allergens to some individuals

Fig. 6.7 Breads may be prepared using a variety of grains (*Source*: Wheat Foods Council)

CULINARY ALERT! For use in bread making, corn needs to be combined with other flour, such as wheat flour, since corn alone yields dense bread. As well, corn does not contain the proteins gliadin and glutenin that form gluten structures.

Other Grains

Other grains, exclusive of the wheat, rice, and corn previously discussed, are *not* abundantly consumed, yet they offer taste variety (Fig. 6.7) and often grow in more adverse environmental conditions where the more common grains will not grow. For some readers, the following grains may be very familiar and frequently utilized. For other readers, these same grains are *not* used at all or may be relatively unheard of! The seeds are used both as forage crops and as food cereals in different parts of the world.

Barley

Barley is "winter-hardy" and is able to survive in the frost of cold climates. It is used for human and animal consumption. Barley is served as a cooked cereal, or the hull of the kernel is removed by abrasion to create pearled barley, i.e., commonly used in soups. Additionally, although barley may not be eaten as a whole grain, it is incorporated into many foods including breads, pilafs or stuffing, or it may be used for ***malt*** production. Barley is the most common malt as it has sufficient enzyme content to hydrolyze the starch efficiently to sugar.

Malt

- In order to create malt, the barley grain is first soaked in water. This soaking causes the germ to sprout and produces an enzyme that hydrolyzes *starch* to a shorter carbon chain, maltose *sugar*. Maltose is a fermentable carbohydrate that is then used to feed yeast and produce CO_2 and ethyl alcohol. (The alcohol and CO_2 are important for brewing alcoholic beverages and for baking.) Dried malt is used in a variety of products including brewed beverages, baked products, breakfast cereal, candies, or malted milks.
- Consumers following a gluten-free diet (no wheat, oats, rye, or barley) must avoid malt. They should read ingredients labels to determine (1) *if* malt is an ingredient in the food, and (2) the *source* of the malt.

Millet

Millet is the general name for *small* seed grass crops. The crops are harvested for food or animal feed (fodder). It is a major crop in some countries and is used as cereal, to make breads or soups. Millet includes proso (the most common) finger, foxtail, and pearl millet. Less common millets include barnyard, browntop, guinea, kodo, and little varieties of millet. Some millet is utilized in birdfeed, for cattle, hogs, poultry, and sheep.

Sorghum is a special type of millet with *large* seeds, typically used for animal feed, so far, yet it is the primary food grain in many parts of the world, where it is ground and made into porridge and cakes. It is also used to yield oil, sugars, and alcoholic beverages. A common variety of sorghum grown in the United States is milo; there are also waxy varieties. Overall, sorghums are resistant to heat and drought, and therefore, are of special value in arid, and hot regions of the world.

Sorghum is useful as a gluten-free way to produce malt. Sorghum and millet seeds are important cereals in semiarid, tropical regions of Asia and Africa.

A *very tiny* millet grain that has been used for centuries in the Ethiopian diet is *teff* or t'ef (*Eragrotis tef*, signifying "love" and "grass"). The seeds are approximately 1/32 of an inch in diameter, with 150 weighing as much as a kernel of wheat! Considering its size, it has a small endosperm in proportion to bran, and therefore is primarily bran, and germ. It is ground for use in flatbread. It grows in tropical climates in Africa, India, and South America. Commercial production of teff as a forage and food crop is also in the United States (Arrowhead Mills, Hereford, TX). It is cultivated in US states including Idaho and South Dakota.

Oats

Oat (referred to singularly when spoken of as a crop) is a significant cereal crop fed to animals such as horses and sheep, and also used by man. It is valued for its high protein content. In milling, the hull is removed and the oats are steamed and "rolled" or flattened for use in food. Oats are incorporated into many ready-to-eat breakfast cereals and snack foods. Oat bran is a soluble fiber that has been shown to be effective in reducing serum cholesterol.

Due to the fact that oats have a fairly high fat content, as far as grains go, rancidity may develop. Lipase activity in the grain is destroyed by the administration of a few minutes of steam treatment.

Quinoa

Quinoa (keen-wa) is the grain highest in protein, although it is *not* an abundantly consumed grain. The small, round, light brown kernels are most often used as a cooked cereal.

Rye

Rye is richer than wheat in lysine, yet it has a relatively low gluten-forming potential.

Therefore, rye does not contribute as good of a structure to dough as is the case with wheat. It is frequently used in combination with wheat flour in breads and quick breads and is made into crackers. There are three types of rye—dark, medium, and light, which may be selected for baking into bread. Rye may be sprouted, producing malt or malt flour.

Triticale

Triticale is a *wheat* and *rye* hybrid, first produced in the United States in the late 1800s. As a crop, it offers the disease resistance of wheat and the hardiness of rye. It has more protein than either grain alone, although the overall crop yield is not high, so its use is not widespread. Triticale was developed to have the baking property of wheat (good gluten-forming potential) and the nutritional quality of rye (high lysine).

With regard to gluten intolerance or celiac disease, many grains are less subject to rigorous testing than wheat, rye, barley, oats, and triticale. Considered to be "safe" on a gluten-restricted diet are grasses such as sorghum, millet, and teff.

Non-cereal "Flours"

Non-cereals, including various *legumes* and *vegetables,* may be processed into *"flour,"* although they do *not* have the composition of grains. For example, soy and garbanzo beans (chick peas) are legumes (from the Leguminosae family) that may be ground and added to baked products. These foods may be found on the list of common food allergens.

Soy "flour" may be incorporated into formulations due to its protein value, or because it aids in maintaining a soft crumb. Cottonseeds (Malvaceae family), and potatoes (tubers), may also be processed into "flour." Buckwheat (fruit of *Fagopyrum esculentum* crop) contains approximately 60 % carbohydrate and may be used in the porridge kasha or as animal feed. Cassava (tuber) is the starch-yielding plant that yields tapioca and is a staple crop in parts of the world.

Cooking Cereals

In cooking, cereal products expand due to retention of the cooking water. *Finely* milled grains such as cornmeal, corn grits, or wheat farina should be *gently* boiled and only *occasionally* be stirred in order to prevent mushy and lumpy textures. *Whole or coarsely* milled grains such as barley, bulgur, rice, and oats (and buckwheat) may be added to *boiling* water and stirred occasionally during cooking.

To control heat while cooking, cereal products may be cooked in the top of a double boiler over boiling water. A disadvantage of this cooking method is that heating time is lengthened compared to direct heating without use of a double boiler.

More later, yet a bit regarding pasta: cooking pasta involves adding it to boiling water and boiling it uncovered until the desired tenderness (typically al dente) is achieved. The addition of a small amount (1/2 teaspoon (2.5 mL) household use) of oil prevents boil-over from occurring.

CULINARY ALERT! Excessive stirring of any milled grain (especially finely milled grains) results in rupturing of the grain contents and is unpalatable, as the cereal forms a gummy, sticky consistency.

Fig. 6.8 Ready-to-eat breakfast cereal (*Source*: Wheat Foods Council)

Breakfast Cereals

Breakfast cereals may perhaps be eaten hot or cold. An American religious group not wanting to consume animal products started the production of ready-to-eat breakfast cereals. The *Western Health Reform Institute* in Battle Creek, Michigan, produced, baked, and then ground a whole meal product to benefit the healthfulness of its institute's patients. A local townsperson, J. H. Kellogg, and his brother W.K. Kellogg started a business with this idea, applying it to breakfast food. A patient, C.W. Post did the same. (Both the Kellogg and Post names are still popular cereal manufacturers today.)

Breakfast cereals (Fig. 6.8) in many forms quickly became popular. Flaking, shredding, puffing, etc. and the production of various forms soon expanded although convenient, some criticize the levels of ingredients, including sugar and fiber, in r-t-e breakfast cereals. *Enrichment* and *fortification* also became a common practice for breakfast cereals that are now in the ranking of one of the most fortified foods available for consumption.

Pasta

Pasta is the paste of milled grains (alimentary paste), extruded through a die or put through a roller. The crushed (not finely ground) endosperm of milled durum spring wheat, known as ***semolina***, is used in the preparation of high-quality pasta products. *Lower-quality* pastas that do *not* use semolina are also available to the consumer. These typically taste "starchy" and are pasty in texture. Although taste may not be affected, rinsing cooked pasta products prior to service may result in the loss of nutrient enrichment.

Pasta frequently appears on restaurant menus and home tables in the form of salads, side dishes, and main dishes. If pasta is processed to include legumes, as part of the formulation, a *complete protein* may be formed in a single food. For instance, pasta may now be commercially formulated to include pureed vegetables, herbs, and spices as well as cheeses. Pasta may also be cholesterol-free or gluten-free, made of non-wheat flour, such as rice. “Technological breakthroughs now make it possible to enjoy rice pasta that tastes, looks, and cooks like regular pasta.”

A variety of products including macaroni, noodles, and spaghetti are created by extrusion. In order to distinguish between macaroni and noodles, *macaroni* does *not* include eggs in its formulation *and noodles must* contain not less than 5.5 % (by weight) of egg solids or yolk (National Pasta Association, Arlington, VA).

Nutritive Value of Grains

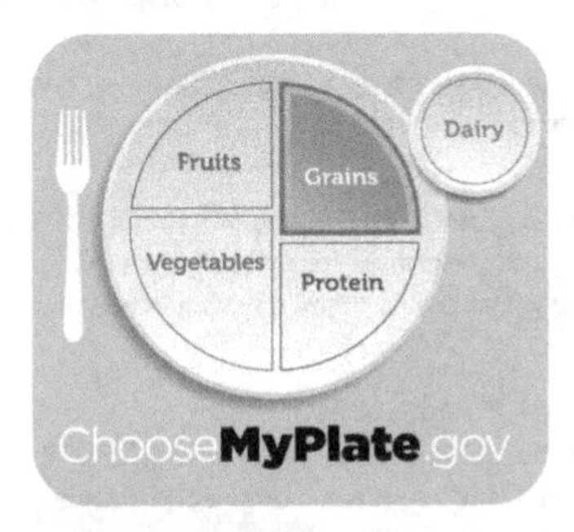

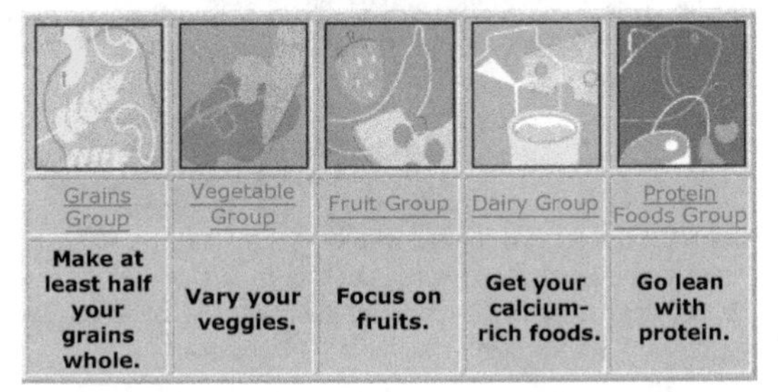

What Foods Are in the Grains Group?

Any food made from wheat, rice, oats, cornmeal, barley, or another cereal grain is a grain product. Bread, pasta, oatmeal, breakfast cereals, tortillas, and grits are examples of grain products.

Grains are divided into two subgroups, Whole Grains and Refined Grains.

Whole grains contain the entire grain kernel—the bran, germ, and endosperm.

Refined grains have been milled, a process that removes the bran and germ. This is done to give grains a finer texture and improve their shelf-life, but it also removes dietary fiber, iron, and many B vitamins.

Most refined grains are *enriched*. This means certain B vitamins (thiamin, riboflavin, niacin, folic acid) and iron are added back after processing. Fiber is not added back to enriched grains. Check the ingredient list on refined grain products to make sure that the word “enriched” is included in the grain name. Some food products are made from mixtures of whole grains and refined grains.

Most of these products are made from refined grains. Some are made from whole grains. Check the ingredient list for the words “whole grain” or “whole wheat” to decide if they are made from a whole grain. Some foods are made from a mixture of whole and refined grains.

Some grain products contain significant amounts of bran. Bran provides fiber, which is important for health.

10 tips
Nutrition Education Series

make half your grains whole

10 tips to help you eat whole grains

Any food made from wheat, rice, oats, cornmeal, barley, or another cereal grain is a grain product. Bread, pasta, oatmeal, breakfast cereals, tortillas, and grits are examples. Grains are divided into two subgroups, **whole grains** and **refined grains**. Whole grains contain the entire grain kernel—the bran, germ, and endosperm. People who eat whole grains as part of a healthy diet have a reduced risk of some chronic diseases.

1 make simple switches

To make half your grains whole grains, substitute a whole-grain product for a refined-grain product. For example, eat 100% whole-wheat bread or bagels instead of white bread or bagels, or brown rice instead of white rice.

2 whole grains can be healthy snacks

Popcorn, a whole grain, can be a healthy snack. Make it with little or no added salt or butter. Also, try 100% whole-wheat or rye crackers.

3 save some time

Cook extra bulgur or barley when you have time. Freeze half to heat and serve later as a quick side dish.

4 mix it up with whole grains

Use whole grains in mixed dishes, such as barley in vegetable soups or stews and bulgur wheat in casseroles or stir-fries. Try a quinoa salad or pilaf.

5 try whole-wheat versions

For a change, try brown rice or whole-wheat pasta. Try brown rice stuffing in baked green peppers or tomatoes, and whole-wheat macaroni in macaroni and cheese.

6 bake up some whole-grain goodness

Experiment by substituting buckwheat, millet, or oat flour for up to half of the flour in pancake, waffle, muffin, or other flour-based recipes. They may need a bit more leavening in order to rise.

7 be a good role model for children

Set a good example for children by serving and eating whole grains every day with meals or as snacks.

8 check the label for fiber

Use the Nutrition Facts label to check the fiber content of whole-grain foods. Good sources of fiber contain 10% to 19% of the Daily Value; excellent sources contain 20% or more.

9 know what to look for on the ingredients list

Read the ingredients list and choose products that name a whole-grain ingredient ***first*** on the list. Look for "whole wheat," "brown rice," "bulgur," "buckwheat," "oatmeal," "whole-grain cornmeal," "whole oats," "whole rye," or "wild rice."

10 be a smart shopper

The color of a food is not an indication that it is a whole-grain food. Foods labeled as "multi-grain," "stone-ground," "100% wheat," "cracked wheat," "seven-grain," or "bran" are usually not 100% whole-grain products, and may not contain **any** whole grain.

Go to www.ChooseMyPlate.gov for more information.

DG TipSheet No. 4
June 2011
USDA is an equal opportunity provider and employer.

10 tips
Nutrition Education Series

choosing whole-grain foods

10 tips for purchasing and storing whole-grain foods

Whole grains are important sources of nutrients like zinc, magnesium, B vitamins, and fiber. There are many choices available to make half your grains whole grains. But whole-grain foods should be handled with care. Over time and if not properly stored, oils in whole grains can cause spoilage. Consider these tips to select whole-grain products and keep them fresh and safe to eat.

1 search the label

Whole grains can be an easy choice when preparing meals. Choose whole-grain breads, breakfast cereals, and other prepared foods. Look at the Nutrition Facts labels to find choices lower in sodium, saturated (solid) fat, and sugars.

2 look for the word "whole" at the beginning of the ingredients list

Some whole-grain ingredients include whole oats, whole-wheat flour, whole-grain corn, whole-grain brown rice, wild rice, and whole rye. Foods that say "multi-grain," "100% wheat," "high fiber," or are brown in color may not be a whole-grain product.

3 kids can choose whole grains

The new school meal standards make it easier for your kids to choose whole grains at school. You can help your child adapt to the changes by slowly adding whole grains into their favorite recipes, meals, and snacks at home.

4 find the fiber on label

If the product provides at least 3 grams of fiber per serving, it is a good source of fiber. If it contains 5 or more grams of fiber per serving, it is an excellent source of fiber.

5 is gluten in whole grains?

People who can't eat wheat gluten can eat whole grains if they choose carefully. There are many whole-grain products, such as buckwheat, certified gluten-free oats or oatmeal, popcorn, brown rice, wild rice, and quinoa that fit gluten-free diet needs.

6 check for freshness

Buy whole-grain products that are tightly packaged and well sealed. Grains should always look and smell fresh. Also, check the expiration date and storage guidelines on the package.

7 keep a lid on it

When storing whole grains from bulk bins, use containers with tight-fitting lids and keep in a cool, dry location. A sealed container is important for maintaining freshness and reducing the possibility of bug infestations or moisture.

8 buy what you need

Purchase smaller quantities of whole-grain products to reduce spoilage. Most grains in sealed packaging can be kept in the freezer.

9 wrap it up

Whole-grain bread is best stored at room temperature in its original packaging, tightly closed with a quick-lock or twist tie. The refrigerator will cause bread to lose moisture quickly and become stale. Properly wrapped bread will store well in the freezer.

10 what's the shelf life?

Since the oil in various whole-grain flours differs, the shelf life varies too. Most whole-grain flours keep well in the refrigerator for 2 to 3 months and in the freezer for 6 to 8 months. Cooked brown rice can be refrigerated 3 to 5 days and can be frozen up to 6 months.

Go to www.ChooseMyPlate.gov for more information.

DG TipSheet No. 22
September 2012
USDA is an equal opportunity provider and employer.

Key Consumer Message: Make at least half of your grains whole grains.
View Grains Food Gallery

Grains make a significant nutritive contribution to the diet (Sebrell 1992). Whole grain products and processed cereal products contribute carbohydrates, vitamins such as B vitamins, minerals such as iron, and fiber to the diet in creative ways. Fortification with vitamin D and calcium are presently under consideration. Ready-to-eat varieties of breakfast cereals are frequently consumed in the more developed countries and many are highly fortified with essential vitamins and minerals, including folate.

Grains are low in fat, high in fiber, and contain no cholesterol, *although* cooked foods, breads, cereals, rice, and pasta dishes may be prepared with added fats, sugars, eggs, and refined flours, which changes the nutritive value profile. Unfortunately, with these additions, many commonly selected r-t-e breakfast cereals lose their original nutritional benefit, as they are manufactured to be high in sugar and/or low in fiber in developed countries.

Safety of Grains

Safety of cereal grains is better assured by proper storage including first-in-first-out (FIFO) rotation. Since whole grains are subjective to rancidity, storage should be kept cold, and not lengthy. All products should be stored off of the floor and a slight distance away from walls due to possible pipe flooding or insect infestation.

Conclusion

Cereals are the edible seeds of cultivated grasses and many cereal foods are prepared from grain. A kernel contains bran, endosperm, and germ, however, if "refined," the refined cereals contain *only* endosperm and are no longer whole grain. For example, *wheat flour* is not the same as *whole wheat flour*. Common cereal grains include wheat, rice, and corn, although other grains such as barley, millet, oats, quinoa, rye, and triticale may be used as a component in meals. Dried grains have a very long storage life, and much of the world depends on them for food.

Over 30,000 varieties of wheat exist, classified according to season, texture, and color. *Hard* wheat is used for bread making, and *soft* wheat for cakes and pastries.

Semolina flour from hard durum wheat is used for pasta production. Pasta is the paste of milled grains, primarily wheat, and increasingly appears in the American diet. It is a complex carbohydrate and low-fat food.

Rice is a staple food of much of the world. It grows as (extra long), long, medium, and short grain rice and grows in a variety of flavors that are used in many entrees, side dishes, even desserts. Corn is also common.

Cereals are included as the base of numerous food guides throughout the world, indicating that they are major foods of a nutritious diet. The USDA recommends that persons avail themselves of the great variety of products that are available in the marketplace.

Notes

CULINARY ALERT!

Glossary

All-purpose wheat flour White flour, not containing the bran or germ. Combining the properties of hard and soft wheat.

Bleached flour Bleaching the pigment to a whiter color, naturally by exposing pigment to air, or by chemical agents.

Bran The layered outer coating of the kernel, offering protection for the seed.

Bread flour Made from a hard wheat kernel, with a high protein–starch ratio; high gluten potential.

Cereal Any edible grain that comes from cultivated grasses.

Endosperm The starch-storing portion of the seed that produces white flour and gluten.

Enrichment Adding back nutrients lost in milling.

Fortification Adding nutrients at levels beyond that present in the original grain.

Germ The embryo; the inner portion of the kernel.

Gluten Protein substances (gliadins, glutenins) left in the flour after the starch have been removed, which when hydrated and manipulated produce the elastic, cohesive structure of dough.

Malt Produced from a sprouting barley germ. Long glucose chains are hydrolyzed by an enzyme to maltose, i.e., involved in both feeding yeast and producing CO_2. May be dried and added to numerous products.

Matured flour Wheat flour, i.e., aged naturally or by chemical agents to improve gluten elasticity and baking properties of dough.

Organic flour Flour from crops grown without the use of chemicals such as herbicides and insecticides.

Patent flour Highest grade of flour from mill streams at the beginning of the reduction rolls. High starch, less protein than mill streams at the end of reduction rolls.

Pasta The paste of milled grains, usually the semolina from durum wheat, extruded through a die to produce a diversity of shaped products. They are dried and then cooked in large amounts of water. Included are macaroni, noodles, spaghetti, ravioli, and the like.

Semolina Flour milled from durum wheat.

Other/Additional Glossary for Cereals, Flour, and Flour Mixtures

Oregon State University Select definitions for a better understanding of cereals, flour, and flour mixtures.

Amylopectin A fraction of starch with a highly branched and bushy type of molecular structure.

Amylose The long-chain or linear fraction of starch.

Baking powder Is a mixture consisting generally of an acid salt and sodium bicarbonate which, when water is added, and possibly heat, will produce carbon dioxide for leavening.

Batter Systems with their relative high water: flour: water is continuous. Structure depends much less on gluten development than on gelatinization of starch.

Bleaching Of flour is the oxidization of the yellow carotenoid pigments in wheat flour. This may be done with either chemicals or during "aging" over a length of time.

Carmelization The development of brown color and caramel flavor as dry sugar is heated to a high temperature; chemical decomposition occurs in the sugar.

Carotenoid pigment Yellow-orange compounds produced by plan cells and found in various fruit, vegetable, and cereal grain tissues; for example, beta-carotene.

Coagulation Change in protein, after it has been denatured, that results in hardening or precipitation and is often accomplished by heating.

Fermentation The transformation of organic substances into smaller molecules by the action of microorganisms; yeast ferments glucose to yield carbon dioxide and alcohol.

Gliadin Is the water-insoluble protein that contributes stickiness and tackiness to gluten structure.

Gluten Is an elastic cohesive mass made up of gliadin, glutenin, water, and a lipoprotein compound.

Glutenin Is the water-insoluble protein that contributes toughness and rubberiness to gluten structure.

Graham flour Is flour essentially from the entire wheat kernel. It may be ground to varying degrees.

Green flour Is flour which has not been aged or matured.

Maturing Of flour is the aging process that affects the flour structural proteins through oxidation of the gliadin and glutenin. Maturing may occur naturally or with chemical additions.

Milling Is the process which generally involves the separation of the bran and germ from the endosperm which is subsequently subdivided.

Oxidation A chemical reaction in which oxygen is added or electrons are lost.

Proofing The last rising of bread dough after it is molded into a loaf and placed in the baking pan.

Reducing substance A molecule that has an effect opposite that of an oxidizing agent: hydrogen or electrons are gained in a reaction involving reducing substances.

Rope Is a bacterial contamination that can originate in the flour bin or in the various constituents used to make bread. It will make a loaf of bread sticky and "ropy" in the interior.

Staling Refers to those changes in quality that occur in baked products after baking. Generally, there is a loss of flavor, softening of the crust or development of a leathery crust, and increased firmness of the crumb.

Starch gelatinization The swelling of starch granules when heated with water, often resulting in thickening.

Straight grade white flour Theoretically should contain all the flour streams resulting from the milling process, but actually 2–3 % of the poorest streams are withheld.

White wheat Flour is a food made by the grinding and sifting of cleaned wheat (definition, FDA).

References

Pszczola DE (2001) Rice: not just for throwing. Food Technol 55(2):53–59

Sebrell WH (1992) A fiftieth anniversary—cereal enrichment. Nutrition Today. (Jan/Feb):20–21

Bibliography

American Association of Cereal Chemists. St Paul, MN

CIMMYT—Centro Internacional de Mejoramiento de Maiz y Trigo—International Maize and Wheat Improvement Center

ConAgra Specialty Grain. Omaha, NE

Cooperative Whole Grain Education Association. Ann Arbor, MI

Fast RB, Caldwell EF (1990) Breakfast cereals and how they are made. American Association of Cereal Chemists, St. Paul, MN

Grain Processing. Muscatine, IA

International Grain Products. Wayzata, MN

International Wheat Gluten Association. Prairie Village, KS

National Barley Foods Council. Spokane, WA

Rice Council of America. Houston, TX

USDA ChooseMyPlate.gov

Wheat Flour Institute. Washington, DC

Wheat Foods Council. Englewood, CO. www.wheatfoods.org

Wheat Gluten Industry Council. Shawnee Mission, KS

Wheat Industry Council. Washington, DC

7 Vegetables and Fruits

Introduction

Vegetables are the edible portion of plants eaten with (or as) the main course. They are in salads and soups. Vegetables may be processed into beverages or vegetable starches, eaten fresh or lightly processed, dried, pickled, or frozen. They impart their own characteristic flavor, color, and texture to diets, and undergo changes during storage and cooking. Ranked next to the cereal crops wheat, rice, and corn, potatoes are the most prolific vegetable crop grown for human consumption.

Fruits are defined in more than one way. *Botanically*, fruits are the mature ovaries of plants with their seeds. Therefore this definition includes all grains, legumes, nuts and seeds, and common "vegetable-fruits" such as cucumbers, olives, peppers, and tomatoes. When defined and considered in a *culinary role*, fruit is the fleshy part of a plant, usually eaten alone or served as a dessert. Fruits are high in organic acids and sugar—higher than vegetables.

The nutritive value of vitamins, minerals, fiber, and other compounds contained in fruits and vegetables is extremely important to the diet. Additional dietary and medicinal benefits of fruits and vegetables are being discovered. "Vary your veggies" and "focus on fruits" is the USDA advice in selecting vegetables and fruits as part of a healthy diet. Also given as USDA advice is "make half your plate fruits and vegetables."

Structure and Composition of Cell Tissue

The structure and composition of vegetables and fruits show that they contain both *simple* and *complex* cells. The *simple* cells are similar to one another in function and structure and include *dermal* tissue and *parenchyma* tissue. Dermal tissue is the single-layer *outside* surface of leaves, young stems, roots, and flowers, while ***parenchyma tissue*** (*see below*) makes up the majority of the plant, and is where basic molecular activity such as the synthesis and storage of carbohydrate by sunlight (photosynthesis) occurs.

Complex tissue includes the vascular, collenchyma, and sclerenchyma supporting tissue. Major vascular tissue consists of the xylem and phloem; *xylem* conducts water from the roots to the leaves, and *phloem* conducts nutrients from the leaves to the roots. These tissues may be located in the center of the vegetable, for example, as is seen in carrots.

A plant is composed *primarily* of simple *parenchyma* tissue (Fig. 7.1). Each cell is bounded by a cell wall produced internally by

V.A. Vaclavik and E.W. Christian, *Essentials of Food Science, 4th Edition*, Food Science Text Series,
DOI 10.1007/978-1-4614-9138-5_7,

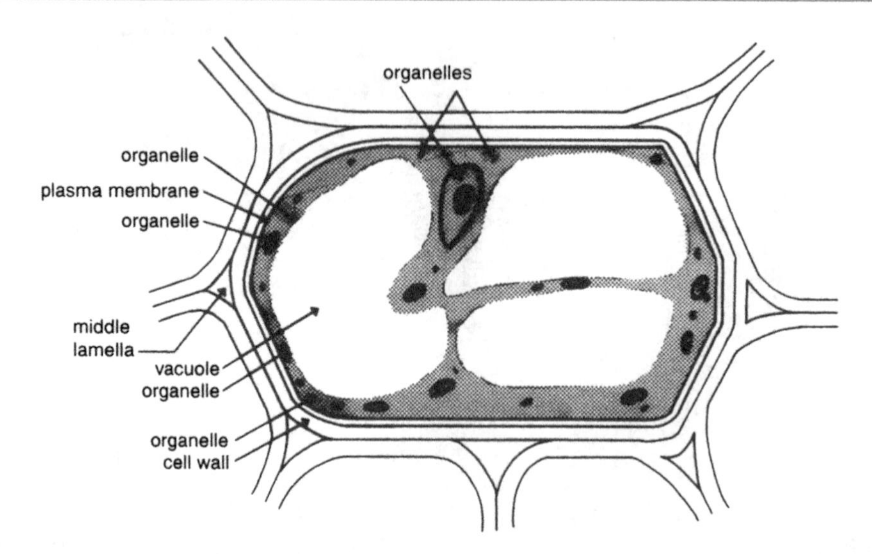

Fig. 7.1 Components of a parenchyma cell (*Source*: Division of Nutritional Sciences, New York State College of Human Ecology)

the protoplast. The wall serves to support and protect cell contents and their retention, influx, or release. When the wall is *firm*, the original shape and texture of the cell are maintained. However, when the wall is *destroyed* (by cutting, dehydration, or cooking), it breaks and spills its contents into the surrounding environment. Thus, water, sugars, or water-soluble vitamins of a cell may be lost.

The *primary* cell wall is made of ***cellulose***, hemicellulose, and pectic substances including pectin. Older, more mature plants may also have a *secondary* cell wall composed of *lignin* (see "Chemical Composition of Plant Material" section), in addition to the primary cell wall.

Inside the parenchyma cell wall is the protoplast, composed of three parts—*plasma membrane, cytoplasm, and organelles*. The *plasma membrane* surrounds the functional cell, while the ***cytoplasm*** of the protoplast includes all of the cell contents inside the membrane, yet outside the nucleus. The *organelles* include nucleus, mitochondria, ribosomes, and plastids. It is the plastids that contain fat-soluble material such as fat-soluble vitamins, and fat-soluble pigments including chlorophyll and carotenoids (each is discussed in a subsequent section of this chapter.)

Outside the cell wall, *between* adjacent cells, is the ***middle lamella***. This is the "cementing" material between adjacent cells and it contains pectic substances, magnesium and calcium, water, and air spaces.

Each parenchyma cell contains an inside cavity known as a ***vacuole***. It may be *large* in size, holding plentiful amounts of water, and comprising the major portion of parenchyma cells, or it may be *small* in size. In an intact, *un*cooked cell, vacuoles hold sufficient water and provide a desirable *crisp* texture to the cell. The opposite effect is noted in *wilted* or cooked cells.

The ***cell sap*** of the vacuole contains the water-soluble materials including vitamins B and C, sugars, inorganic salts, organic acids, sulfur compounds, and the water-soluble pigments. These cell sap components may escape into the surrounding soaking/rehydrating/cooking water.

Chemical Composition of Plant Material

Carbohydrate

Carbohydrate constitutes the *largest* percentage of the dry weight of plant material. It is the basic molecule formed during *photosynthesis* when water (H_2O) and carbon dioxide (CO_2) combine to yield carbohydrate (CHO) and oxygen (O_2).

Carbohydrate is present in both the simple and complex forms. For example, *simple* carbohydrates are the *mono*saccharides—including *glucose* and *fructose*; and *di*saccharides, such as *sucrose*, that may increase during the fruit ripening process. *Complex* carbohydrates or *poly*saccharides are synthesized from simple carbohydrates and include *cellulose* and *starch.*

Various complex carbohydrates and the effect of heat on those carbohydrates are discussed in the following (see also Chap. 3)

Starch is the storage component of carbohydrate located in roots, tubers, stems, and seeds of plants. When subjected to heat and water, starch absorbs water and gelatinizes (Chap. 4). Vegetables vary in their starch content. Some vegetables such as potatoes are starchy, some moderate, and others such as parsley are less starchy. Starch is digestible as the bonds between the glucose units are $\alpha-1,4$.

Cellulose is water-insoluble fiber that provides *structure* to plant cell walls. The molecular bonds between glucose units are β-1,4; therefore, cellulose remains *in*digestible to humans, although it may be softened in cooking.

Hemicellulose fiber provides structure in cell walls, and the majority is *in*soluble. It is softened when heated in an alkaline environment, such as, if baking soda is added to cooking water for the purpose of green color retention.

Pectic substances (Chap. 5) are the firm, intercellular "cement" between cell walls, the gel-forming polysaccharide of plant tissue, and are hydrolyzed by cooking. Large *in*soluble forms of pectin become *soluble* pectin with ripening of the plant material.

In addition to carbohydrates, there is a noncarbohydrate fiber material present in the complex vascular and supporting tissue. It is ***lignin*** and is found in older vegetables. Lignin remains unchanged by heat and may exhibit an unacceptable "woody" texture.

Protein

Protein makes up less than 1 % of the composition of a *fruit*, and protein *is low* in most *vegetables*. Protein is most prevalent in legumes—peas and beans—yet, even then, it is an *incomplete* protein as it lacks the essential amino acid, methionine.

Protein that is present as enzymes may be extracted from plants and used in other foods. Examples include the proteolytic enzymes that contribute the beneficial tenderizing effects to meats such as *papain* (derived from papaya), *ficin* (obtained from figs), and *bromelain* (extracted from pineapple).

Fat

Fat composes approximately 5 % of the dry weight of roots, stems, and leaves of vegetables. It makes up less than 1 % of the dry weight of a fruit, except for fruits such as avocados and olives that contain 16 % and 14 % fat, respectively. Fat is instrumental in development during the early growth of a plant.

Vitamins

The vitamins present in vegetables and fruit are primarily *carotene* (a vitamin A precursor) and *vitamin C*. *Beta-carotene*, is present in dark orange fruits, vegetables, and as an underlying pigment in green vegetables. Vitamin B_1 (thiamin) is also present. Fruits supply more than 90 % of the water-soluble vitamin C and a major percentage of the fat soluble vitamin A in a diet.

Water-soluble vitamin losses may occur upon soaking when vitamins leach out, and also in heating. Losses occur primarily in heating. In addition to soaking and heating, enzymatic action may negatively affect the nutritive quality of fruits and vegetables. Specifically, the enzymes, ascorbic acid oxidase and thiaminase, can cause nutritional changes in vitamins C and B_1, respectively, during storage. Therefore, retention of these vitamins is controlled by deactivating the enzymes in blanching prior to freezing.

CULINARY ALERT! It is interesting to note that vitamins A and C, so plentiful in fruits and vegetables, are both listed on Nutrient Facts labels as vitamins that Americans lack, therefore increase consumption of fruits and vegetables.

Minerals

Minerals are more prolific in vegetables than in fruits and are notably calcium, magnesium, and iron. Calcium ions are *added* to some canned vegetables in order to promote firmness and lessen softening of pectic substances. Yet, since the oxalic acid in spinach and the phytates in peas *bind calcium*, decreasing its bioavailability, calcium is *not* added to these canned vegetables.

Water

Water is found *in* and *between* plant cell walls. Some of its functions in the plant are to transport nutrients, to promote chemical reactions, and to provide plants with a crisp texture if cell membranes are intact.

Water constitutes a small percentage (10 %) of seeds and is a substantially larger percent of leaves. It makes up 80–90 % of a plant, as is evidenced by the drastic size reduction of a measure of vegetables that is subject to dehydration.

CULINARY ALERT! Think about how the volume of plant material changes significantly when a food dehydrator is used to remove water from food.

Phytochemicals (More in Appendices)

Phytochemicals are plant chemicals. They are non-nutrient materials that may be especially significant in preventing disease and controlling cancer. These chemicals are the focus of much research concerning their importance to human health.

The list is long of the many examples of such plant chemicals. It includes the beta-carotene of carotenoid pigments, the flavonoid group of pigments, as well as the sulfur-containing allyl sulfide and sulforaphane. Additionally, dithiolthiones, indoles, and isothiocyanates in cruciferous ("cross-shaped blossom," cabbage family) vegetables, isoflavones, phytosterols, protease inhibitors, saponins in legumes, and limonene and the phenols of citrus fruit are among the plant chemicals that may be effective in disease prevention.

Turgor Pressure

A plant's ***turgor pressure*** is the pressure that water-filled vacuoles exert on the cytoplasm and the partially elastic cell wall. A raw product still attached to the plant prior to harvesting is generally crisp because the vegetable or fruit contains a large percentage of water, which provides turgidity to the plant. As previously mentioned, the structure of plant material is, to a large degree, dependent on the water content of the parenchyma cell.

Shortly after the fruit or vegetable is "picked" from the plant, water is lost to the air (evaporation) due to air flow with its evaporation, or due to low humidity storage. As a result, there is a loss of turgor pressure. The product becomes limp, wilted, and dehydrated. If the parenchyma cell is still *intact* (*not cooked* or otherwise destroyed), water may reenter the cell, and turgor of this wilted, limp product may be restored. Soaking is an example of rehydration.

CULINARY ALERT! It is possible to rehydrate or recrisp by storage in high humidity (refrigerator's hydrator box or crisper) or by minimal soaking in warm, 70–90 °F (21–32 °C), water.

Subsequent to soaking, when plant pores open and take up water, the plant pores then re-close, and hold the absorbed water for approximately 6 h if the plant is refrigerated (Produce Marketing Association, Newark, DE). Soaking raw plant material may be discouraged though, as water-soluble nutrients and pigments may, by purely physical means, escape into the soaked water. [To rehydrate lettuce, it is suggested that it should not be soaked, except rather, placed in only 2 in. or so of warm water (Produce Marketing Association, Newark, DE).] Sprays of dips to make the produce waterproof may also be employed.

Once the parenchyma cell is subjected to cooking, *osmosis* ceases and *diffusion* occurs, which changes the texture, flavor, and shape of fruits. ***Osmosis*** represents water movement across a semipermeable membrane. ***Diffusion*** signifies water *and solute* movement across a *permeable* membrane.

Pigments and Effects of Additional Substances

Plant pigments enhance the aesthetic value of fruits and vegetables for humans, as well as attract insects and birds, which fosters pollination. These pigments are subject to change with ripening and processing of the raw vegetables or fruits. The four pigments found in plants are *chlorophyll*, the green pigment; *carotenoids*, a yellow, red, or orange pigment; and the flavonoids, both *anthocyanin*, the red, blue, or purple pigment, and *anthoxanthin*, the white pigment. "... a variety of different colors of non-starchy vegetables and fruits, including red, green, yellow, white, purple and orange, as well as tomato-based products and allium vegetables, such as garlic, are recommended daily." (Food Product Design 2012)

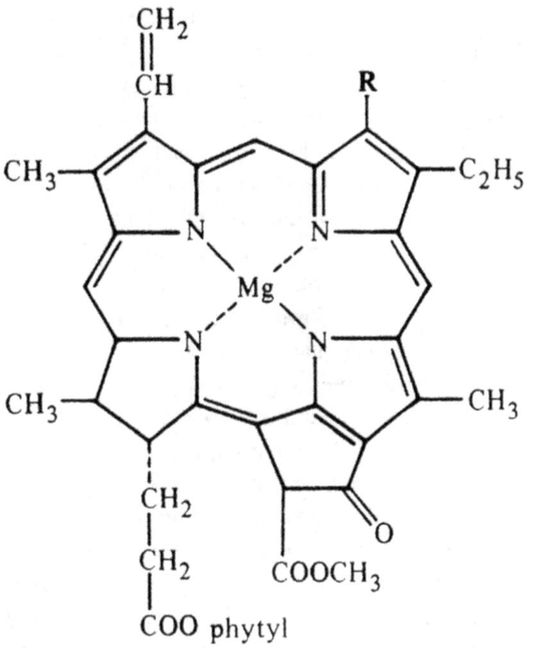

Fig. 7.2 Chlorophyll

High-performance liquid chromatography (HPLC) is generally used for plant pigment analysis. A discussion of the major pigments and a description of how they may change appears in the following material.

Chlorophyll

Chlorophyll is perhaps the most well-recognized plant pigment. It is the green pigment found in the cell chloroplast, and it is responsible for photosynthesis (i.e., converting sunlight to chemical energy). It is *fat-soluble* and may appear in vegetable cooking water if the water also contains fat.

Chlorophyll is structurally a porphyrin ring containing magnesium at the center of a ring of four pyrrole groups (Fig. 7.2). Phytol alcohol is esterified to one of the pyrrole groups, and it confers solubility to fat and fat solvents. Methyl alcohol is attached to another pyrrole group.

If the magnesium in chlorophyll is *displaced* from its central position on the porphyrin ring, an *irreversible* pigment change occurs. A number of factors cause this pigment color change, including prolonged storage, the heat of *cooking*, changes in *hydrogen ion* concentration (pH), and the presence of the minerals, zinc and copper. These factors are responsible for producing a drab, olive-green colored pigment in the *cooked* product. In a *raw* form the cell membrane does not allow H to contact/change a pigment.

Initially, as green vegetables are *heated*, air is removed from *in* and *between* the cell, and a *bright green* color becomes apparent. Then internal organic acids are released and hydrogen displaces magnesium, producing pheophytins. Either magnesium-free *pheophytin a*, which is a gray-green pigment, or *pheophytin b*, an olive-green pigment, is formed. These changes to the chlorophyll pigment become more marked with time, so a short cooking time is recommended.

As well, cooking the product *uncovered* for the first 3 minutes allows the escape of volatile plant acids that would otherwise remain in the cooking water and react to displace magnesium. Using a cover while cooking allows less change of chlorophyll to occur. (This is not true of all vegetable pigments as seen later.)

When *heated*, green-pigmented vegetables that are *high* in acid content undergo *more* color change than green vegetables *low* in acid, and green vegetables show less color change than fruits with their *high* acid. Even *raw* green vegetables, such as *raw* broccoli, change color to the underlying yellowish color as the chlorophyll degrades.

CULINARY ALERT! Pigments may change from the natural color due to extended heating and release of the plant's *internal* organic acids; therefore, minimum cooking is preferred. In addition to the internal organic acids, an *external* acid environment (i.e., acid added to cooking water) causes the natural green color to change into olive-green pheophytin.

The preceding discussion has been on the effect of *acids* on pigment color. As opposed to an acid environment, an *alkaline* environment also affects the green pigment. As the professional or home chef knows, the addition of the alkaline material, sodium bicarbonate (baking soda), produces and maintains a desirable green color. The soda reacts with chlorophyll, displacing the phytyl and methyl groups on the molecule, and the green pigment forms a bright-green, water-soluble chlorophyllin.

Nonetheless, although producing a desirable *appearance* with pH change of added soda, the benefit is accompanied by an unacceptable 1. *loss of texture*, due to softening of hemicellulose. Sodium bicarbonate also 2. *destroys ascorbic acid* (*vitamin C*) and *thiamin* (*vitamin B_1*). Therefore, due to these texture and nutrient losses the addition of this alkali substance is *not* recommended.

CULINARY ALERT! Sodium bicarbonate (baking soda) has a positive effect on color. However, it negatively affects texture and nutritive value.

In food preparation, the minerals, copper and zinc, may be released in the process of cutting or chopping. Also, some knives, copper bowls, or colanders may produce *un*desirable color changes in chlorophyll by displacing magnesium.

Regardless of the manner in which chlorophyll is changed, when the chlorophyll is destroyed, a second underlying carotenoid pigment may become apparent. Carotenoids are discussed below.

Carotenoids

The ***carotenoids*** are red, orange, and yellow *fat-soluble* pigments in fruits and vegetables, including *carotenes* (the hydrocarbon classification) and *xanthophylls* (the oxygenated class). They are found in chloroplasts along with chlorophyll, where the green pigment dominates, and also in chromoplasts without chlorophyll. The carotenoid pigment is seen especially in flowers, fruits, including tomatoes, peppers, and citrus fruits, as

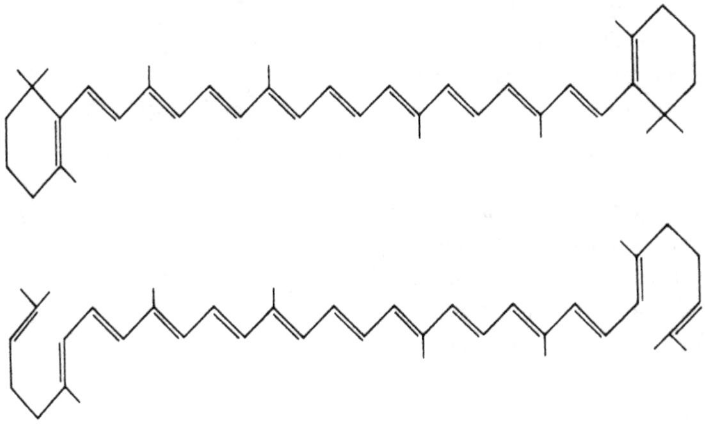

Fig. 7.3 Beta-carotene, lycopene

well as roots, including carrots and sweet potatoes.

Carotenes are unsaturated hydrocarbons containing many carbon atoms. The conjugated double bonds (i.e., double bonds alternating with single bonds) are responsible for the color; the greater the number of conjugated double bonds, the deeper the color. For example,

- *Beta*-carotene is naturally orange in color and contains a six-membered ring at each end of the chain (Fig. 7.3) In comparison to beta-carotene:
  - *Alpha*-carotene has one *less* conjugated double bond and is *paler* in color.
  - *Lycopene*, found in tomatoes and watermelon, has the *deepest red-orange color* because it has two *more* double bonds than beta-carotene, and it has two open rings (Fig. 7.3) at each end of the chain.

There exist hundreds of types of carotenes—40 or more carotenoids are known to be precursors of vitamin A. The most well-known carotene is the aforementioned beta-carotene, cleaved by an enzyme in the intestinal mucosa to yield vitamin A.

Xanthophylls are the yellow-orange colored derivatives of carotenes containing carbon, hydrogen, and oxygen. Xanthophylls include lutein and zeaxanthin.

Autumn leaves show evidence of destruction of the green chlorophyll pigment, as the carotenes, and "autumn xanthophylls" that existed along with the chlorophyll become visible. Corn contains the xanthophyll *cryptoxanthin*, and green leaves contain *lutein*. Paprika also contains xanthophyll pigment.

The carotenoid pigment may undergo autoxidation due to the large number of double bonds. This oxidation may result in "off-flavor" and color loss, yielding unsatisfactory products. Antioxidants such as butylated hydroxy anisole (BHA), butylated hydroxy toluene (BHT), or tertiary butylated hydroxy quinone (TBHQ) are frequently added to a wide variety of foods containing fruits and vegetables, herbs, or spices to prevent this detrimental oxidation.

> The FDA does not allow health claims for spices. However supportive research into the health benefits of spices fits nicely into two consumer trends: movement toward natural remedies and a growing appetite for spicy foods. (Hazen 2012)

Whereas *oxidation* causes development of a *lighter-color* cooked vegetable, *caramelization* of plant sugar may result in a *darker-color* cooked vegetable. It is recommended that carotene-pigmented vegetables should be either covered during cooking, or cooked quickly, as in stir-frying. Since the pigment is fat soluble, table fat such as butter or margarine should be minimized or omitted in cooking as the pigment may become paler.

The length of cooking time does *not* negatively affect carotenoid pigments as much as it does for chlorophyll, and changes are *not* as noticeable. However, upon *heating*, and in the presence of *acid*, some molecular isomerization occurs. Specifically, in carotenoids, the predominant *trans* molecular form, naturally present in plants, is changed to *cis* configuration in a matter of a few minutes, and the pigment becomes less bright. Unlike the case with chlorophyll pigments, *alkali* environments do *not* produce a color change.

Carotenes provide color in food. Food technologists have developed annatto, carrot, paprika, and tomato *extracts* to provide color in foods. (Pinkish-white flowers of the annatto plant with their small reddish-orange seeds inside offer dye used to color foods such as cheddar cheese.)

In addition to the plant pigments, added herbs and spices also provide carotene coloring and flavor. Albeit in *small* amounts in foods, they contribute to vitamin A values that appear on nutrition labels. They supply advantageous nutrients such as beta-carotene. This addition offers the same nutrients as a diet of yellow, green, and leafy vegetables, although in significantly lesser amounts.

Carotene from vegetables or fruits may prevent oxidation of body tissues, and development of *cancer*, although much remains unknown about possible benefits of *supplements* of this biologically active component of plant material. The Academy of Food and Nutrition advocates *foods* in the diet as the best source of good nutrition (see "Nutritive Value of Vegetables and Fruits" section rather than supplements).

CULINARY ALERT! Cooking change is minimal for carotenoids.

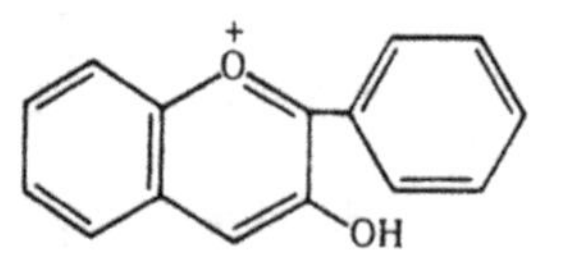

Fig. 7.4 Anthocyanin

Another group of pigmented compounds consisting of anthocyanins and anthoxanthins are the flavonoids.

Anthocyanin

Anthocyanin (Fig. 7.4) is the red, blue-red, blue, or purple pigment in fruits and vegetables such as blueberries, cherries, raspberries, red cabbage, red plums, and rhubarb (not beets; see "Betalaines" section). The *skins* of radishes, red apples, red potatoes, grapes, and eggplant also contain anthocyanin pigment. It is prevalent in buds and young shoots, and is an underlying pigment of chlorophyll, that becomes apparent as a purplish pigment in autumn leaves when chlorophyll decomposes.

Anthocyanins contain a *positively charged oxygen* in the central group of the molecule and belong to the flavonoid group of chemicals. Thus they are distinguished from the orange-red found in carotenoids. These pigments are *water-soluble* and are found in the cell sap of plants. They may be released into the cooking water with soaking or prolonged heat exposure.

In the following is a discussion regarding anthocyanin and pH.

pH and color: Care must be taken when working with the anthocyanin pigments. Mixed fruit juices for a punch drink or fruits incorporated into baked goods with alkaline leavens may produce undesirable color. Either the addition of *alkali* or an alkaline cooking medium produces unwelcome violet-blue or turquoise color.

In an *acidic* environment, the anthocyanin pigment exhibits a more characteristic

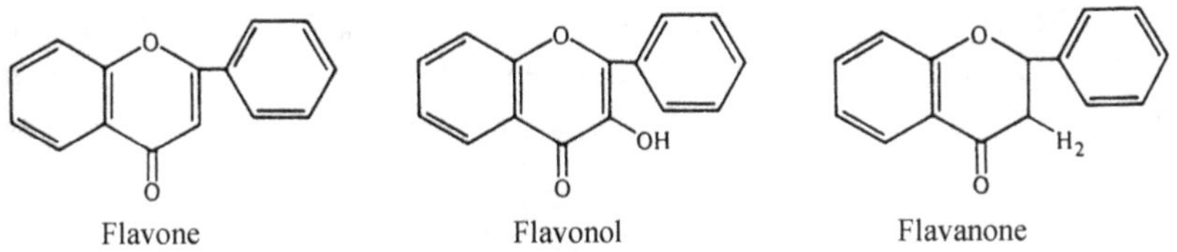

Fig. 7.5 Anthoxanthin

red color. A tart, acidic apple is often added to red cabbage while cooking in order to produce a more appealing finished product.

pH and texture: Textural characteristics are *also* affected by pH. If acids such as lemon juice or vinegar are added to fruits and vegetables (anthocyanin pigment) for better color, it should be *after* desired softening has occurred because acid prevents softening (see "Cooking Effect" section).

Recall the *negative* changes to the chlorophyll pigments when it is cooked *covered*, and still *retaining* plant acids. The opposite is true of anthocyanin pigmented vegetables. In fact, cooking the anthocyanin-pigmented vegetable *with* a cover is recommended for better color retention since *plant acids* are then *retained* yielding acidic cooking water. If fruits containing anthocyanins are added to batters and dough, such as in the preparation of blueberry muffins, *acidic* buttermilk is also incorporated to assist in preserving color. Color changes that occur in cooking are reversible.

Metals, such as iron from non-stainless-steel preparation tools, also change pigments. Metals can change the natural purplish pigment to a blue-green color. Therefore, food products containing the anthocyanin pigment are often canned in lacquer-coated (enamel-lined) metal cans to prevent the product acid from interacting with the can metal and causing undesirable color changes.

CULINARY ALERT! Anthocyanin color is subject to reversible changes in cooking.

Anthoxanthin

A fourth major pigment, ***anthoxanthin*** pigment (Fig. 7.5) is also a flavonoid, and is similar to anthocyanin, yet it exists in a *less* oxidized state, as the oxygen on the central group is *uncharged*. (Recall: anthocyanins contain a *positively charged oxygen*.) Anthoxanthins are white, or pale yellowish, *water-soluble* pigments found in a plant's cell sap. This classification represents flavone, flavonol, flavanone, and flavanol pigments, and includes fruits such as apples, or vegetables such as cauliflower, onions, and potatoes.

A *short* cooking time is desired for the anthoxanthin pigment. Otherwise, with *prolonged* heat, the pigment turns into an undesirable brownish gray color. For example, white, anthoxanthin potatoes, with their low organic acid content, may become unfavorably dark colored after prolonged cooking due to formation of an iron–chlorogenic acid complex. Some anthoxanthins may *change* to anthocyanins and exhibit a pinkish tinge if vegetables are overcooked.

In *acid* environments, anthoxanthin becomes lighter. (Therefore, in household use, cooking in an acid environment, incorporating one teaspoon of cream of tartar per quart of water may be useful in lightening the color.) If cooking water is *alkaline* or contains traces of iron salts, the result may be a *yellow* or *brown* discoloration of white cooked vegetables. Cooking in aluminum cookware also causes the same negative discoloration.

CULINARY ALERT! Short cooking of the anthoxanthin pigment is advised. It remains whiter in acid.

Betalaines

Betalaines are a minor group of pigments that contain a nitrogen group in the molecular structure. They are similar to, yet *not* categorized as anthocyanins or anthoxanthins. These pigments differ in color.

For example,

- Betacyanines are **red-colored** like anthocyanins at a pH of 4–7
- Betaxanthines are **yellow-colored** like anthoxanthins at a pH above 10
- Betalaine below a pH of 4 is **violet-colored**

A lacquered can (as with anthocyanins) is used in order to prevent color changes that may result from metals in the can reacting with the betalaine pigment.

Tannins

Tannins (tannic acid) are polyphenolic compounds that add both color and astringent flavor to foods. They may be responsible for the *unwanted* brown discoloration of fruits and vegetables, as well as for the *desirable* changes that provide tea leaves with their characteristic color. They range in color from pale yellow to light brown, and due to their acidic nature, they tend to cause the mouth to pucker. (Astringents shrink mucous membranes, extract water, and dry up secretions.)

The term tannin represents a broad group of compounds found in plants—normally in bark, fruit, leaves, and roots. Tannins, such as the brownish pigment found in tea leaves may be used as the brown colored dye in dyeing fabrics or tanning leather. Food tannins found in wines, and teas, contain antioxidant properties correlated with good health. Tannins precipitate proteins causing them to become a solid and "*settle*" out to the lower edge of a solute phase (however, proteins will *float* if they are less dense than the solvent).

Flavor Compounds

The flavor of cooked vegetables is greatly influenced by the presence of sulfur both *allium* and *brassica* compounds, although aldehydes, ketones, organic compounds, and alcohols are contributors to the flavor profile. Some of the favorable sulfur compounds, including allyl sulfides may increase excretion of carcinogens from the body, according to the American Cancer Society (ACS).

Allium

Vegetables of the genus ***allium*** include chives, garlic, leeks, onions, and shallots, each different members of the *lily* family. Onions, for example, contain strong sulfur compounds, and exhibit enzyme activity when cut, causing the eyes to tear (lachrymatory effect). Similarly, garlic undergoes an enzymatic change to sulfur compounds, precursor (+)-*S-allyl-L-cysteine sulfoxide*, producing the identifiable garlic odor.

While these flavor compounds in plants are *water soluble*, they may be lost from the vegetable to the water, then volatized as steam during cooking. It follows then, that if a *mild* flavor is desired in cooked onions, a *large* amount of boiling water and cooking *uncovered* for a long time period is recommended. In that manner, sulfur flavor compounds are degraded and vaporized. Inversely, a *sweeter*, *more concentrated* flavor is produced if *less* water and a *cover* is used. The most *intense* flavor results from cooking in fat where flavor is simply *not* lost.

CULINARY ALERT! Allium: is mild if vegetable is cooked in a large amount of water, uncovered. A concentrated, stronger flavor is apparent if less water, and covered cooking is chosen. Intensity is greatest when cooking in oil.

Brassica

Vegetables of the genus ***Brassica*** include broccoli, Brussels sprouts, cabbage, cauliflower, kale, kohlrabi, mustard greens, rutabaga, and turnips. They are of the *mustard* family and are known as *cruciferous* vegetables that have a cross-shaped blossom on the young, growing plant. As opposed to allium, the *naturally mild* flavor of the *raw* vegetables with brassica *becomes quite strong* and objectionable with prolonged *cooking* as hydrogen sulfide is produced.

That is why, for optimal flavor of *brassica* flavored vegetables, a *small* amount of briskly boiling cooking water and *short* cooking is advised. Also, in order to allow the volatile organic acids to escape, it is recommended that the vegetables should be *uncovered* at the beginning of the cooking process. Then they may be *covered* to keep the cooking time short.

Several vegetables of the Brassica genus, such as cabbage, contain a sulfur compound known as *sinigrin*. Sinigrin may interact with an enzyme, myrosinase released from the cut or bruised cell and produce potent mustard oil. *The* (+)-*S-methyl-L-cysteine sulfoxide* compound may convert to the more desirable dimethyl disulphide.

CULINARY ALERT! Brassica: cooking in a small amount of water, for a short time, with a cover, prevents development of an unacceptable strong flavor.

Organic Acids

Organic acids give the tart, sour taste of fruits, and they include some of the following acids—citric acid, malic acid, or tartaric acid. *Vegetables* contain a greater variety of organic acids, yet maintain a less acidic pH level than fruits.

Concentrates, Extracts, Oils, Spices, and Herbs

Concentrates, extracts, oils, spices, and herbs incorporate flavor into food as it is processed. These may be used as an alternative to fresh, frozen, or dehydrated vegetables in a product formulation and they provide products with a pure, consistent quality of flavor when they are added. *Concentrates* impart the vegetable's characteristic flavor. *Natural plant extracts* may be used to yield the flavors and aroma of fresh herbs and spices. *Essential oils* are also removed from a plant and concentrated to produce flavoring oils. These may be the replacement for some spices and herbs.

Although there may not be a clear-cut distinction between an herb and spice, a herb is generally from the *herbaceous* part of plants. According to the American Spice Trade Association (ASTA), a spice is "any dried plant product used primarily for seasoning purposes." Spices may come from fruits, flowers, roots, or seeds, as well as from shrubs and vines. They enhance color, flavor, palatability, and they exhibit antimicrobial properties (Sherman & Flaxman 2001). (The Food and Drug Administration (FDA) does not include dehydrated vegetables in its definition of spices, but rather they are "flavors.")

An immense amount of folklore goes along with herbs and spices, which may be used for medicinal as well as culinary purposes. While *traditional medicine* in practice for centuries, includes the use of herbs and spices, such traditional medicine may be combined in practice *today* with *Western or modern medicine*. In fact, the *National Center for Complementary and Alternative Medicine*, established by the *National Institutes of Health*, has as its mission to seek out effective and alternative medical treatment, to evaluate the outcomes, and report findings to the public!

Vegetable Classifications

Vegetable classifications demonstrate the parts of the plant eaten as food. This varies throughout the world. The eight common parts, beginning with underground parts of the plant and progressing to those parts growing above ground, are as follows:

- **Roots**—underground; beet, carrot, jicama, parsnips, radish, rutabaga, sweet potato, turnip, yam ("Sweet potatoes" are a yellow to orange color flesh, either dry or moist. In the USA, they may be known as "yams," and both names are stated on a label. The non-orange, true "yam" is yellow, white, or purple-pigmented flesh root vegetables)
- **Tubers**—underground; enlarged fleshy stem; starch storage area after leaves manufacture carbohydrates; buds or eyes form new plants; Irish potato, Jerusalem artichoke
- **Bulbs**—stems with an underground bulb of food reserve; garlic bulb, leeks, onions, shallots, spring onions (green onions or scallions do not possess a real bulb)
- **Stems**—a plant's vascular system, nutrient pathway; a lot of cellulose; asparagus, celery, kohlrabi, rhubarb
- **Leaves**—the manufacturing organ for carbohydrate that is then stored elsewhere in the plant; Brussels sprouts, cabbage, lettuce, parsley, spinach; also seaweed and "greens" such as beet, collards, kale, and mustard greens
- **Flowers**—clusters on the stem; artichoke, broccoli, cauliflower
- **Fruits**—the mature ovaries with seeds, generally sweet, and fleshy; apple, banana, berry, and orange; although including vegetable-fruits, such as avocado, cucumber, eggplant, okra, olive, pepper, pumpkin, snap beans, squashes, and tomato that are not sweet, and seeds
- **Seeds**—in fruit of a plant; may be in pods; includes legumes such as dried beans, peas, and peanuts, and, in the USA, sweet corn (although it is a grain, not vegetable); may be sprouted

Harvesting and Postharvest Changes

Harvesting and postharvest processing *schedules* and *procedures* should be strictly followed to ensure fruits and vegetables with the highest possible quality. Crops are harvested at *different* stages of maturity prior to storage, and they are likely to be larger and less tender with age. It may be ideal to harvest *less mature* fruits and vegetables, *or* to allow them to "ripen on the vine."

Another ideal practice is that of *cooling* fresh produce *in the field*, and then canning *close* to the field, prior to transport. This practice minimizes negative changes in quality.

After harvest, fruits and vegetables continue to undergo respiration, the metabolic process of taking in oxygen (O_2) and giving off carbon dioxide (CO_2), moisture, and heat. The maximum rate of respiration occurs just before full ripening. *Climacteric* fruits, such as the apple, apricot, avocado, banana, peach, pear, plum, and tomato ripen *after* harvesting. Tropical fruits such as the papaya and mango are also climacteric, as is the avocado.

On the other hand, *non-climacteric fruits*, such as the cherry, citrus fruit, grapes, melon, pineapple, and strawberry ripen *prior* to harvest.

- Climacteric—ripens AFTER harvesting

- Non-climacteric—ripens BEFORE harvesting

Natural postharvest sunlight, artificial, or fluorescent light exposure may form a green chlorophyll pigment and *solanine* (bitter, and toxic at high levels) in some vegetables such as onions or potatoes. Green colored spots may appear just below the skin, and if small, these small amounts may easily be cut away.

Proper packaging for shipping is significant. Storage conditions that retain plant's moisture or heat reduce negative changes in the fruit or vegetable, such as undesirable mold or rot.

Ripening

Evidence of ripening can be *seen* and felt to the touch in a physical evaluation. For example, changes from the green color (due to chlorophyll degradation) allow more carotenoid pigment to be visible in the fruit as it ripens. Flavor changes are noted with an *increase* in the sugar and *decrease* in the acid content.

Between the produce maturity and ripening, there is a lot of unseen enzymatic activity. Although ripening may be *unseen*, there is internal hormonal and enzyme activity prior to change in the physical appearance.

A *noticeable* ripening change that occurs is due to the production of odorless and colorless ethylene gas. For example, the emission of this gas generates a softening of the plant cell wall. *Ethylene gas* is a naturally occurring hydrocarbon produced by some vegetables and fruits, especially apples, bananas, citrus fruit, melon, and tomatoes. In particular, lettuce and leafy vegetables as well as any bruised fruits are especially susceptible to undesirable respiration due to the presence of ethylene gas. Storage conditions should *separate* ethylene producers from other fruits that do not require ripening. ("*One bad apple spoils the whole bunch*!")

In addition to *natural* ethylene gas, there is also *artificially* produced ethylene gas, made by the burning of hydrocarbons. Food distributors may introduce measured doses of ethylene gas into a closed food chamber for the purpose of ripening unripened fruits before they are sold to retailers. The effectiveness of ethylene in achieving faster and more uniform ripening is dependent on the pulp temperature and stage of maturity of the fruit, and the relative humidity of the ripening room (SYSCO Foods).

CULINARY ALERT! A technique for ripening fruit at home is to place unripened fruit in a closed paper bag, which then traps ethylene gas and speed up desirable ripening.

There is some control for the unwanted effects of natural ethylene gas, which may overripen the fruit and result in poor quality. Specifically, *gibberellic acid* may be added as a control to the external storage environment of fruits and vegetables. A preharvest application of this plant growth regulator delays ripening, and retains firmness in a fruit, both of which are important considerations in postharvest handling, storage, and transportation.

In the process of *senescence* (overripening), the intracellular protopectin develops into water-soluble pectin. *Overripe* fruits and vegetables become soft or mushy as the once-firm cells separate from one another. To control unwanted ripening and extend shelf life, edible waxes and other treatments, including irradiation treatment, may be applied to fruits and vegetables. The enzymes, sucrose synthetase and pectinase, are used in measuring maturity of *some* potatoes and fruits.

Refrigeration may reduce adverse chemical reactions. As well, manipulation of CO_2 and O_2 through controlled atmosphere storage (CA), controlled atmosphere packaging (CAP), and modified atmosphere packaging (MAP) offers control of ripening (Chap. 19).

Enzymatic Oxidative Browning

Enzymatic oxidative browning (EOB) occurs when the plant's phenolic compounds react with enzymes in the presence of oxygen. When bruised or cut during preparation, discoloration of *some* fruits or vegetables may occur. For example, when some varieties of apples, apricots, bananas, cherries, peaches, pears, eggplant, or potatoes are bruised or cut, the product *enzymes* are exposed to *oxygen* in the atmosphere, and the produce is subject to undesirable *browning or EOB*.

Control measures to prevent EOB may not be easy. For example, there may be more than one substrate existing in a fruit or vegetable, also, oxygen may come from intercellular spaces, not solely surface air, and then the responsible enzyme must be denatured. Damaging enzymes spread in storage, and as mentioned earlier, it is true that "one bad apple spoils the whole bunch!"

One effective control of browning is to avoid contact between the substrate and oxygen. In order to achieve this, food may be covered with a *sugar syrup* in order to block oxygen, or it may be covered with a film wrap that limits oxygen permeability. *Another* control is the application of a commercially prepared *citric acid powder* or *ascorbic acid* to the cut fruit surface. Lemon juice in a ratio of 3:1 with water may be applied to the surface of the fruit, according to the Produce Marketing Association. In this manner, the vitamin C juice is oxidized *instead* of the pigment, and the acidic pH inhibits enzymatic action.

Pineapple juice, because of its sulfhydryl groups (–SH) acts as an antioxidant, and is an *additional*, effective means of protection against browning. (As with lemon juice, the Produce Marketing Association recommends dipping cut fruits in pineapple juice [3:1 ratio, pineapple juice to water] for controlling EOB.) Sulfur compounds in the juice interfere with the darkening of various foods, such as cut fruit, cut lettuce leaves, and white wine. However, due to health concerns of a small percentage of the population allergic to sulfites, the use of sulfiting agents to prevent browning is restricted in raw products. Other available agents may be used.

Home gardeners usually blanch fruits or vegetables prior to freezing. *Blanching* destroys the polyphenol oxidase enzyme and enables the product to withstand many months of freezer storage *without* degradation. Blanching entails the placement of (usually) cut-up fruit or vegetable pieces in boiling water for a precise period of time prior to freezing. The exact length of time depends on the volume and texture of the product.

CULINARY ALERT! To control browning, avoid contact between the substrate and oxygen—cover susceptible fruit with a sugar syrup or film wrap. Cover or immerse cut fruit in lemon juice, orange juice, pineapple juice, or a commercial treatment to control browning.

Cooking Effect

Cooking has many effects on food—its water retention, color, texture, flavor, and nutritive value to name a few of the effects. When short cooking periods and cooking methods such as steaming are selected, the effect is minimal loss of both flavor and nutritive value. Also, steaming retains the natural color as it does not allow contact between released internal acids and the food.

Vegetables and fruits may be consumed raw, without cooking, or are made ready for consumption by methods such as baking, boiling, frying, pressure-cooking, sautéing, steaming, stir-frying, and so forth. Cooking introduces appearance and texture changes, as well as flavor and nutritive value changes, as shown in the following.

Water Retention/Turgor

Water retention and turgor are changed once a fruit or vegetable is cooked. Once cooked, the cell membranes *lose* their selective permeability, and unlike the simple movement of water/osmosis that occurs in *raw* produce, the *cooked* cell membranes allow the additional movement of sugars and some nutrients as well as water. *Diffusion* occurs as substances move from an area of higher concentration to an area of lower concentration and the plant cell loses its form, water, and turgor.

Color

The natural color of *raw* fruits and vegetables varies, and the color of *cooked* fruits and vegetables is influenced by a number of factors as previously discussed in this chapter. These factors include the natural plant pigment and pH, age, duration of cooking, use of a pan lid, cooking method employed, and surrounding environment. *Blanching* serves to inactivate enzymes and expel intercellular air that may negatively affect color.

Cooking in aluminum or cast-iron cookware may *negatively discolor* cooked products, therefore, instead, the use of stainless steel may be recommended for cooking vegetables or fruits. Another color change accompanies the use of sodium bicarbonate, which yields a *brighter green* color. However, as earlier mentioned, this usage is not recommended, as vitamin and texture losses occur.

Texture

Without doubt, the *texture* of a fruit or vegetable changes upon cooking. The texture of the cooked vegetable depends on a number of factors. These factors include pH, age, duration of cooking, and water composition. For example, lengthy cooking in boiling *alkaline* water drastically softens texture as hemicelluloses break down; cellulose is softened, and pectins degrade. The addition of *acid*, such as the addition of a tomato to another vegetable recipe, yields a firm cooked vegetable because tissues do not soften, and pectin precipitates.

Helping to retain texture are calcium ions. These calcium ions are *naturally* present in hard water or may be *added* to many canned vegetables in commercial processing. For example, canning tomatoes with the addition of calcium *retains* the texture of cooked plant tissue forming insoluble salts with pectic substances. For a similar reason, brown sugar or molasses are common additives that are useful in retaining texture as well—e.g., Boston Baked Beans. Of course, the texture is also related to maturity of the plant, which may become tougher and "woody" due to the presence of lignin in older plants.

Flavor

The flavor of cooked vegetables is dependent on factors such as the classification as either *Allium* or *Brassic*a, and loss of both water-soluble organic acids and sugars from the vacuole. Additionally, recipe ingredients including sugar, fat, herbs, and spices vary the flavor of vegetables and may actually encourage a wary person to eat the vegetables!

Nutritive Value

Nutritive value is presented in much more detail *later* in this chapter. For now, discussion is limited to *cooking effect* on nutritive value. The nutritive value of cooked fruits and vegetables is influenced by factors such as nutrients naturally present in the food, the type of cooking medium, duration of cooking and added substances. Through diffusion, water-soluble *vitamins and sugars* in the cell sap are lost from parenchyma cells and may be oxidized. On the other hand, *minerals* present in plant material are inorganic substances that cannot be destroyed (although they may be discarded in fruit or vegetable trimmings).

Of nutritional benefit in cooking is a *short cooking* time in a *minimal* amount of water or *steaming* the vegetables. Yet, there are times when just the opposite, that is, *lengthy cooking*, with *plentiful* water may be desirable to achieve mild taste—foods such as mild tasting cooked onions may benefit from lengthy cooking and plentiful water.

Regarding the use of lids, it may be beneficial to cook *with a pan lid on* since it speeds up cooking and leads to the *desirable* retention of acids, flavor, or nutrients. Recall however, cooking with a *lid on* for the entire duration of cooking is *detrimental* to the green chlorophyll pigment and *Brassica* flavored vegetables as has been described.

CULINARY ALERT! Cooking produces changes in the turgor, appearance, texture, flavor, and nutritive value. Some changes are desirable, some not! Cook vegetables minimally.

Fruits—Unique Preparation and Cooking Principles

In this portion of the chapter, attention is given to some of the unique aspects of cooking and preparing *fruits*. Further discussion of "fruits" in this section may include vegetable-fruits such as avocados and peppers, yet most typically, fruit is referring to sweet, fleshy fruits containing seeds. It should be kept in mind that bananas and seedless grapes are examples of *fruits without seeds*.

To repeat a previously mentioned concept, the *botanical* definition of a fruit includes all grains, legumes (beans and peas), nuts, as well as some plant parts commonly eaten as "vegetables" (i.e., tomatoes) and thus is different from the culinary definition. According to its *culinary* role, fruit is the sweet, fleshy part of a plant, usually eaten alone or served as dessert. Grains, legumes, and nuts do not fit into this culinary definition of fruit; neither do the "vegetable-fruits" such as avocadoes, cucumbers, eggplant, okra, olives, peppers, pumpkin, snap beans, squash, and tomatoes, which are typically considered as *vegetables* in dietary regimes. The following is interesting:

> A 1893 tax dispute led to the ruling by the United States Supreme Court that a tomato was a vegetable. "Botanically, tomatoes are considered a fruit of the vine, just as are cucumbers, squashes, beans, and peas. But in common language of people, whether sellers or consumers of provisions, all these are vegetables which are grown in kitchen gardens, and which, eaten cooked or raw, are, like potatoes, carrots, parsnips, turnips, beets, cauliflower, cabbage, celery, and lettuce, usually served at dinner in, with, or after the soup, fish, or meats which constitute the principal part of the repast, and not like fruits, generally, as dessert" (United States Supreme Court) (Cunningham 2002).

Fruit Preparation

During fruit preparation, *water loss* may occur. For instance, when fresh-cut strawberries are sprinkled with sugar for added flavor, water is lost from the fruit through osmosis, and red (sweetened) liquid can be seen collecting in the bowl of strawberries. Other fruits may show the same effect or undergo *discoloration* due to EOB.

Whether prepared commercially by Industry, by a foodservice establishment or at home, cooking fruit in different manners/mediums may occur as follows:

Water: When fruits are cooked in *plain water*, water moves into the tissues (***osmosis***), and sugar, at a 12–15 % level naturally, diffuses out (***diffusion***). The fruit, including dried fruit, such as raisins, becomes plump. Pectins become soluble and diffuse into water; cells become less dense, and the product becomes tenderer. Cellulose is softened, and lignins remain unchanged. The fruit loses its shape.

Sugar addition: Sugar may be utilized in cooking. It offers flavor and some preservation. When *large* amounts of sugar (amounts greater than that found naturally in fruits) are added to the cooking water at the *beginning* of cooking, the tenderization is diminished and the *shape* will be *maintained*. This is because the water moves out, and the higher concentration of sugar outside of the piece of fruit moves in by diffusion. As well, the sugar interferes with plant pectin solubility. It also dehydrates cellulose and hemicellulose resulting in shrunken, tough walls.

Timing for the addition of sugar is significant. If sugar is added to fruit *early* in cooking, then that is desirable for berries or slices, where retaining shape is important. *Conversely*, when fruits are cooked in plain water and sugar is added *late*, after cooked fruit *loses its shape* and softens, desirable fruit sauces such as applesauce are formed.

Flavor changes: There are flavor changes that occur in a fruit preparation method such as cooking fruit. Water-soluble sugars and other small molecules, escape to the surrounding water in cooking. Consequently, the cooked fruit tastes blander, unless sugar is added during cooking.

Fruit Juices and Juice Drinks

Fruit "Juices" are 100 % fruit by definition, while "juice drinks" must only contain 10 % or more of real juice. Each may be formulated from a variety of fruits. Data on yield and amounts of produce needed to extract juice becomes important in studies on diet and disease (Newman et al. 2002). The FDA requires that commercial juices be pasteurized to control microbial growth. Treatment with ultraviolet (UV) irradiation is given in order to reduce the pathogens and other detrimental microorganisms.

Grading Vegetables and Fruits

Grading by the United States government (USDA) is a *voluntary* function of packers and processors. It is *not* an indication of safety, nutritive value, or type of packs (e.g., "packed in heavy syrup" and so forth). Wholesalers, commercial, and institutional food service, including restaurants and schools, may purchase according to grade using written specifications, although consumers may be unaware of grading.

Dried and *frozen* forms of fruits and vegetables are graded, although grading indications appear *less* commonly than on *canned* or *fresh* products that often show grade. In the highly competitive wholesale food-service market, *canned* fruits and vegetables receive US Grade A, B, or C.

US Grade A is the *highest* rating and indicates the best appearance and texture, including clarity of liquid, color, shape, size, absence of blemishes or defects, and maturity. US Grade C is the *lowest* grade. *Fresh fruits* and vegetables are rated US Fancy, US No. 1, and US No. 2.

Private labeling by some companies may have specifications that state a narrow range within a grade. Proprietary names may be assigned to various grades.

Organically Grown Vegetables and Fruits

"Organically grown" was formerly a term without a federal standard for the foods' production, handling, and processing. Finally, in February 2001, the USDA provided a federal definition for "organically grown." Rules for implementing The Organic Foods Production Act of 1990 took several years to go into effect and proposals were released for feedback several years prior to the final ruling. A tremendous amount of public input was obtained in an attempt to satisfy both the organic farmer and the consumer.

The intent of the final comprehensive Organic Foods standard was to *clarify* for the consumer. As well, it was to *ease potential confusion* in domestic and export sales, and make use of just *one* product label, eliminating the need for individual state and/or private standards. The USDA Organic Seal was also redesigned for better consumer understanding and became effective for use in August 2002.

Subsequent to legislation, foods labeled "organic" must be grown *without* the use of chemical pesticides, herbicides, or fertilizers (Wardlaw & Smith 2011) and have *verifiable records* of their system of production. Organic products must be 95 % organically produced; processed foods may be labeled "made with organic ingredients." If organic production and handling is *not* followed, yet a product *is* offered for sale as organic, a large monetary fine may be imposed.

Even though there is the *absence* of chemical pesticides, herbicides, and fertilizers used during growth, which would be desirable to some individuals, there is *no* evidence that organically grown foods are *higher* in *nutrient* content than conventionally grown foods. A poor soil may yield a lower crop, yet *not* one of lesser nutritive value (Newman et al. 2002).

While the *pesticide* residue would certainly be lowered or nonexistent, *bacterial counts* of organically grown plant material may be *higher* than conventionally grown foods. This is especially true if animal manure was used as a fertilizer, and care in washing was overlooked. Organically grown is *not* synonymous with *food safety* either, therefore, as with all produce, care must be taken to wash contaminants off all fruits and vegetables.

Of note in this discussion is the reminder that the National Organic Program (NOP) applies to *more than fruits and vegetables*. Crop standards, livestock standards, and handling standards are all addressed by the Act.

Biotechnology (More in Appendices)

Biotechnology (biotech) advocates say that biotech assists in providing a less expensive, safer, and better tasting food supply. Several years of conventional breeding techniques may be shortened by gene manipulation, possibly by half for some foods. Growers have strived to increase availability and yield of their crops, despite factors such as weather conditions in the growing region, insect infestation, and the lengthy time period of conventional breeding. It could be said that biotechnology goes back many centuries as a tool in breeding crops.

Biotech represents a combination of (a) conventional breeding, by plant breeders including selection, gene-crossing, and mutation, with (b) biotechnology, including recombinant DNA, and gene transfer. Continued collaboration between scientists using both approaches is needed in order to improve product quality and meet consumers' demands. Many consumers want to have genetically altered food products so-labeled.

The FDA ensures the safety of genetically altered foods and food ingredients in two ways, by regulating: *adulteration*, and by the *food additive* provision of the rules. These two FDA regulating methods provide the same safety standards of any other *non-bioengineered* food product.

The following is a statement by the FDA Biotechnology Coordinator regarding food biotechnology:

Food biotechnology

First, let me explain what we mean when we refer to food biotechnology or genetically engineered foods. Many of the foods that are already common in our diet are obtained from plant varieties that were developed using conventional genetic techniques of breeding and selection. Hybrid corn, nectarines (which are genetically altered peaches), and tangelos (which are a genetic hybrid of a tangerine and grapefruit) are all examples of such breeding and selection. Food products produced through modern methods of biotechnology such as recombinant DNA techniques and cell fusion are emerging from research and development into the marketplace. It is these products that many people refer to as "genetically engineered foods." The European Commission refers to these foods as Genetically Modified Organisms. The United States uses the term genetic modification to refer to all forms of breeding, both modern, i.e., genetic engineering, and conventional.

The new gene splicing techniques are being used to achieve many of the same goals and improvements that plant breeders have sought through conventional methods. Today's techniques are different from their predecessors in two significant ways. First, they can be used with greater precision and allow for more complete characterization and, therefore, greater predictability about the qualities of the new variety. These techniques give scientists the ability to isolate genes and to introduce new traits into foods without simultaneously introducing many other undesirable traits, as may occur with traditional breeding. This is an important improvement over traditional breeding.

Second, today's techniques give breeders the power to cross biological boundaries that could not be crossed by traditional breeding. For example, they enable the transfer of traits from bacteria or animals into plants.

In conducting its safety evaluations of genetically engineered foods, FDA considers not only the final product but also the techniques used to create it. Although study of the final product ultimately holds the answer to whether or not a product is safe to eat, knowing the techniques used to create the product helps in understanding what questions to ask in reviewing the product's safety. That is the way FDA regulates both traditional food products and products derived through biotechnology.

Statement of:
James H. Maryanski, Ph.D., Biotechnology Coordinator
Center for Food Safety and Applied Nutrition Food and Drug Administration

Before:
The Subcommittee on Basic Research, House Committee on Science

FDA

Providing human and environmental *safety*, as well as *high-quality* foods, is of great significance to public. The FDA requires that all bioengineered foods be labeled *if* they are significantly *different* from the original conventional food in nutritive value, or in posing food allergies.

Areas of research continue to focus on improving the areas previously mentioned. Certainly, the nutritive content of plant foods, such as improving the protein content of plants, and increasing their resistance to pests, or improving their storage is researched. In addition to providing the consumer with greater economy, convenience, and improved nutritive value, safety is a factor that is important to both the grower and consumer. *Safety* of biotechnology has been debated and discussed by the public, educators, environmentalists, and scientists. The future may hold more such debate.

Historically, the safety of the first genetically engineered food designed for human consumption was demonstrated to the FDA and approval was granted for use of the Flavr-Savr tomato (in *May 1994*). Its shelf life was 10 days longer than other tomatoes. Due to the polygalacturonase (PG) enzyme, it stayed on the vine *longer*, thus it could be vine ripened with enhanced flavor. Then, in 1996, the planting of corn, potato, soybeans, and tomato varieties developed through *biotechnology* began following FDA decisions on safety. Currently many more food varieties are being developed through advances in biotechnology.

According to the International Food Information Council (IFIC), a significant component of the US harvest is produced by biotechnology (IFIC). In 10 years, during 2005, over 1,400 biotech notifications were acknowledged, and over 500 permits were approved (USDA).

The USDA's Agricultural Research Service (ARS) along with private industry and Academic research centers maintain the goal of developing improved genetic engineering. To date, there are some food companies that have ceased using, or announced that they will not use GMO's due to negative consumer reaction. The debate continues.

1. What is Agricultural Biotechnology?
Agricultural biotechnology is a range of tools, including traditional breeding techniques, that alter living organisms, or parts of organisms, to make or modify products; improve plants or animals; or develop microorganisms for specific agricultural uses. Modern biotechnology today includes the tools of genetic engineering.

2. How is Agricultural Biotechnology being used?

Biotechnology provides farmers with tools that can make production cheaper and more manageable. . ..

Researchers are at work to produce hardier crops that will flourish in even the harshest environments and that will require less fuel, labor, fertilizer, and water, helping to decrease the pressures on land and wildlife habitats. . . .

In addition to genetically engineered crops, biotechnology has helped make other improvements in agriculture not involving plants. Examples of such advances include making antibiotic production more efficient through microbial fermentation and producing new animal vaccines through genetic engineering for diseases such as foot and mouth disease and rabies.

USDA

For a more in-depth report on biotechnology and foods, see reports by the Institute of Food Technologists.

Irradiation

Irradiation is reported elsewhere in this text and in other writings. The aim is to control pathogens. There is much information available to learn beyond the scope of material in this text. Some fresh fruits, juices, and sprouts have also been treated in this manner. Plant seeds may be irradiated to control pathogens. On the horizon are the results of further studies seeking suitable methods to control pathogens in products other than fruits and vegetables.

According to the USDA "Food irradiation is a technology for controlling spoilage and eliminating foodborne pathogens." The result is similar to pasteurization. The fundamental difference between food irradiation and pasteurization is the source of the energy used to destroy the microbes. While conventional pasteurization relies on heat, irradiation relies on the energy of ionizing radiation.

"Food irradiation is a process in which approved foods are exposed to radiant energy, including gamma rays, electron beams, and x-rays. In 1963, the Food and Drug Administration (FDA) found the irradiation of food to be safe. . . . Irradiation is not a substitute for good sanitation and process control in meat and poultry plants. It is an added layer of safety".

Radura Symbol

Vegetarian Food Choices

Vegetarian foods are chosen by a growing number of vegetarians, whether it is for religious, political, health, or other reasons. To clarify "vegetarian" is not simple, one must realize that it may indicate something different to various individuals. The meaning varies. However, true *vegans* are vegetarians who omit all animal products from their diet. If other types of vegetarian cuisines are followed, vegetarians might consume milk, or eggs, white meat, or fish. Persons adhering to consumption of minimal animal products are classified as "flexitarians."

In view of the fact that *animal* products are the only significant source of vitamin B_{12}, vegans consuming a meat-less diet may be wise to obtain reliable, vitamin B_{12} fortified foods. Vitamin B_{12} supplementation may be chosen in order to maintain the myelin sheath surrounding the nerves and prevent permanent nerve damage and paralysis. It is valuable to note that microwave heating *inactivates* vitamin B_{12} in foods (Chap. 9).

Labeling of Vegetables and Fruits

Nutrition Facts

Nutrition Facts labeling on foods in the USA must report on four items—vitamin A and C and the minerals, calcium and iron. These are identified below the solid line on all Nutrition Facts food labels. These four nutrients in particular are listed as nutrients that fall short of adequate levels for the population. Many Americans would do well to increase their intake of these two vitamins that are simultaneously so prevalent in fruits and vegetables. The label provides the consumer with information regarding the percentage of Daily Value that they are consuming in each serving. Individual fresh fruits and vegetables do not have labels, yet supermarket brochures, posters, or plastic bags relate the nutrient contribution.

Label Terms

Labeling terms that apply to fruits and vegetables include the following and must appear as a product descriptor after the product name, for example, "green beans, fresh"

- A "***Fresh***" food must be a raw food, alive, and respiring. Some skin surface treatment is acceptable, such as application of wax, or pesticides. Treatment with less than 1 kGy irradiation, to inactivate pathogenic and spoilage microorganisms is allowed. (The FDA is considering use of the term "fresh" for alternative nonthermal technologies that function to protect the US food supply, and clearly convey food characteristics to consumers.)
- "Freshly prepared" is food that has not been frozen, heat processed, or preserved.
- A "Good Source of" must contain 10–19 % of the Daily Value of that nutrient per serving.
- If an item states "Fat-free," it must have less than 1/2 g of fat per serving. "Lowfat" indicates that the product must contain 3 g of fat or less per serving.
- Calorie level is important to many consumers. If an item states "Low-calorie," it must contain less than 40 cal per serving.
- "Sodium-free" signifies that a product contains less than 5 mg of sodium (Na) per serving. "Very-low-sodium" is used for a product that contains less than 35 mg of Na per serving, and "Low-sodium" is less than 140 mg Na per serving.
- "High-fiber" is 5 mg or more of fiber per serving.

The 1991 nutrition labeling produce regulations were amended by the FDA. Regulations exist for labeling nutritive value of the 20 most frequently consumed vegetables and fruits. In addition to the top 20, other vegetables and fruits *must* be labeled if nutritional claims are made. Such labeling is *voluntary* and will continue to be voluntary if there is sufficient compliance noted by the FDA.

Nutritive Value of Vegetables and Fruits

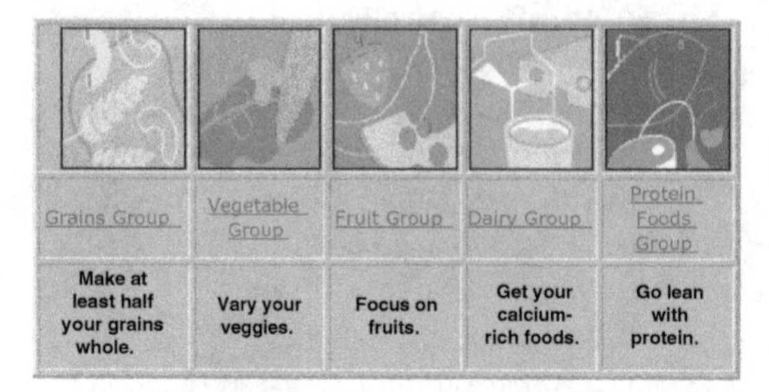

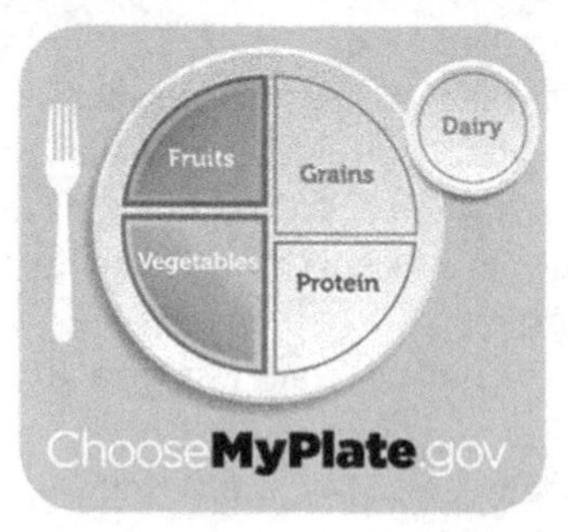

(USDA)

ChooseMyPlate.gov

http://www.choosemyplate.gov/food-groups/vegetables.html

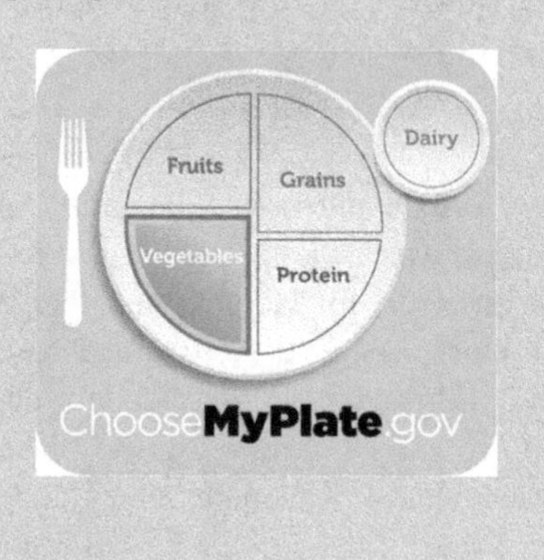

What Foods Are in the Vegetable Group?

Any vegetable or 100 % vegetable juice counts as a member of the Vegetable Group. Vegetables may be raw or cooked; fresh, frozen, canned, or dried/dehydrated; and may be whole, cut-up, or mashed.

Vegetables are organized into five subgroups, based on their nutrient content.

Key Consumer Message: Vary your veggies. Make half your plate fruits and vegetables.

View Vegetables Food Gallery

The *nutritional value* of vegetables and fruits is important in the diet. This text section is lengthy! Due to a worldwide supply and international purchasing potential, vegetables and fruits have year-round availability. Achieving good nutrition is enhanced by availability of the nutrients present in fruits and vegetables.

Vitamins, notably vitamins A and C, minerals (calcium and iron), and dietary fiber, are among the great benefits of a high fruit and vegetable diet, whether foods are canned, frozen, or fresh. As well, there are antioxidant properties (beta-carotene, vitamin C, and vitamin E), and anticarcinogenic properties, and fat is low for the majority of fruits and vegetables.

http://www.choosemyplate.gov/food-groups/fruits.html

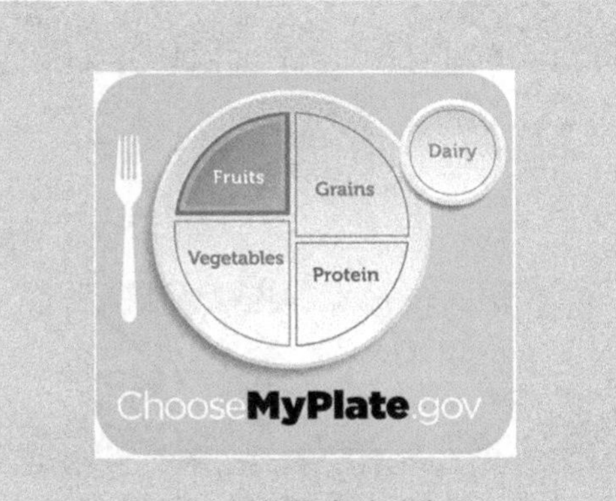

Any fruit or 100 % fruit juice counts as part of the Fruit Group. Fruits may be fresh, canned, frozen, or dried, and may be whole, cut-up, or pureed. Some commonly eaten fruits are identified on the website.

Key Consumer Message: Focus on fruits. Make half your plate fruits and vegetables

View Fruits Food Gallery

Further dietary and medicinal benefits of fruits and vegetables are shown. For example, non-nutrients, such as the ***phytochemicals*** (phyto = plant) in fruits and vegetables, may function in the prevention of human disease. This further supports the idea that nutrition is obtained from food rather than isolated compounds. Isolated compounds of fruits, vegetables, and other foods that are thought to provide health and medicinal benefits to the diet are ***nutraceuticals***. The FDA has not recognized the term nutraceuticals or allowed health claims on products beyond those that are supported by the scientific community (Chap. 20, Appendix).

Additional evaluation and research is needed in order to address the many potential health benefit/disease-preventing properties of plant material. Some nutrition facts are included in Figs. 7.6 and 7.7.

Unfortunately, the USDA Department of Health and Human Services has noted: "In this land of plenty, millions of Americans aren't eating wisely. Not because they haven't had enough to eat, but because they eat too many of the wrong things or too little of the right."

According to the American Diabetic Association Exchange List, one serving of vegetables contains 25 cal and one serving of fruit contains 60 cal.

"Vary your veggies" and "focus on fruits" is the USDA advice in selecting vegetables and fruits as part of a healthy diet.

Citrus fruits contain antioxidants, vitamin C, and relatively good amounts of folic acid that has been shown to prevent reoccurrence of neural tube defect in pregnant women. The FDA allows a label claim regarding foods with dietary fiber and a reduction of cancer incidence.

Taste is the most important factor that influences food choices; positive messages about benefits of diets with plenty of fruits and vegetables help with making choices. On a regular basis, the American Public eats too little of fruits and vegetables containing nutrients, such as vitamins A and C (on all Nutrition Facts labels), or the antioxidant vitamin E, all of which have an important role in preventing or delaying major degenerative diseases of Americans.

VEGETABLES

NUTRI-FACTS UPDATE

NUTRITION FACTS FOR RAW VEGETABLES[1]

Vegetables, Serving portion (gram weight/ounce weight)	Calories	Calories From Fat	Total Fat (g/%DV)	Sodium (mg/%DV)	Potassium (mg/%DV)	Total Carbohydrate (g/%DV)	Dietary Fiber (g/%DV)	Sugars (g)	Protein (g)	Vitamin A (%DV)	Vitamin C (%DV)	Calcium (%DV)	Iron (%DV)
Asparagus, 5 spears (93 g/3.3 oz)	25	0	0 / 0	0 / 0	230 / 7	4 / 1	2 / 8	2	2	10	15	2	2
Bell Pepper, 1 medium (148 g/5.3 oz)	30	0	0 / 0	0 / 0	270 / 8	7 / 2	2 / 8	4	1	8	190	2	2
Broccoli, 1 medium stalk (148 g/5.3 oz)	45	0	0.5 / 1	55 / 2	540 / 15	8 / 3	5 / 20	3	5	15	220	6	6
Carrot, 7" long, 1 1/4" diameter (78 g/2.8 oz)	35	0	0 / 0	40 / 2	280 / 8	8 / 3	2 / 8	5	1	270	10	2	0
Cauliflower, 1/6 medium head (99 g/3.5 oz)	25	0	0 / 0	30 / 1	270 / 8	5 / 2	2 / 8	2	2	0	100	2	2
Celery, 2 medium stalks (110 g/3.9 oz)	20	0	0 / 0	100 / 4	350 / 10	5 / 2	2 / 8	0	1	2	15	4	2
Cucumber, 1/3 medium (99 g/3.5 oz)	15	0	0 / 0	0 / 0	170 / 5	3 / 1	1 / 4	2	1	4	10	2	2
Green (Snap) Beans, 3/4 cup cut (83 g/3.0 oz)	25	0	0 / 0	0 / 0	200 / 6	5 / 2	3 / 12	2	1	4	10	4	2
Green Cabbage, 1/12 medium head (84 g/3.0 oz)	25	0	0 / 0	20 / 1	190 / 5	5 / 2	2 / 8	3	1	0	70	4	2
Green Onion, 1/4 cup chopped (25 g/0.9 oz)	10	0	0 / 0	5 / 0	70 / 2	2 / 1	1 / 4	1	0	2	8	0	0
Iceberg Lettuce, 1/6 medium head (89 g/3.2 oz)	15	0	0 / 0	10 / 0	120 / 3	3 / 1	1 / 4	2	1	4	6	2	2
Leaf Lettuce, 1 1/2 cups shredded (85 g/3.0 oz)	15	0	0 / 0	30 / 1	230 / 7	4 / 1	2 / 8	2	1	40	6	4	0
Mushrooms, 5 medium (84 g/3.0 oz)	20	0	0 / 0	0 / 0	300 / 9	3 / 1	1 / 4	0	3	0	2	0	2
Onion, 1 medium (148 g/5.3 oz)	60	0	0 / 0	5 / 0	240 / 7	14 / 5	3 / 12	9	2	0	20	4	2
Potato, 1 medium (148 g/5.3 oz)	100	0	0 / 0	0 / 0	720 / 21	26 / 9	3 / 12	3	4	0	45	2	6
Radishes, 7 radishes (85 g/3.0 oz)	15	0	0 / 0	25 / 1	230 / 7	3 / 1	0 / 0	2	1	0	30	2	0
Summer Squash, 1/2 medium (98 g/3.5 oz)	20	0	0 / 0	0 / 0	260 / 7	4 / 1	2 / 8	2	1	6	30	2	2
Sweet Corn, kernels from 1 medium ear (90 g/3.2 oz)	80	10	1 / 2	0 / 0	240 / 7	18 / 6	3 / 12	5	3	2	10	0	2
Sweet Potato, medium, 5" long, 2" diameter (130 g/4.6 oz)	130	0	0 / 0	45 / 2	350 / 10	33 / 11	4 / 16	7	2	440	30	2	2
Tomato, 1 medium (148 g/5.3 oz)	35	0	0.5 / 1	5 / 0	360 / 10	7 / 2	1 / 4	4	1	20	40	2	2

Most fruits and vegetables provide negligible amounts of saturated fat and cholesterol.

[1] Raw, edible weight portion.
Percent Daily Values are based on a 2,000 calorie diet.

Developed by: Food Marketing Institute, American Dietetic Association, American Meat Institute, Food Distributors International, National Broiler Council, National Cattlemen's Beef Association, National Fisheries Institute, National Grocers Association, National Turkey Federation, Produce Marketing Association, United Fresh Fruit and Vegetable Association

Data Source: U.S. Food and Drug Administration

(7/96)

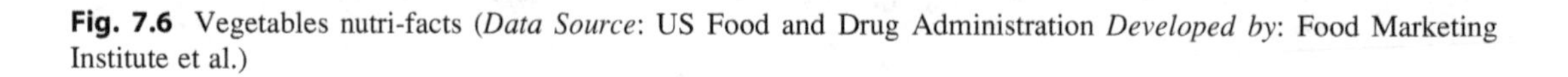

Fig. 7.6 Vegetables nutri-facts (*Data Source*: US Food and Drug Administration *Developed by*: Food Marketing Institute et al.)

FRUITS

NUTRI-FACTS UPDATE

NUTRITION FACTS FOR RAW FRUITS[1]

FRUIT, Serving portion (gram weight/ounce weight)	Calories	Calories From Fat	Total Fat (g/%DV)	Sodium (mg/%DV)	Potassium (mg/%DV)	Total Carbohydrate (g/%DV)	Dietary Fiber (g/%DV)	Sugars (g)	Protein (g)	Vitamin A (%DV)	Vitamin C (%DV)	Calcium (%DV)	Iron (%DV)
Apple, 1 medium (154 g/5.5 oz)	80	0	0/0	0/0	170/5	22/7	5/20	16	0	2	8	0	2
Avocado, California, 1/5 medium (30 g/1.1 oz)	55	45	5/8	0/0	170/5	3/1	3/12	0	1	0	4	0	0
Banana, 1 medium (126 g/4.5 oz)	110	0	0/0	0/0	400/11	29/10	4/16	21	1	0	15	0	2
Cantaloupe, 1/4 medium (134 g/4.8 oz)	50	0	0/0	25/1	280/8	12/4	1/4	11	1	100	80	2	2
Grapefruit, 1/2 medium (154 g/5.3 oz)	60	0	0/0	0/0	230/7	16/5	6/24	10	1	15	110	2	0
Grapes, 1 1/2 cups (138 g/4.9 oz)	90	10	1/2	0/0	270/8	24/8	1/4	23	1	2	25	2	2
Honeydew Melon, 1/10 medium melon (134 g/4.8 oz)	50	0	0/0	35/1	310/9	13/4	1/4	12	1	2	45	0	2
Kiwifruit, 2 medium (148 g/5.3 oz)	100	10	1/2	0/0	480/14	24/8	4/16	16	2	2	240	6	4
Lemon, 1 medium (58 g/2.1 oz)	15	0	0/0	5/0	90/3	5/2	1/4	1	0	0	40	2	0
Lime, 1 medium (67 g/2.4 oz)	20	0	0/0	0/0	75/2	7/2	2/8	0	0	0	35	0	0
Nectarine, 1 medium (140 g/5.0 oz)	70	0	0.5/1	0/0	300/9	16/5	2/8	12	1	4	15	0	2
Orange, 1 medium (154 g/5.5 oz)	70	0	0/0	0/0	260/7	21/7	7/28	14	1	2	130	6	2
Peach, 1 medium (98 g/3.5 oz)	40	0	0/0	0/0	190/5	10/3	2/8	9	1	2	10	0	0
Pear, 1 medium (166 g/5.9 oz)	100	10	1/2	0/0	210/6	25/8	4/16	17	1	0	10	2	0
Pineapple, 2 slices, 3" diameter, 3/4" thick (112 g/4 oz)	60	0	0/0	10/0	115/3	16/5	1/4	13	1	0	25	2	2
Plums, 2 medium (132 g/4.7 oz)	80	10	1/2	0/0	220/6	19/6	2/8	10	1	6	20	0	0
Strawberries, 8 medium (147 g/5.3 oz)	45	0	0/0	0/0	270/8	12/4	4/16	8	1	0	160	2	4
Sweet Cherries, 21 cherries; 1 cup (140 g/5.0 oz)	90	0	0.5/1	0/0	300/9	22/7	3/12	19	2	2	15	2	2
Tangerine, 1 medium (109 g/3.9 oz)	50	0	0.5/1	0/0	180/5	15/5	3/12	12	1	0	50	4	0
Watermelon, 1/18 medium melon; 2 cups diced pieces (280 g/10.0 oz)	80	0	0/0	10/0	230/7	27/9	2/8	25	1	20	25	2	4

Nutrient / % Daily Value of Nutrient

Most fruits and vegetables provide negligible amounts of saturated fat and cholesterol; avocados provide 1g of saturated fat per ounce.

[1] Raw, edible weight portion.
Percent Daily Values are based on a 2,000 calorie diet.

Developed by: Food Marketing Institute, American Dietetic Association, American Meat Institute, Food Distributors International, National Broiler Council, National Cattlemen's Beef Association, National Fisheries Institute, National Grocers Association, National Turkey Federation, Produce Marketing Association, United Fresh Fruit and Vegetable Association

Data Source: U.S. Food and Drug Administration

(7/96)

Fig. 7.7 Fruits nutri-facts (*Data Source*: US Food and Drug Administration *Developed by*: Food Marketing Institute et al.)

The Academy of Nutrition and Dietetics (Position of The Academy of Nutrition and Dietetics) states that eating a wide variety of foods, including an emphasis on grains, vegetables, and fruits is the best way to obtain adequate amounts of beneficial food constituents: "It is the position of The American Dietetic Association that the best nutritional strategy for promoting optimal health and reducing the risk of chronic disease is to obtain adequate nutrients from a variety of foods. Vitamin and mineral supplementation is appropriate when well-accepted, peer-reviewed scientific evidence shows safety and effectiveness." (Position of The Academy of Nutrition and Dietetics)

> Nutrition continues to drive decision making in supermarket aisles across the country, according to *Shopping for Health* 2012, the 20th in a yearly study released today by the Food Marketing Institute (FMI) and *Prevention*, and published by Rodale Inc. (Prevention Magazine and Food Marketing Institute 2012)
>
> [FMI conducts programs in public affairs, food safety, research, education, and industry relations on behalf of its nearly 1,250 food retail and wholesale member companies in the USA and around the world. FMI's US members operate more than 25,000 retail food stores and almost 22,000 pharmacies with a combined annual sales volume of nearly $650 billion. FMI's retail membership is composed of large multi-store chains, regional firms, and independent operators. Its international membership includes 126 companies from more than 65 countries. FMI's nearly 330 associate members include the supplier partners of its retail and wholesale members].

It is interesting to note that *The American Dental Association* recommends eating fruits such as apples and oranges and many uncooked vegetables such as carrots and celery. These act as "detergent" foods, cleaning teeth, and gums of food debris that may otherwise lead to the major nutrition-related problem of tooth decay.

Nutrient Losses

Nutrient losses may result from:

- Ascorbic acid (vitamin C) and thiamin (B_1) diffused to the water and oxidized.
- Mineral salts lost in soaking or cooking water.
- Excessive peel removal.
- Excessive chopping.
- Prolonged or high temperature storage.

Storage:

- Succulents, and leafy fruits and vegetables—stored covered in the refrigerator.
- Tubers—stored in a dark, cool place for quality.

10 tips
Nutrition Education Series

add more vegetables to your day

10 tips to help you eat more vegetables

It's easy to eat more vegetables! Eating vegetables is important because they provide vitamins and minerals and most are low in calories. To fit more vegetables in your meals, follow these simple tips. It is easier than you may think.

1 discover fast ways to cook

Cook fresh or frozen vegetables in the microwave for a quick-and-easy dish to add to any meal. Steam green beans, carrots, or broccoli in a bowl with a small amount of water in the microwave for a quick side dish.

2 be ahead of the game

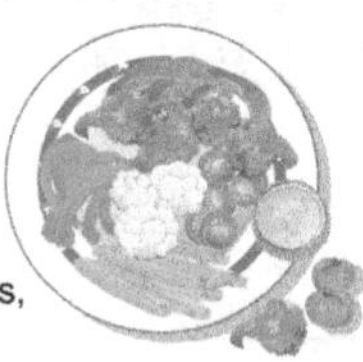

Cut up a batch of bell peppers, carrots, or broccoli. Pre-package them to use when time is limited. You can enjoy them on a salad, with hummus, or in a veggie wrap.

3 choose vegetables rich in color

Brighten your plate with vegetables that are red, orange, or dark green. They are full of vitamins and minerals. Try acorn squash, cherry tomatoes, sweet potatoes, or collard greens. They not only taste great but also are good for you, too.

4 check the freezer aisle

Frozen vegetables are quick and easy to use and are just as nutritious as fresh veggies. Try adding frozen corn, peas, green beans, spinach, or sugar snap peas to some of your favorite dishes or eat as a side dish.

5 stock up on veggies

Canned vegetables are a great addition to any meal, so keep on hand canned tomatoes, kidney beans, garbanzo beans, mushrooms, and beets. Select those labeled as "reduced sodium," "low sodium," or "no salt added."

6 make your garden salad glow with color

Brighten your salad by using colorful vegetables such as black beans, sliced red bell peppers, shredded radishes, chopped red cabbage, or watercress. Your salad will not only look good but taste good, too.

7 sip on some vegetable soup

Heat it and eat it. Try tomato, butternut squash, or garden vegetable soup. Look for reduced- or low-sodium soups.

8 while you're out

If dinner is away from home, no need to worry. When ordering, ask for an extra side of vegetables or side salad instead of the typical fried side dish.

9 savor the flavor of seasonal vegetables

Buy vegetables that are in season for maximum flavor at a lower cost. Check your local supermarket specials for the best-in-season buys. Or visit your local farmer's market.

10 try something new

You never know what you may like. Choose a new vegetable—add it to your recipe or look up how to fix it online.

Go to www.ChooseMyPlate.gov for more information.

DG TipSheet No. 2
June 2011
USDA is an equal opportunity provider and employer.

10 tips
Nutrition Education Series

focus on fruits

10 tips to help you eat more fruits

Eating fruit provides health benefits. People who eat more vegetables and fruits as part of an overall healthy diet are likely to have a reduced risk of some chronic diseases. Fruits provide nutrients vital for health, such as potassium, dietary fiber, vitamin C, and folate (folic acid). Most fruits are naturally low in fat, sodium, and calories. None have cholesterol. Any fruit or 100% fruit juice counts as a part of the Fruit Group. Fruits may be fresh, canned, frozen, or dried, and may be whole, cut-up, or pureed.

1 keep visible reminders

Keep a bowl of whole fruit on the table, counter, or in the refrigerator.

2 think about taste

Buy fresh fruits in season when they may be less expensive and at their peak flavor. Add fruits to sweeten a recipe.

3 think about variety

Buy fruits that are dried, frozen, and canned (in water or 100% juice) as well as fresh, so that you always have a supply on hand.

4 don't forget the fiber

Make most of your choices whole or cut-up fruit, rather than juice, for the benefits that dietary fiber provides.

5 be a good role model

Set a good example for children by eating fruit every day with meals or as snacks.

6 include fruit at breakfast

At breakfast, top your cereal with bananas, peaches, or strawberries; add blueberries to pancakes; drink 100% orange or grapefruit juice. Or, try a fruit mixed with fat-free or low-fat yogurt.

7 try fruit at lunch

At lunch, pack a tangerine, banana, or grapes to eat, or choose fruits from a salad bar. Individual containers of fruits like peaches or applesauce are easy and convenient.

8 experiment with fruit at dinner, too

At dinner, add crushed pineapple to coleslaw, or include orange sections, dried cranberries, or grapes in a tossed salad.

9 snack on fruits

Dried fruits make great snacks. They are easy to carry and store well.

10 keep fruits safe

Rinse fruits before preparing or eating them. Under clean, running water, rub fruits briskly to remove dirt and surface microorganisms. After rinsing, dry with a clean towel.

Go to www.ChooseMyPlate.gov for more information.

DG TipSheet No. 3
June 2011
USDA is an equal opportunity provider and employer.

Safety of Vegetables and Fruits

Safety is a food characteristic that the public expects. Foods should be safe, and in fact, the public is encouraged to eat more fruits and vegetables for health. Fruits and vegetables are *not* considered "potentially hazardous foods" that allow the "rapid and progressive growth of infectious or toxigenic microorganisms" (Model FDA Food Code).

In comparison to animal-based foods, there are few problems with plant-based products, yet, unfortunately, plant-based products can carry disease. Recently fresh, bagged spinach was pulled off the market nationwide, due to E-coli bacteria. Health Departments across the USA advised that washing the bagged spinach could not guarantee safety. One death and illness in many states followed ingestion of the spinach.

Pathogenic microorganisms are found in the environment and can contaminate food, causing illness. Imports from less-developed regions of the world may be implicated as a contributing factor in the increase in fruit- and vegetable-related foodborne illness.

Check your steps at FoodSafety.gov. Also see the chapter on Food Safety.

Some foods are more frequently associated with foodborne illness. With these foods, it is especially important to:

- CLEAN: Wash hands and food preparation surfaces often. And wash fresh fruits and vegetables carefully.
- SEPARATE: Don't cross-contaminate! When handling raw meat, poultry, seafood, and eggs, keep these foods and their juices away from ready-to-eat foods.
- COOK: Cook to proper temperature. See the Minimum Cooking Temperatures chart for details on cooking meats, poultry, eggs, leftovers, and casseroles.
- CHILL: At room temperature, bacteria in food can double every 20 min. The more bacteria there are, the greater the chance you could become sick. So, refrigerate foods quickly because cold temperatures keep most harmful bacteria from multiplying.

Regardless of its source, it is recommended that "ready-to-eat" value-added fresh produce be *washed* prior to consumption, and then refrigerated in order to maintain food safety. Washing is recommended despite the label statement that the product is washed and ready-to-eat.

Cross-contamination from other foods, such as meats, should be avoided, pull dates should be adhered to, and assembly/preparation areas should be sanitary. Of course personal hygiene is crucial to food safety.

Hydrogen peroxide is a generally recognized as safe (*GRAS*) substance that has also been used as a bleaching agent (as in milk used for cheese), and as an antimicrobial agent in foods. Some antimicrobials are effective due to their low pH, yet are not usable due to the unacceptable flavor that they impart. Other substances having antimicrobial properties include essential oils from citrus, coriander, mint, parsley, and vanillin juice peels.

Conclusion

Plant tissue is composed primarily of parenchyma tissue. The structure and composition of a fresh fruit or vegetable changes as the cell is destroyed. As fruits and vegetables typically contain a very large percentage of water, the maintenance of turgor pressure is an important factor in determining plant material quality.

The desirable pigments and flavor compounds contained in fruits and vegetables may undergo unacceptable changes upon preparation and cooking. Discoloration of some cut vegetables or fruits is known as EOB, which must be controlled. Improper storage or cooking can result in quality losses.

The nutritive value of vitamins, pro-vitamins (carotene) minerals, fiber, and other compounds contained in fruits and vegetables are extremely important to the diet, and there are medicinal benefits of fruits and vegetables. Many are low in fat content. Vegetarian food choices may be met with consumption of a variety of fruits and vegetables. "Vary your veggies" and "focus on fruits" is the USDA advice in selecting vegetables and fruits as part of a healthy diet.

Biotechnology provides the consumer with greater economy and convenience. Coupled with an understanding of the role of phytochemicals in disease prevention, vegetables and fruits may provide a greater nutrient contribution to the human diet. Irradiation is utilized as a means of ensuring food safety. Items of high nutritional value that were once unfamiliar and not used, as well as new items from around the world are now available on grocery shelves.

Notes

CULINARY ALERT!

Glossary

Allium Flavor compounds in the genus *Allium* that contain sulfur compounds and offer phytochemical value.

Anthocyanin Red-blue pigmented vegetables of the Flavone family.

Anthoxanthin Whitish pigmented fruits and vegetables of the Flavone group of chemicals.

Biotechnology Biogenetic engineering of animals, microorganisms, and plants to alter or create products that have increased resistance to pests, improved nutritive value, and shelf life.

Brassica Flavor compound of *Brassica* genus including cruciferous vegetables with sulfur compounds.

Carotenoid The group of red-orange pigmented fruits and vegetables; some are precursors of vitamin A and also have antioxidant value.

Cellulose Glucose polymer joined by β-1,4 glycosidic linkages; cannot be digested by human enzymes, thus it provides insoluble dietary fiber.

Cell sap Found in the plant vacuole; contains water-soluble components such as sugars, salts, and some color and flavor compounds.

Chlorophyll The green pigment of fruits and vegetables.

Cytoplasm Plant cell contents inside the cell membrane, but outside the nucleus.

Diffusion Movement of solute across a permeable membrane from an area of greater concentration to lesser concentration in heated products that do not have an intact cell membrane.

Enzymatic oxidative browning Browning of cut or bruised fruits and vegetables due to the presence of phenolic compounds, enzymes, and oxygen.

Fresh Alive and respiring as evidenced by metabolic and biochemical activities.

Fruit The mature ovaries of plants with their seeds.

Hemicellulose The indgestible fiber in cell walls that provides bulk in the diet; may be soluble, but primarily insoluble.

Lignin The noncarbohydrate component of fiber of plant tissue that is insoluble and excreted from the body. It provides the undesirable woody texture of mature plants.

Middle lamella The cementing material between adjacent plant cells, containing pectic substances, magnesium, calcium, and water.

Nutraceuticals The name given to a proposed new regulatory category of food components that may be considered a food or part of a food and may supply medical or health benefits including the treatment or prevention of disease. A term not recognized by the FDA.

Osmosis The movement of water across semipermeable membranes from an area of greater concentration to lesser concentration in products with an intact cell membrane.

Parenchyma tissue Majority of plant cells containing the cytoplasm and nucleus.

Pectic substances The intercellular "cement" between cell walls; the gel-forming polysaccharide of plant tissue.

Phytochemicals Plant chemicals; natural compounds other than nutrients in fresh plant material that help in disease prevention. They protect against oxidative cell damage and may facilitate carcinogen excretion from the body to reduce the risk of cancer.

Turgor pressure Pressure exerted by water-filled vacuoles on the cytoplasm and the partially elastic cell wall.

Vacuole Cavity filled with cell sap and air.

References

Cunningham E (2002) Is a tomato a fruit *and* a vegetable? J Am Diet Assoc 102:817

Food Product Design (2012) Fruits and vegetables versus cancer. Food Product Design (October):20–22

Hazen C (2012) A taste of healthy spices and seasonings. Food Product Design (June):73–82

Newman V, Faerber S, Zoumas-Morse C, Rock CL (2002) Amount of raw vegetables and fruits needed to yield 1 c juice. J Am Diet Assoc 102:975–977

Prevention Magazine and Food Marketing Institute (2012) The 20th annual "Shopping For Health" survey results

Sherman PW, Flaxman SM (2001) Protecting ourselves from food. Am Sci 89:142–151

SYSCO Foods. Ethylene gas and applications

Wardlaw G, Smith A (2011) Contemporary nutrition, 8th edn. McGraw Hill, New York

Bibliography

A recent web-based seminar offered to industry and media reps http://www.foodseminars.net/product.sc?productId = 111 included the following seminar: organic and conventional foods: safety and nutritional comparisons

Academy of Nutrition and Dietetics (formerly American Dietetic Association.) Chicago, IL

American Cancer Society (ACS)

American Soybean Association. St. Louis, MO

Arrowhead Mills. Hereford, TX

Centers for Disease Control and Prevention (CDC)

Fresh-cut Produce Association (1976) How to buy canned and frozen fruits. Home and Garden Bulletin no. 167. Consumer and Marketing Service, USDA, Washington, DC

Fresh-cut Produce Association (1977a) How to buy dry beans, peas, and lentils. Home and garden bulletin no. 177. USDA, Washington, DC

Fresh-cut Produce Association (1977b) How to buy fresh fruits. Home and garden bulletin no. 141. USDA, Washington, DC

Fresh-cut Produce Association (1980) How to buy fresh vegetables. Home and Garden Bulletin no. 143. USDA, Washington, DC

McCormick and Co. Hunt Valley, MD

Model FDA Food Code

Shattuck D (2002) Eat your vegetables: make them delicious. J Am Diet Assoc 102:1130–1132

USDA ChooseMyPlate.gov

Vaclavik VA, Pimentel MH, Devine MM (2010) Dimensions of food, 7th edn. CRC Press, Boca Raton, FL

Van Duyn MAS, Pivonka E (2000) Overview of the health benefits of fruit and vegetable consumption for the dietetics professional. Selected literature. J Am Diet Assoc 100:1511–1521

Part III

Proteins in Food

8 Proteins in Food: An Introduction

Introduction

Proteins are the most abundant molecules in cells, making up 50 % or more of their dry weight. Every protein has a unique structure and conformation or shape, which enables it to carry out a specific function in a living cell. Proteins comprise the complex muscle system and the connective tissue network, and they are important as carriers in the blood system. All enzymes are proteins; enzymes are important as catalysts for many reactions (both desirable and undesirable) in foods.

All proteins contain carbon, hydrogen, nitrogen, and oxygen. Most proteins contain sulfur, and some contain additional elements; for example, milk proteins contain phosphorus, and hemoglobin and myoglobin contain iron. Copper and zinc are also constituents of some proteins.

Proteins are made up of amino acids. There are at least 20 different amino acids found in nature, and they have different properties depending on their structure and composition. When combined to form a protein, the result is a unique and complex molecule with a characteristic structure and conformation and a specific function in the plant or animal where it belongs. Small changes, such as a change in pH, or simply heating a food, can cause dramatic changes in protein molecules. Such changes are seen, for example, when cottage cheese is made by adding acid to milk or when scrambled eggs are made by heating and stirring eggs.

Proteins are very important in foods, both nutritionally and as functional ingredients. They play an important role in determining the texture of a food. They are complex molecules, and it is important to have an understanding of the basics of protein structure to understand the behavior of many foods during processing. This chapter covers the basics of amino acid and protein structure. Individual proteins, such as milk, meat, wheat, and egg proteins, are covered in the chapters relating to these specific foods.

Amino Acids

General Structure of Amino Acids

Every ***amino acid*** contains a central carbon atom, to which is attached a carboxyl group (COOH), an amino group (NH_2), a hydrogen atom, and another group or side chain R specific to the particular amino acid. The general formula for an amino acid is

$$\begin{array}{c} \text{H} \\ | \\ \text{COOH} - \text{C} - \text{NH}_2 \\ | \\ \text{R} \end{array}$$

Glycine is the simplest amino acid, with the R group being a hydrogen atom. There are more than 20 different amino acids in proteins. Their

For use with subsequent Protein food chapters.

DOI 10.1007/978-1-4614-9138-5_8,

properties depend on the nature of their side chains or R groups.

In a solution at pH 7, all amino acids are ***zwitterions***; that is, the amino group and carboxyl groups are both ionized and exist as COO^- and NH_3^+, respectively. Therefore, amino acids are ***amphoteric*** and can behave as an acid or as a base in water depending on the pH. When acting as an acid or proton donor, the positively charged amino group donates a hydrogen ion, and when acting as a base the negatively charged carboxyl group gains a hydrogen ion, as follows:

Acid. $R-CH(NH_3^+)-COO^- \rightleftharpoons R-CH(NH_2)-COO^- + H^+$

Base: $R-CH(NH_3^+)-COO^- + H^+ \rightleftharpoons R-CH(NH_3^+)-COOH$

Categories of Amino Acids

Amino acids can be divided into four categories, according to the nature of their side chains, as shown in Fig. 8.1. The first category includes all the amino acids with ***hydrophobic*** or **nonpolar** side chains. The hydrophobic (water-hating) amino acids contain a hydrocarbon side chain. Alanine is the simplest one, having a methyl group (CH_3) as its side chain. Valine and leucine contain longer, branched, hydrocarbon chains. Proline is an important nonpolar amino acid. It contains a bulky five-membered ring, which interrupts ordered protein structure. Methionine is a sulfur-containing nonpolar amino acid. The nonpolar amino acids are able to form hydrophobic interactions in proteins; that is, they associate with each other to avoid association with water.

The second group of amino acids includes those with **polar uncharged** side chains. This group is ***hydrophilic***. Examples of amino acids in this group include serine, glutamine, and cysteine. They either contain a hydroxyl group (OH), an amide group ($CONH_2$), or a thiol group (SH). All polar amino acids can form hydrogen bonds in proteins. Cysteine is unique because it can form ***disulfide bonds*** (—S—S—), as shown below:

$$X-CH_2-SH + HS-CH_2-X \rightarrow X-CH_2-S-S-CH_2-X + H_2$$

cysteine → cystine

$X = NH_2-CH(COOH)-$

A disulfide bond is a strong covalent bond, unlike hydrogen bonds, which are weak interactions. Two molecules of cysteine can unite in a protein to form a disulfide bond. A few disulfide bonds in a protein have a significant effect on protein structure, because they are strong bonds. Proteins containing disulfide bonds are usually relatively heat stable, and more resistant to unfolding than other proteins. The presence of cysteine in a protein therefore tends to have a significant effect on protein conformation.

The third and fourth categories of amino acids include the charged amino acids. The **positively charged (basic)** amino acids include lysine, arginine, and histidine. These are positively charged at pH 7 because they contain an extra amino group. When a basic amino acid is part of a protein, this extra amino group is free (in other words, not involved in a peptide bond) and, depending on the pH, may be positively charged.

The negatively charged (acidic) amino acids include aspartic acid and glutamic acid. These are negatively charged at pH 7 because they both contain an extra carboxyl group. When an acidic amino acid is contained within a protein, the extra carboxyl group is free and may be charged, depending on the pH.

Oppositely charged groups are able to form ionic interactions with each other. In proteins, acidic and basic amino acid side chains may interact with each other, forming ionic bonds or salt bridges.

Protein Structure and Conformation

All proteins are made up of many amino acids, joined by ***peptide bonds*** as shown below:

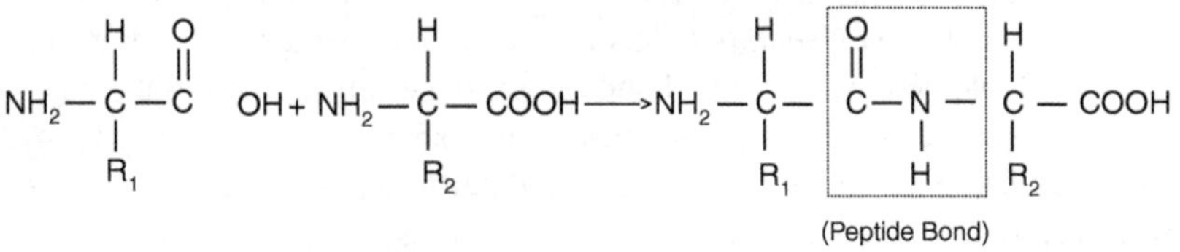

$$NH_2-CH(R_1)-C(=O)-OH + NH_2-CH(R_2)-COOH \longrightarrow NH_2-CH(R_1)-C(=O)-NH-CH(R_2)-COOH$$

(Peptide Bond)

Fig. 8.1 Examples of amino acids classified according to the nature of their R groups (only the side groups are shown)

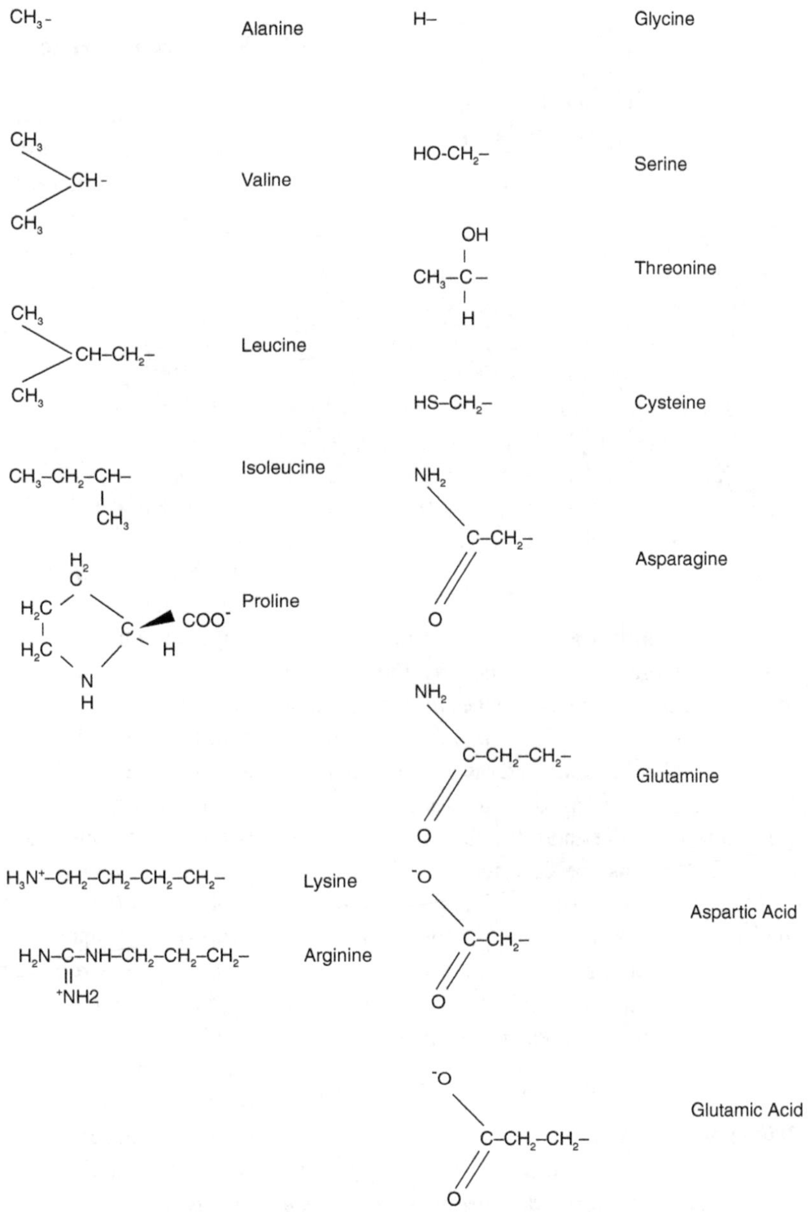

Side group	Amino acid	Side group	Amino acid
CH_3-	Alanine	H–	Glycine
$(CH_3)_2CH$-	Valine	$HO-CH_2$–	Serine
		$CH_3-CH(OH)$–	Threonine
$(CH_3)_2CH-CH_2$–	Leucine	$HS-CH_2$–	Cysteine
$CH_3-CH_2-CH(CH_3)$–	Isoleucine	$NH_2-C(=O)-CH_2$–	Asparagine
H_2C, H_2C, H_2C, NH ring; C–H, COO^-	Proline		
		$NH_2-C(=O)-CH_2-CH_2$–	Glutamine
$H_3N^+-CH_2-CH_2-CH_2-CH_2$–	Lysine	$^-O-C(=O)-CH_2$–	Aspartic Acid
$H_2N-C(={}^+NH2)-NH-CH_2-CH_2-CH_2$–	Arginine		
		$^-O-C(=O)-CH_2-CH_2$–	Glutamic Acid

Peptide bonds are strong bonds and are not easily disrupted. A ***dipeptide*** contains two amino acids joined by a peptide bond. A ***polypeptide*** contains several amino acids joined by peptide bonds. Proteins are usually much larger molecules, containing several hundred amino acids. They can be hydrolyzed, yielding smaller polypeptides, by enzymes or by acid digestion.

The sequence of amino acids joined by peptide bonds forms the backbone of a protein, as shown below:

```
 O   R   H   O
 ‖   |   |   ‖
 C   C  N H  C
\ / \ /|\ / \|/ \ /
 C  N H C   C   N
 |  |   ‖   |   |
 R  H   O   R   H
```

- The protein backbone consists of repeating N—C—C units.
- The amino acid side chains (R groups) project alternately from either side of the protein chain.
- The nature of the R groups determines the structure or ***conformation*** of the chain. (In other words, the shape the protein assumes in space.)

Each protein has a complex and unique conformation, which is determined by the specific amino acids and the sequence in which they occur along the chain. To understand the function of proteins in food systems and the changes that occur in proteins during processing, it is important to understand the basics of protein structure. Proteins are described as having four types of structure—primary, secondary, tertiary, and quaternary structure—and these build on each other. The primary structure determines the secondary structure and so on. The different types of protein structures are outlined below.

Primary Structure

The primary structure (***protein primary structure***) of a protein is the specific sequence of amino acids joined by peptide bonds along the protein chain. This is the simplest way of looking at protein structure. In reality, proteins do not exist simply as straight chains. However, it is the specific sequence of amino acids that determines the form or shape that a protein assumes in space. Therefore, it is essential to know the primary structure if a more detailed understanding of the structure and function of a particular protein is desired.

Secondary Structure

The secondary structure (***protein secondary structure***) of a protein refers to the three-dimensional organization of segments of the polypeptide chain. Important secondary structures include the following:

- Alpha helix—ordered structure
- Beta-pleated sheet—ordered structure
- Random coil—disordered structure

The ***alpha (α) helix*** is a corkscrew structure, with 3.6 amino acids per turn. It is shown in Fig. 8.2. It is stabilized by intrachain hydrogen bonds; that is, the hydrogen bonds occur within a single protein chain, rather than between adjacent chains. Hydrogen bonds occur between each turn of the helix. The oxygen and hydrogen atoms that comprise the peptide bonds are involved in hydrogen bond formation. The α-helix is a stable, organized structure. It cannot be formed if proline is present, because the bulky five-membered ring prevents formation of the helix.

The ***beta (β)-pleated sheet*** is a more extended conformation than the α-helix. It can be thought of as a zigzag structure rather than a corkscrew. It is shown in Fig. 8.3. The stretched protein chains combine to form β-pleated sheets. These sheets are linked together by interchain hydrogen bonds. (Interchain hydrogen bonds occur between adjacent sections of the protein chains rather than within an individual chain.) Again, the hydrogen and oxygen atoms that form the peptide bonds are involved in hydrogen bond

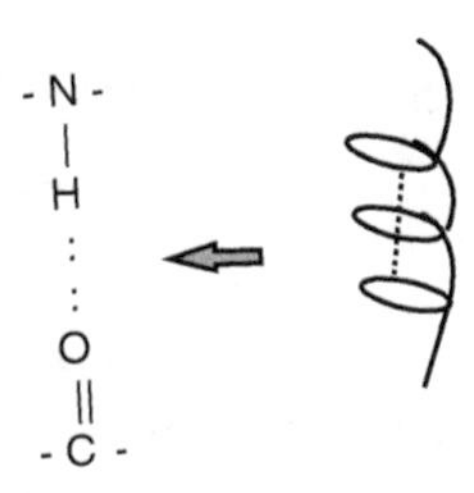

Fig. 8.2 Schematic three-dimensional structure of an α-helix

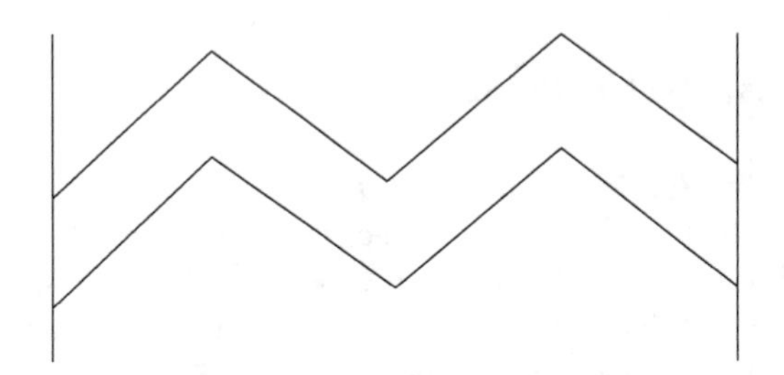

Fig. 8.3 Schematic three-dimensional structure of β-pleated sheets

formation. Like the α-helix, the β-pleated sheet is also an ordered structure.

The ***random coil*** is a secondary structure with no regular or ordered pattern along the polypeptide chain. This is a much more flexible structure than either the α-helix or β-pleated sheet. It is formed when amino acid side chains prevent formation of the α-helix or β-sheet. This may occur if proline is present or if there are highly charged regions within the protein.

A protein may contain regions of α-helix, β-sheet, and random coil at different places along the chain. How much of each type of secondary structure it contains depends on the sequence of amino acids or, in other words, on the primary structure of the protein.

Tertiary Structure

The ***tertiary structure*** of a protein refers to the three-dimensional organization of the complete protein chain. In other words, it refers to the spatial arrangement of a protein chain that contains regions of α-helix, β-sheet, and random coil. So, this structure is really an overview of a protein chain, rather than a detailed look at a small section of it. Again, the tertiary structure is built on the secondary structure of a specific protein.

There are two types of protein tertiary structure:
- Fibrous proteins
- Globular proteins

Fibrous proteins include structural proteins such as collagen (connective tissue protein), or actin and myosin, which are the proteins that are responsible for muscle contraction. The protein chains are extended, forming rods or fibers. Proteins with a fibrous tertiary structure contain a large amount of ordered secondary structure (either α-helix or β-sheet).

Globular proteins are compact molecules and are spherical or elliptical in shape, as their name suggests. These include transport proteins, such as myoglobin, which carry oxygen to the muscle. The whey proteins and the caseins, both of which are milk proteins, are also globular proteins. Globular tertiary structure is favored by proteins with a large number of hydrophobic amino acids. These orient toward the center of the molecule and interact with each other by hydrophobic interactions. Hydrophilic amino acids orient toward the outside of the molecule and interact with other molecules; for example, they may form hydrogen bonds with water. The orientation of the hydrophobic amino acids toward the center of the molecule produces the compact globular shape that is characteristic of globular proteins.

Quaternary Structure

Protein quaternary structure, or the quaternary protein structure, involves the noncovalent association of protein chains. The protein chains may or may not be identical. Examples of quaternary structure include the actomyosin system of muscles and the casein micelles of milk. For more information on these structures, the reader is referred to the chapters on meat and milk, respectively.

Interactions Involved in Protein Structure and Conformation

Protein primary structure involves only peptide bonds, which link the amino acids together in a specific and unique sequence. Secondary and tertiary structures may be stabilized by hydrogen bonds, disulfide bonds, hydrophobic interactions, and ionic interactions. ***Steric*** or spatial effects are also important in determining protein conformation. The space that a protein molecule occupies is determined partially by the size and shape of the individual amino acids along the protein chain. For example, bulky side chains such as proline prevent formation of the α-helix and favor random coil formation. This prevents the protein from assuming certain arrangements in space.

Quaternary structures are stabilized by the same interactions, with the exception of disulfide bonds. As has already been mentioned, disulfide bonds are strong, covalent bonds, and so the presence of only a few disulfide bonds will have a dramatic effect on protein conformation and stability. Hydrogen bonds, on the other hand, are weak bonds, yet they are important because there are so many of them.

Each protein takes on a unique native conformation in space, which can almost be considered as a "fingerprint." As has already been mentioned, the exact folding of the protein into its natural conformation is governed by the amino acids that are present in the protein and the bonds that the side chains are able to form in a protein. The amino acid sequence is also important, as the location of the amino acids along the chain determines which types of bonds will be formed and where, and thus determines how much α-helix, β-sheet, or random coil will be present in a protein. This, in turn, determines the tertiary and quaternary structure of a protein, all of which combine to define its native conformation. Knowledge of protein conformation and stability is essential to understanding the effects of processing on food proteins.

Reactions and Properties of Proteins

Amphoteric

Like amino acids, proteins are ***amphoteric*** (being able to act as an acid or a base) depending on the pH. This enables them to resist small changes in pH. Such molecules are said to have buffering capacity.

Isoelectric Point

The ***isoelectric point*** of a protein is the pH at which the protein is electrically neutral (it is denoted by pI). At this pH, the global or overall charge on the protein is 0. This does not mean that the protein contains no charged groups. It means that the number of positive charges on the protein is equal to the number of negative charges. At the isoelectric point, the protein molecules usually precipitate because they do not carry a net charge. (Molecules that carry a like charge repel each other, and thus form a stable dispersion in water. Removal of the charge removes the repulsive force and allows the molecules to interact with each other and precipitate, in most cases.)

The pH of the isoelectric point differs for each protein. It depends on the ratio of free ionized carboxyl groups to free ionized amino groups in the protein.

The isoelectric point is important in food processing. For example, cottage cheese is made by adding lactic acid to milk to bring the pH to the isoelectric point of the major milk proteins (the caseins). The proteins precipitate at this pH, forming curds. These are separated from the rest of the milk and may be pressed and/or mildly salted before being packaged as cottage cheese.

Water-Binding Capacity

Water molecules can bind to the backbone and to polar and charged side chains of a protein. Depending on the nature of their side chains, proteins may bind varying amounts of water—they have a ***water-binding capacity***. Proteins with many charged and polar groups bind water readily, whereas proteins with many hydrophobic groups do not bind much water. As proteins get closer to their isoelectric point, they tend to bind less water, because reduced charge on the protein molecules results in reduced affinity for water molecules.

The presence of bound water helps to maintain the stability of protein dispersion. This is due to the fact that the bound water molecules shield the protein molecules from each other. Therefore, they do not associate with each other or precipitate as readily, and so the dispersion tends to be more stable.

Salting-in and Salting-out

Some proteins cannot be dispersed in pure water yet are readily dispersed in dilute salt solutions. When a salt solution increases the dispersibility of a protein, this is termed "***salting-in***." It occurs because charged groups on a protein bind the anions and cations of the salt solution more strongly than water. The ions, in turn, bind water; thus, the protein is dispersed in water more easily.

Salting-in is important in food processing. For example, brine may be injected into ham to increase the dispersibility of the proteins. This has the effect of increasing their water-binding capacity, and so the ham is moister and its weight is increased. The same is true for poultry to which polyphosphates are added.

Salting-out occurs at high salt concentrations, when salts compete with the protein for water. The result is that there is insufficient water available to bind to the protein, and so the protein precipitates. This is not normally a problem in food processing. However, it may be a contributing factor to the deterioration of food quality during freezing of foods; during the freezing process, water is effectively removed as ice crystals, and so the concentration of liquid water decreases and the solute concentration increases dramatically. This is discussed in Chap. 17.

Denaturation

Denaturation is the change in the secondary, tertiary, and/or quaternary structure of a protein. There is no change in the primary structure. In other words, denaturation does not involve breaking of peptide bonds. The protein unfolds, yet there is no change in its amino acid sequence.

Denaturation may occur as a result of the following:

- Heat
- pH change
- Ionic strength change (changes in salt concentration)
- Freezing
- Surface changes (occurring while beating egg whites)

Any of these factors may cause breaking of hydrogen bonds and salt bridges. As a result, the protein unfolds and side chains that were buried in the center of the molecule become exposed. They are then available to react with other chemical groups and, in most cases, the denatured protein precipitates. This reaction is usually irreversible; it is not possible to regain the original conformation of the denatured protein.

The changes that produce denaturation are usually mild changes. In other words, mild heat treatment, such as pasteurization or blanching, or small changes in pH are sufficient to change the conformation of a protein.

Denatured proteins normally lose their functional properties; that is, they are unable to perform their normal function in a food. Enzymes are inactivated and so the reactions that they catalyzed can no longer take place. This has important implications in food processing.

Denaturation may be desirable and can be deliberately brought about by food processing. Examples of desirable denaturation include heating beaten egg white foams to form meringues, adding acid to milk to form cottage cheese, or inactivating enzymes by heat, as occurs when vegetables are blanched before freezing.

Blanching is a mild heat treatment that denatures and inactivates enzymes that would cause rancidity or discoloration during frozen storage.

Sometimes denaturation is undesirable. For example, frozen egg yolks are lumpy and unacceptable when thawed because the lipoproteins denature and aggregate. Overheating of foods can also cause unwanted denaturation. Food processors must be careful to utilize processing methods that do not cause unnecessary deterioration of food quality due to protein denaturation.

Hydrolysis of Peptides and Proteins

Hydrolysis of proteins involves breaking peptide bonds to form smaller peptide chains. This can be achieved by acid digestion, using concentrated acid. This may be appropriate in protein research, but it is not an option in food processing. Hydrolysis is also catalyzed by ***proteolytic*** enzymes. Examples of such enzymes used in foods include ficin, papain, and bromelain, which are used as meat tenderizers. They hydrolyze muscle protein or connective tissue, making meat more tender. It is important to control the duration of time that they are in contact with the meat so that too much hydrolysis does not occur. Too much hydrolysis would make the texture of the meat soft and "mushy" (see Chap. 9).

Another example of a proteolytic enzyme is rennet, which is used to make cheese (see Chap. 11). This enzyme is very specific in its action, hydrolyzing a specific peptide bond in the milk protein. The result of this hydrolysis reaction is aggregation of the milk proteins to form curds, which can then be processed into cheese. (This enzyme continues to act as a proteolytic agent during cheese aging in conjunction with natural enzymes from milk and the starter cultures. Their combined action results in flavor and texture development in aged cheeses.)

Maillard Browning

Maillard browning is the reaction that is responsible for the brown color of baked products. A free carbonyl group of a reducing sugar reacts with a free amino group on a protein when heated, and the result is a brown color. The reaction is highly complex and has a significant effect on the flavor of foods as well as the color. It is known as nonenzymatic browning, because the reaction is not catalyzed by an enzyme. (Maillard browning must be distinguished from enzymatic browning, which is the discoloration of damaged fruits or vegetables and is catalyzed by an enzyme such as phenol oxidase; enzymatic browning is discussed in Chap. 7.)

The Maillard Reaction is favored by the following:

- High sugar content
- High protein concentration
- High temperatures
- High pH
- Low water content

Maillard browning is responsible for the discoloration of food products such as powdered milk and powdered egg. Before drying, eggs are usually "desugared" enzymatically to remove glucose and prevent Maillard browning (see Chap. 10).

The reaction causes loss of the amino acids lysine, arginine, tryptophan, and histidine, as these are the amino acids with free amino groups that are able to react with reducing sugars. With the exception of arginine, these are essential amino acids. (The body cannot make them, and so they must be included in the diet.) Therefore, it is important to retard the Maillard Reaction, particularly in susceptible food products (such as food supplies sent to underdeveloped countries) in which the nutritional quality of the protein is very important.

Enzymes

All enzymes are proteins. Enzymes are important in foods, because they catalyze various reactions that affect color, flavor, or texture, and hence quality of foods. Some of these reactions may be desirable, whereas others are undesirable, and produce unwanted discoloration or off-flavors in foods.

Each enzyme has a unique structure or conformation, which enables it to attach to its specific substrate and catalyze the reaction. When the reaction is complete, the enzyme is released to act as a catalyst again. All enzymes have an optimal temperature and pH range, within which the reaction will proceed most rapidly. Heat or changes in pH denature the enzymes, making it difficult or impossible for them to attach to their respective substrates, and thus inactivating them.

If an enzymatic reaction is required in food processing, it is important to ensure that the optimal pH and temperature range for that enzyme is achieved. Outside the optimal range, the reaction will proceed more slowly, if at all. Heat treatment must therefore be avoided. If this is not possible, the enzyme must be added after heat treatment and subsequent cooling of the food.

On the other hand, if enzyme action is undesirable, the enzymes must be inactivated. This is usually achieved by heat treatment, although it may also be accomplished by adding acid to change the pH.

Examples of desirable enzymatic reactions include the clotting of milk by *rennet*, which is the first step in making cheese (Chap. 11). Ripening of cheese during storage is also due to enzyme activity. Ripening of fruit is also due to enzyme action (Chap. 7). Other desirable enzymatic reactions include tenderizing of meat by proteolytic enzymes such as *papain*, *bromelain*, and *ficin* (Chap. 9). These enzymes are obtained from papaya, pineapple, and figs, respectively.

As was mentioned earlier, these enzymes catalyze hydrolysis of peptide bonds in proteins. They are added to the meat and allowed to work for a period of time. The reaction must be controlled, to prevent too much breakdown of the proteins. The optimum temperature for these enzymes occurs during the early cooking stages. (Hydrolysis proceeds very slowly at refrigeration temperatures.) As meat is cooked, the enzymes promote hydrolysis. However, as the internal temperature continues to rise, the enzymes are inactivated and the reaction is stopped.

Although useful as meat tenderizers, proteolytic enzymes may be undesirable in other circumstances. For example, if a gelatin salad is made with raw pineapple, the jelly may not set, due to action of *bromelain*, which is contained in pineapple. This can be prevented by heating the pineapple to inactivate the enzyme before making the gelatin salad.

Additional examples of unwanted enzymatic reactions include enzymatic browning, which occurs when fruits and vegetables are damaged, due to the action of *polyphenol oxidase*, and produces undesirable discoloration (Chap. 7). Development of off-flavors in fats and fat-containing foods may also be a problem in some circumstances, and this may be caused by *lipase* or *lipoxygenase* (Chap. 12).

Enzymes are inactivated in fruits and vegetables prior to freezing by a mild heat process known as blanching (Chap. 17). The fruits or vegetables are placed in boiling water for a short time, in order to inactivate the enzymes that would cause discoloration or development of off-flavors during frozen storage.

CULINARY ALERT! Do not add fresh pineapple, papaya, kiwi, or other fruits containing proteolytic enzymes to a gelatin gel, or it will not set! Canned fruit of these fruit varieties is not generally available yet it yields better results than fresh.

Functional Roles of Proteins in Foods

Proteins have many useful ***functional properties*** in foods. A functional property is a characteristic of the protein that enables it to perform a specific role, or function, in a food. For example, a protein with the ability to form a gel may be used in a food with the specific intention of forming a gel, as in use of gelatin to make jelly.

Functional properties or roles of proteins in foods include solubility and nutritional value. They may also be used as thickening, binding, or gelling agents, and as emulsifiers or foaming agents.

The functional properties of a specific protein depend on its amino acid composition and sequence since these determine the conformation and properties of the protein.

Although no single protein exhibits all the functional properties, most proteins may perform several different functions in foods, depending on the processing conditions. Some proteins are well known for specific functional properties in foods.

Whey protein is an example of a protein that is used for its *solubility* (Chap. 11). Whey is soluble at acid pH, because it is relatively hydrophilic and able to bind a lot of water, and so, unlike many proteins, it does not precipitate at its isoelectric point. Because of its solubility, whey protein is used to fortify acidic beverages such as sports drinks. Whey protein may also be used as a *nutritional fortifier* in other products including baked goods.

Egg proteins are used as *thickening* or *binding* agents in many food products (Chap. 10). Meat proteins are also good binding agents.

Gelatin and egg white proteins are examples of *gelling* agents (Chap. 10). When egg whites are heated, they form a firm gel as can be seen in a boiled egg. Gelatin is used to make jelly and other congealed products. Gelatin gels are formed when the protein molecules form a three-dimensional network due to association by hydrogen bonds. Gelatin gels can be melted by heating, and reformed on cooling. Egg white gels, on the other hand, are formed due to association by hydrophobic interactions and disulfide bonds, and they do not melt on heating. The proteins of gluten are another example of proteins that are able to associate to form a three-dimensional network. The gluten network is formed during kneading of bread dough, and is responsible for the texture and volume of a loaf of bread. Soy protein may also be used to form food gels.

This "structure-forming protein found in wheat, barley, rye and triticale is also used as a food additive for stabilizing or thickening foods. Yet because an estimated 1 in every 133 people is afflicted with celiac disease, an autoimmune condition in which gluten damages the lining of the small intestine when ingested, food manufacturers are continually looking for ways to remove gluten from some foods, while maintaining product appeal." (Grain Processing Corp 2012) (Diarrhea, constipation, migrains, weight loss, and more may be symptoms that occur.)

Many proteins are used as either *emulsifiers* or *foaming agents*, as discussed in Chap. 13. Proteins are ***amphiphilic***, containing both hydrophobic and hydrophilic sections in the same molecule. This allows them to exist at an interface between oil and water, or between air and water, rather than in either bulk phase. They are able to adsorb at an interface and associate to form a stable film, thus stabilizing emulsions or foams. Egg white proteins are the best foaming agents, whereas egg yolk proteins are the best emulsifying agents. The caseins of milk are also excellent emulsifiers.

Proteins are used in many foods to control texture, due to their ability to thicken, gel, or emulsify. Such food products must be processed, handled, and stored with care, to ensure that the proteins retain their functional properties. Some protein denaturation is usually necessary to form an emulsion, a foam, or a gel. However, too much denaturation due to incorrect processing conditions or poor handling and storage may result in undesirable textural changes (such as breaking of emulsions, loss of foam volume, or syneresis in gels) and must be avoided.

Conjugated Proteins

Conjugated proteins are also known as heteroproteins. They are proteins that contain a prosthetic group that may be an organic or an inorganic component. Examples of conjugated proteins include the following:

- Phosphoproteins—for example, casein (milk protein); phosphate groups are esterified to serine residues.
- Glycoproteins—for example, κ-casein; a carbohydrate or sugar is attached to the protein.
- Lipoproteins—for example, lipovitellin, in egg yolk; a lipid is attached to the protein.
- Hemoproteins—for example, hemoglobin and myoglobin; iron is complexed with the protein.

Protein Quality

Press Release on New Protein Quality Measurement FAO proposes new protein quality measurement.

The Food and Agriculture Organization of United Nations (FAO) has released a report recommending a new, advanced method for assessing the quality of dietary proteins. The report, "Dietary protein quality evaluation in human nutrition," recommends that the Digestible Indispensable Amino Acid Score (DIAAS) replace the Protein Digestibility Corrected Amino Acid Score (PDCAAS) as the preferred method of measuring protein quality.

The report recommends that more data be developed to support full implementation, but in the interim, protein quality should be calculated using DIAAS values derived from fecal crude protein digestibility data. Under the current PDCAAS method, values are "truncated" to a maximum score of 1.00, even if scores derived are higher.

Protein is vital to support the health and well-being of human populations. However, not all proteins are alike as they vary according to their origin (animal, vegetable), their individual amino acid composition and their level of amino acid bioactivity. "High quality proteins" are those that are readily digestible and contain the dietary essential amino acids in quantities that correspond to human requirements.

"Over the next 40 years, three billion people will be added to today's global population of 6.6 billion. Creating a sustainable diet to meet their nutritive needs is an extraordinary challenge that we won't be able to meet unless we have accurate information to evaluate a food's profile and its ability to deliver nutrition," said Paul Moughan, Co-director of the Riddet Institute, who chaired the FAO Expert Consultation.

"The recommendation of the DIAAS method is a dramatic change that will finally provide an accurate measure of the amounts of amino acids absorbed by the body and an individual protein source's contribution to a human's amino acid and nitrogen requirements. This will be an important piece of information for decision makers assessing which foods should be part of a sustainable diet for our growing global population."

Using the DIAAS method, researchers are now able to differentiate protein sources by their ability to supply amino acids for use by the body. For example, the DIAAS method was able to demonstrate the higher bioavailability of dairy proteins when compared to plant-based protein sources. Data in the FAO report showed whole milk powder to have a DIAAS score of 1.22, higher than the DIAAS score of 0.64 for peas and 0.40 for wheat.

DIAAS determines amino acid digestibility, at the end of the small intestine, providing a more accurate measure of the amounts of amino acids absorbed by the body and the protein's contribution to human amino acid and nitrogen requirements. PDCAAS is based on an estimate of crude protein digestibility determined over the total digestive tract, and values stated using this method generally overestimate the amount of amino acids absorbed. Some food products may claim high protein content, but since the small intestine does not absorb all amino acids the same, they are not providing the same contribution to a human's nutritional requirements.

Site: http://www.fao.org/ag/humannutrition/35978-02317b979a686a57aa4593304ffc17f06.pdf—2011

This FAO recommendation "overcomes some of the criticisms of the PDCAAS method, such as considering high-quality proteins, antinutritional factors, amino acid bioavailability and the effect of limiting amino acid." (Kuntz 2013)

Nutrition: See More in Specific Food Commodity Chapters

Nutrition comes into play with new product introductions—both original products and reformulations. "... nearly every fat has been implicated in some sort of dietary brouhaha. Carbohydrates, well, steer clear of sugars and starches, and be wary of fibers that might cause digestive upset. Water seems safe—for now.

And then there's protein .... Protein is a key factor in satiety, so it can help battle the bulge. ... one of the strongest nutritional trends for 2013 and beyond. And ... not just by adding high-protein ingredients like meat, eggs or beans, but purified sources, like dairy proteins, and vegetable proteins, including soy, canola and even the dreaded gluten." (Kuntz 2013)

What Foods Are in the Protein Foods Group?
All foods made from meat, poultry, seafood, beans and peas, eggs, processed soy products, nuts, and seeds are considered part of the Protein Foods Group. Beans and peas are also part of the Vegetable Group.

Vegetarian Choices in the Protein Foods Group
Vegetarians get enough protein from this group as long as the variety and amounts of foods selected are adequate. Protein sources from the Protein Foods Group for vegetarians include eggs (for ovo-vegetarians), beans and peas, nuts, nut butters, and soy products (tofu, tempeh, veggie burgers).
Choosemyplate.gov

Since the Great Depression adequate protein intake has not been a concern for most Americans, as meat, poultry and other forms of animal protein are readily available and even typically, overconsumed. (Berry 2012)

Conclusion

Proteins are complex molecules that are widely distributed in all foodstuffs. It is important to understand their conformation and reactions in order to know how they will behave during food processing and to understand how to maximize their functional properties. This is especially true of protein-rich foods, where the quality of the final product depends to a large extent on the treatment of the protein during processing and handling. This chapter has focused on general properties of food proteins. More details of the composition and functional properties of some specific food proteins are given in the ensuing chapters.

Notes

10 tips
Nutrition Education Series

with protein foods, variety is key

10 tips for choosing protein

Protein foods include both animal (meat, poultry, seafood, and eggs) and plant (beans, peas, soy products, nuts, and seeds) sources. We all need protein—but most Americans eat enough, and some eat more than they need. How much is enough? Most people, ages 9 and older, should eat 5 to 7 ounces* of protein foods each day.

1 vary your protein food choices
Eat a variety of foods from the Protein Foods Group each week. Experiment with main dishes made with beans or peas, nuts, soy, and seafood.

2 choose seafood twice a week

Eat seafood in place of meat or poultry twice a week. Select a variety of seafood—include some that are higher in oils and low in mercury, such as salmon, trout, and herring.

3 make meat and poultry lean or low fat
Choose lean or low-fat cuts of meat like round or sirloin and ground beef that is at least 90% lean. Trim or drain fat from meat and remove poultry skin.

4 have an egg
One egg a day, on average, doesn't increase risk for heart disease, so make eggs part of your weekly choices. Only the egg yolk contains cholesterol and saturated fat, so have as many egg whites as you want.

5 eat plant protein foods more often
Try beans and peas (kidney, pinto, black, or white beans; split peas; chickpeas; hummus), soy products (tofu, tempeh, veggie burgers), nuts, and seeds. They are naturally low in saturated fat and high in fiber.

6 nuts and seeds
Choose unsalted nuts or seeds as a snack, on salads, or in main dishes to replace meat or poultry. Nuts and seeds are a concentrated source of calories, so eat small portions to keep calories in check.

7 keep it tasty and healthy
Try grilling, broiling, roasting, or baking—they don't add extra fat. Some lean meats need slow, moist cooking to be tender—try a slow cooker for them. Avoid breading meat or poultry, which adds calories.

8 make a healthy sandwich

Choose turkey, roast beef, canned tuna or salmon, or peanut butter for sandwiches. Many deli meats, such as regular bologna or salami, are high in fat and sodium—make them occasional treats only.

9 think small when it comes to meat portions
Get the flavor you crave but in a smaller portion. Make or order a smaller burger or a "petite" size steak.

10 check the sodium
Check the Nutrition Facts label to limit sodium. Salt is added to many canned foods—including beans and meats. Many processed meats—such as ham, sausage, and hot dogs—are high in sodium. Some fresh chicken, turkey, and pork are brined in a salt solution for flavor and tenderness.

* What counts as an ounce of protein foods? 1 ounce lean meat, poultry, or seafood; 1 egg; ¼ cup cooked beans or peas; ½ ounce nuts or seeds; or 1 tablespoon peanut butter.

Go to www.ChooseMyPlate.gov for more information.

DG TipSheet No. 6
June 2011
USDA is an equal opportunity provider and employer.

CULINARY ALERT!

Glossary

Amino acid Building block of proteins; contains an amino group, a carboxyl group, a hydrogen, and a side chain, all attached to a central carbon atom.

Amphiphilic A molecule that contains both hydrophobic and hydrophilic sections.

Amphoteric Capable of functioning as either an acid or as a base depending on the pH of the medium.

Alpha helix Ordered protein secondary structure: corkscrew shape, stabilized by intrachain hydrogen bonds.

Beta-pleated sheet Ordered protein secondary structure; zigzag shape, stabilized by interchain hydrogen bonds.

Conformation The specific folding and shape that a protein assumes in space.

Denaturation Changes in the conformation (secondary, tertiary, or quaternary structure) of a protein caused by changes in temperature, pH, or ionic strength, or by surface changes.

Dipeptide Two amino acids joined by a peptide bond.

Disulfide bond Strong covalent bond formed by the reaction of two thiol (SH) groups.

Functional property Characteristic of the molecule that enables it to perform a specific role in a food. Examples of functional properties of proteins include solubility, thickening, binding, gelation, foaming, and emulsifying capacity.

Hydrolysis Breaking of one or more peptide bonds in a protein to form smaller polypeptide chains.

Hydrophilic Water-loving; characteristic of polar and charged groups.

Hydrophobic Water-hating; characteristic of nonpolar groups.

Isoelectric point pI; the pH at which the overall charge on a protein is zero; the number of positive charges is equal to the number of negative charges; the protein is most susceptible to denaturation and precipitation at this pH.

Maillard browning The free carbonyl group of a reducing sugar and the free amino group of a protein react to form a brown color; complex nonenzymatic reaction that is favored by high temperatures.

Peptide bond Bond formed by the reaction of the amino group of one amino acid and the carboxyl group of another.

Polypeptide Several amino acids joined together by peptide bonds.

Protein primary structure Specific sequence of amino acids along the protein chain, joined by peptide bonds; the covalently bonded protein backbone.

Protein quaternary structure The noncovalent association of protein chains to form a discrete unit.

Protein secondary structure Three-dimensional arrangement of sections of the protein chain; secondary structures include the α-helix, β-pleated sheet, and random coil.

Protein tertiary structure Three-dimensional arrangement of the whole protein chain; the shape that a protein chain assumes in space; includes fibrous and globular structures.

Proteolytic Breaks down or hydrolyzes proteins.

Random coil A protein secondary structure that exhibits no regular, ordered pattern.

Salting-in Addition of a dilute salt solution to improve the dispersibility of a protein.

Salting-out Addition of a concentrated salt solution to precipitate a protein.

Steric effects Effects caused by the size and shape of the amino acids comprising the protein chain; spatial effects; for example, bulky amino acids can prevent a protein from folding upon itself in certain ways.

Water-binding capacity The ability of a protein to bind water; this ability depends on the number of charged and polar groups along the protein chain.

Zwitterion Contains a positively charged group and a negatively charged group within the molecule.

References

Berry D (2012) Pumping up protein. Food Product Design. (May):66

Grain Processing Corp. (2012) Gluten-free goodness. Food Product Design (May):81

Kuntz LA (2013) In terms of protein. Food Product Design. (March/April):10

Bibliography

Coultate T (2009) Food. The chemistry of its components, 5th edn. RSC, Cambridge

Damodaran S (2007) Amino acids, peptides and proteins. In: Damodaran S, Parkin K, Fennema O (eds) Fennema's food chemistry, 4th edn. CRC, Boca Raton

McWilliams M (2012) Foods: experimental perspectives, 7th edn. Prentice-Hall, Upper Saddle River

Potter N, Hotchkiss J (1999) Food science, 5th edn. Springer, New York

The Academy of Nutrition and Dietetics (AND)

The Food and Agriculture Organization of United Nations (FAO)

Vieira ER (1999) Elementary food science, 4th edn. Chapman & Hall, New York

Meat, Poultry, Fish, and Dry Beans 9

Introduction

Meat is the edible portion of mammals—the flesh of animals used for food. "Meat" may include rabbit, venison, and other game, as well as the *nonmammals* poultry and fish. The flesh from various animals may be used as food throughout the world.

Red meat is the meat from mammals including beef and veal, lamb, mutton, and pork. White meat refers to meat from poultry. Addressing the question of pork as a white meat, it is determined that its myoglobin content is lower than beef, and yet significantly higher than chicken or turkey white meat. The USDA treats pork as a red meat. In 1987, the US National Pork Board began a successful advertising campaign stating that pork was "the other white meat." This was intended to give the perception that, similar to chicken and turkey (white meat), it was more healthy than red meat.

Other than the red or white meats, seafood is derived from fish, and game is from nondomesticated animals. These may be sold fresh or frozen. Meat is also available in processed or manufactured products.

Meat is composed of three major parts: *muscle, connective tissue, and adipose tissue* (fat). Lean meats contain less adipose tissue than well-marbled cuts of meat. The location of the cut of meat on the animal, muscle contraction, and postmortem changes all influence the degree of meat tenderness. Individual cuts vary in inherent tenderness, requiring different cooking methods

All meat is subject to mandatory *inspection* by the USDA and voluntary *grading*. After inspection alteration may occur due to processing methods including curing, smoking, restructuring, and tenderizing. Kosher and Halal inspections mean much more than having a religious official blessing.

Incomplete plant proteins of animal feed are resynthesized in meat, and it is important to know that only animal protein is a *complete* protein. Thus, if meat consumption is minimized or omitted from the diet, for any number of reasons, an individual must obtain similar nutrients from a nonmeat source, such as combination of various plants (Chap. 7).

The USDA estimates 2011 US per capita beef consumption at 57.4 lb, down 13 % from 10 years ago and down about 25 % from 1980. The 2012 USDA prediction was that Americans would eat less beef than they ate in 2011 (only 54.1 lb of beef on average). Reuters News Service reports that this low amount "an opportunity for beef companies and retailers to promote ... higher-end cuts in supermarkets but in smaller portions (National Cattleman's Beef Association (NCBA)".

V.A. Vaclavik and E.W. Christian, *Essentials of Food Science, 4th Edition*, Food Science Text Series,
DOI 10.1007/978-1-4614-9138-5_9,

Beef demand has been up and down—depending on such things as health news and the economy. Yet, some individuals may have environmental, religious, vegetarian/flexitarian beliefs, or other concerns related to the consumption of meat, thus they might choose to avoid meat products, or consume meat minimally. The USDA recommends to "go lean with protein." See your personal intake recommendation for daily consumption by utilizing choosemyplate.gov.

Meat must satisfy the requirements of appearance, texture, and flavor, as well as nutrition, safety, and convenience. Therefore, if eating meat, in order to keep it safe, it becomes important to know the effects of storing and cooking meat on its various components.

Characteristics of Meat

Physical Composition of Meat

The physical composition of meat is composed of three tissues: muscle tissue, connective tissue, and adipose or fatty tissue. Each is discussed in the text below.

Muscle Tissue. Muscle tissue is referred to as the *lean* tissue of meat. It includes *cardiac, skeletal, and smooth muscle. Cardiac* muscle is located in the heart. *Skeletal* muscle, the primary component of the carcass, provides support for the weight of the body, and its movement, or locomotion. When a muscle is used, it serves to strengthen the bone to which it is attached (true in humans too). *Smooth* muscle is the visceral muscle, located for example, in the digestive tract, reproduction system, and throughout the blood vessels of the circulatory system.

Within the muscle cell membrane (Fig. 9.1), there are ***myofibrils*** containing alternating thin and thick protein filaments, namely the *actin* and *myosin*, which contract and relax in the living animal. They are varied in length, perhaps 1 or 2 in. long, and are very small in diameter. Each fiber is cylindrical, with tapered ends, and is covered by a thin connective tissue sheath called ***endomysium***. Small bundles of 20–40 fibers make up one primary bundle that represents the "***grain***" of meat. This primary bundle is surrounded by ***perimysium*** connective tissue.

CULINARY ALERT! In carving meats it is often recommended to cut "across the grain," thus shortening the fibers for enhanced tenderness.

Collectively, several primary bundles form a larger, *secondary* bundle that also contains blood vessels and nerves. As is the case with the primary bundles that make it up, each secondary bundle is also surrounded by ***perimysium*** connective tissue. In turn, several of the secondary bundles are surrounded by ***epimysium*** connective tissue dividing one skeletal muscle from another. In *between* the muscle bundles, there are blood vessels (capillaries) and small pockets of fat cells.

Connective Tissue. (mostly collagen and elastin) The connective tissue is made up of protein and mucopolysaccharides. It is located throughout the muscle (Fig. 9.1) and determines the degree of meat tenderness. Lesser amounts of connective tissue equates to meat that is more tender. Various types of connective tissue—the endomysium, perimysium, and epimysium, bind the muscle fibers in bundles to form the muscle.

The connective tissue extends beyond the muscle fibers to form *tendons*, which attach the muscle to bones and holds and connects various parts of the body. It also forms *ligaments*, attaching one bone to another. Additionally, the tough skin, or hide of an animal, is connected to underlying animal tissue by connective tissue.

It follows that meat containing a *high* degree of muscle tissue naturally has a *greater* amount of connective tissue to hold myofibrils and bundles in the muscle. ***Collagen*** is the most abundant protein found in mammals—in bone, cartilage, tendons, and ligament, enveloping muscle groups and separating muscle layers. It is also in horns, hooves, and skin.

Collagen is a triple-coil protein structure that is white in color and contracts to a thick mass when heated. Yet, it becomes tenderized when cooked with *moist* heat. This tenderization may be referred to in several manners. For example,

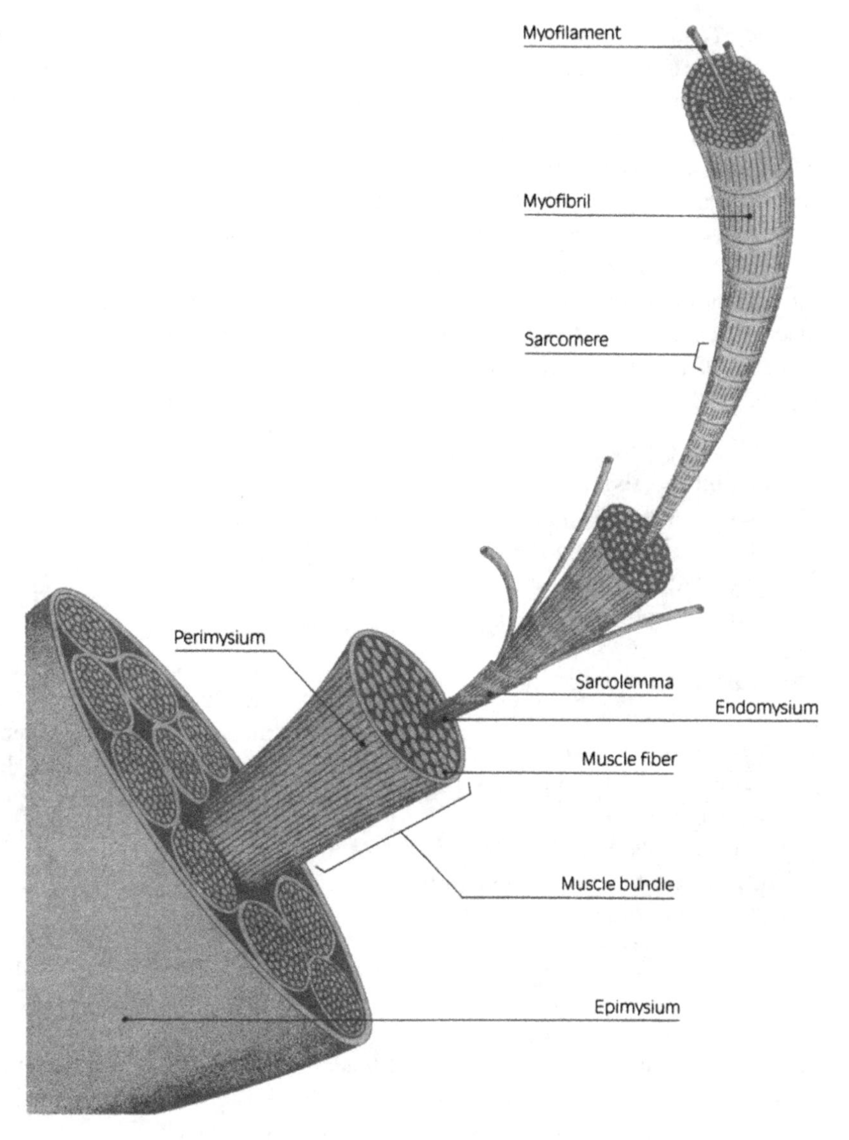

Fig. 9.1 Diagram of a lean muscle and its connective tissue (*Source*: National Cattlemen's Beef Association)

collagen may be "*converted to*," "*solubilized to*," or "*gelatinized to*" water-soluble ***gelatin***. (This is the same gelatin that may in turn be used for edible gels in the diet.) In older animals, the collagen is increased and may form many cross-linkage, thus *preventing* solubilization of collagen to the more tender gelatin. The meat from older animals is therefore tough.

A second, *lesser* component of the meat's connective tissue is the yellow-colored ***elastin*** protein, which is more elastic than collagen. It is found in the flexible walls of the circulatory system and throughout the animal body, assisting in holding bone and cartilage together. Elastin is extensive in muscles used in locomotion, such as legs, neck, and shoulders. Unlike collagen, it is *not* softened in cooking.

Another minor connective tissue component is ***reticulin***. This is a protein found in *younger* animals. It may be the precursor of collagen or elastin.

Usually, connective tissue is present to a greater degree in the muscle of *older* animals. Meat high in connective tissue may be mechanically *ground* to break the connective tissue and increase tenderization of the meat.

Fatty Tissue. A third meat component in addition to the muscle tissue and connective tissue is fatty tissue. Cuts of meat may vary substantially in composition and appearance due to the presence of ***adipose*** or ***fatty tissue***. Animal fat stores energy, and its content is dependent on factors such as animal feed, hormone balance, age, and genetics.

Fat is held by strands of connective tissue throughout the body and is deposited in several places such as around organs, under the skin, and between and within muscles as described below.

- **Adipose tissue**—fat that is stored around the heart, kidney organs, and in the pelvic canal areas. (*Suet* refers to the hard fatty tissue around the kidneys and other glandular organs of cattle and sheep.)
- **Subcutaneous fat** (finish)—fat that is visible after the skin is removed. (This is also referred to as *cover fat*. If well trimmed, the visible fat layer is less apparent.)
- **Intermuscular fat**—fat *between* muscles (also known as seam fat).
- **Intramuscular fat**—fat *within* muscles (marbling) (Fig. 9.2).

Upon cooking, the melted fat component contributes to juiciness, the sensation of tenderness, and flavor. Thus, well-***marbled*** meat with intramuscular fat may be desirable (despite the high level of fatty tissues). *Lean* meat is primarily muscle tissue and is *lower* in fat. The percentage of fat stores in an animal will generally increase with the animal's age.

Chemical Composition of Meat

The *chemical* composition of various meat cuts varies to a large degree from one cut to another. Meat may contain a range of 45–70 % water, 15–20 % protein, and anywhere from 5 to 40 % fat, depending on the cut and trim. Meat contains no carbohydrate (except for the liver, which stores glycogen). These meat constituents are described in the following text.

Water

Water is the major constituent of meat, and the *greatest percentage* is found in *lean* meat and young animals where *fatty tissue* is *low*. Then, as an animal becomes more mature and fatter, with more adipose tissue, the water forms a *smaller proportion* of the entire makeup compared to young, lean animals.

Water exists in muscle fibers and, to a lesser degree, in connective tissue. It is *released* from the protein structure in a number of ways. For example, water loss occurs as the muscle coagulates during cooking. Loss occurs as muscle fibers are broken (due to chemical, enzymatic, or mechanical tenderization), by salting, and if the pH changes. Inversely, water may be *added* to meats such as cured ham, with a notation appearing by law, on the ham label.

CULINARY ALERT! A recent repercussion of labeling that meets the government's food safety requirements of raw meats and poultry is that an identification of water retention must be stated. Thus, according to the USDA, processors must list either the maximum percentage of absorbed water, or retained water on applicable food labels.

Protein

Protein of *animal* sources is of *high* biological value. It is known as a *complete* protein, indicating that it contains all of the essential amino acids in amounts and proportions that can be used in synthesizing body proteins. The three primary types of proteins in meats are *myofibril*, *stromal proteins*, and *sarcoplasmic proteins* as described in the following:

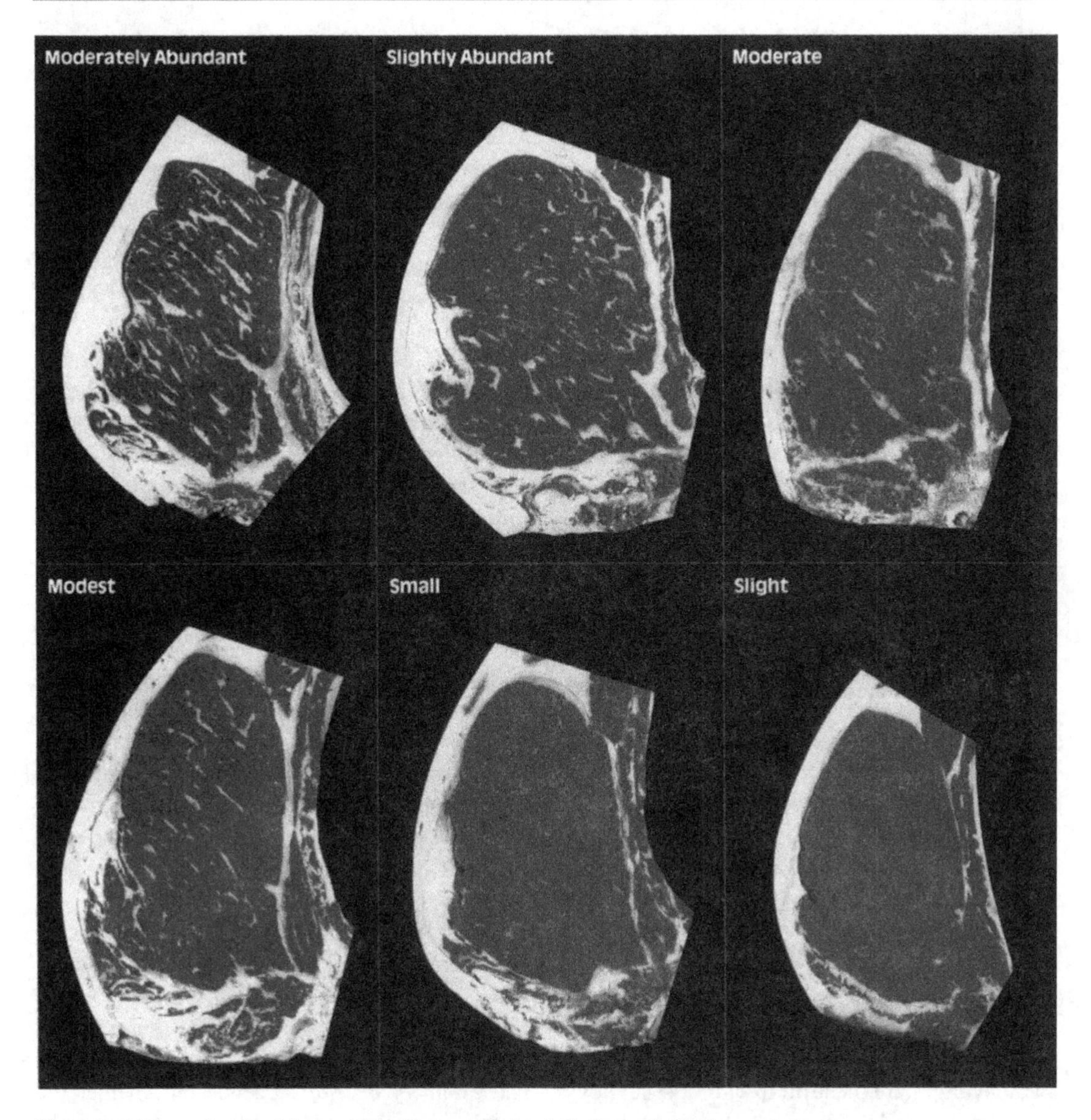

Fig. 9.2 Different levels of fat marbling (*Source*: National Cattlemen's Beef Association)

Myofibril proteins: Muscle bundles are groups of *myofibrils* composed of several protein molecules including actin and myosin that may form an overlap complex called ***actomyosin***.

Stromal proteins (connective tissue proteins): The watery connective tissue contains fibrils of *stromal* proteins: collagen, elastin, and reticulin (discussed earlier).

Sarcoplasmic proteins: *Sarcoplasmic* proteins are a third general classification of meat proteins. They include the *pigments* and *enzymes*. For example, the hemoglobin *pigment* stores oxygen in the red blood cells bringing it to tissues, including the muscles, while myoglobin stores oxygen in the muscle where it is needed for metabolism.

Enzymes are found in meat proteins. They may be *proteolytic*—degrading protein during the aging of meat, *amylolytic*—degrading carbohydrates, or *lypolytic*—degrading fats. There are also numerous enzymes in fluid of the muscle cell.

Fat

Fat may be a major component of meat. Fat varies in its degree of saturation (see Fig. 9.2). For example, subcutaneous fats are generally more *unsaturated* than fat around glandular organs. *Saturated* fat promotes less oxidation, and therefore less rancidity. In the animal, fat contributes to the survival of the living animal at low environmental temperatures.

In the diet, fat allows the fat-soluble vitamins A, D, E, and K to be carried. As well, fats contain some essential fatty acids that are the precursor material used in the synthesis of phospholipids for every cell membrane

Cholesterol, a sterol, is present in the cell membranes of all *animal* tissue. Typically, lean meats have a lower cholesterol content than higher fat meats. An exception to this level of existence of fat and cholesterol in lean meat is veal (young, lean calf meat), which is *low* in fat, yet *high* in cholesterol.

Carbohydrates

Carbohydrates are plentiful in *plant* tissue; however, are negligible in *animal* tissue. Approximately half of the small percentage of carbohydrates in animals is stored in the *liver* as *glycogen*. The other half exists throughout the body as glucose, especially in muscles, and in the blood. A small amount is found in other glands and organs of an animal. If an animal is exercised or not fed prior to slaughter, low stores of glycogen appear in the liver and muscles.

Vitamins and Minerals

Both vitamins and significant minerals are present in meats. The *water-soluble* B-complex vitamins function as cofactors in many energy-yielding metabolic reactions. The liver stores the four *fat-soluble* vitamins - vitamins A, D, E, and K. The minerals iron (in heme and myoglobin pigments), zinc, and phosphorus are present in meat.

Muscle Contraction in Live Animals

Muscle tissue of slaughtered animals undergoes several changes *after* slaughter. In order to better understand the reactions that occur in meat and their effects on tenderness and quality, it is necessary to have a basic understanding of the structure and function of muscle in a live animal.

Structure of the Myofilaments of Muscle

As previously mentioned, muscle fibers contain bundles of myofibrils. The myofibrils themselves are composed of bundles of protein filaments as shown in Fig. 9.3. These include *thin* filaments, made mostly of ***actin***, and *thick* filaments, which contain ***myosin***. They are arranged in a specific pattern within a repeating longitudinal unit called a ***sarcomere***.

The thin filaments occur at each end of the sarcomere, and they are held in place by ***Z-lines***. The Z-lines define the ends of each sarcomere. The thick filaments occur in the center of the sarcomere, and they overlap the thin filaments. The *extent of overlap* depends on whether the muscle is contracted or relaxed. In a *relaxed* muscle, the sarcomeres are extended, and there is *not* much overlap of thick and thin filaments. However, a *contracted* muscle has a *lot* of overlap because the sarcomeres shorten as part of the contraction process.

The thin and thick filaments are interspersed between each other in the regions where they overlap. A cross-section of the myofibrils shows that each thick filament is surrounded by six of the thin filaments, and every thin filament is surrounded by three of the thick filaments. This facilitates interaction between the thin and thick filaments when contraction occurs (see "Post-mortem Changes in the Muscle" section).

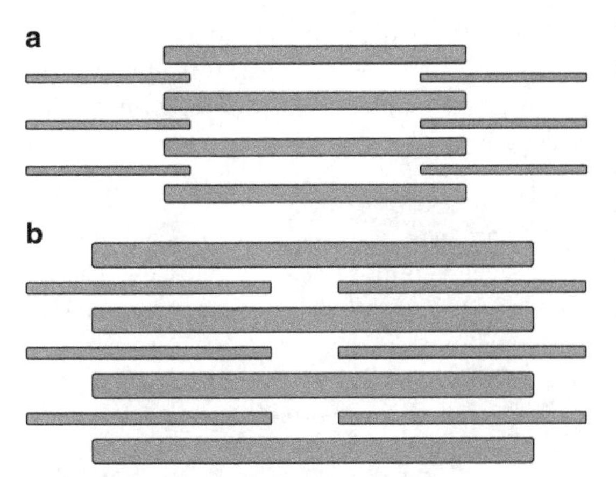

Fig. 9.3 Sarcomere and protein filaments (**a**) relaxed (**b**) contracted

Muscle Contraction

Muscle contraction starts when a nerve impulse causes release of calcium ions from the sarcoplasmic reticulum into the sarcoplasm, which is a jelly-like substance surrounding the thin and thick filaments of the myofibrils. The calcium ions bind to a specific site on the thin filaments, causing the active site on actin to be exposed. Actin molecules are then able to react with myosin, forming actomyosin. Adenosine triphosphate (ATP) is necessary as the energy source for this reaction.

The myosin then contracts and pulls the actin-containing filaments further in toward the center of the sarcomere. The actomyosin complex then breaks, and myosin forms another cross-link with a different actin molecule. As the cycle continues, the sarcomere continues to shorten, due to the formation of more cross-links, and contraction occurs.

When the nerve impulse ceases, calcium ions are pumped out of the sarcoplasm and returned to the sarcoplasmic reticulum. Actin and myosin cannot interact without calcium ions, and so the actomyosin complex breaks. The muscle relaxes and returns to its original extended state.

Energy for Contraction

The energy for contraction comes primarily from aerobic respiration, which enables glucose to be broken down completely to yield CO_2 and 36 molecules of ATP. In animals, glucose is stored as glycogen, which is broken down as needed to supply energy. When short bursts of extreme muscle activity are necessary, aerobic respiration does not supply adequate amounts of ATP, and so energy is also obtained by anaerobic glycolysis. This is a more rapid yet less efficient way of producing energy, as only two molecules of ATP are produced for every glucose molecule.

Glycolysis converts glucose to lactic acid, which builds up in the muscle. (It is the buildup of lactic acid that makes muscles sore and stiff after strenuous exertion. When the strenuous activity ceases, lactic acid is oxidized and removed from the muscle.)

Both aerobic respiration and glycolysis can take place in a *live* animal. *After* slaughter, aerobic respiration ceases, yet glycolysis continues for a while.

Postmortem Changes in the Muscle

Postmortem changes in the muscle make several meat characteristics differ. Some time *after* slaughter (from 6 to 24 h), muscle *stiffens* and becomes hard and inextensible. (Perhaps you have seen this stiffness in deer or other deceased animals in roadside accidents, or hunting sites.) *Prior* to slaughter, muscle tissue in the living animal is *soft and pliable*. Then there follows a time period for stiffening. It is species-specific, and it is known as ***rigor mortis***, which literally means "the stiffness of death." This stiffening is due to loss of extensibility by the myofibril proteins, actin, and myosin, once *energy reserves* become nonexistent, and oxygen does not reach the cells.

If meat is cooked at this stage, it is extremely *tough*. In fact, most meat is *aged or conditioned* to allow the muscles to relax and become *soft and pliable* again before it is cooked. This "*resolution of rigor*" is due to the enzymatic breakdown of proteins that hold muscle fibers together. This stiffness is temporary.

Subsequent to slaughter, a sequence of events takes place in muscle that leads to the onset of rigor mortis. When the animal is killed, *aerobic* respiration ceases, blood flow stops, and the muscles are no longer supplied with oxygen. Therefore, *anaerobic* conditions soon prevail. Glycolysis continues, and glycogen stores are converted to lactic acid with the formation of ATP. The reaction continues until glycogen stores are depleted or until a pH of 5.5 is reached. At this pH, the enzymes that are responsible for glycolysis are denatured, and so the reaction *stops*. If glycogen is in short supply, glycolysis may stop due to depletion of glycogen, before the pH drops as low as 5.5.

When glycolysis stops, the ATP supply is quickly depleted. Lack of ATP prevents calcium ions from being pumped out of the sarcoplasm, and so the active site on the actin molecules of the thin myofilaments is available to bind with the myosin of the thick filaments. Actin and myosin unite, forming *actomyosin cross-links*. This cross-link formation is irreversible, as there is no available ATP. (In a live animal, actomyosin cross-links are formed and broken repeatedly, as part of contraction, though the cycle requires ATP.)

Formation of these irreversible *actomyosin cross-links* causes the muscle to become rigid. This is *rigor mortis*, and it correlates with the depletion of ATP in the muscle. Once formed, *actomyosin cross-links* do not break down, even during aging of meat, and their presence makes meat tough (Fig. 9.3b).

Accordingly, the stiffness of the muscle at rigor depends on the extent of actomyosin formation, which, in turn, depends on the extent of *overlap* of the thin and thick myofilaments. Recall that the *greater* is the overlap of thin and thick myofilaments, the more *extensive* the formation of actomyosin, and the *stiffer* the muscle. This results in *tough* meat:

- Little overlap—few actomyosin cross-links (tender meat)
- Substantial overlap—many actomyosin cross-links (tough meat)

Since the extent of actomyosin formation affects the toughness of meat, it is important to *minimize* the number of *cross-links* formed. This is done in two ways:

1. The meat is *hung* on the carcass after slaughter to *stretch* the muscles. This minimizes shortening of the sarcomeres and results in formation of *fewer* actomyosin cross-links. (more later).
2. *Pre-rigor temperature* is controlled to minimize fiber shortening. The optimum temperature is between 59 °F and 68 °F (15–20 °C). *Above* this temperature, increased shortening occurs. *Below* it, "cold shortening" occurs. At low temperatures, the sarcoplasmic reticulum pump is unable to pump calcium ions out of the sarcoplasm, and so contraction occurs.(more later).

Both hanging the carcass and controlling pre-rigor temperature minimize contraction before the onset of rigor mortis, result in fewer actomyosin cross-links, and increase meat tenderness.

Ultimate pH

The *ultimate pH* is the pH that is reached when glycolysis ceases and is usually around 5.5. After slaughter, the pH drops due to the buildup of lactic acid, which is normally removed from the blood of the living animal. As mentioned already, glycolytic enzymes are close to their isoelectric point and are inactivated at this pH, thus preventing glycolysis from continuing. Therefore, a pH of 5.5 is the lowest possible ultimate pH. It is possible to obtain a higher ultimate pH if the animal is starved or stressed before slaughter. This depletes the glycogen reserves, thus glycolysis stops before sufficient lactic acid has been formed to bring the pH to 5.5.

Meat with a high ultimate pH has excellent water-holding capacity, because many of the proteins are not as close to their isoelectric point and, therefore, are able to bind more water. However, a low ultimate pH is desirable from a microbiological point of view, because it inhibits microbial growth. A high ultimate pH results in poor resistance to microbial growth.

The *rate of change* of pH after slaughter also has a significant effect on the quality of meat. A *rapid* pH change while the temperature is still high causes considerable denaturation of contractile and/or sarcoplasmic proteins and loss of water-holding capacity. Lysozomal enzymes are also released at high temperatures, and these cause hydrolysis of proteins. Such *undesirable* changes may happen if the carcass is *not* cooled rapidly after slaughter (e.g., if the pH drops to 6.0 before the temperature of the carcass drops below 95 °F (35 °C)).

Aging or Conditioning of Meat

Natural ***aging*** or conditioning of meat involves holding meat for several days, beyond rigor mortis. Under controlled storage conditions of temperature and humidity, (and perhaps light) the muscles become soft and pliable again, making the meat tender. Meat aging occurs as muscles become tender due to (protein and) actomyosin breakdown. A protease, which is active at around pH 5.5, breaks down the thin myofilaments at the Z-lines. This causes the muscle to become pliable again, and meat to be tender. The sarcoplasmic proteins denature and there is some denaturation of the myofibril proteins, with a resultant loss of water-holding capacity, and so the meat drips. Collagen and elastin do not denature significantly during aging.

CULINARY ALERT! Natural, proteolytic enzymes in meat may sufficiently tenderize meat in the time between slaughter and retail sale; however, controlled aging is sometimes induced.

As mentioned above, actoyosin formation affects the toughness of meat, and it is important to minimize the number of cross-links formed. This is done in two ways. Aging is achieved by *hanging* the carcass in a cold room, at 34–38 °F (2 °C) for 1–4 weeks. Although the meat regains tenderness after about a week, the best flavor and tenderness develops in about 2–4 weeks. Humidity levels of approximately 70 % are controlled, and the meat may be wrapped in vacuum bags to minimize dehydration and weight loss.

Higher temperatures for shorter times, such as 68 °F (20 °C) for 48 h, have also been used to age beef. However, development of surface bacterial slime tends to be a problem for meat aged by such methods. It is shown that exposing the meat to ultraviolet light during the aging is of help in this regard.

Aging requirements differ among meat types. For example, pork and lamb do *not* require aging such as occurs with beef, since the animals are slaughtered while they are young and inherently tender. They are usually processed the day following slaughter.

Meat Pigments and Color Changes

Various meat pigments and color changes are seen in meat tissue. Meat may appear as red meat or white meat, depending upon the predominant pigment and its concentration contained in the meat. The two major pigments in meat responsible for the red color are myoglobin and hemoglobin. *Myoglobin* (with one heme group as part of its structure) is 80–90 % of the total meat pigment (see "Chemical Composition of Meat" section). It allows oxygen to be stored in the muscles. *Hemoglobin* (with four heme groups in its structure) is present at levels of 10–20 % of well-bled meat. It carries oxygen in the bloodstream.

Myoglobin, the primary pigment contributor of meat, is purplish-red. It is present in frequently exercised portions of the animal that expend great amounts of oxygen, such as muscles of a chicken leg. It produces the "dark meat" of turkey for example. The specific myoglobin level is influenced by the species, age, sex, and specific muscle. There is *more* myoglobin in the muscles of cows than pigs, *more* in older sheep than young lambs, and *more* in bulls (adult males) than cows.

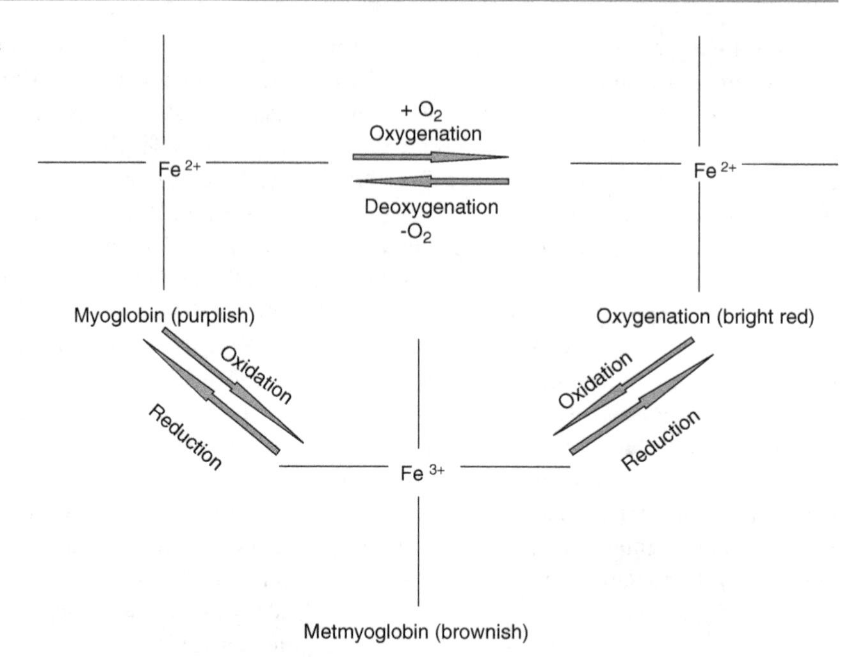

Fig. 9.4 Pigment change

When *myoglobin* is exposed to oxygen in the air, bright red, oxygenated *oxymyoglobin* is produced. With time, *metmyoglobin* is apparent. It is the *undesirable*, brownish-red colored pigment found in meat due to oxidation of the iron molecule. This unwelcome metmyoglobin pigment is found in meat that is *not* fresh, that contains significant levels of bacteria, and in meat exposed to light or exposed to low levels of oxygen (Fig. 9.4).

In processed meats, such as lunchmeats, nitrites may be added in order to both preserve the desirable pink color and control the growth of *C. botulinum*.

Meat-Handling Process

USDA Inspections

"*United States Inspected and Passed*" and the packaging/processing plant number is specified in the round *stamp* (Fig. 9.5) found on the primal cut of inspected meat. The stamp is made of a nontoxic purple vegetable dye. Packaged, *processed* meat must show a somewhat *similar stamp* on the packaging or *carton* of the meat. It is common for specific *state* inspection stamps (as opposed to USDA) to display a stamp with the shape of the state. Inspections for *wholesomeness* and accurate *labeling* are a service of the US government and are paid for with tax dollars.

The *Federal* Meat Inspection Act of 1906 requires inspection of all meat packing plants slaughtering and processing meat for *inter*state commerce. The Wholesome (Wholesale) Meat Act of 1967 required the same inspection program for *intra*state transport.

Trained veterinarians and agents of the USDA **Food Safety and Inspection Service** (FSIS) inspect the *health* of the animal, as well as *the sanitation* of the physical meat plant. The inspection for meat is *mandatory* (Chap. 20). Meat's inspection for ***wholesomeness*** indicates that it is safe to eat, without adulteration, and that examination of the carcass and viscera of the animal did *not* indicate the presence of *disease*. (It is *not* meant to imply freedom from all disease-causing microorganisms.)

Inspections occur *before*, *after*, and *throughout* meat processing. Diseased and unwholesome animals may *not* be used; harmful ingredients may *not* be added; misleading names or labels may *not* be used, and there must be established sanitation codes for the plant. Safety in meat processing is of

Fig. 9.5 The USDA inspection, quality grade, and yield grade stamps (*Source*: USDA)

utmost importance to the processor and customer alike. Violations of the Meat Act are fined and persons committing violations have been imprisoned! The Meat Inspection Program also controls and monitors *imported* meat.

In view of the fact that the pathogenic microorganism E. coli 0157:H7 may be undetected if *only visual inspections* are used for inspection, actual *bacterial counts* are included in inspections. Meat processing inspections with the inclusion of bacterial count checks now include the *Hazard Analysis and Critical Control Point* (HACCP) method of food safety (Chap. 16). It is the current program for inspecting meat.

> As far as USDA is concerned, its FSIS holds that a meat or poultry product can claim to be "natural" as long as it does not contain any artificial flavor or flavoring, color ingredient, chemical preservative or any other artificial or synthetic ingredient, and that the product and its ingredients are not more than minimally processed. Reports Hoffman at Solae regarding USDA's stance on "natural". (Decker 2013)

Kosher Inspection

Kosher inspection indicates that the meat is "*fit and proper*" for consumption or "*properly prepared*". Following Mosaic and Talmudic Laws, a specially trained rabbi slaughters the animal, e.g., beef, lamb, goat, and the meat is well-bled, and then salted. All processing is done under the supervision of individuals authorized by the Jewish faith. According to Mosaic Law, meat *must* come from an animal that has split hooves and chews its cud. Therefore, hogs and all pork products *cannot* be Kosher (see "Why Americans Buy Kosher" below).

The Kosher stamp (Fig. 9.6) does not indicate grade or wholesomeness. Meat is *still* subject to federal or state inspection.

Kosher Industry Facts

Why Americans Buy Kosher

55 %—Health and safety
38 %—Vegetarians
16 %—Eat halal
35 %—Taste or flavor
16 %—Guidelines that they were produced
8 %—Good products
8 %—Keep kosher all the time
8 %—Looking for vegetarian products, either for religious or dietary reasons
(Note: Respondents gave multiple answers)
Compiled by Lubicom Marketing and Consulting (2009)

Overall, there is an "estimated sales of kosher meat with 45 % being glatt or strictly kosher: $550,000,000" (Lubicom Marketing and Consulting, LLC, Brooklyn, NY http://www.lubicom.com/kosher/statistics/). Muslims and of other religious groups comprise a large percent of the Kosher market. Of course, not all persons of Jewish faith keep Kosher in their diet.

Kosher is not considered an acceptable substitute for proper and permitted Muslim foods. Only *some* foods are acceptable in also meeting Halal certification for Muslim requirements.

Fig. 9.6 Kosher Symbol (*Source*: National Cattlemen's Beef Association)

Halal Certification

The ***Halal*** certification indicates "*proper and permitted*." Only foods prepared and processed under Halal standards are to be consumed by Muslims, although not all persons of the Muslim faith keep to a diet of solely Halal-certified foods.

Certification uses trained Muslim inspectors who assist, participate in, and supervise food production in companies complying with Halal standards. A *crescent M* symbol on the product package indicates that the product meets the Halal standards of the Islamic Food and Nutrition Council of America (Food Technol 2000).

The types of foods permitted, including use of additives, slaughtering, packaging, labeling, shipping, and other aspects of food handling, are *regulated*. For example, the halal production does *not* accept alcohol, gelatin prepared from swine (for use as a food ingredient or packaging ingredient), or meat from animals that was *not* individually blessed.

Halal does not employ the same processing as Kosher food, although some products bear certification that they *are both* Kosher and Halal (Eliasi and Dwyer 2002). As per the founder and president of a company producing both Kosher and Halal refrigeration-free meals, a Kosher or Halal food product means much more than having the product blessed by a representative religious official.

In brief, there exist over 13 million persons of the Jewish faith, and one billion Muslims in the world. Their dietary laws are *not* interchangeable, yet are *similar*. A look at dietary restrictions of other religions is included in another reference (Food Technol 2000). As is true for Kosher, the Halal Certification does not indicate grade or wholesomeness. Meat is *still* subject to federal or state inspection.

Fig. 9.7 Halal certification (*Source*: Islamic Food and Nutrition Council of America)

The Halal stamp of certification (Fig. 9.7).

Grading of Meat

Voluntary grading of meat is part of the processing cost and is *not* paid for by tax dollars. The task of *grading* of meat (as opposed to the inspection of carcasses) was established by the USDA in 1927. It reports on both *quality* and *yield*, as described below (Fig. 9.5).

Voluntary *quality* grading evaluates various characteristics of animals. Evaluation includes age; color of lean, external fat quality and distribution, marbling; shape of animal carcass; and firmness of the muscle and meat texture—coarseness of muscle fiber bundles. So it looks at an evaluation of marbling, maturity, texture, and appearance.

Beef grades are according to the grades listed below. *Other* meats have *different* standards with less categories of grades.

- Prime
- Choice
- Select
- Standard
- Commercial
- Utility
- Cutter
- Canner

Prime grades of beef are very *well*-marbled. Prime is followed by Choice, Select, and Standard rankings with *less* marbling. The younger animal is more likely to be tender and receive a grade of

Prime, Choice, Select, or Standard. Older, more mature aged beef typically qualify for Commercial, Utility, Cutter, or Canner grades.

As well as voluntary *quality* grading, meat is also graded for *yield*, which is useful at the wholesale level. The highest percentage of lean, boneless yield (usable meat) on the carcass is given an assigned yield grade of "1." If yield of a carcass is *less*, the yield grade may be assigned a value as low as "5." Sales and marketing of meat products are based on both grades and yields.

Hormones and Antibiotics

Hormone and antibiotic use in animals are both monitored by the FDA. *Hormones* may be used in animal feeds to promote growth and/or to increase lean tissue growth and reduce fat content. All hormone-use in animal feeding must be *discontinued* for a specified time period prior to slaughter and must be *approved* by the FDA. A random sampling of carcasses provides tests for and monitors growth hormone residues (growth promotants).

Antibiotics in animal feed, when not utilized properly, may also pose a food safety concern. Antibiotics have been used for well over half of a century to treat disease and the FDA monitors their use in animal feeds to prevent their transfer to man. *Subtherapeutic* doses of antibiotics do more than treat *already-existing* diseases; such doses may be used to *prevent* disease and promote growth of animals. With this practice, the *therapeutic* administration of antibiotics to humans may be rendered ineffective if antibiotic-resistant strains of bacteria are passed from the livestock to man. The National Academy of Sciences (NAS) concluded that they were "unable to find data directly implicating the subtherapeutic use of feed antibiotics in human illness."

In order to protect human health, the FDA (even recently) has urged that specific antibiotics used to treat animals be removed from the market, if they have been found to compromise other drugs used in treating animals or humans. "There are many countries [such as Denmark] that do not allow antibiotics as growth promoters in farm animals" (Peregrin 2002a). See Question below:

Question of the Week

Do Hormones and Antibiotics Cause Health Problems in Humans?

Myth:

The use of antibiotics and hormone growth implants in livestock production is causing hazardous residues in beef and contributing to the development of health problems in humans.

Fact:

1. No residues from feeding antibiotics are found in beef, and there is no valid scientific evidence that antibiotic use in cattle causes illness resulting from the development of antibiotic-resistant bacteria.
2. Scientific authorities agree that use of hormone implants results in the efficient production of beef that is safe. (Montana State University)

Antibiotics that are used *solely* for the purpose of *animal* growth have often been debated. Unfortunately, as has been reported in the press involving *human* growth hormone, a small percentage of illegal users may stay just one step ahead of regulatory inspectors.

Animal Welfare Approval

Meat may be certified by Animal Welfare Approved, one of the leading advocates for humane livestock treatment. The first Animal Welfare Approved restaurant opened in Hudson, NY in 2011.

Cuts of Meat

Primal or Wholesale Cuts

A ***primal cut*** is also known as a *wholesale* cut of an animal. Meat cutting separates cuts into *tender* and *less tender* cuts, and *lean* and *fatty*. Cuts differ with species, and primal cuts of *beef* are identified below. They are listed according to tenderness. *Less exercised* skeletal muscles that provide support (as in: cuts of meat along the backbone, such as the loin) are usually more tender than other skeletal muscles that are used in *locomotion*. Ultimately though, tenderness is a function of how meat is torn by the teeth, not just the cut, age, and so forth.

Most tender	Medium tender	Least tender
Rib	Chuck	Flank (<brisket)
Short loin	Round	Short plate
Sirloin		Brisket
		Foreshank
		Tip

Subprimal Cuts

Subprimal cuts are divisions of *primal* cuts, often sent to the grocery market for further cutting. They may be boneless. If they are vacuumpacked, they are considered as "beef-in-a-bag", if boxed, "boxed meat". Subprimal cuts are further divided into individual *retail* cuts such as roasts, steaks, and chops.

Retail Cuts

Retail cuts are those available in the *retail* market, cut from primal or subprimal cuts. They may be named for the primal cut in which they are located *or* for the bones they contain. (Fig. 9.8).

In most cases, the cuts from the neck, legs, and lower belly are *least tender* for the reason that, as mentioned, they are the most exercised portions of the animal. These cuts are made more palatable when cooked with *moist heat* to soften connective tissue, although the same less tender cuts may be cooked with *long*, dry heat cooking at low temperatures and produce a satisfactory product. *Tender* cuts are cooked with *dry heat*.

CULINARY ALERT!

- Less tender cuts—best: moist heat or long, low heat cooking
- Tender cuts—best: dry heat cooking, quick

The wholesale and retail cuts of beef are identified in Fig. 9.8. In the1970s, The National Livestock and Meat Board (now National Cattlemen's Beef Association (NCBA)) coordinated a committee of retail and meat industry representatives and federal agencies, which standardized names for 314 retail cuts of meat. They published the *Uniform Retail Meat Identity Standards* (URMIS). URMIS labels include the kind of meat (beef, veal, pork, or lamb), the primal cut from which the meat originated (chuck, rib, loin, or round of the animal), and the name of the retail cut.

Beef (Fig. 9.11) is *most commonly* obtained from carcasses of the following:
- Steer: young, castrated male carcass
- Heifers: young, females before breeding, beyond veal and calf age

Beef is *less frequently* obtained from the carcasses of the following:
- Cows: females that have had a calf
- Bulls: adult male
- Baby beef: young cattle, 8–12 months of age
- Calves: young cattle, 3–8 months of age, beyond veal classification

Veal is from the carcass of:
- Beef calves, generally 3 weeks to 3 months or more.
- Veal is milk-fed, not grass-fed, thus is low in iron and pale in color.
- Young calf meat is normally lighter pin-gray than older calf meat.

Pork is the flesh of swine (pig).

Lamb is the flesh of *young* sheep, not more than 14 months old.

Mutton is the flesh of sheep *older* than 2 years.

Cooking Meat

To better assure a successful cooked meat product, it is critical to know the effects of cooking meat on

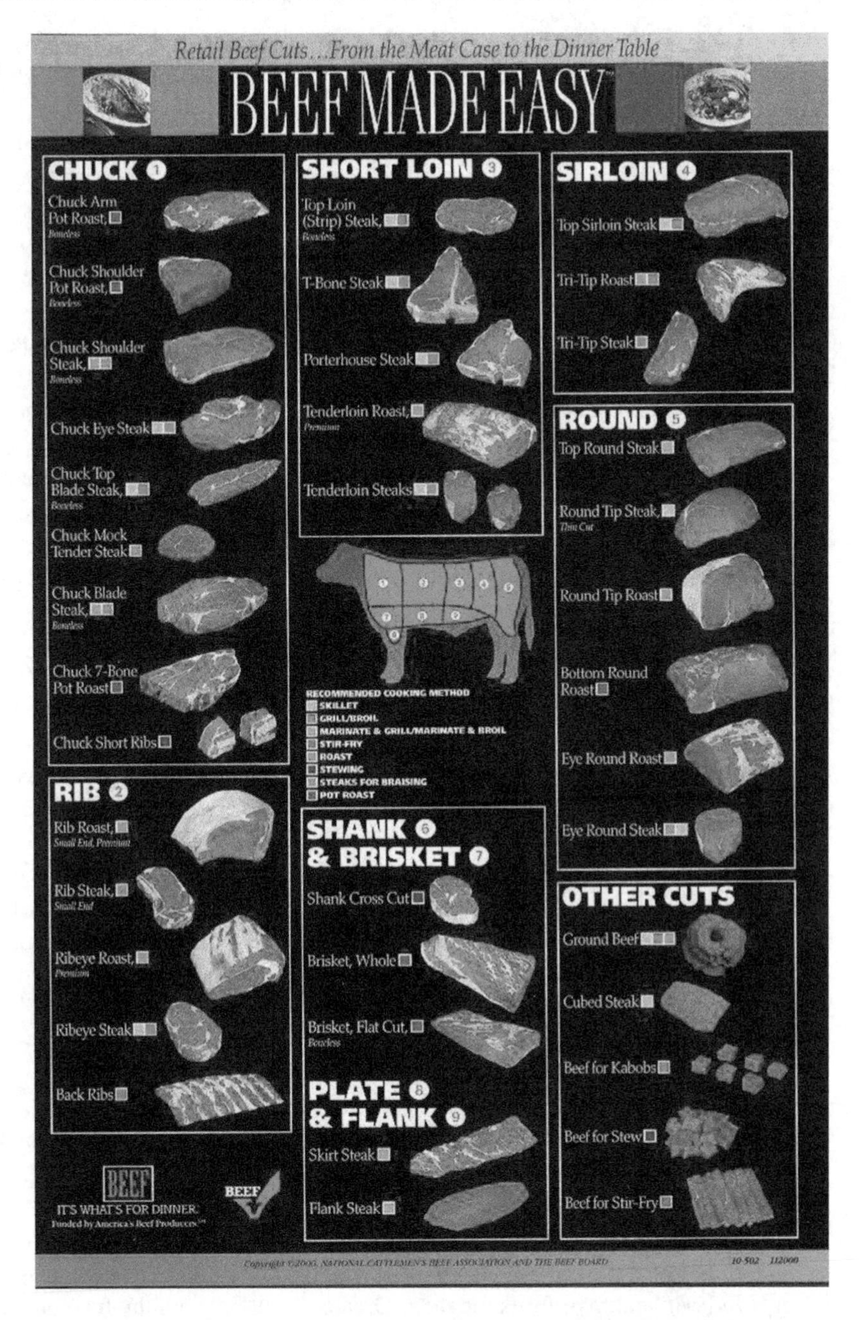

Fig. 9.8 Wholesale and retail cuts of beef (*Source*: USDA)

its various components. The processing plant and consumer both require a familiarity with types of cuts and cooking methods for meat. Although it may be a lot to ask, meat must satisfy the requirements of appearance, texture, and flavor, as well as nutrition, safety, and convenience.

Certainly, some individuals may have environmental, religious, vegetarian, health, or other concerns related to the consumption of meat, thus they might choose to *avoid* meat products, or consume meat minimally.

CULINARY ALERT! The purpose of cooking is to improve appearance, tenderness, and flavor while destroying pathogenic microorganisms.

In cooking, the peptide chains of amino acid chains uncoil (denature) and reunite or coagulate, releasing water and melted fat (Chap. 8). Consequently, the meat shrinks. When heat is applied, muscle fibers toughen and connective tissue becomes tender. These reactions are opposing effects and the method, time, and temperature of cooking differ.

Effects of Cooking on Muscle Proteins

Cooking has an effect, both desirable, and more negative on muscle proteins. The muscles used in locomotion (muscles for physical movement), as well as muscles of older animals each contain a large number of myofibrils in each muscle. As opposed to this, the less-used muscles and muscles of young animals have less myofibrils in each muscle. Thus, in the former case, muscle bundles are physically larger, and the large size indicates that they have a more coarse grain.

Temperatures around 131 °F (55 °C) precipitate myosin, and 158–176 °F (70–80 °C) precipitates actin. This precipitation denatures, shortens, toughens, and shrinks the surrounding connective tissue, causing a loss of water-holding capacity. The longer the cooking time, the more serious this effect will be, although at temperature of 170 °F (77 °C) tenderness may improve.

Tender cuts of meat contain *small* amounts of connective tissue and should be cooked for a short time at a *high* temperature with dry heat. Such cooking minimizes coagulation and shrinkage of muscle fibers and prevents loss of water-holding capacity. Toughening of the muscle fibers is minimized if tender cuts are cooked to *rare* rather than the well-done stage.

Tender meat that is *overcooked* tends to be *dry and tough* because the protein coagulates, water is squeezed out, and myofibrils toughen. Greater cooking intensity for a short time is advantageous for tender cuts of meat, and prolonged moist heat cooking is recommended for *less tender* cuts of meat.

Effects of Cooking on Collagen

As the collagen in meat is subject to the heat of cooking, the effect is that hydrogen bonds and some heat-sensitive cross-links are broken. Collagen, as mentioned earlier, is the major component of connective tissue. At temperatures between 122 °F and 160 °F (50–71 °C), connective tissue begins shrinking. Some of the tough-structured collagen is then solubilized and converted to gelatin. As the collagen fibers are weakened, the meat becomes more tender. Breakdown (or "melting," solubilization, gelatinization) of collagen is faster as collagen reaches higher temperatures.

Young animals contain few cross-links in collagen thus, it is readily converted to gelatin and meat tends to be tender. Collagen from *older* animals, on the other hand, contains many more covalent cross-links, most of which are *not* broken down by cooking; therefore, older animals yield tough meat unless it is heated in a *moist* atmosphere.

Cuts of meat *low* in collagen, such as rib or loin steaks, are inherently tender and do *not* benefit from slow, moist heat cooking. These cuts are more tender when cooked *quickly* and served to the *rare* or *medium-rare* stage.

On the other end of the spectrum, when collagen levels in a cut of meat are *high*, slow, moist heat cooking, to achieve a *well-done* stage, is recommended as it gelatinizes the collagen. Some tenderization of the meat surface (1/4 in.) occurs as meats are placed in marinades.

Effect of Cooking on Fat

The effect of cooking on fat is seen as fat melts throughout the meat with cooking. This melted fat produces a perception of a tender product. If a cut is high in fat content, or well-marbled, it yields a more tender cooked meat. In cooked, left-over meat, fat oxidation contributes to flavor deterioration.

Methods of Cooking

Normally, as mentioned previously, there are various "best" methods of cooking—dry heat and moist heat.

- ***Dry heat* methods** of cooking include broiling, frying, pan-frying, roasting, sautéing, and stir-frying. Meat is cooked *uncovered. Tender* steaks, chops, ground meat, and thin cuts are cooked this way
- ***Moist heat* methods** of cooking include braising, pressure-cooking, simmering, steaming stewing, or using a slow cooking pot to simmer. Meat is cooked *covered. Less tender* cuts such as chuck, flank, or round may be cooked in this manner

Meat contains water and therefore, to some extent, *all* meat if it is covered, provides moist heat cooking. This reduces surface drying that occurs with dry heat cooking and gives time for collagen to become gelatin.

The effect of *dry* heat (uncovered) and *moist* heat (covered) on two *identical*, less tender cuts of meat such as the chuck or round is seen. When the two roasts are removed from the oven at the *same* time, the *covered* roast temperature shows a lower temperature and less weight loss than the *uncovered* roast. Covering is preferable for these less tender cuts.

Other Factors Significant in Cooking

Cooking *methods* have previously been discussed. However, additional factors are significant in cooking. For example inherent pigments, the reading of thermometers, searing, and removal temperature are important factors in cooking. They are discussed in the following:

Pigments. Color may be an indication of the degree of doneness. When cooked, the myoglobin pigment is denatured, and therefore meat changes color from a red or purple, to pale gray-brown.

Use of Thermometers. The use of a calibrated thermometer to measure temperature provides the necessary assurance that the recommended safe cooking temperature is achieved. The cooking thermometer may be designed to detect *multiple* food temperatures along its stem and show an "average" temperature of those multiple readings. Depending upon where/how the thermometer is inserted—into the fat, muscle, or placement near a bone, the reported temperature varies and may not accurately reflect doneness.

CULINARY ALERT! The thermometer will give a false reading if placed in the fat or touching a bone. Thermometers provide a more accurate reading of meat doneness when they are inserted at an angle, rather than vertically where fat could drip down the stem, and again, give a false reading.

Searing. Initial high heat, or the short practice of "searing" with *dry* heat imparts flavor; however, further *moist* heat cooking is used for the continued cooking of less tender cuts of meat, high in collagen-containing connective tissue.

Removal temperature. Large roasts increase in internal temperature for 15–45 min or more *after* removal from the oven. Roasts removed from the oven at the *rare* stage (more moisture) exhibit a *greater* temperature rise following removal from the oven, than roasts removed at the drier, *well-done* stage. This should be kept in mind when a specific doneness is desired.

Specific temperatures for doneness of meat are as follows:

Doneness	Temperature
Rare	140 °F (60 °C)
Medium-rare	150 °F (65 °C)
Medium	160 °F (71 °C)
Well done	170 °F (77 °C)

CULINARY ALERT! Roasts increase in temperature following removal from the oven. Use a calibrated thermometer correctly.

Alterations to Meat

Processed Meat

Processed meat is defined as meat that has been changed by any mechanical, chemical, or enzymatic treatment, altering the taste, appearance, and often keeping the quality of the product (NCBA). It may be cured, smoked, or cooked, and it includes cold cuts (lunch meats), sausage, ham, and bacon. Processed meats may be available in low-fat formulations. Meat that has been processed is subject to the same USDA inspection as other meat.

Of all the meat produced in the United States, about one third is processed meat. Most of processed meat is *pork*, approximately one quarter of it is *beef*, and a small amount is *lamb or mutton*. If formulated with meat trimmings and variety meats, that fact must be stated on the label.

Processed meats may contain salt, phosphates, nitrate (NO_3), or nitrite, which provide beneficial *microbial control*. Additionally, these ingredients supply flavor, texture, and protein-binding contributions (see Restructured Meats) to foods. Processed meats may contain a reduction or replacement of sodium. Sectors of the population watching sodium or some additives may desire to reduce their intake of processed meat. Lower sodium processed meat is also discussed under the Nutrition section of this chapter.

The FSIS allows specific additions to meat. *Carrageenan*, and locust bean gum, may be utilized as meat additives. *Xanthan* gum at maximum levels of 0.5 % to prevent escape of the brine solution added to cured pork products is allowed. *Nitrite* is added to processed meat to preserve the color of meat and control the growth of *C. botulinum*.

A health concern regarding the addition of *nitrites* is that they could combine with amines (the by-product of protein breakdown) in the stomach and form carcinogenic "nitrosamines." This was addressed in a report by the NAS that stated that neither sodium nitrate nor nitrite was carcinogenic. Nitrites remain at levels less than 50 ppm in processed meat at the point of consumption. Many processors add ascorbic acid (vitamin C), erythorbic acid and their salts, sodium ascorbate, and sodium erythorbate, to cured meat to maintain processed meat color. These same additives also inhibit the production of nitrosamines from nitrites.

What Nitrite Does in Meat

Nitrite in meat greatly delays development of the botulinum toxin (which causes botulism), develops cured meat flavor and color, retards development of rancidity and off-odors and off-flavors during storage, inhibits development of warmed-over flavor, and preserves flavors of spices, smoke, etc.

Adding nitrite to meat is only part of the curing process. Ordinary table salt (sodium chloride) is added because of its effect on flavor. Sugar is added to reduce the harshness of salt. Spices and other flavorings are often added to achieve a characteristic "brand" flavor. Most, but not all, cured meat products are smoked after the curing process to impart a smoked meat flavor.

Sodium nitrite, rather than sodium nitrate, is most commonly used for curing (although in some products, such as country ham, sodium nitrate is used because of the long aging period). In a series of normal reactions, nitrite is converted to nitric oxide. Nitric oxide combines with myoglobin, the pigment responsible for the natural red color of uncured meat. They form nitric oxide myoglobin, which is a deep red color (as in uncooked dry sausage) that changes to the characteristic bright pink normally associated with cured and smoked meat (such as wieners and ham) when heated during the smoking process. (http://www.extension.umn.edu/distribution/nutrition/DJ0974.html, University of Minnesota Extension)

Curing and Smoking of Meat

Curing is a modification of meat that increases shelf life, forms a pink color, and produces a salty flavor. As mentioned, cured meats contain nitrite that controls the growth of *C. botulinum*. Additional color changes to the cured meat may result in fading as the pigment *oxidizes* when exposed to oxygen. Exposure to *fluorescent light* may actually give cured hams a fluorescent sheen and also causes a graying or fading of the color. Therefore, cured meats are packaged so that they are *minimally* exposed to the deleterious effects of oxygen and light.

Very popular ***cured meats*** include ham (pork) and corned (cured) beef as well as bacon and pastrami. Corned beef was given the name for the reason that beef was preserved with "corns" (grains) of salt.

Smoked meats prepared on a smoker are very popular in parts of the United States; they are relatively unheard of in other localities. Commercially or at home, beef, ham, and turkey are smoked (heat processed) to impart flavor. Liquid smoke could also be used to impart flavor.

CULINARY ALERT! Smoking treats meat by exposure to the aromatic smoke of hardwood and smoking also dehydrates, thus offering microbial control to the meat.

FDA Ruling on Curing and Smoking

(A) Introduction

Meat and poultry are cured by the addition of salt alone or in combination with one or more ingredients such as sodium nitrite, sugar, curing accelerators, and spices. These are used for partial preservation, flavoring, color enhancement, tenderizing, and improving yield of meat. The process may include dry curing, immersion curing, direct addition, or injection of the curing ingredients. Curing mixtures are typically composed of salt (sodium chloride), sodium nitrite, and seasonings. The preparation of curing mixtures must be carefully controlled ... It is important to use curing methods which achieve uniform distribution of the curing mixture in the meat or poultry product.

FDA—Smoking

Smoking is the process of exposing meat products to wood smoke. Depending on the method, some products may be cooked and smoked simultaneously, smoked and dried without cooking, or cooked without smoking. Smoke may be produced by burning wood chips or using an approved liquid smoke preparation. Liquid smoke preparations may also be substituted for smoke by addition directly onto the product during formulation in lieu of using a smokehouse or another type of smoking vessel. As with curing operations, a standard operating procedure must be established to prevent contamination during the smoking process.

Ham is cured pork from the hind leg of the hog. Picnic shoulder or picnic ham is made from the front leg of the hog. Bacon is cured and/or smoked hog meat from the pig belly. The University of Georgia, Cooperative Extension Service

Restructured Meat

Restructured meat contains muscle tissue, connective tissue, and adipose tissue of a natural cut of meat; however, proportions of each may differ. In the process of restructuring, meat is (1) flaked, ground, or chunked to a small particle size, (2) reformed, and (3) shaped—perhaps into roasts or steaks.

Myosin in the meat muscle may be instrumental in causing meat particles to bind together. As well, salts, phosphates, and other *nonmeat* binders such as egg albumen, gelatin, milk protein, wheat, or textured vegetable protein may be added for the purpose of holding protein particles together. Generally, the restructuring process provides a less expensive menu item that *resembles* a whole meat portion. It offers consistency in serving size and appearance. Most boneless hams and some breakfast meats are restructured meats.

Tenderizing, Artificial Tenderizing

Tenderizing meats may be desirable prior to cooking. *Young* animals are *naturally* tender and do not need *artificial* tenderizing. Recall that the connective tissue of *older* animals contains more covalent cross-links, is less soluble, and less readily converted to gelatin. Therefore, meat from older animals may require tenderization. In addition to the age, the origin of the specific cut on the animal is also a factor influencing tenderness.

The less tender cuts of meat may be *artificially* tenderized to break down the proteins of muscle or connective tissue. This may be achieved by mechanical, electrical, or enzymatic treatment as discussed below.

The *mechanical* tenderization includes chopping, cubing, and grinding. Meat may cubed, ground, or pounded prior to stuffing or rolling or use in a recipe. These techniques break the surface muscle fibers and connective tissue. A special instrument that pierces the meat with multiple thin, tenderizing needles is involved in the "needling" or "blade tenderizing" of meat.

Electrical stimulation such as ultrasonic vibration indirectly tenderizes meat by the physical vibrations that stimulate muscles to break down ATP to lactic acid. It also decreases the pH. The electrical stimulation of a carcass tenderizes without degrading the muscle fibers and texture of meats to a mushy state.

Natural *enzyme* tenderizers derived from tropical plants are available as powders or seasoning compounds that may be applied by dipping or spraying meat. They are more effective in tenderizing than marinades, which only penetrate approximately one fourth inch into the interior of the meat. Enzymes include papain from the papaya plant, bromelain from pineapple, and ficin from figs.

Various enzymes treat the muscle tissue, and others, the connective tissue. For example,

- The enzymes chymopapain, or papain, and ficin exert a greater effect on tenderizing *muscle fibers* than tenderizing connective tissue.
- The enzyme bromelain degrades *connective tissue* more than the myofibrils.

Any overapplication of natural enzyme tenderizers to meat surfaces, or allowing the treated meat to remain at temperatures conducive to enzymatic activity, could produce an *overly* soft meat consistency.

The natural enzyme tenderizer papain may also be injected into the jugular vein (bloodstream) of an animal a few minutes before slaughter. It is distributed throughout the animal tissue. The enzyme is heat-activated (by cooking at 140–160 °F (60–71 °C)) and eventually denatured in cooking.

CULINARY ALERT! With the addition of acid marinades to meat, collagen is softened to gelatin. The collagen fibers exhibit swelling and retain more water. Tomato and vinegar are acids that cause meat to respond in this manner and become tender.

There is a more *recent* development utilized in tenderizing meat without affecting appearance and taste. This is a *noninvasive process* used to tenderize meat: The process employs a 3 min cycle in a high pressure, water-filled, closed tank. A 4-ft diameter stainless steel tank, sealed with a stainless steel domed lid, creates a high-pressure wave as a small explosive charge is detonated within *the* tank. Lower grades of meat, especially cuts that are low in fat content, may increase in value as they are made more tender for consumer use (Morris 2000).

Poultry

Poultry (bird) sales increased in times when beef sales declined. All poultry is subject to *inspection* under the Wholesome Poultry Products Act of 1968 and is *graded* US Grade A, B, or C quality based on factors including conformation, fat, and freedom from blemishes and broken bones. The inspection, labeling, and handling of poultry products is similar to the meat inspection process (Chap. 19).

Chicken is the *primary* poultry consumed in the US diet. It is classified according to weight, age, and condition of the bird as follows:

Broilers/fryers	2–2.5 lb	3–5 months of age
Roasters	3–5 lb	9–12 weeks of age
Capons	4–8 lb	Less than 8 months of age
Hens, stewing hens, or fowls	2.5–5 lb	Less than 1 year
Rock Cornish game hen	1–2 lb	5–7 weeks of age

Turkey is the *second* most frequently consumed poultry in the United States and is classified as follows:

Fryers/roasterst	10 weeks of age
Mature roasting birds	20–26 weeks of age
Tom turkey (male)	greater than 5 months of age

Duck, geese, guinea, and pigeon provide variety to the diet; however, are consumed less frequently than chicken or turkey. Each is subject to inspection by the USDA's Food Safety and Inspection Service (FSIS).

The *dark meat* of poultry represents portions of meat from more exercised parts of the animal. Dark meat contains *more* myoglobin, and fat, and more iron and zinc than *white meat*. It also contains *less* protein. However, regarding fat content, any poultry *without* skin on during cooking allows less drippage of fat into meat than poultry cooked with the skin. It is possible that poultry *with* skin may contain slightly more of the lipid cholesterol than an equal portion of lean beef.

In addition to the *whole bird*, individual pieces of breast meat, legs, or thighs are sold separately, and there are many *processed* poultry products on the American market. For example, many *lunch meats* contain turkey or chicken which provides the benefit of poultry in place of beef or pork and may reduce fat content. *Ground turkey* may serve as a replacement for *ground beef* in cooked dishes, and many *formed entrees* such as nuggets, patties, or rolls are available to the consumer.

Proper poultry cooking is imperative as it may carry *Salmonella* bacteria. It must be adequately cooked (165° F) to assure destruction of this living pathogen. The FSIS allows the use of trisodium phosphate as an antimicrobial agent on raw, chilled poultry carcasses that have passed inspection for wholesomeness.

Fish

Fish is consumed by many people. It includes both edible *finfish* and *shellfish* (both appear in charts below) obtained from marine and freshwater sources. Fish is *softer* and *flakier* than either mammals or poultry, because muscle fibers exist as short bundles, which contain thin layers of connective tissue (see Fig. 9.9).

Worldwide, there are several thousand species of seafood, and with current processing, preservation, and marketing methods, a greater variety of species are consumed. Yet, only a few species are used as edible fish and shellfish (Fig. 9.9).

Fish are classified as follows:

Finfish (vertebrate with fins): Finfish are fleshy fish with a bony skeleton and are covered with scales. They may be lean or fat. For example:

Fig. 9.9 Fish (Courtesy of SYSCO® Incorporated)

Lean

- *Lean* saltwater fish—cod, flounder, haddock, halibut, red snapper, whiting
- *Lean* freshwater fish—brook trout and yellow pike

Fat

- *Fat* saltwater fish—herring, mackerel, and salmon
- *Fat* freshwater fish—catfish, lake trout, and whitefish

Shellfish (invertebrates): Shellfish are either *crustaceans or mollusks*—the former with a crustlike shell and segmented bodies, the latter with soft structures in a partial or whole, hard shell. Some examples are as follows:

Crustacea

- Crab, crayfish, lobster, and shrimp
- Crustlike shell and segmented bodies

Mollusks

- Abalone, clams, mussels, oysters, and scallops
- Soft structures in a partial or whole, hard shell

The physical, nutritive components of fish are similar to mammals—for example,

- Carbohydrate: negligible content, as is the case with mammals
- Fat: variable percentage, primarily liquid (hence, fish *oil*) and *not* saturated.
  - Fish feed on marine and freshwater *plants* that contribute to their high content of omega-3 polyunsaturated fatty acids. (Eicosapentaenoic acid—EPA, and docosahexaenoic acid—DHA, both demonstrated to be 'protective against diseases, such as heart disease.)
- Protein: high quality, complete proteins including:
  - Myofibril proteins—actin and myosin;
  - Connective tissue—collagen; and sarcoplasmic proteins—enzymes and myoglobin.

In the past, the "R-Month Rule" stated that the months of September through April indicated safe harvesting of shellfish. The letter "R" is in their spelling and these are also colder months. Today, adequate refrigeration and care dispels this belief.

Fish is classified as inherently tender because fish contains *less* connective tissue than beef, and *more* of it converts to gelatin during cooking. The *flakes* that appear in a cooked fish are due to a change in connective tissue that occurs with heating and are a sign that cooking is complete.

CULINARY ALERT! The appearance of flakes in cooked fish indicates doneness.

Restructured, or "formed" fish of various types that have been minced prior to cooking will *not* show such flakes. *Minced fish* may be

10 tips
Nutrition Education Series

eat seafood twice a week

10 tips to help you eat more seafood

Twice a week, make seafood—fish and shellfish—the main protein food on your plate.* Seafood contains a range of nutrients, including healthy omega-3 fats. According to the *2010 Dietary Guidelines for Americans*, eating about 8 ounces per week (less for young children) of a variety of seafood can help prevent heart disease.

1 eat a variety of seafood
Include some that are higher in omega-3s and lower in mercury, such as salmon, trout, oysters, Atlantic and Pacific mackerel, herring, and sardines.

2 keep it lean and flavorful
Try grilling, broiling, roasting, or baking—they don't add extra fat. Avoid breading or frying seafood and creamy sauces, which add calories and fat. Using spices or herbs, such as dill, chili powder, paprika, or cumin, and lemon or lime juice, can add flavor without adding salt.

3 shellfish counts too!
Oysters, mussels, clams, and calamari (squid) all supply healthy omega-3s. Try mussels marinara, oyster stew, steamed clams, or pasta with calamari.

4 keep seafood on hand
Canned seafood, such as canned salmon, tuna, or sardines, is quick and easy to use. Canned white tuna is higher in omega-3s, but canned "light" tuna is lower in mercury.

5 cook it safely
Check oysters, mussels, and clams before cooking. If shells don't clamp shut when you tap them, throw them away. After cooking, also toss any that didn't open. This means that they may not be safe to eat. Cook shrimp, lobster, and scallops until they are opaque (milky white). Cook fish to 145°F, until it flakes with a fork.

*This recommendation does not apply to vegetarians.

6 get creative with seafood
Think beyond the fish fillet. Try salmon patties, a shrimp stir-fry, grilled fish tacos, or clams with whole-wheat pasta. Add variety by trying a new fish such as grilled Atlantic or Pacific mackerel, herring on a salad, or oven-baked pollock.

7 put it on a salad or in a sandwich
Top a salad with grilled scallops, shrimp, or crab in place of steak or chicken. Use canned tuna or salmon for sandwiches in place of deli meats, which are often higher in sodium.

8 shop smart
Eating more seafood does not have to be expensive. Whiting, tilapia, sardines, canned tuna, and some frozen seafood are usually lower cost options. Check the local newspaper, online, and at the store for sales, coupons, and specials to help save money on seafood.

9 grow up healthy with seafood
Omega-3 fats from seafood can help improve nervous system development in infants and children. Serve seafood to children twice a week in portions appropriate for their age and appetite. A variety of seafood lower in mercury should also be part of a healthy diet for women who are pregnant or breastfeeding.

10 know your seafood portions
To get 8 ounces of seafood a week, use these as guides: A drained can of tuna is about 3 to 4 ounces, a salmon steak ranges from 4 to 6 ounces, and 1 small trout is about 3 ounces.

Go to www.ChooseMyPlate.gov for more information.

DG TipSheet No. 15
December 2011
USDA is an equal opportunity provider and employer.

produced from less popular varieties of fish, or from the fish-flesh remains of the filet process.

The washed, minced fish, coupled with heating, produces gel-like properties in the flesh and it may be "*formed*" for use as various products. For example, with centuries of production in Japan, and developing technology in handling, the minced fish is used in the production of fish sticks, nuggets, patties, or other unbreaded, "formed" fish items.

In the production of *surimi*, for example, minced fish such as pollack is washed to remove both oil and water-soluble substances such as colors and flavor compounds, leaving only protein fibers as the remains. The washing also removes sarcoplasmic proteins that interfere with the necessary gelling. (Thus some oil and sarcoplasmic enzyme residue remaining in the fibers.)

Subsequent to washing, the flesh (protein fibers) is mixed with salt to solubilize the myofibril proteins—actin and myosin. Other characteristic flavors and pigments, as well as ingredients that promote the elastic texture and stability of the product, are added to the fish so that it may be incorporated into chowders, resembles crabmeat, lobster meat, or sausage-type products.

If surimi is used to create these crabmeat, lobster meat, or sausage-type products, they are called "imitation" (e.g., "imitation crabmeat."). Two of the more common raw fish dishes are sashimi and sushi. Sashimi is sliced and prepared fish and sushi if vinegared rice, rolled with raw fish, and covered with seaweed. Care in handling is required of raw fish dishes.

In addition to the aforementioned meat, poultry, and fish, other protein sources in the diet are listed below.

Dry Beans and Peas (Legumes) as Meat Alternatives

Legumes offer great variety of diet. As plant material, legumes are *incomplete* proteins, while *complete* proteins are animal proteins that contain all the essential amino acids present at superior levels—for example, meat, poultry, fish, or milk and eggs.

In order to obtain the same essential amino acid profile as complete proteins, two or more *plant* protein foods are typically *combined* and eaten in the same day. The requirement is that they are combined in a *day* (not needed in the same *meal*) in order to provide the body with *essential amino acids*.

Examples of plant foods' amino acid composition may be seen in the chart below:

Legumes—for example, soybean, black-eyed peas, pinto beans; good source of lysine poor source of (limited) tryptophan and sulfur-containing (S–C) amino acids (soybeans contain tryptophan)
Nuts–Seeds—for example, peanuts, sesame seeds; good source of tryptophan and S–C amino acids; poor source of lysine (peanuts contain less S–C amino acids)
Cereals–Grains (whole grains)—corn, rice, whole grains (Chap. 4); good source of tryptophan and S–C amino acids; poor source of lysine (corn is a poor source of tryptophan and good source of S–C amino acids; wheat germ is poor in tryptophan and S–C amino acids but a good source of lysine)

"Combining" (see above—combination of two or more complementary sources of incomplete proteins in order to provide a complete amino profile) may include serving beans with rice, tofu with vegetables on rice, black-eyed peas served with cornbread, tofu and cashews stir-fry, chick peas and sesame seeds (hummus) or peanut butter on whole wheat bread. As a common example, a vegetarian diet may frequently combine legumes (beans or peas) with either nuts–seeds or grains.

Mutual supplementation is the name given to this combination of two or more complementary sources of incomplete proteins in order to provide a complete amino profile. The requirement is that they are combined in a *day* (not needed in the same *meal*) in order to provide the body with *essential amino acids*.

CULINARY ALERT! Combine the appropriate incomplete proteins in order to create a complete protein.

Legumes

Legumes (Fig. 9.10) are the seeds of a pod of the *Leguminosae* family. The seed, found inside the pod, splits into two distinct parts attached to each other at the lower edge. They include edible peas that may be green, yellow, white, or variegated in color. They include sugar peas with edible pods, black-eyed peas, and more.

Legumes may exist as elongated, flattened, spherical, or kidney-shaped beans or peas. Notable are the edible, podded string beans/snap beans or green beans, kidney beans, or soybeans. Various beans, such as mung beans, are sprouted for culinary use, and others may be used for animal fodder. Frequently, legumes may be referred to as a pulse, and part of the bean, or pea family.

In addition to beans and peas, carob pods and lentils are legumes. Peanuts, despite their name, are *not* true nuts, however, legumes. They are the high protein seeds of a brown pod that appears contracted between the seeds (humped and inverted). Their ripening occurs underground.

Conspicuous changes occur in cooking legumes such as softening due to the gelatinization of starch and flavor improvement. The protein is coagulated and its availability is higher following cooking.

> Edible bean products contain both soluble and insoluble fiber which may help to slow digestion. These fibers can contribute to providing a feeling of fullness, or satiety, which can help in weight management . . . also contribute to a low glycemic index . . . a decreased level of glucose in blood following a meal. (Foster 2012)

Legumes may be the origin of intestinal distress and gas in some consumers. For that reason, an enzyme derived from *A. niger* has been processed for addition to foods such as these and is commercially available to consumers for their dietary use. As well, some individuals exhibit an *allergic* response with antibody production following ingestion of various legumes, thus Physicians and dietitians would recommend avoidance.

CULINARY ALERT! Beans and peas [along with cruciferous (cabbage family) vegetables and whole grains in the diet] are wholesome food choices recommended to many by physicians and dieticians for healthy eating. They are low in fat, contain no cholesterol, and are good sources of fiber.

Usually, soybeans are derived from an autumn harvest and are processed into oil, tofu, frozen dessert, "flour", or textured vegetable protein.

- **Soybean oil**, pressed from the bean, is the highest volume vegetable oil in the United States and is commonly a constituent of margarine.
- **Tofu** is soy milk that has been coagulated to make the gel. Tofu is available in various types, ranging from soft to extra firm, depending on the water content. Extra firm tofu may be cut into small pieces and used in stir-fry cooking.
- **Frozen dessert** the curd is further processed and sweetened, it may be served as a frozen tofu- based dessert, similar to ice cream or ice milk. The soft tofu may be an ingredient of "shakes" or frozen, sweetened dessert mixtures.
- **Soy flour** is made of dehulled beans with the oil (that was 18 %) pressed out. It is useful by consumers who cannot consume wheat or flours with gluten-forming proteins. It is non-gluten. Although a soybean is not a

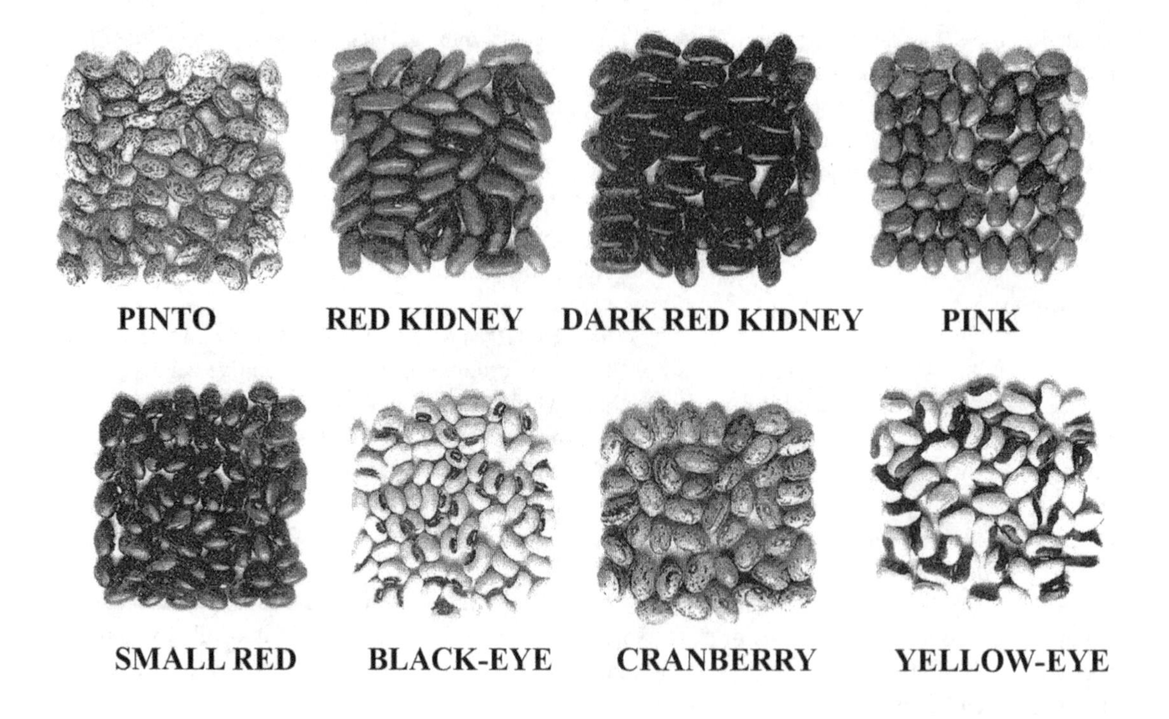

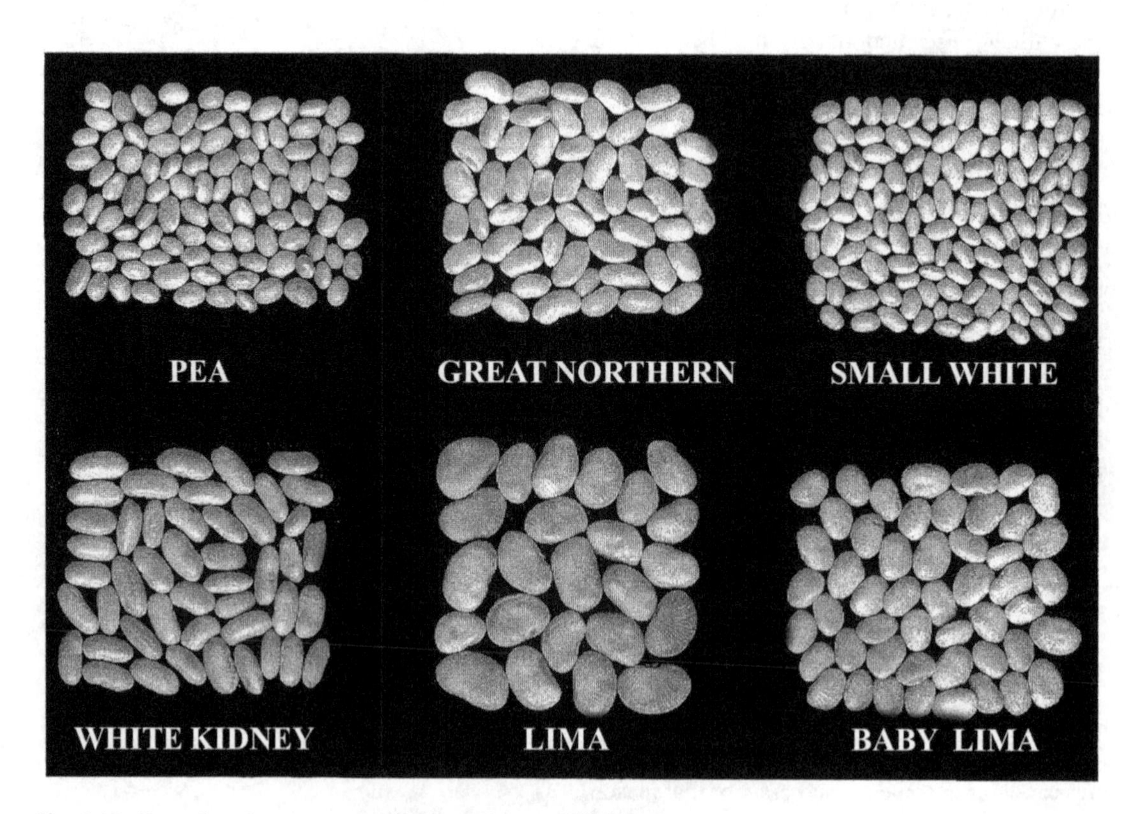

Fig. 9.10 Examples of some common legumes (*Source*: USDA)

cereal, it may be a source of "flour" in recipes.

- **Textured vegetable protein (TVP)** is used by food-service establishments, including school lunch programs that use soy protein on their menus. TVP may simulate a ground form of meat or meat flakes, it resembles the texture of meat, and is a good source of protein in meatless diets. It is the principal ingredient in artificial bacon sprinkles used for salads and other foods. In combination with colors, flavors, and egg binders (for ease of fabrication), the often-unpleasant characteristic flavor of soybeans can be covered.
- **Fermented soybeans** produce soy sauce, miso, and tempeh. Soy sauce is a combination of the fermented soy and wheat; miso is fermented soybean and/or rice used in oriental sauces and soup bases. Tempeh is similar to tofu; however is inoculated with different bacteria.
  - A soy protein *concentrate* is soy that has been defatted, with soluble carbohydrate removed. It is *70 %* protein. An even higher quality soy ingredient may be manufactured using a soy protein isolate.
  - An *isolate* is *90 %* protein, with even more of the nonprotein material extracted, and with the addition of flavors and colors, it may be satisfactorily included in numerous foods. Nuts are addressed in the discussion of fruits and vegetables (Chap. 7).

Quorn as a Meat Alternative

Quorn (pronounced "kworn") is a meat alternative that became available to American consumers in early 2002 after being sold in Europe for over a decade. It is not a vegetable however, it is a fungus made into many meat-type products including patties and nuggets, and casseroles. A former FDA food safety chief (Sanford Miller, PhD.) and senior fellow at the Center for Food and Nutrition Policy states that "This product meets what the nutrition community thinks a product should be and in addition, it tastes good! Modern science can fabricate anything. We can imitate anything, but we always run into problems on how to have it taste good. Not taste alright, but taste good. This product does that."

The mycoprotein was approved after undergoing a 5-year approval process with extensive animal and human testing. This process included a close look at possible allergens, which showed less than the allergens of mushrooms or soy (Peregrin 2002b).

According to an earlier report by the Center for Science in the Public Interest (CSPI), claims on some labels that the key ingredient in Quorn is "mushroom in origin," were not true. "Quorn products contain no mushrooms. Rather, the so-called "mycoprotein" in these products is actually grown in large fermentation vats from *Fusarium venenatum*, a non-mushroom fungus. On other Quorn packages, the source of mycoprotein is omitted altogether."

CSPI executive director Michael F. Jacobson said. "But Quorn's mycoprotein has nothing to do with mushrooms, plants, or vegetables. It is a fungus and should be labeled as such. Saying that Quorn's fungus is in the mushroom family is like saying that jellyfish are in the human family. If an obscure term like "mycoprotein" is to be used in Quorn's ingredient listings, says CSPI, packages should be required to disclose clearly the product's fungal origins." (CSPI)

Nutritive Value of Meat, Poultry, and Fish

Selected nutritive value aspects of meats (beef, veal, pork, and lamb), poultry (chicken and turkey), and fish and shellfish are shown in Figs. 9.11, 9.12, 9.13, and 9.14. These NutriFacts figures are the most recent ones reviewed by the USDA. Nutritive values for calories, calories

BEEF & VEAL

NUTRI-FACTS UPDATE

BEEF & VEAL NUTRITION FACTS

(⅛" fat trim / trimmed of visible fat)

	Calories	Calories From Fat	Total Fat	Saturated Fat	Cholesterol	Sodium	Protein	Iron
BEEF, 3 oz cooked serving			g	g	mg	mg	g	%DV
Ground Beef, broiled, well done (10% fat*)	210	100	11	4	85	70	27	15
Ground Beef, broiled, well done (17% fat*)	230	120	13	5	85	70	24	15
Ground Beef, broiled, well done (27% fat*)	250	150	17	6	85	80	23	15
Brisket, Whole, braised	290 / 210	190 / 100	21 / 11	8 / 4	80 / 80	55 / 60	22 / 25	10 / 15
Chuck, Arm Pot Roast, braised	260 / 180	160 / 60	18 / 7	7 / 3	85 / 85	50 / 55	24 / 28	15 / 20
Chuck, Blade Roast, braised	290 / 210	190 / 100	21 / 11	9 / 4	90 / 90	55 / 60	23 / 26	15 / 15
Rib Roast, Large End, roasted	300 / 200	220 / 100	24 / 11	10 / 4	70 / 70	55 / 60	20 / 23	10 / 15
Rib Steak, Small End, broiled	280 / 190	190 / 90	21 / 10	9 / 4	70 / 70	55 / 60	20 / 24	10 / 10
Top Loin, Steak, broiled	230 / 180	130 / 70	15 / 8	6 / 3	65 / 65	55 / 60	22 / 24	10 / 10
Loin, Tenderloin Steak, broiled	240 / 180	150 / 80	16 / 9	6 / 3	75 / 70	50 / 55	22 / 24	15 / 15
Loin, Sirloin Steak, broiled	210 / 170	110 / 60	12 / 6	5 / 2	75 / 75	55 / 55	24 / 26	15 / 15
Eye Round, Roast, roasted	170 / 140	60 / 40	7 / 4	3 / 2	60 / 60	50 / 55	24 / 25	10 / 10
Bottom Round, Steak, braised	220 / 180	110 / 60	12 / 7	5 / 2	80 / 80	40 / 45	25 / 27	15 / 15
Round, Tip Roast, roasted	190 / 160	90 / 50	10 / 6	4 / 2	70 / 70	55 / 55	23 / 24	15 / 15
Top Round, Steak, broiled	180 / 150	70 / 40	7 / 4	3 / 1	70 / 70	50 / 50	26 / 27	15 / 15
VEAL, 3 oz cooked serving			g	g	mg	mg	g	%DV
Shoulder, Arm Steak, braised	200 / 170	80 / 40	9 / 5	3 / 1	125 / 130	75 / 75	29 / 30	6 / 6
Shoulder, Blade Steak, braised	190 / 170	80 / 50	9 / 6	3 / 2	130 / 135	85 / 85	27 / 28	6 / 6
Rib Roast, roasted	190 / 150	110 / 60	12 / 6	5 / 2	95 / 95	80 / 80	20 / 22	4 / 4
Loin Chop, roasted	180 / 150	100 / 50	10 / 6	4 / 2	85 / 90	80 / 80	21 / 22	4 / 4
Cutlets, roasted	140 / 130	35 / 25	4 / 3	2 / 1	85 / 90	60 / 60	24 / 24	4 / 4

Not a significant source of total carbohydrate, dietary fiber, sugars, vitamin A, vitamin C, and calcium.

*Before cooking

Serving Size: 3 oz. cooked portion, without added fat, salt or sauces.

Developed By: Food Marketing Institute, American Dietetic Association, American Meat Institute, National-American Wholesale Grocers' Association, National Broiler Council, National Fisheries Institute, National Grocers Association, National Live Stock and Meat Board, National Turkey Federation, United Fresh Fruit and Vegetable Association.

Reviewed By: United States Department of Agriculture

Data Source: USDA Handbook 8-13, revised 1990 and Bulletin Board, 1994 (beef) and USDA Handbook 8-17, 1989 (veal)

3/95

Fig. 9.11 Nutrifacts of beef and veal in 3-ounce cooked portions (*Source*: Food Marketing Institute)

from fat, total fat, saturated fat, cholesterol, sodium, protein, and iron are reported for beef and veal, pork and lamb, and chicken and turkey. Similarly, calories, calories from fat, total fat, saturated fat, cholesterol, sodium, potassium, total carbohydrate, protein, vitamin A, vitamin C, calcium, and iron are also reported for seafood.

PORK & LAMB

NUTRI-FACTS

UPDATE

PORK & LAMB NUTRITION FACTS

⅛" fat trim / trimmed of visible fat

PORK, 3 oz cooked serving	Calories	Calories From Fat	Total Fat	Saturated Fat	Cholesterol	Sodium	Protein	Iron
			g	g	mg	mg	g	%DV
Ground Pork, broiled	250	160	18	7	80	60	22	6
Shoulder, Blade Steak, broiled	220 / **190**	130 / **100**	14 / **11**	5 / **4**	80 / **80**	60 / **65**	22 / **23**	6 / **8**
Loin, Country Style Ribs, roasted	280 / **210**	190 / **110**	22 / **13**	8 / **5**	80 / **80**	45 / **25**	20 / **23**	6 / **6**
Loin, Rib Chop, broiled	220 / **190**	120 / **80**	13 / **8**	5 / **3**	70 / **70**	55 / **55**	24 / **26**	4 / **4**
Center Chop, Loin, broiled	200 / **170**	100 / **60**	11 / **7**	4 / **3**	70 / **70**	50 / **50**	24 / **26**	4 / **4**
Top Loin, Chop, boneless, broiled	200 / **170**	90 / **60**	10 / **7**	3 / **2**	70 / **70**	55 / **55**	25 / **26**	4 / **4**
Top Loin, Roast, boneless, roasted	190 / **170**	90 / **60**	10 / **6**	4 / **2**	65 / **65**	40 / **40**	24 / **26**	4 / **6**
Loin, Tenderloin Roast, roasted	150 / **140**	45 / **35**	5 / **4**	2 / **1**	65 / **65**	45 / **50**	24 / **24**	6 / **6**
Loin, Sirloin Roast, roasted	220 / **180**	120 / **80**	14 / **9**	5 / **3**	75 / **75**	50 / **55**	23 / **25**	4 / **6**
Spareribs, braised	340	230	26	9	105	80	25	8
LAMB, 3 oz cooked serving			g	g	mg	mg	g	%DV
Shoulder, Arm Chop, broiled	230 / **170**	140 / **70**	15 / **8**	7 / **3**	80 / **80**	65 / **70**	21 / **24**	10 / **10**
Shoulder, Blade Chop, broiled	230 / **180**	140 / **90**	16 / **10**	6 / **3**	80 / **80**	70 / **75**	20 / **22**	8 / **8**
Shank, braised	210 / **160**	100 / **45**	11 / **5**	5 / **2**	90 / **90**	60 / **65**	24 / **26**	10 / **10**
Rib Roast, roasted	290 / **200**	210 / **100**	23 / **11**	10 / **4**	80 / **75**	65 / **70**	19 / **22**	8 / **8**
Loin Chop, broiled	250 / **180**	160 / **80**	18 / **8**	7 / **3**	85 / **80**	65 / **70**	22 / **25**	10 / **10**
Leg, Whole, roasted	210 / **160**	110 / **60**	12 / **7**	5 / **2**	80 / **75**	55 / **60**	22 / **24**	10 / **10**

Not a significant source of total carbohydrate, dietary fiber, sugars, vitamin A, vitamin C, and calcium.

Serving Size: 3 oz. cooked portion, without added fat, salt or sauces.

Developed By: Food Marketing Institute, American Dietetic Association, American Meat Institute, National-American Wholesale Grocers' Association, National Broiler Council, National Fisheries Institute, National Grocers Association, National Live Stock and Meat Board, National Turkey Federation, United Fresh Fruit and Vegetable Association.

Reviewed By: United States Department of Agriculture

Data Source: *USDA Handbook 8-10*, 1992 (pork) and *USDA Handbook 8-17*, 1989 and *Bulletin Board*, 1994 (lamb)

3/95

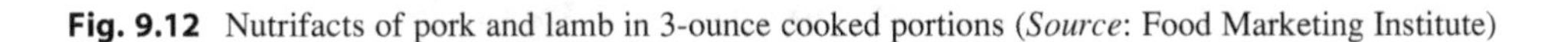

Fig. 9.12 Nutrifacts of pork and lamb in 3-ounce cooked portions (*Source*: Food Marketing Institute)

POULTRY NUTRITION FACTS

With skin / Skinless

	Calories	Calories From Fat	Total Fat	Saturated Fat	Cholesterol	Sodium	Protein	Iron
Chicken, 3 oz cooked serving			g	g	mg	mg	g	%DV
Whole*, roasted	200 / 130	100 / 35	12 / 4	3 / 1	75 / 75	70 / 75	23 / 23	6 / 6
Breast, baked	170 / 120	60 / 15	7 / 1.5	2 / .5	70 / 70	60 / 65	25 / 24	6 / 4
Wing, baked	250 / 150	150 / 50	17 / 6	5 / 1.5	70 / 70	70 / 80	23 / 23	6 / 6
Drumstick, baked	180 / 130	90 / 35	9 / 4	3 / 1	75 / 80	75 / 80	23 / 23	6 / 6
Thigh, baked	210 / 150	120 / 60	13 / 7	4 / 2	80 / 80	70 / 75	21 / 21	6 / 6
Turkey, 3 oz cooked serving			g	g	mg	mg	g	%DV
Whole*, roasted	180 / 130	70 / 25	8 / 3	2 / 1	70 / 65	60 / 60	24 / 25	8 / 8
Breast, baked	160 / 120	60 / 10	6 / 1	2 / 0	65 / 55	55 / 45	24 / 26	6 / 8
Wing, baked	200 / 140	100 / 25	11 / 3	3 / 1	70 / 60	50 / 75	23 / 26	6 / 8
Drumstick, baked	170 / 140	70 / 40	8 / 4	2 / 1	70 / 65	75 / 80	23 / 24	10 / 15
Thigh, baked	160 / 140	60 / 40	7 / 5	2 / 1.5	70 / 65	70 / 70	22 / 23	10 / 15

*without neck or giblets

Not a significant source of total carbohydrate, dietary fiber, sugars, vitamin A, vitamin C, and calcium.

Serving Size: 3 oz. cooked portion, without added fat, salt or sauces.

Developed By: Food Marketing Institute, American Dietetic Association, American Meat Institute, National-American Wholesale Grocers' Association, National Broiler Council, National Fisheries Institute, National Grocers Association, National Live Stock and Meat Board, National Turkey Federation, United Fresh Fruit and Vegetable Association.

Reviewed By: United States Department of Agriculture

Data Source: USDA Handbook 8-5 and research conducted in cooperation with USDA.

3/95

Fig. 9.13 Nutrifacts of chicken and turkey in 3-ounce cooked portions (*Source*: Food Marketing Institute)

SEAFOOD

NUTRI-FACTS

UPDATE

NUTRITION FACTS FOR COOKED SEAFOOD[1]

Nutrient / % Daily Value of Nutrient	Calories	Calories From Fat	Total Fat	Saturated Fat	Cholesterol	Sodium	Potassium	Total Carbohydrate	Protein	Vitamin A	Vitamin C	Calcium	Iron
SEAFOOD (84 g/3 oz)			(g/%DV)	(g/%DV)	(mg/%DV)	(mg/%DV)	(mg/%DV)	(g/%DV)	(g)	(%DV)	(%DV)	(%DV)	(%DV)
Blue Crab	100	10	1 / 2	0 / 0	90 / 30	320 / 13	360 / 10	0 / 0	20	0	0	8	4
Catfish	140	80	9 / 14	2 / 10	50 / 17	40 / 2	230 / 7	0 / 0	17	0	0	0	0
Clams, about 12 small	100	15	1.5 / 2	0 / 0	55 / 18	95 / 4	530 / 15	0 / 0	22	10	0	6	60
Cod	90	0	0.5 / 1	0 / 0	45 / 15	60 / 3	450 / 13	0 / 0	20	0	0	2	2
Flounder/Sole	100	14	1.5 / 2	0.5 / 3	60 / 20	90 / 4	290 / 8	0 / 0	21	0	0	2	2
Haddock	100	10	1 / 2	0 / 0	80 / 27	85 / 4	340 / 10	0 / 0	21	0	0	2	6
Halibut	110	20	2 / 3	0 / 0	35 / 12	60 / 3	490 / 14	0 / 0	23	2	0	4	4
Lobster	80	0	0.5 / 1	0 / 0	60 / 20	320 / 13	300 / 9	1 / 0	17	0	0	4	2
Mackerel, Atlantic/Pacific	210	120	13 / 20	1.5 / 8	60 / 20	100 / 4	400 / 11	0 / 0	21	0	0	0	5
Ocean Perch	110	20	2 / 3	0 / 0	50 / 17	95 / 4	290 / 8	0 / 0	21	0	0	10	6
Orange Roughy	80	10	1 / 2	0 / 0	20 / 7	70 / 3	330 / 9	0 / 0	16	0	0	0	0
Oysters, about 12 medium	100	35	3.5 / 5	1 / 5	115 / 38	190 / 8	390 / 11	4 / 1	10	0	0	6	45
Pollock	90	10	1 / 2	0 / 0	80 / 27	110 / 5	360 / 10	0 / 0	20	0	0	0	2
Rainbow Trout	140	50	6 / 9	2 / 10	60 / 20	35 / 1	370 / 11	0 / 0	21	4	4	6	2
Rockfish	100	20	2 / 3	0 / 0	40 / 13	70 / 3	430 / 12	0 / 0	21	4	0	0	2
Salmon, Atlantic/Coho	160	60	7 / 11	1 / 5	50 / 17	50 / 2	490 / 14	0 / 0	22	0	0	0	4
Salmon, Chum/Pink	130	35	4 / 6	1 / 5	70 / 23	65 / 3	410 / 12	0 / 0	22	2	0	0	2
Salmon, Sockeye	180	80	9 / 14	1.5 / 8	75 / 25	55 / 2	320 / 9	0 / 0	23	4	0	0	2
Scallops, about 6 large or 14 small	120	10	1 / 2	0 / 0	55 / 18	260 / 11	280 / 8	2 / 1	22	0	0	2	2
Shrimp	80	10	1 / 2	0 / 0	165 / 55	190 / 8	140 / 4	0 / 0	18	0	0	2	15
Swordfish	130	35	4.5 / 7	1 / 5	40 / 13	100 / 4	310 / 9	0 / 0	22	2	2	0	4
Whiting	110	25	3 / 5	0.5 / 3	70 / 23	95 / 4	320 / 9	0 / 0	19	2	0	6	0

Seafood provides negligible amounts of dietary fiber and sugars.

[1] Cooked, edible weight portion.
Percent Daily Values are based on a 2,000 calorie diet.

Serving Size: 3 oz. skinless cooked portion, without added fat, salt or sauces.

Developed by: Food Marketing Institute, American Dietetic Association, American Meat Institute, Food Distributors International, National Broiler Council, National Cattlemen's Beef Association, National Fisheries Institute, National Grocers Association, National Turkey Federation, Produce Marketing Association, United Fresh Fruit and Vegetable Association

Data Source: U.S. Food and Drug Administration

(7/96)

Fig. 9.14 Nutrifacts of seafood in 3-ounce cooked portions (*Source*: Food Marketing Institute)

Meats are excellent sources of complete protein, many B vitamins, including B_{12} that is only found in animal products, and also the minerals iron and zinc. For additional information on the nutritive value of meat and its many vitamins and minerals, the reader is referred to Figs. 9.11,

9.12, 9.13, and 9.14 that follow as well as other nutrition textbooks.

"Most health and nutrition authorities believe the majority of Americans consume too much sodium." … The CDC "believes that if manufacturers of the top-10 categories of food responsible for 44 % of people's sodium intake were to reduce the sodium content of these foods by 25 %, they could help prevent an estimated 28,000 deaths annually. With CDC having identified cold cured cuts and cured meats as well as fresh and processed poultry as two of the top-10 categories, processed-meat manufacturers stepping up to the challenge (Berry 2013)".

In 2000, The American Heart Association announced the organization's official recommendation for daily consumption of soy protein. Soybeans have antioxidant properties and contain saponins noted for their disease-fighting potential.

What Foods Are in the Protein Foods Group?

All foods made from meat, poultry, seafood, beans and peas, eggs, processed soy products, nuts, and seeds are considered part of the Protein Foods Group. Beans and peas are also part of the Vegetable Group. For more information on beans and peas, see Beans and Peas Are Unique Foods.

Select a variety of protein foods to improve nutrient intake and health benefits, including at least 8 ounces of cooked seafood per week. Young children need less, depending on their age and calorie needs. The advice to consume seafood does not apply to vegetarians. Vegetarian options in the Protein Foods Group include beans and peas, processed soy products, and nuts and seeds. Meat and poultry choices should be lean or low-fat.

All foods made from meat, poultry, seafood, beans and peas, eggs, processed soy products, nuts, and seeds are considered part of the Protein Foods Group. Beans and peas are also part of the Vegetable Group. For more information on beans

10 tips
Nutrition Education Series

healthy eating for vegetarians

10 tips for vegetarians

A vegetarian eating pattern can be a healthy option. The key is to consume a variety of foods and the right amount of foods to meet your calorie and nutrient needs.

1 think about protein

Your protein needs can easily be met by eating a variety of plant foods. Sources of protein for vegetarians include beans and peas, nuts, and soy products (such as tofu, tempeh). Lacto-ovo vegetarians also get protein from eggs and dairy foods.

2 bone up on sources of calcium

Calcium is used for building bones and teeth. Some vegetarians consume dairy products, which are excellent sources of calcium. Other sources of calcium for vegetarians include calcium-fortified soymilk (soy beverage), tofu made with calcium sulfate, calcium-fortified breakfast cereals and orange juice, and some dark-green leafy vegetables (collard, turnip, and mustard greens; and bok choy).

3 make simple changes

Many popular main dishes are or can be vegetarian—such as pasta primavera, pasta with marinara or pesto sauce, veggie pizza, vegetable lasagna, tofu-vegetable stir-fry, and bean burritos.

4 enjoy a cookout

For barbecues, try veggie or soy burgers, soy hot dogs, marinated tofu or tempeh, and fruit kabobs. Grilled veggies are great, too!

5 include beans and peas

Because of their high nutrient content, consuming beans and peas is recommended for everyone, vegetarians and non-vegetarians alike. Enjoy some vegetarian chili, three bean salad, or split pea soup. Make a hummus-filled pita sandwich.

6 try different veggie versions

A variety of vegetarian products look—and may taste—like their non-vegetarian counterparts but are usually lower in saturated fat and contain no cholesterol. For breakfast, try soy-based sausage patties or links. For dinner, rather than hamburgers, try bean burgers or falafel (chickpea patties).

7 make some small changes at restaurants

Most restaurants can make vegetarian modifications to menu items by substituting meatless sauces or non-meat items, such as tofu and beans for meat, and adding vegetables or pasta in place of meat. Ask about available vegetarian options.

8 nuts make great snacks

Choose unsalted nuts as a snack and use them in salads or main dishes. Add almonds, walnuts, or pecans instead of cheese or meat to a green salad.

9 get your vitamin B_{12}

Vitamin B_{12} is naturally found only in animal products. Vegetarians should choose fortified foods such as cereals or soy products, or take a vitamin B_{12} supplement if they do not consume any animal products. Check the Nutrition Facts label for vitamin B_{12} in fortified products.

10 find a vegetarian pattern for you

Go to www.dietaryguidelines.gov and check appendices 8 and 9 of the *Dietary Guidelines for Americans, 2010* for vegetarian adaptations of the USDA food patterns at 12 calorie levels.

Go to www.ChooseMyPlate.gov for more information.

DG TipSheet No. 8
June 2011

and peas, see Beans and Peas Are Unique Foods.

Select a variety of protein foods to improve nutrient intake and health benefits, including at least 8 ounces of cooked seafood per week. Young children need less, depending on their age and calorie needs. The advice to consume seafood does not apply to vegetarians. Vegetarian options in the Protein Foods Group include beans and peas, processed soy products, and nuts and seeds. Meat and poultry choices should be lean or low-fat.

Key consumer message: Go lean with protein View Protein Food Gallery

A nutrition article on the Paleo Diet is not limited to a discussion on red meat, yet is included in this meat chapter.

The Paleo Diet

Should We Eat Like Our Caveman Ancestors?

By Lauren Innocenzi

"The Paleolithic (Paleo) diet, also called the "Caveman" or "Stone Age" diet, centers around the idea that if we eat like our ancestors did 10,000 years ago, we'll be healthier, lose weight and curb disease. "A quick and pithy definition of the Paleo diet is—if the cavemen didn't eat it then you shouldn't either," says Academy Spokesperson Jim White, RD, ACSM/HFS. That means foods that can be hunted, fished or gathered: meat, fish, shellfish, poultry, eggs, veggies, roots, fruits, and berries. No grains, no dairy, no legumes (beans or peas), no sugar, no salt. Why? "According to proponents, our bodies are genetically predisposed to eat this way. They blame the agricultural revolution and the addition of grains, legumes and dairy to the human diet for the onset of chronic disease (obesity, heart disease, and diabetes)," says White.

On one hand, this way of eating encourages including more fruits and vegetables and cutting out added sugar and sodium—which aligns with the 2010 *Dietary Guidelines for Americans*. The combination of plant foods and a diet rich in protein can help control blood sugar, regulate blood pressure, contribute to weight loss and prevent Type 2 diabetes, says White.

But a typical plan also exceeds the *Dietary Guidelines* for daily fat and protein intake and falls short on carbohydrate recommendations, according to a review from *US News & World Report*. The exclusion of whole grains, legumes, and dairy can be risky as well. "These foods are nutrient-rich and contain important vitamins and minerals such as calcium and vitamin D. Without these foods, supplementation is necessary," says White. "Eating this way … can be very healthy but the lack of certain foods may result in certain deficiencies."

Eliminating whole grains and dairy is not necessarily the ticket to ending disease and ensuring weight loss. Whole grains contain dietary fiber, which may help reduce your risk of heart disease, cancer and diabetes, and other health complications. And studies suggest that dairy may play a role in weight loss. "The crux of the problem, with respect to grains and dairy, stem from over consumption, and as with anything, excess quantities will become problematic," explains White.

The Paleo diet might also be hard to sustain. "We live in a society where it is not possible to eat exactly as our ancestors ate. For example, wild game is not readily available as most of the meat we consume has been domesticated. And the plant food we eat has also been processed rather than grown and gathered in the wild," says White. "While strict conformity is not realistic, it is possible to modify the plan, eating only wild caught fish, grass-fed meat, and organic fruits and vegetables." But even that can be hard to follow because of lack of variety, need for planning, supplementation, and cost, White adds.

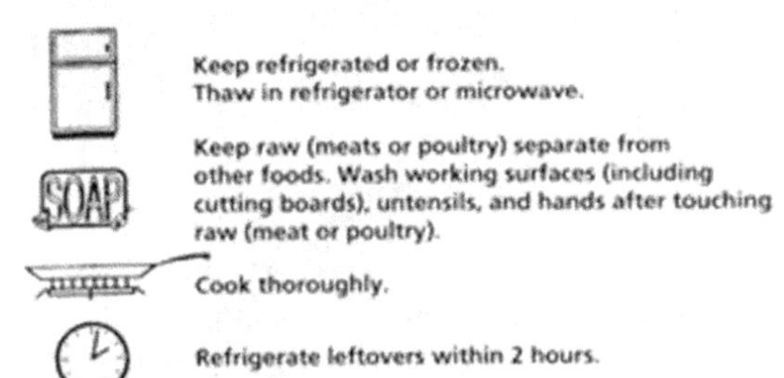

Fig. 9.15 Safe handling instructions

Lauren Innocenzi is an online content manager for the Academy of Nutrition and Dietetics.

Safety of Meat, Poultry, and Fish

The safety of meat, poultry, and fish is of utmost concern. Meat is a potentially hazardous food that supports rapid bacterial growth if contaminated and then stored at improper temperatures (Chap. 16). Because all meat contains bacteria, it should be maintained in a clean and covered condition, at temperatures that retard the growth of microorganisms which may both contaminate and spoil meat, producing changes in the color, odor, and safety. *Safe handling* instructions appearing on meat packages is shown in Fig. 9.15.

A former USDA deputy undersecretary of food safety has said "We are strongly encouraging specific interventions for raw meat and poultry in order to further reduce the level and incidence of pathogens such as *Salmonella* in these products. We feel that there is a whole arsenal of potentially effective interventions that could be utilized" (CDC).

Regarding irradiation to control pathogens, according to the USDA, "If irradiated meat is used in another product, such as pork sausage, then the ingredients statement must list irradiated pork, but the radura does not have to appear on the package."

Radura

Restaurants are not required to disclose the use of irradiated products to their customers; however, some restaurants voluntarily provide irradiation information on menus.

How should I handle irradiated meat and poultry?

Food irradiation is not a substitute for good sanitation and does not replace safe cooking and handling. Consumers should handle irradiated foods just like any other food and always follow safe food handling practices.

"Food irradiation is a technology for controlling spoilage and eliminating foodborne pathogens. The result is similar to pasteurization. The fundamental difference between food irradiation and pasteurization is the source of the energy used to destroy the microbes. While conventional pasteurization relies on heat, irradiation relies on the energy of ionizing radiation."

Table 9.1 Minimum safe internal temperature for selected meats

Meat	Temperature
Beef steaks (rare)	155 °F (68 °C) (upon consumer order)
Roast beef (rare)	130 °F (54 °C) (time dependent)
Pork	155 °F (68 °C) or 170 °F (77 °C) in a microwave oven
Ground beef	155 °F (68 °C)
Poultry	165 °F (74 °C)

Source: FDA

Check your steps at FoodSafety.gov. Also see Chap. 19.

- **CLEAN.** Wash hands and surfaces often.
- **SEPARATE.** Separate raw meats from other foods.
- **COOK.** Cook food to the right temperature.
- **CHILL.** Refrigerate food promptly.

CULINARY ALERT! Adherence to specific temperatures is necessary for the prevention of growth and the destruction of harmful microorganisms in meat. Adequate refrigeration, cooking, and holding, as well as reheating are all important in controlling bacteria. Personal hygiene and sanitation are also important in preventing the spread of bacteria.

The bacteria, *Clostridium botulinum,* is an *anaerobic* bacteria that causes the disease botulism. It is a deadly form of food poisoning that may result from consuming improperly processed canned or vacuum-packed meats. To control this, nitrite may be added to processed meat to inhibit the reproduction of bacterial spores.

A more widespread, less deadly bacteria such as *Staphylococcus aureus* may grow in contaminated meat products. Subsequently the bacteria may be destroyed in cooking; however, the toxin that the bacteria secretes survives cooking and may cause food illness in as little as one hour after consumption of contaminated meat.

Additional bacteria and agents are problematic as well. Poultry without sufficient cooking may contain the live, infection-causing *Salmonella* bacteria, which is the most common cause of foodborne infections in the United States. Most *Salmonella* are destroyed at 161 °F (72 °C) for 16 s, or 143 °F (62 °C) for 30 min. Another Clostridium, *Clostridium perfringens*, is found in meats especially those that were allowed to cool slowly following cooking. Undercooked pork may contain the parasite *Trichinella spiralis*, which is killed at temperatures of 155 °F (68 °C).

Ground beef, the *combined* meat from *many* cattle, is more likely to have contamination with *E. coli 0157:H7* than is a *single* cut of meat (such as steak) coming from a *single* animal. Bacteria may spread during processing and handling, and then cooking temperatures of 155 °F (68 °C) are necessary to destroy any E. coli that might be in the meat. A major challenge to the safety of ready-to-eat (r.t.e.) products includes *Listeria monocytogenes.* This may grow under refrigeration, yet is destroyed by thermal processing.

The USDA gave approval for *steam pasteurization* as an antimicrobial treatment of beef carcasses (see below). This treatment reduces the risk of *E. coli 0157:H7* by exposing the entire surface of the carcass to steam that kills the bacteria. Meat processors must avoid subsequent recontamination of the product, and the consumer must handle the meat with care. The use of steam pasteurization for *pork and poultry* is subject to further research.

The American Meat Institute Foundation (AMIF) speaks for the industry in saying that it *sanitizes* fresh meat, as well as the r.t.e meat products, including hams, and hot dogs. This sanitization is achieved either by steam pasteurization, which (1) exposes the carcass to a steam filled cabinet, or (2) uses of a handheld device in steam vacuuming, whereby steam is sprayed directly onto carcass spots where contamination is suspected. With the added food protection provided by the uses of multiple intervention strategies, sprays and organic acid (lactic and acetic acids) and hot water treatment are also used widely (Mermelstein 2000).

Further FDA-approved treatments include high-intensity pulsed-light treatment for the control of microorganisms on the surface of food (61 FR42381-42382). Irradiation is a process often used to destroy the pathogens that are present in meat and extend refrigerator shelf life. The FDA has approved *radiation* of fresh, frozen meats.

A US patent has been awarded to a company that uses electricity as the energy source to pasteurize processed and packaged foods, including r.t.e. meats such as hot dogs and luncheon meats (Food Eng 2000). The use of *ozone* to disinfect poultry processing water is reviewed on a case-by-case basis.

CULINARY ALERT! Meat must be kept safe in the defrosting process. The FDA advises thawing below temperatures of 45 °F (7 °C), under cold, running water, or by microwave, if immediately cooked. Slow thawing, with intact wrappers, is the defrosting method that allows the least moisture loss. The USDA recommends refreezing only in the case of properly thawed and cooked meats.

The FDA-recommended cooking temperatures to control bacterial growth and prevent foodborne illness are listed in Table 9.1 (check local jurisdiction).

Concern exists over *Bovine Spongiform Encephalopathy* (BSE) or "*Mad Cow Disease*." Further understanding of the disease and vigilance is needed to protect the food supply.

At this time, "meat recalls are down and safety is up" according to the data from the CDC (Decker 2012). The usual suspects include *E. coli 0157:H7* in beef and dairy cattle, *Salmonella* and *Campylobacter* in swine and poultry, *Listeria monocytogenes* in r-t-e foods. As well, *staphylococcus aureus* is a common contaminant. "... ground products may have pathogens spread throughout the product", so ordering a 'rare" burger may pose risk.

Handling the foods properly is also a priority, as it is reported "ultimately, the riskiest meat product is the one that's not handled properly, but is abusively stored, handled or undercooked, allowing pathogens to be transmitted to the consumer."(Decker 2012)

Reported by the Meat Science Department at a leading university,

> While not all lessons are easily learned, and some might come at great costs, the meat and poultry industries have actively updated their practices with new information as it has become available. (Decker 2012)

Also of current concern is the unique disease known as Alpha-Gal or *galactose alpha 1,3-galactose*. In this disease, an *IgE antibody* binds onto the *carbohydrate* present in mammal meat *galactose-alpha 1,3-galactose*. Affected persons eating this carbohydrate (not a protein) in meat show a *delayed anaphylaxis* response of several hours, instead of the typical minutes.

Mammalian products including beef, pork, lamb, rabbit, goat, or deer meat cause this disease. Alpha-Gal is *not* found in the nonmammalian poultry or fish.

It may be noted that previous tick bites in an individual trigger this reaction (tiny ticks are often known as chiggers).

Conclusion

Meat is the edible portion of animals used for food. Beef, pork, lamb, and veal are included in the definition of meat, and other animal products such as poultry and fish are commonly considered to be "meats". The amount and type of meat consumption varies throughout the world. Meat is primarily a muscle tissue and also contains connective tissue with a greater variance in the amount of adipose tissue held inside. Water is present to a greater degree in lean meats and young animals. The protein is a complete protein and contains all the essential amino acids.

Cuts of meat include primal or wholesale, subprimal, and retail cuts, with the latter being more familiar to consumers, as it is what they may purchase at their grocery market. The inherent tenderness of a particular cut depends on such factors as location on the carcass, postmortem changes in the muscle, including the stage of rigor mortis, aging, and the method of cooking. Meat color such as red or white, or "dark meat" is dependent on myoglobin and hemoglobin pigments. Changes in the color of meat may result from exposure to oxygen, acidity, and light.

Meat is subject to inspections and grading in order to provide the consumer with safe, more consistent, and reliable meat products. Meat is a potentially hazardous food and adherence to specific temperatures (cold and hot) is necessary for the prevention of growth and the destruction of harmful microorganisms.

Cooking meat causes the uncoiling or denaturation of peptide protein chains to occur. Tender cuts of meat remain tender when cooked by dry heat for a short time at high temperatures. Overcooking tender cuts of meat produces tough, dry meat, because water is released during denaturation. Less tender cuts of meat become increasingly tender as collagen solubilizes during lengthy exposure to moist heat cooking.

Beef, veal, pork, and lamb may be altered by various processing methods. This includes restructuring, and artificial tenderizing. Ham, corned beef, and bacon are examples of cured meat. Beef, ham, and turkey may be smoked to impart flavor and offer microbial control by dehydration. An alteration to meat occurs as meat is artificially tenderized and includes mechanical, electrical, and enzymatic treatment.

Poultry makes a significant contribution to the US diet and is classified according to age and condition of the bird. Many processed poultry products, including ground turkey, lunchmeats, and formed entrees, are available for use by consumers. Edible fish and shellfish including restructured fish such as surimi provide high-quality protein food to the diet.

Various legumes are consumed. They are incomplete proteins and when eaten in combination according to amino acid profiles, they form a complete protein and function as meat alternatives.

Make informed choices of your protein selection!

Notes

CULINARY ALERT!

Glossary

Actin The protein of muscle that is contained in the thin myofilaments and is active in muscle contraction.

Actomyosin The compound of actin and myosin that forms in muscle contraction.

Adipose tissue Fatty tissue; energy storage area in an animal.

Aging Process in which muscles become more tender due to protein breakdown.

Collagen Connective tissue protein; the largest component that gives strength to connective tissue; is solubilized to gelatin with cooking.

Connective tissue The component of animal tissue that extends beyond the muscle fibers to form tendons which attach the muscle to bones; it connects bone to bone; endomysium, perimysium, and epimysium connective tissue surrounds muscle fibers, muscle bundles, and whole muscles, respectively.

Cured meat Contains nitrite to form the pink color and control the growth of *Clostridium botulinum.*

Dry heat Method of cooking tender cuts of meat, including broiling, frying, pan-frying, and roasting.

Elastin Connective tissue protein; the yellow component of connective tissue that holds bone and cartilage together.

Endomysium Connective tissue layer that surrounds individual muscle fibers.

Epimysium Connective tissue layer that surrounds an entire muscle.

Gelatin Formed from the tenderization of collagen, used for edible gels in the human diet.

Grain Primary bundle containing 20–40 muscle fibrils.

Halal "Proper and permitted" food under jurisdiction of trained Muslim inspection.

Kosher "Fit and proper" or "properly prepared" food under jurisdiction of the Jewish faith; following the Mosaic or Talmudic Law.

Marbled Intermuscular and intramuscular fatty tissue distributed in meat.

Moist heat Method of cooking less tender cuts of meat, including braising, pressure-cooking, simmering, or stewing.

Muscle tissue The lean tissue of meat.

Myofibril The contractile actin and myosin elements of a muscle cell.

Myosin Protein of a muscle contained in the thick myofilaments that reacts with actin to form actomyosin.

Perimysium The connective tissue layer that surrounds muscle bundles.

Primal cut Wholesale cut of meat; it contains the subprimal and retail cuts.

Retail cut Cuts of meat available in the retail market; cut from primal cuts.

Reticulin Minor connective tissue found in younger animals; it may be the precursor of collagen or elastin.

Rigor mortis Postmortem state 6–24 h after death in which muscles stiffen and become less extensible; onset of rigor mortis correlates with depletion of ATP in the slaughtered animal.

Sarcomere Repeating unit of the muscle myofibrils.

Sarcoplasmic protein The hemoglobin and myoglobin pigments, and enzymes in the cytoplasm of a muscle fiber.

Smoked meat Meat that has been treated to impart flavor by exposure to aromatic smoke of hardwood; smoking preserves by dehydrating, thus offering microbial control.

Stromal protein Proteins including collagen, elastin, and reticulin of the connective tissue and supporting framework of an animal organ.

Subprimal cut Division of a primal cut.

Wholesome Inspection does not indicate the presence of illness.

Z-lines Boundaries of the sarcomere; holds thin filaments in place in the myofibril.

References

Decker KJ (2013) A natural approach to fortification. Food Product Design (Jan/Feb):66–73

My own meals. Food Technol. 2000;54(7):60–62

Eliasi J, Dwyer JT (2002) Kosher and Halal: Religious observances affecting dietary intakes. J Am Diet Assoc 102:911–913
Peregrin T (2002a) Limiting the use of antibiotics in livestock: Helping your patients understand the science behind this issue. J Am Diet Assoc 74(6):768
Morris CE (2000) Bigger buck for the bang. Food Eng 72 (1):25–26
Foster RJ (2012) Bean there, done that. Food Product Design (Sept):14–15
Peregrin T (2002b) Mycoprotein: Is America ready for a meat substitute from fungus? J Am Diet Assoc 102:628
Berry D (2013) Lower-sodium processed meats—Is it possible? Food Product Design (Jan/Feb):48–52
Mermelstein NH (2000) Sanitizing meat. Food Technol 55(3):64–65
Portable pasteurization on the way. Food Eng 2000;72 (July/Aug):18
Decker KJ (2012) Lessons learned: A new era for meat and poultry safety. Food Product Design (Sept):18–28

Bibliography

Academy of Nutrition and Dietetics (formerly The American Dietetic Association, Chicago, IL)
Academy of Nutrition and Dietetics. Lauren Innocenzi—online content manager
American Heart Association, Dallas, TX
American Meat Institute (AMI), Washington DC
Center for Science in the Public Interest (CSPI)
Centers for Disease Control and Prevention (CDC)
Model FDA Food Code
Montana State University
National Academy of Sciences (NAS)
National Cattlemen's Beef Association (NCBA)—a merger of the National Livestock and Meat Board, and National Cattlemen's Association, Chicago, IL
National Cholesterol Education Program (NCEP)
National Heart, Lung and Blood Institute
National Restaurant Association (1992) The Educational Foundation. Applied Foodservice Sanitation, 4th ed. New York, NY: Wiley
The American Meat Institute Foundation (AMIF)
The University of Georgia, Cooperative Extension Service
The University of Minnesota Extension
TX A&M University. Meat Science Dept
Uniform Retail Meat Identity Standards (URMIS)
USDA Choosemyplate
USDA's Meat and Poultry Hotline (1-800-535-4555), Food Safety and Inspection Service, Washington, DC. For more information on food irradiation and other food safety issues, contact USDA's Meat and Poultry Hotline at 1-888-MPHotline (1-888-674-6854) or visit www.fsis.usda.gov
Vegetarian nutrition—http://fnic.nal.usda.gov/lifecycle-nutrition/vegetarian-nutrition Food and Nutrition Information Center (FNIC)
National Agricultural Library (NAL). http://www.nal.usda.gov/fnic/pubs/bibs/gen/vegetarian.pdf

Eggs and Egg Products 10

Introduction

The eggs of various birds are consumed throughout the world; however, the discussion that follows in this chapter is regarding *hen* eggs. Eggs are a natural biological structure with shells offering protection for developing chick embryos. They have numerous functions in food systems and must be protected against becoming or offering contamination. Eggs provide nutritive value and culinary variety to the diet, while being an economical source of food. Today, we see a reversal of dietary limitations, and healthy persons can enjoy eggs as long as they form part of a healthy, balanced diet.

Eggs are considered by the World Health Organization (WHO) to be the reference protein worldwide, to which all other protein is compared. A vegetarian diet that includes eggs is an ovo-vegetarian diet.

The quality and freshness of eggs is important to regulatory agencies, processors, and consumers, and is determined by a number of factors. The age, temperature, humidity, and handling of eggs determine freshness. Egg safety is significant.

Physical Structure and Composition of Eggs

The Whole Egg

An average hen egg weighs about 2 ounces (57 g), which includes the weight of the yolk, white, and shell. Each component differs in composition as shown in Tables 10.1 and 10.2. See structure of a hen's egg, Fig. 10.1 (California Egg Commission, Upland, CA).

CULINARY ALERT! Egg protein includes the enzyme *alpha-amylase*. This enzyme must be inactivated by heat in order to have desirable cooked egg mixtures. Undercooked egg mixtures may not show a deleterious effect until after the egg has been refrigerated.

The Yolk

An egg yolk comprises approximately 31 % of the weight of an egg, *all* of the egg's cholesterol, and almost all of the fat. Generally, it has a *higher* nutrient density than the white, containing all of the vitamins known, except vitamin C.

V.A. Vaclavik and E.W. Christian, *Essentials of Food Science, 4th Edition*, Food Science Text Series,
DOI 10.1007/978-1-4614-9138-5_10, © Springer Science+Business Media New York 2014

Table 10.1 Chemical composition of the hen's egg by percentage

Component	%	Water	Protein	Fat	Ash
Whole egg	100	65.5	11.8	11.0	11.7
Egg white	58	88.0	11.0	0.2	0.8
Egg yolk	31	48.0	17.5	32.5	2.0
Shell	11				

Source: USDA

Table 10.2 Protein and fat content of egg components in grams

Component	Protein	Fat
Whole egg	6.5	5.8
Egg white	3.6	—
Egg yolk	2.7	5.2

Source: USDA

Additionally, yolks supply flavor and mouthfeel that consumers find acceptable; they have many culinary uses.

A cluster of developing yolks, each within its own sac, is present in the hen ovary.

Egg yolks contain all three *lipids*—triglycerides—fats and oils, phospholipids, and sterols in large spheres, granules, and micelles. The primary phospholipid is phosphatidyl choline, or *lecithin*; the most well known sterol is *cholesterol* found in the yolk.

Protein in the yolk represents 40 % of the eggs protein. Primarily vitellin is present in a lipoprotein complex as lipovitellin and lipovitellinin. The phosphorus-containing phosvitin and sulfur-containing livetin are also present in yolks. Scientists measure dietary protein quality by (1) its amino acid composition—quality and quantity, and (2) its digestibility—how well the human body absorbs and uses the ingested protein. Eggs are often used as the "gold standard" for measuring the protein quality of other foods (see PDCAAS).

The yolk *pigments*—mainly xanthophylls, also carotene and lycopene—come from animal feed such as the green plants and yellow corn that the hen eats. If yolks have a higher carotenoid content, they are *darker* (although not necessarily of vitamin A potential); however, chickens producing eggs with *pale* yolks may be fed supplements that darken the yolk. Concentric rings of slightly different colors appear in the yolk, beginning in the center with a very small white spot. Green color of boiled eggs is discussed later.

There is a higher concentration of *solids* in the yolk than in the white, and thus water movement into the yolk occurs as the egg ages. This water movement causes the egg yolks enlarge and become less viscous.

Surrounding the yolk is a colorless sac, the **vitelline membrane** (Fig. 10.1). It is continuous with the opaque-colored chalazae (kah-lay-za) cord structure. The chalazae is a ropelike cord that attaches to the yolk vitelline membrane yet is actually found in the albumen, or white. It holds the yolk in place at the center of the egg, preventing it from the damaging effects of hitting the shell (similar to a bungi jumping cord!).

The White

The egg white, also known as the **albumen** (Fig. 10.1), comprises approximately 58 % of the weight of an egg. As with the yolk, the white too consists of concentric layers. There are four layers - two thick and two thin whites separated by inner and outer thin whites. In lower grade or older eggs, the thick albumen becomes *indistinguishable* from the thin whites. The chalazae is located within these layers of the albumen and is continuous with the vitelline membrane that surrounds the yolk. A fresh egg has a more prominent chalazae than older eggs.

Eggs contain a high biological value protein, which is a *complete* protein, with all of the essential amino acids in a well-balanced proportion. Over half of the protein in whites is *ovalbumin*, although conalbumin, ovomucid, and globulins (including lysozyme, which is able to lyse some bacteria) contribute lesser percentages of protein in the egg whites. Whites provide *more* protein than the yolk and are often cooked and eaten alone, or incorporated into a recipe. The addition of egg whites in place of an entire egg adds protein while limiting fat and cholesterol. Avidin is another egg white protein. If consumed *raw*,

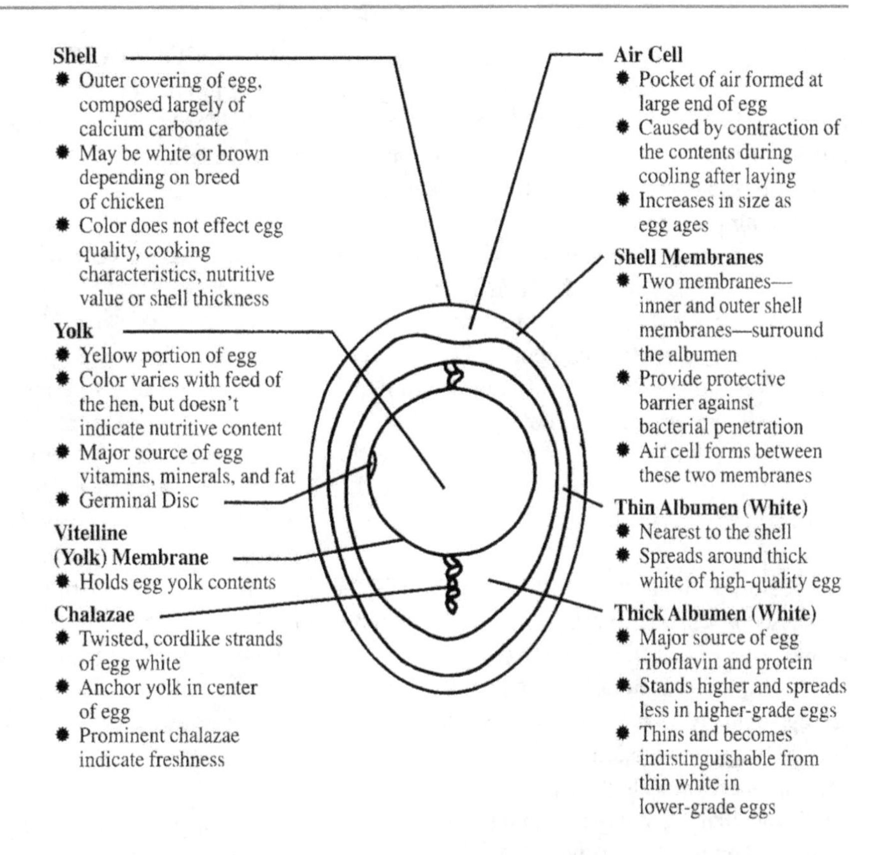

Fig. 10.1 Structure of a hen's egg (*Source*: California Egg Commission

the avidin binds with the vitamin biotin and renders biotin ineffective when consumed. Approximately 60 % of egg protein is located in the egg white.

Egg white proteins attach themselves to the yolk as it descends down the oviduct of the hen. Other constituents of the white are a negligible amount of fat; the vitamins riboflavin (which imparts a greenish tint to the white), niacin, biotin; and minerals including magnesium and potassium.

The Shell

The shell contributes the remaining 11 % weight of the whole egg. The dry shell contains the following:

- 94 % Calcium carbonate
- 1 % Magnesium carbonate
- 1 % Calcium phosphate
- 4 % Organic matrix made primarily of protein

Layers of the shell consist of a *mammillary* or inner layer, a *spongy* layer, and the outer *cuticle* (which may erroneously be referred to as "bloom"). The cuticle blocks the pores and protects the egg against outside contamination entering the egg.

Thousands of pores run throughout these layers of the shell, with a greater number at the large end. A shell is *naturally* porous for a potentially developing chick inside. As a result of the pores, CO_2 and moisture losses occur and O_2 enters the shell. The shell also functions as a barrier against harmful bacteria and mold entry, as a protein layer of keratin partially seals the shell pores.

"Sweating" or moisture condensation on the shell may produce stains. The presence of animal droppings may also stain. However, simply washing is *not* recommended as it may remove the shell's outer cuticle lining or open its pores resulting in a diminished shelf life. Once the outside protection is violated, microorganisms from the outside can travel to the inside contents and contaminate the egg.

Two thin shell *membranes* (Fig. 10.1) are inside of the shell, *one* of which is attached to the shell, and the *other* is not attached, but rather, moves with the egg contents. The air cell (discussed later) develops as the two membranes separate at the large end of the egg.

Color

The color of both the *shell* and *yolk* will be addressed in this section. *Egg shell color* depends on the *breed* of hen and has *no* known effect on egg flavor or quality, including the nutritive quality of the egg contents. *White Leghorn* hens are the chief breed for egg production in the United States and they produce *white* shells. Upon a closer look, it is significant that this *White* Leghorn breed of hen has *white* ears under their feathers.

Brown eggs (brown-colored egg shells) are popular in some regions of the United States, and with some individuals. The eggs are from slightly larger birds (requiring more feed), and they are not as prevalent as white shell eggs; therefore, for those reasons, brown eggs are usually more expensive than white. Brown eggs are produced from a *different* breed of hen than white eggs—notably hens with *reddish-brown* ears, such as *Rhode Island Red hens, Plymouth Rock hens, and New Hampshire breeds.*

Brown eggs are more difficult to classify by candling as to interior quality than are white eggs (United States Department of Agriculture (USDA)). In addition to the white and brown eggs, some egg shells are bluish or greenish. (Yes, the ears of the chicken are of that same color tinge!)

The yolk color depends on the *feed* given to the hen. As mentioned earlier, yolks may be a *deep* yellow pigment due to carotene, xanthophyll, or lycopene in the feed (not necessarily of vitamin A potential), or they may be *pale* yolks.

CULINARY ALERT! Color is not an indication of quality or nutritive value.

- Shell color—due to breed
- Yolk color—due to feed

Changes Due to Aging

Changes to the egg that occur with age are numerous. For example, contents inside the shell *shrink* and the air cell enlarges due to water loss (Jordan et al. Bulletin #612). The yolk flattens as the vitelline membrane thins, and the surrounding thick white becomes thinner, no longer holding the yolk centered in the egg. Also, the thick white thins as sulfide bonds break, and it loses CO_2 with age. Subsequently the pH rises to a more alkaline level—from 7.6 to 9.6, which allows bacterial growth. Along with these changes, another alteration with age is that the chalazae cord appears less prominent.

Abnormalities of an Egg Structure and Composition

Abnormalities in the structure and composition of eggs may be detected with or without candling (see "Candling" section). Consumers with first-hand experience may be familiar with some of these abnormalities. The USDA cites examples:

- Double-yolked egg—results when two yolks are released from the ovary about the same time or when one yolk is lost into the body cavity and then picked up when the ovary releases the next day's yolk.
- Yolkless eggs—usually formed around a bit of the tissue that is sloughed off the ovary or oviduct. This tissue stimulates the secreting glands of the oviduct and a yolkless egg results.
- Egg within an egg—one day's egg is reversed in direction by the wall of the

oviduct and is added to the next day's egg. A shell is formed around both.

- Blood spots—rupture of one or more small blood vessels in the yolk follicle at the time of ovulation, although chemically and nutritionally they are fit to eat.
- Meat spots—either blood spots that have changed in color due to chemical action, or tissue sloughed off from the reproductive organs of the hen.
- Soft-shelled eggs—generally occur when an egg is prematurely laid and insufficient time in the uterus prevents the deposit of the shell (e.g., minerals).
- Thin-shelled eggs—may be caused by mineral deficiencies, heredity, or disease.
- Glassy- and chalky-shelled eggs—caused by malfunctions of the uterus of the laying bird. Glassy eggs are less porous and will not hatch but may retain their quality.
- Off-colored yolks—due to substances in feed that cause off-color.
- Off-flavored eggs—may be due to certain feed flavors, such as fish oil or garlic. Eggs stored near some fruits and vegetables or chemicals readily absorb odors from these products.

Egg Function

The function of eggs is important to the processing facilities, retail foodservice operations, and the consumer alike, who depend on eggs for many uses in food preparation. Due to the any number of functions of an egg, a recipe formulation without eggs may not exhibit the same qualities as one that contains eggs. "Eggs supply aeration and provide structure resulting in moist, flavorful and tender baked foods. And eggs give you a clean ingredient label naturally" (American Egg Board (AEB), Park Ridge, IL).

Eggs are multifunctional products. Perhaps the contents of eggs are not ingested! Or in various ethnic holiday celebrations—the egg shell may be filled with confetti.

CULINARY ALERT! Some of the functions of eggs are listed in Table 10.3.

A high-quality egg that is fit for the consumer is one without blemishes and with a shell that is intact and clean.

Inspections and Grading for Egg Quality

Eggs are subject to inspections and are graded for quality. The USDA grades eggs on a fee-for-service basis in order to assign grades. Grading involves an evaluation of the exterior shell, its shape, texture, soundness (it should not be broken), and cleanliness, as well as the interior white and yolk, and air-cell size. Lesser grades and older eggs may be used successfully in other applications than high-grade, fresh eggs.

The 1970 Federal Egg Products Inspection Act provides the assurance that egg products are *wholesome and unadulterated* and that plants processing egg products are *continuously inspected*. Grading though is voluntary, although most eggs on the retail market are graded under federal inspections (USDA), according to established standards.

Candling

Candling is a technique that allows a view of the shell and *inside* of eggs without breaking the shell—double yolks and so forth may be seen. *Candlelight* was once used for inspecting the interior of eggs, where egg contents could be seen when held up to a candle while being rapidly rotated—thus the name *candling*. Today, commercial eggs may be scanned in mass, with bright

Table 10.3 Some of the functions of eggs in food systems

• ***Binder***
Eggs are viscous and they coagulate (to a solid or semisolid state); therefore, they bind ingredients such as those in meatloaf or croquettes, and they bind breading
• ***Clarifying agent***
Raw egg whites coagulate around foreign particles in a hot liquid. For example, when added to liquid, eggs whites seize loose coffee grounds in a coffee pot, and they clarify broth and soups, bringing the stray material to the surface for subsequent removal
• ***Emulsifier***
Egg yolks contain phospholipid emulsifiers, including lecithin. Emulsifiers allow two ordinarily immiscible liquids, such as oil and water to mix in the preparation of mayonnaise
• ***Foaming, leavening agent, aeration***
Egg whites increase 6–8 times in volume when beaten to a ***foam***. As the egg white foam is heated, the protein coagulates around air cells, maintaining a stable foam structure. Egg white foams leaven angel food cake and are created for meringues and desserts
• ***Gel***
A two-phase system of liquids in solids forms as eggs coagulate, forming a gel in custards
• ***Thickening agent***
Eggs coagulate and thicken mixtures such as custards and hollandaise sauce
• ***Other***: color, flavor, nutritive value, surface drying and crisping, etc. Eggs serve numerous other roles in foods
Egg yolk carotenoids add yellowish color to baked products, or yolks may be spread on dough to brown, dry, glaze, and impart a crusty sheen
Fat provides flavor, *inhibits crystal formation* in sugars, and inhibits staling
Eggs provide *nutritional value* in cooked or baked food mixtures

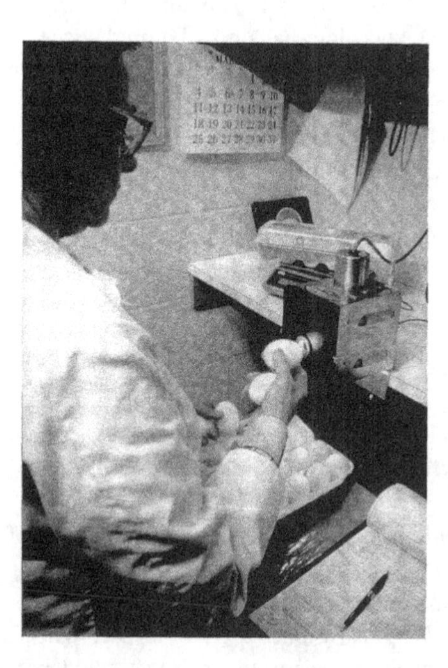

Fig. 10.2 Candling eggs by hand (*Source*: USDA)

lights under trays of eggs. The USDA bases grades on candling quality, evaluated either by hand (Fig. 10.2) or by mass scanning (Fig. 10.3).

Candling may be completed *either* at the farm or at the egg distributor before eggs are sold to the consumer. *External* observation of the shape and cleanliness of the shell may occur prior to or subsequent to candling. A candler will also form occasional comparisons of the broken-out, internal appearance evaluation with candled appearance.

CULINARY ALERT! Blood spots may be undesirable to some consumers; however, they pose no health hazard.

Letter Grades

Letter grades are issued voluntarily. Letter grades are based on candled quality and may appear as shields on the egg cartons. Grade shields on the carton indicate that the eggs were graded for quality and checked for size under the supervision of a trained packer. Packers who do *not* choose to use the federal USDA grading service are monitored by *state* agencies, and may *not* use the federal USDA grade shield.

Fig. 10.3 Candling by mass screening (*Source*: USDA)

Fig. 10.4 USDA emblem certifying quality (*Source*: USDA)

The USDA grade shields are shown in Fig. 10.4. The USDA assigns a grade of "AA" to the highest quality egg. Even this high quality may quickly diminish if eggs are exposed to improper storage conditions (USDA).

Referencing a recent *Wall Street Journal* article, the USDA shield of approval is no guarantee of safety (http://online.wsj.com/article/SB10001424052748704791004575466014072143010.html).

Occasional micrometer measurements of thick albumen egg height may also be carried out in a grading office where samples are tested (see Figs. 10.5 and 10.6).

> In the grading process, eggs are examined for both interior and exterior quality before they're sorted according to weight (size). Grade quality and weight (size) are not related to one another. Eggs of any quality grade may differ in weight (size). In descending order of quality, grades are designated AA, A and B. (American Egg Board (AEB), Park Ridge, IL)

See more on egg grades

Air Cell

The ***air cell***, also known as the *air sac* or *air pocket*, is the empty space formed at the large end of the egg. By definition, it holds oxygen. Initially, there is either *no* air cell or a small one. Then it becomes large and apparent to the eye when the warm egg cools, the egg contents shrink, and the *inner* membrane pulls away from the *outer* membrane. The air cell increases in size with *age, cooling, and moisture loss.* It could result in microbial spoilage due to the plentiful oxygen it supplies to microorganisms.

Oftentimes a large air cell is noted in *older* eggs as they are shelled for consumption. As mentioned above, due to the fact that oxygen is available, microbial spoilage may follow development of a large air cell, as O_2 migrates to the yolk. It is recommended, therefore, that eggs should be packed with the large, blunt end of the egg *up*. If packed and stored in this manner, air movement from the cell to the yolk is minimized.

According to the American Egg Board, "Although the air cell usually forms in the large end of the egg, it occasionally moves freely toward the uppermost point of the egg as the egg is rotated. It is then called a free or floating air cell. If the main air cell ruptures, resulting in one or more small separate air bubbles floating beneath the main air cell, it is known as a bubbly air cell." (American Egg Board (AEB), Park Ridge, IL)

An acceptable air-cell size for the different grades is as follows: 1/8 in. for Grade AA, 3/16 in. for Grade A, and no limit in air-cell size for Grade B quality eggs.

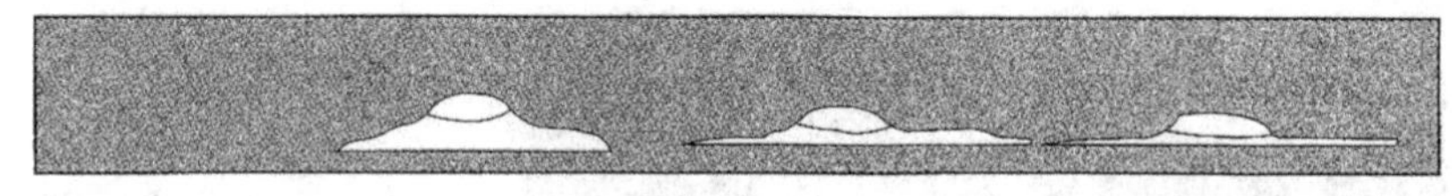

	GRADE AA	GRADE A	GRADE B
Break Out Appearance	Covers a small area.	Covers a moderate area.	Covers a wide area.
Albumen Appearance	White is thick and stands high; chalaza prominent.	White is reasonably thick, stands fairly high, chalaza prominent.	Small amount of thick white; chalaza small or absent. Appears weak and watery.
Yolk Appearance	Yolk is firm, round, and high.	Yolk is firm and stands fairly high.	Yolk is somewhat flattened and enlarged.
Shell Appearance	Approximates usual shape; generally clean,* unbroken; ridges/rough spots that do not affect the shell strength permitted.		Abnormal shape; some slight stained areas permitted; unbroken; pronounced ridges/ thin spots permitted.
Usage	Ideal for any use, but are especially desirable for poaching, frying, and cooking in shell.		Good for scrambling, baking, and use as an ingredient in other foods.

*An egg may be considered clean if it has only very small specks, stains or cage marks. Source: USDA

Fig. 10.5 Grades of eggs (*Source*: California Egg Commission)

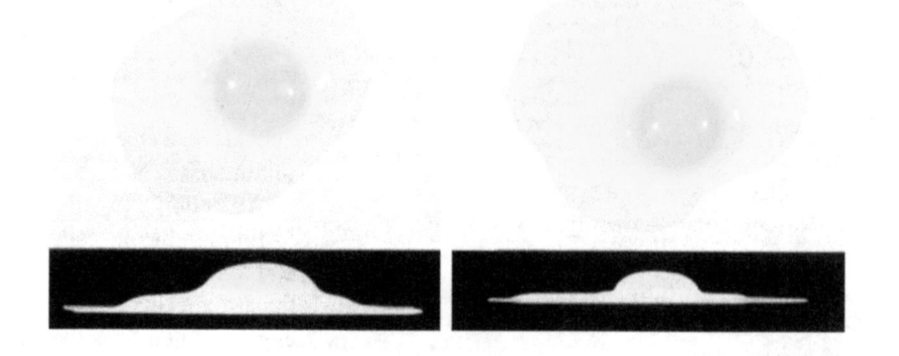

Fig. 10.6 Quality standards for grades (*Source*: USDA)

CULINARY ALERT! As a consequence of formation of large air cells, older eggs will float if placed in a bowl of water. Floating is an indication of less desirable eggs. The consumer may be familiar with the "floating" test.

Egg Size

Egg-size comparisons are shown in Fig. 10.7. The USDA does *not* include an evaluation of egg *size* as a part of egg *quality*. USDA classifications according to size and weight (minimum weight per dozen) are as follows:

• Jumbo	30 ounces (4 per cup)
• Extra large	27 ounces
• Large	24 ounces (5 per cup)
• Medium	21 ounces
• Small	18 ounces
• Pee wee	15 ounces

There is a difference of 3 ounces per dozen between each size class. Knowing the ounces in the various sizes assists with calculating pricing

Fig. 10.7 Egg sizes (*Source*: California Egg Commission)

a

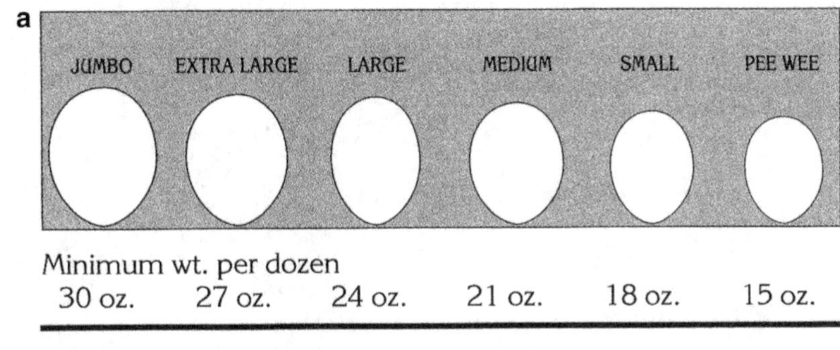

Minimum wt. per 30 dozen case

56 lbs. 50 1/2 lbs. 45 lbs. 39 1/2 lbs. 34 lbs. 28 lbs.

b

JUMBO	X-LARGE	LARGE	MEDIUM	SMALL
1	1	1	1	1
2	2	2	2	3
5	5	6	7	8
9	10	12	13	15
18	21	24	27	28
37	44	50	56	62

as the best value may be computed by comparing price *per ounce*. Of course, pricing of *individual* eggs may still be costed out also. Although undesignated in many recipes for the consumer at home, *large* eggs are the standard size egg used in published recipes.

The primary factor in determining egg size is the *age* of the hen; an *older* hen produces a larger egg. Secondary factors influencing egg size are the *breed* and *weight* of the hen. The quality of the feed, as well as henhouse overcrowding and stress, all impact size, perhaps negatively.

USDA services available to volume purchases of eggs appear in Fig. 10.8.

Processing/Preservation of Eggs

Processing or preservation treatments for eggs may occur both for food safety purposes and to keep the egg fresher, longer, therefore limiting negative quality changes. Eggs are laid at a hen's body temperature and require subsequent refrigeration. It is possible to hold an egg for 6 months in cold [29–32 °F (0 °C)] storage if the shell pores are closed. Fresh eggs have thick whites and thus do not run when cracked. They contain a very pronounced chalazae cord. Over time, eggs lose moisture and CO_2.

Shell eggs or egg products may be preserved in the subsequent manners.

Mineral Oil

Mineral oil application is one means of preserving eggs. When oil is applied, it partially closes the shell pores and allows less microorganism permeability. It also allows an egg to hold more moisture within, retain its CO_2, and be protected against a pH rise in storage. Shell eggs may be sprayed or dipped in mineral oil on the same day they are laid, and washing off by consumers is unnecessary. Mineral oil dips or sprays may cause a hard cooked egg to be more difficult to peel.

Pasteurization

Pasteurization is a process required by the Food and Drug Administration (FDA) for all

Large		Small	Medium	X-Large	Jumbo
1	=	1	1	1	1
2	=	3	2	2	2
3	=	4	3	3	2
4	=	5	5	4	3

Fig. 10.8 USDA shell egg certification (*Source*: USDA)

commercial liquid, dry, or frozen egg products that are out of the shell. This treatment destroys microorganisms such as *Salmonella* bacteria that can travel from the digestive tract and droppings of birds into the egg, causing foodborne illness infection. The USDA requires a process of pasteurization that achieves a temperature of 140–143 °F (60–62 °C), held for 3-1/2 min or longer. This is less time than fluid milk and its typical 30 min pasteurization.

Pasteurization must allow maintenance of the functional properties of the egg. For example, following pasteurization, egg *whites* can still be whipped for use in a meringue although they need a longer time period to beat to a foam, and *yolks* or *whole eggs* remain functional when used as emulsifiers. Prior to pasteurization, *aluminum sulfate* may be added to egg whites in order to stabilize conalbumin protein that becomes unstable at a pH of 7.0.

Ultrapasteurization of liquid whole eggs combined with aseptic packaging creates a commercial product with numerous advantages over frozen or shell eggs. According to a market leader in refrigerated ultrapasteurized liquid whole eggs and scrambled egg mixture, the eggs have a shelf life of 10 weeks when stored between 33 and 40 °F (1–4 °C). The eggs are *Salmonella*-, *Listeria*-, and *E. coli*-negative. The eggs are not frozen, so that they are not subjected to freezer-to-refrigerator storage, which can result in a loss of functional properties.

Freezing

Freezing is a means of preservation. Since the eggs are broken open they must first be pasteurized prior to freezing. *Uncooked whites* retain their functional properties after freezing and thawing, whereas *cooked* whites exhibit ***syneresis*** (water leakage) upon thawing.

Whole eggs and *yolks* may *gel* and become gummy upon thawing, as a result of an aggregation of low-density lipoproteins in the yolk. Gumminess is controlled by sugar, corn syrup, or salt addition. Processors may add the enzyme papain in order to hydrolyze the protein. As water is bound to the enzyme, the defrosted product exhibits less gel formation.

CULINARY ALERT! A 10 % sugar solution (1 tablespoon of sugar per cup of eggs, household measure), a 5 % inclusion of corn syrup, or 3 % salt (1 tablespoon per cup of eggs) may be added to yolks before freezing as a control against aggregation. Choose the solution according to egg usage.

Dehydration

Egg dehydration is a simple process of preservation that began in the 1870s. Over the years dehydration has been much improved. It offers microbial control to egg products when water levels are reduced by techniques such as spray drying or drying on trays (producing a flaked, granular form). The dehydrated whole egg, white, yolk, or blend is then packaged in various sized packages or drums. Subsequently, it may be reconstituted and cooked, or added as an ingredient to packaged foods such as cake mixes or pasta.

Egg *whites* require the removal of glucose prior to dehydration in order to improve storage stability, because, otherwise glucose in the whites leads to unacceptable browning and flavor changes. The browning is a result of the Maillard reaction (nonenzymatic) of proteins and sugars in long or hot storage. Glucose may be removed by *lactobacillus* microbial fermentation or by enzymatic fermentation with commercial enzymes such as glucose oxidase or catalase.

Egg *yolks* undergo irreversible changes in their lipoprotein structure when dehydrated, losing some functional and desirable sensory characteristics. Dried eggs should be kept cold to meet food safety guidelines.

Storing Eggs

Storage of eggs requires cold temperatures as well as other significant conditions. For example, it is recommended that the consumer should store eggs on an *inside* shelf of the refrigerator, large end up, *not* on the door where the temperature is *warmer*. Whether it is 1 dozen eggs or flats of 30 dozen or more, eggs should be kept *in the carton* in which they were obtained, in order to prevent moisture loss and the absorption of odors and flavors from other refrigerated ingredients.

CULINARY ALERT! Hard cooked (boiled) eggs may be retained in a refrigerated unit for up to 1 week. Any break-out portions of egg may be safely stored under refrigeration in this manner: yolks in water, for 1–2 days, whites in a covered container for up to 4 days.

The USDA-graded eggs are washed, sanitized, oiled, graded, and packaged soon after they are laid, and it is usually a matter of days between the egg leaving the hen house and reaching the supermarket. Cold temperatures, high humidity, and proper handling are required in storage, and, when kept cold, eggs may be safely stored for 45 days past the pack date.

According to the American Egg Board, "Egg cartons from USDA-inspected plants must display a Julian date—the date the eggs were packed. Although not required, they may also carry an expiration date beyond which the eggs should not be sold. In USDA-inspected plants, this date cannot exceed 30 days after the pack date. It may be less through choice of the packer or quantity purchaser such as your local supermarket chain. Plants not under USDA inspection are governed by laws of their states." (American Egg Board (AEB), Park Ridge, IL)

Many eggs reach stores only a few days after the hen lays them. Egg cartons with the USDA grade shield on them must display the "pack date" (the day that the eggs were washed, graded, and placed in the carton). The number is a three-digit code that represents the consecutive day of the year (the "Julian Date") starting with January 1 as 001 and ending with December 31 as 365. When a "sell-by" date appears on a carton bearing the USDA grade shield, the code date may not exceed 45 days from the date of pack.—USDA

Denaturation and Coagulation: Definitions and Controls

Denaturation may be mild or extensive. It occurs when a protein molecule (helical shape) unfolds, changing its nature (thus the word denaturing). This is an *irreversible* change in the specific folding and shape that a protein assumes in space.

Denaturation of the protein in an egg may occur due to *heat*, *mechanical action* such as beating or whipping, or an *acidic pH*. Regardless of the cause, the helical chains with intramolecular bonds uncoil and align in a parallel fashion, forming intermolecular bonds, and the protein chains shrink.

In the *raw* state, eggs appear *translucent* because light is refracted and passed between individual proteins. As the egg denatures, the egg changes in appearance from translucent to *opaque* or *white*. Once *cooked*, light is no longer able to pass between the newly formed protein mass.

Coagulation represents the *further* process that occurs when denatured protein molecules form a solid mass. The liquid/fluid egg (which is a sol) is converted into a solid or semisolid state (which is a gel). Water escapes from the structure as unfolded helixes attach to each other.

This coagulation occurs over a wide temperature range and is influenced by factors previously mentioned such as heat, beating, pH, and also use of sugar and salt. Coagulation results in the precipitation of the protein and is usually a *desirable* characteristic.

Curdling may occur next. Beyond denaturation and coagulation, *undesirable curdling* of egg mixtures results in an egg mixture shrinking or becoming tough. Some factors involved in denaturation, subsequent coagulation, and possible curdling are as follows:

Heat

- Heat should be slow and mild. The egg *white* denatures, coagulates, and becomes solid at temperatures of 144–149 °F (62–65 °C). Egg *yolks* begin to coagulate at 149 °F (65 °C) and become solid at 158 °F (70 °C). *Whole eggs* coagulate at an intermediate temperature. In the preparation of an egg mixture such as an egg custard, the rate of heating and intensity of heat must be controlled. These heating characteristics are discussed below.

- **Rate and coagulation**:
  A *slow* rate of heating safely coagulates the egg mixture at a lower temperature than a *rapid* rate of heating. A slow rate provides the "margin of error" or *extra time* (for possibly interrupting cooking) between the coagulation temperature and undesirable, fast approaching curdling. A *rapid* rate of heating may quickly exceed the desired temperature and result in undesirable curdling.
- **Intensity and coagulation**:
  A *mild* heating intensity denatures and coagulates with desirable molecular associations. As opposed to this, *intense* heating applies too much heat, too quickly, and causes undesirable curdling with negative changes such as water loss and shrinkage (Chap. 8).

- **Water bath and coagulation**:
  Using a water bath controls *both* the rate and intensity of coagulation. It is therefore an advisable baking strategy for baking egg dishes, commercially and at home. The reason it works is that the egg product is placed in an external water medium that cannot exceed the boiling temperature of water.

Additional factors influencing the denaturation and coagulation of eggs include the following:

- **Surface changes**. Beating, and so forth, denatures the helical protein structure. This is readily seen in the white color (explained above) and increased volume of egg white foams used for the preparation of meringues.
- **Acid pH**. An acid pH coagulates egg protein. For example, adding acid to the water used for poaching eggs coagulates the egg white so that it remains small and compact. As well, acid in the cooking water offers control, by immediately coagulating undesirable strands of leakage escaping from cracks in eggs that are hard cooked.

CULINARY ALERT! Acidic cooking water may cause difficulty in peeling an older, more alkaline egg. Thus older eggs, which have become alkaline with age, may be cooked in salted water.

Effect of Added Ingredients on Denaturation and Coagulation

In addition to the aforementioned surface changes and acidic pH of the water, extra ingredients in an egg mixture may affect both the denaturation and coagulation processes.

Sugar. The addition of sugar exerts a protective effect on the egg by controlling the rate of denaturation and ultimate formation of intermolecular bonds. This is seen in the preparation of meringues. The foam in the meringue will not be as large if sugar is added early, *prior* to denaturation. For larger foams, sugar should be added late, *after* the egg white has denatured.

Sugar also raises the temperature required for coagulation. A custard prepared *with* sugar has a *higher* coagulation temperature than a similar egg–milk mixture *without* sugar but produces no change in the finished gel (Chap. 14).

Salt. When salt is added, it promotes denaturation, coagulation, and gelation. Salt may be a constituent of food, such as the *milk salts* in milk, or it may be *added* to a product formulation. Milk salts contribute to custard gelation, whereas the addition of water to eggs does *not* promote gelation.

Acid level. As the pH *decreases* and becomes more acidic, coagulation of the egg white occurs more readily. An older, more alkaline egg will result in *less* coagulation than a fresh, neutral pH egg. Vinegar may be added to the water of poached and hard cooked egg to aid in denaturation and coagulation *and* to prevent spreading of egg strands. Coagulation depends on which egg protein is involved, and its isoelectric point (pI)—the point at which a protein is least soluble and usually precipitates.

Other ingredients. Ingredients such as fat vary and therefore all specifics cannot be addressed (Fig. 10.9). Egg is often diluted by the addition of other substances in a food system. For example, the coagulation temperature is *elevated* if an egg mixture is made dilute by water or milk. If a mixture is diluted, a less firm finished product results.

Fig. 10.9 Pan-fried egg

Cooking/Baking Changes

Cooking typically produce noticeable egg changes, hopefully while keeping a tender, high-quality product. With the expected/continued rise in gluten-free baked products, eggs offer nutritional and functional property benefits for those foods. "Egg products contribute humectancy, that helps optimize moisture for better density and rise, and prevents dry, crumbly characteristics often associated with gluten-free formulations" (Foster 2013).

Several cooking methods include the following:

Pan Frying:

- Method: Eggs placed in a preheated pan coagulate the egg proteins.
- Heated pan: A preheated pan allows coagulation before the egg has an opportunity to spread. However, an overheated pan may overcoagulate the egg and produce a tough product.
- Use of fat: Pan Frying in a measurable amount of fat and basting the top of the egg with fat produce a tender egg, but may not be desirable in terms of the calorie and fat contribution that is offered.

CULINARY ALERT! Eggs may be pan-"fried" in liquids other than fat or oil, and the pan lid may remain in use to create steam that cooks the egg's upper surface.

Hard Cooked Eggs:

"Hard boiled" is another, yet less appropriate, term for these "hard cooked" eggs. (*boiling* the eggs is not desirable)

- Method: It is recommended that eggs be placed one layer deep in a covered saucepan of boiling water, and then simmered, not boiled, for 15–18 min for a hard cooked egg, or just 2–5 min for a soft "boiled" egg. More than one layer deep, or placing eggs in cold water at the start of cooking, may retard the "doneness" of hard cooked eggs.

  Alternative method: Place the eggs one layer deep in an uncovered saucepan of cold water. Heat water to boiling. Then, remove the pot from the stove burner and cover. Allow standing time of 9 min for medium eggs and 12 min for large eggs.
- Peeling: Eggs should be cooled *rapidly* to facilitate easier peeling. Fresh eggs may be difficult to peel, in part because an alkaline pH has not yet been achieved.
- Cracking: In order to prevent cracking from an expansion of air in the air cell, and the buildup of internal pressure, it may be recommended that the egg be punctured at the large end. However, this seemingly logical step has *not* been shown to prevent cracking of the shell. For prevention, the egg may be warmed slightly prior to cooking.
- Color: Green discoloration of hard cooked eggs occurs with long and high heat exposure. The green color is due to the formation of *ferrous sulfide* from *sulfur* in the egg white protein combining with *iron* from the yolk. "Greenish yolks can best be avoided by using the proper cooking time and temperature and by rapidly cooling the cooked eggs." (American Egg Board (AEB), Park Ridge, IL)

CULINARY ALERT! Hard *cooked* is the term of choice when referring to "hard *boiled* eggs." Eggs are more tender when they reach a simmering, not boiling temperature.

Custard

- Method: Custards (served plain or incorporated into cream desserts, flan, or quiche) are cooked with a *slow rate* of heating. This provides a margin of error that protects against a rapid temperature elevation from the point of coagulation to undesirable curdling where the protein structure shrinks and releases water. Custards cooked with the addition of a starch white sauce are able to withstand higher heat because starch exerts a protective effect on the denatured proteins.
- *Stirred* custard: Custards may be stirred or baked. Soft, *stirred* custard will cling to a stirring spoon, as it thickens. It remains pourable and does not form a gel. If overheated, or heated too quickly, the mixture curdles and separates into curds and whey. Therefore, the use of a double boiler is recommended in order to control temperature and the rate of cooking. As mentioned, starch may be added to the formulation in order to prevent curdling.
- *Baked* custard (see Fig. 10.10) reaches a *higher* temperature than stirred custard and gels. Baking in a *water bath* is recommended in order to control the rate and intensity of heat and prevent the mixture from burning. With the addition of starch in a recipe, this is not required. Cooking/holding for an extended period of time, even in a water bath, could cause syneresis.
- Texture: The texture of an egg custard is dependent on a number of factors, including the extent of egg coagulation and added ingredients. A well-coagulated custard is fine textured; a curdled custard is extremely porous, tough, and watery.

Fig. 10.10 Custard baked in water bath (*Source*: American Egg Board)

CULINARY ALERT! Milk salts and added sugar raise the coagulation temperature; custards prepared with starch (such as arrowroot, cornstarch, flour, and tapioca) control curdling.

CULINARY ALERT! Cooking with a double boiler or water bath in the oven often brings success to egg custards.

Scrambled eggs

- Method: Cook with short cooking, at medium high.
- Dilution: This may result in less solid coagulation.
- Discoloration: Negative coloring may appear in eggs as ferrous sulfide forms. Avoid direct heat when holding eggs. Water may be placed between the pan of eggs and the source of heat.

Egg White Foams and Meringues

Egg white foam is created as the liquid egg whites are *beaten or whipped* to incorporate air. The egg white volume expands with beating as the protein denatures and coagulates around the many newly formed air cells.

Beaten whites are used in numerous food applications, as meringues, or incorporated into a recipe to lighten the structure. The volume and stability of egg white foams is dependent on conditions such as the humidity in the air, temperature of the egg, and other added substances, as shown in Table 10.4.

CULINARY ALERT! Care should be taken to gently *fold*, not *stir*, the beaten egg white foam into the other recipe ingredients. After all, it was work to create the foam, and the air cells should not be roughly treated!

A variety of food products are created using egg white foams, including cakes, dessert shells, sweet or savory soufflés, and pies. A sweet egg white foam is known as a meringue and may be either soft or hard, the latter prepared with more sugar. Examples of sweet meringue confections include pies, cookies, and candies.

The preponderance of meringues require that egg whites be beaten to either the soft or stiff peak stage, and then immediately be added to the recipe. Processors use egg white foams to create special appearance and volume for their products.

CULINARY ALERT! Use *super-fine* sugar in order to create glossy meringues. Ordinary *granulated* sugar may be successfully utilized if it is processed in the food processor for 1 min, prior to use.

Possible unsuccessful egg white foams or meringues may result if the foam is *not immediately* incorporated into a formulation or if eggs are *overbeaten*. A brief explanation: when not incorporated immediately, the recipe may lose some of its characteristic elasticity, and, upon standing, become stiff and brittle. If overbeaten, the foam is not able to expand with heat since the eggs have now become inelastic.

A further error results from using *cold* temperature eggs. These cold eggs have a high surface tension and do not beat to as high a volume as *room temperature* eggs. It is recommended to allow eggs to reach room temperature for better whipping, although this practice carries with it the increased risk of salmonella growth.

CULINARY ALERT! Rather than setting eggs out to slowly warm to room temperature, and encouraging bacterial growth, egg whites may

Table 10.4 Some factors affecting the volume and stability of egg white foams

Temperature—The temperature of eggs influences beating ability. At room temperature, eggs have less surface tension and are more easily beaten than if they were cold. Yet, at warm temperatures, Salmonella may grow and cause illness in susceptible individuals
pH—Acid should be added in the whipping process *after* eggs reach the foamy stage and have large air cells. If acid substances such as cream of tartar are added to raw egg whites at the *beginning* of the beating process, there is less volume although greater stability due to intramolecular bond coagulation
Salt—Salt adds flavor. Its presence delays foam formation, and, if added early in the beating process, produces a drier foam with less volume and stability. Salt should be added to egg white foams at the foamy stage or later if flavor is needed
Sugar—The protective effect of sugar on eggs has been discussed
Early addition
The early addition of sugar causes less intermolecular bonding of the egg proteins than would occur in the absence of sugar. Therefore, the addition of sugar results in an egg foam that is stable, but has less volume. A fine-textured, more stable foam develops if finely ground sugar is added *early* in the beating process
Late addition
Sugar (2–4 tablespoons per egg white, respectively, for soft or hard meringues) should be added to foams gradually, at the soft peak or stiff peak stage of development, *after* large air cells have formed and denaturation has begun. On a damp day, the preparation area may contain a lot of humidity that is absorbed by the sugar, and this results in a softer meringue (hydroscopicity Chap. 14)
Fat—Traces of fat may remain in the equipment used for beating egg white foams, or it may originate from the egg yolk, or be introduced by another added ingredient in the product formulation. If fat enters the egg white, there will be substantially less foaming, and less volume. Fat interferes with the foaming that would occur if protein aligned itself around the air cell and coagulated
Liquid—The addition of liquid dilutes the egg white. A benefit is that added liquid, such as water, will increase volume and tenderness of foams, yet it results in a less stable, softer foam and an increased likelihood of syneresis. Dried egg white that has been reconstituted with a liquid requires longer beating time than fresh egg whites, due to some protein breakdown in the drying process
Starch—Starch assists in controlling coagulation in proteins; starch is of benefit to soft meringues. A starch should first be cooked and then incorporated into the meringue

be slightly warmed by placing the appropriate number of separated egg whites in a bowl over warm water. This allows the egg whites to warm up prior to successful whipping.

Leavening is diminished if *older* eggs are used for the creation of foams. While older eggs whip up more easily than fresh eggs, protein does not coagulate well around the air cells and there is a higher percentage of thin whites that create large, unstable foams.

Egg *yolks* contain fat that physically interferes with the alignment of protein around air cells. Therefore, the yolks should be *completely* separated from the whites, not allowing *stray* yolk to enter the white in separation. Separation works easier when eggs are cold. Although egg yolks cannot form foams, they may be beaten to become thicker. They may be used in other cooking applications.

It may be common practice to separate the yolk and white by passing the egg contents between the two broken halves of the shell. Repeatedly passing the egg contents from one shell to the other releases the white and retains the yolk. A warning though: The American Egg Board offers information about separating eggs. "Bacteria are so very tiny that, even after washing and sanitizing, it's possible that some bacteria may remain in the shell's pores. The shell might also become contaminated from other sources. When you break or separate eggs, it's best to avoid mixing the yolks and whites with the shells. Rather than broken shell halves or your hands, use an inexpensive egg separator or a funnel when you separate eggs to help prevent

a

b

Fig. 10.11 Unbeaten (*top*) and beaten egg whites (*bottom*) after addition of acid and sugar (*Source*: American Egg Board)

introducing bacteria. Also use a clean utensil to remove any bits of eggshell that fall into an egg mixture and avoid using eggshells to measure other foods." (American Egg Board (AEB), Park Ridge, IL)

Commercial egg *substitutes* may be successfully used in the preparation of foams since they consist primarily of whites and contain no fat. They are similar to shell egg whites, aside from imparting the yellowish color.

A further point to address in egg meringues is the use of copper bowls for creating meringues. Copper bowl usage for beating egg whites has been a recommendation over the years. However, it turns out that conalbumin protein from the egg white combines with traces of copper from the bowl, producing *copperconalbumin*. There is no noticeable effect in the *unbaked* foam; however, due to toxicity issues, the use of copper bowls is no longer recommended.

Hard meringues may be a key ingredient of some cookies or candy. Soft meringues are used most notably on pie (Fig. 10.11). The special problems that may arise with soft meringues are shrinking, weeping, and beading. A hot oven and cold pie filling may be responsible for these problems in the same meringue.

Weeping is the release of water from *under*coagulated (perhaps underbeaten or undercooked) egg white foam. A release of water at the interface of meringue and filling may form a water layer causing the meringue to slide off. This occurs if the meringue is placed on a *cold* filling.

Consequently, to prevent weeping, prepare the meringue first and then the already-prepared meringue may be placed on a *hot* filling and *immediately* baked. Both the filling and the oven should be hot. Another method used for control is the addition of ½ to 1 teaspoon of cornstarch to the sugar prior to beating it into the eggs.

Beading is apparent in *over*coagulated (overcooked) meringues. Beading appears as drops of amber-colored syrup on top of meringue. It may be the result of (1) adding too much sugar, or the insufficient incorporation of the sugar into the beaten egg whites. It may also be the result of (2) baking too long, at a low temperature. For control, a high temperature for a short time is needed.

CULINARY ALERT! Placing one layer of meringue at a time, after the previous layer adheres to the filling, is helpful in maintaining a meringue. As well, a fine layer of breadcrumbs may be sprinkled on the hot filling prior to topping with the meringue.

A brief look at the stages of denaturation when egg whites are beaten to foam appears in Table 10.5

CULINARY ALERT! Slicing a gummy, sticky, beading meringue surface is better with the use of a sharp serrated knife dipped in cold water prior to each slice.

Egg Products and Egg Substitutes

Egg products and egg substitutes on the market include pasteurized, processed, refrigerated liquid, frozen, and dried eggs that are available for commercial and retail users.

If in *liquid* form, eggs may be ultrapasteurized or aseptically packaged to extend shelf life. ***Egg substitutes*** have *no* yolks and may contain 80 % egg white. Generally, the "yolk" is made of corn oil, nonfat milk solids (NFMS), calcium caseinate, soy protein isolate, soybean oil, and other substances, including vitamins and minerals. The egg substitute also contains no cholesterol, less fat, and more unsaturated fat than whole egg. Many US egg patents have been issued relating to low-fat and low or decholesterized egg products.

CULINARY ALERT! Egg substitutes although yellowish in color may be beaten for use in egg white foams.

Nutritive Value of Eggs

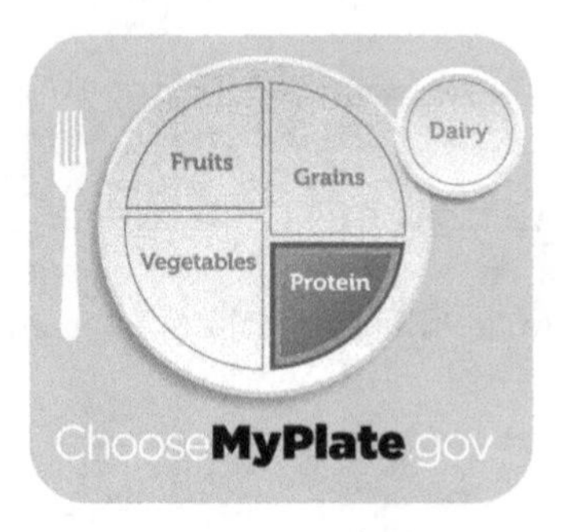

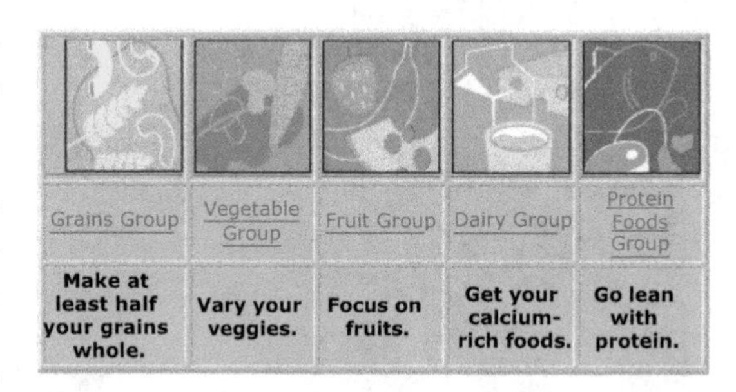

The nutritive value of eggs includes vitamins A, D, E, the water-soluble Bs, and minerals such as iron, phosphorus, and zinc as well as iodine, potassium, and sulfur. Eggs are low in calories—75 cal per large egg—and are used to fortify other foods that may originally be low in protein.

Eggs are a complete protein, with a ***biological value*** of 100, which indicates that all of the protein is retained by the body. All other protein sources are evaluated against this standard. That is not to say that eggs are "the perfect food." Persons who follow an ovo-vegetarian diet include eggs in their diet and assist in meeting essential protein requirements.

Egg whites are given the highest ***protein-digestibility-corrected amino acid score*** (***PDCAAS***) of 1.0, which corrects the amino acid composition with its digestibility.

For FDA labeling purposes, the PDCAAS method of determining protein quality is used. The % Daily Value for protein that appears on labels reflects both the quantity (in grams) and quality of protein (Table 10.6).

New, Digestible Indispensable Amino Acid Score (DIAAS) as a replacement for PDCAAS has been proposed. See this press release:

FAO Proposes New Protein Quality Measurement

The Food and Agriculture Organization of United Nations (FAO) has released a report recommending a new, advanced method for assessing the quality of dietary proteins. The report, "Dietary protein quality evaluation in human nutrition," recommends that the Digestible Indispensable Amino Acid Score (DIAAS) replace the Protein Digestibility Corrected Amino Acid Score

Table 10.5 Beaten egg white foam

Stage	Description
Unbeaten raw egg white	• Small volume of thick and thin whites
	• No initial additives
Foamy	• Unstable, large air-cell volume, transparent
	• Bubbles coalesce if beating is halted
	• Acid coagulates protein around air cell
	• Add cream of tartar (acid) now
Soft rounded peaks	• Air cells subdivide in size and are whiter
	• Volume is increased
	• Add sugar now
	• May be used for food applications
	• Used for soft meringue
Stiff pointed peaks	• Many small air cells, volume is increased
	• Egg protein coagulates around fine air cells
	• Ready for most food applications
	• Used for hard meringue
Dry peak foam	• Brittle, inelastic; less volume as air cells break
	• Denatured, water escapes, ***flocculated***
	• Not as effective as a leavening agent
	• Overcoagulated, curdled appearance

Table 10.6 PDCAAS of selected foods

Egg white	1	Chick peas	0.66
Casein (milk)	1	Pinto beans	0.63
Soybean isolate	0.99	Rolled oats	0.57
Beef	0.92	Lentils	0.52
Kidney beans	0.68	Whole wheat	0.40

(PDCAAS) as the preferred method of measuring protein quality.

The report recommends that more data be developed to support full implementation, but in the interim, protein quality should be calculated using DIAAS values derived from fecal crude protein digestibility data. Under the current PDCAAS method, values are "truncated" to a maximum score of 1.00, even if scores derived are higher...

http://www.fao.org/ag/humannutrition/35978-02317b979a686a57aa4593304ffc17f06.pdf

Egg yolks contain cholesterol, and consequently are restricted by some individuals with known heart disease. Over the past years, the American Heart Association (AHA) has changed its recommendations to where the current recommendation for intake is now consuming seven eggs per week. Dietary sources of cholesterol do not equate to personal levels of blood cholesterol (American Egg Board (AEB), Park Ridge, IL)

How much the cholesterol in a person's diet increases blood cholesterol varies from person to person. See the 2012 published research with references http://www.incredibleegg.org/health-and-nutrition

Statement by the Egg Nutrition Center and American Egg Board on Recently Published Research on Egg and Cholesterol Consumption

Park Ridge, Ill. (August 14, 2012)—Eggs have been shown to have a wide range of health benefits, providing 13 essential vitamins and minerals, high-quality protein and International contamination of eggs with SE is very low, even from a known positive flock (American Egg Board (AEB), Park Ridge, IL). Yet, the safety of eggs must be ensured. For example, only clean, uncracked eggs should be purchased from a reputable supplier. Exterior surface bacteria can enter shells of dirty eggs, or even clean ones, especially through cracks causing the egg to be unsafe.

A Harvard study with more than one hundred thousand subjects found no significant difference in cardiovascular disease risk between those consuming less than one egg per week and those consuming one egg per day. The researchers concluded that consumption of up to one egg per day is unlikely to have substantial overall impact on the risk of heart disease or stroke among healthy men and women. Another study published in Risk Analysis estimates that eating one egg per day is responsible for less than one percent of the risk of coronary heart disease in healthy adults. Alternatively, lifestyle factors including poor diet, smoking, obesity, and physical inactivity contribute to 30–40 % of heart disease risk, depending on gender...

Additionally, research has shown that saturated fat may be more likely to raise a person's serum cholesterol than dietary cholesterol...

Also see about reevaluating eggs' cholesterol risk:

http://www.sciencenews.org/view/generic/id/7301/description/Reevaluating_Eggs_Cholesterol_Risk

The American Heart Association cautions, for people with existing "coronary heart disease, diabetes, high-LDL cholesterol or other cardiovascular disease, your daily cholesterol limit is less than 200 mg."

The Egg Nutrition Center site is for anyone who wants to learn about eggs and good nutrition. The truth is, an egg a day is OK! http://www.enc-online.org/.

"Egg Nutrition Center (ENC) monitors scientific findings and regulatory developments, and serves as a resource for health practitioners in need of current nutrition information to share with their patients. Key vehicles for disseminating information are the Nutrition Close-Up newsletter, various educational brochures and tool kits, published scientific articles, and symposia presented at health professional conferences and the eggnutritioncenter.org website."

See also:

http://www.incredibleegg.org/health-and-nutrition/cracking-the-cholesterol-myth

Cracking the Cholesterol Myth

More than 40 Years of Research Supports the Role of Eggs in a Healthy Diet

Many Americans have shied away from eggs—despite their taste, value, convenience, and nutrition—for fear of dietary cholesterol. However, more than 40 years of research have shown that healthy adults can eat eggs without significantly impacting their risk of heart disease.

And now, according to new **United States Department of Agriculture** (USDA) nutrition data, eggs are lower in cholesterol than previously recorded. The USDA recently reviewed the nutrient composition of standard large eggs and results show the average amount of cholesterol in one large egg is 185 mg, a 14 % decrease. The analysis also revealed that large eggs now contain 41 IU of vitamin D, an increase of 64 %.

For more information, contact the Egg Nutrition News Bureau at 312-233-1211 or info@eggnutrition.org.

"Research shows diets with increased protein are beneficial for weight loss, specifically to reduce body fat, protect lean tissue, increase satiety and stabilize glycemic indices like blood sugar and triglycerides." Suggested is "a high protein-rich breakfast reduces hunger, boosts satiety and reduces brain responses involved with food cravings to a greater degree than a typical breakfast that's lower in protein." (Lockwood et al. 2006)

Below is a Nutrition Facts Label Printed Directly on an Egg still subject to on-going testing:

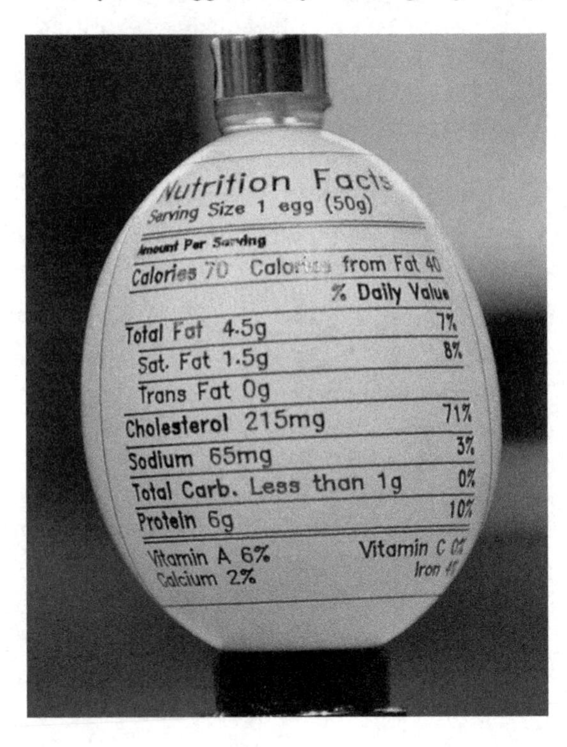

See: TipSheet no. 6 (in Meats)

American Egg Board		
Realities/misconceptions		
Issue	Misconception	Fact
Salmonella	Eggs pose a serious risk of foodborne illness due to salmonella contamination	Eggs used in processed foods are from further processed eggs which are pasteurized and free from *Salmonella*. Keep in mind that proper food handling is still important with further processed eggs
Cholesterol	For many years, consumers and the media have viewed eggs as a high cholesterol food associated with high blood cholesterol levels	Studies have shown that saturated fat, not dietary cholesterol from eggs, is the real culprit. Most healthy people can include eggs in their diet
Egg substitutes and replacers	Food processors sometimes believe that ingredients provide better functions than real eggs in formulations	Eggs provide superior polyfunctional properties. Food processors find that replacers do not function as well as real eggs in various food applications. Often a replacer provides one function only. For example, the substitute may act as a foaming agent but exhibits poor coagulation properties

"Egg protein contributes essential functional and nutritional properties" and "is well suited for a variety of gluten-free applications." (Foster 2013)

Safety of Eggs

"SAFE HANDLING INSTRUCTIONS" on egg labels assists in preventing illness from bacteria. This product warning provides instructions that include keep eggs refrigerated, cook eggs until yolks are firm, and cook foods containing eggs thoroughly.

The contents of freshly laid eggs are generally sterile, although ***Salmonella enteritidis*** (***SE***) has been found inside some eggs. Eggs are *usually* protected from bacteria by the shell and the two shell membranes; however, the surface of shell eggs may contain a high level of bacteria. These bacteria may enter the shell through the pores, especially when it is soiled and washed. If bacteria travel to the internal portion of the egg, it is typically the *egg yolk membrane* (i.e., the vitellin membrane), *not* the yolk itself nor the egg white that harbors the bacteria.

According to the American Egg Board, "… protective barriers include the shell and yolk membranes and layers of the white which fight bacteria in several ways. The structure of the shell membranes helps prevent the passage of bacteria. The shell membranes also contain *lysozyme*, a substance that helps prevent bacterial infection. The yolk membrane separates the nutrient-rich yolk from the white." (American Egg Board (AEB), Park Ridge, IL)

The Egg Safety Action Plan, a joint effort by the FDA and USDA, was announced in the late 1999. Its intent was to reduce the incidence of SE, and it contains two important requirements:

1. The refrigeration requirement in this Plan is that eggs delivered to a retail establishment (restaurants, hospitals, schools, nursing homes, grocery stores, delis, and vending operations) be quickly stored at an ambient temperature of 45 °F (7 °C) or less upon receipt
2. There is a required statement on shell egg cartons that reads as follows:
   According to the FDA raw eggs should not be used, especially by population groups such as the young, elderly, or immune-compromised.

Food Safety and Pregnancy

Safe Eats—Eating Out & Bringing In

Welcome to Safe Eats, your food-by-food guide to selecting, preparing, and handling foods safely throughout your pregnancy and beyond!

Meat, Poultry & Seafood|Dairy & Eggs|Fruits, Veggies & Juices|Ready-to-Eat Foods|Eating Out & Bringing In

For a printable chart of cooking temperatures, see the Apply the Heat (PDF | 20.3KB) chart

The Food and Drug Administration (FDA) cautions, "During pregnancy, your immune system is weakened, which makes it harder for your body to fight off harmful foodborne microorganisms that cause foodborne illness." How can you safeguard the health of yourself and your baby? See some tips from the FDA.

Raw eggs or undercooked eggs are the most common source of Salmonella illness (salmonellosis). The U.S. National Pasteurized Eggs, Inc. (NPE), producer of Safest Choice™ pasteurized eggs, has received the 2011 NSF Food Safety Leadership Award in Breakthrough Technologies for its in-shell egg pasteurization process. "In-shell pasteurized eggs may be used safely without cooking."—USDA

Some USDA and FDA rulings follow:

As Recommended by the USDA: Dates on Egg Cartons

Use of either a "Sell-by" or "Expiration" (EXP) date is not federally required, but may be State required, as defined by the egg laws in the State where the eggs are marketed. Some State egg laws do not allow the use of a "sell-by" date. Many eggs reach stores only a few days after the hen lays them. Egg cartons with the USDA grade shield on them must display the "pack date" (the day that the eggs were washed, graded, and placed in the carton). The number is a three-digit code that represents the consecutive day of the year starting with January 1 as 001 and ending with December 31 as 365. When a "sell-by" date appears on a carton bearing the USDA grade shield, the code date may not exceed 45 days from the date of pack. (The day that the eggs were washed, graded, and placed in the carton.)

Always purchase eggs before the "Sell-by" or "EXP" date on the carton. After the eggs reach home, refrigerate the eggs in their original carton and place them in the coldest part of the refrigerator, not in the door. For best quality, use eggs within 3–5 weeks of the date you purchase them. The "sell-by" date will usually expire during that length of time, but the eggs are perfectly safe to use.

fsis.usda.gov

Check your steps at FoodSafety.gov. Also see the chapter on Food Safety.

Some foods are more frequently associated with food poisoning or foodborne illness. With these foods, it is especially important to:

- CLEAN: Wash hands and food preparation surfaces often. And wash fresh fruits and vegetables carefully.
- SEPARATE: Don't cross-contaminate! When handling raw meat, poultry, seafood, and eggs, keep these foods and their juices away from ready-to-eat foods.
- COOK: Cook to proper temperature. See the Minimum Cooking Temperatures chart for details on cooking meats, poultry, eggs, leftovers, and casseroles.
- CHILL: At room temperature, bacteria in food can double every 20 min.

The FDA prohibits the use of raw or lightly cooked eggs in food production or manufacturing facilities. Eggs must reach an internal temperature of 145 °F (63 °C) or higher to be considered safe for consumption (check local jurisdiction). Egg products must be pasteurized. A previous FDA Commissioner added ". . . no sunny side up, no over easy." (Dr. Jane E. Henney, FDA Commissioner)

International contamination of eggs with SE is very low, even from a known positive flock (California Egg Commission, Upland, CA). Yet, the safety of eggs must be ensured. For example, only clean, uncracked eggs should be purchased from a reputable supplier. Exterior surface bacteria can enter shells of dirty eggs, or even clean ones, especially through cracks causing the egg to be unsafe.

Also, The President's Council on Food Safety has encouraged developments in science and technology by companies and universities to reduce the incidence of SE. For example, methods are being investigated/employed to bring an egg temperature down from 109 °F (43 °C) (the internal temperature of hens) to a cold temperature of 45 °F (7 °C) to control SE. One such method utilizes cryogenic carbon dioxide; another uses a clean warm-water bath to kill bacteria without cooking (Praxair Inc. 2000; Mermelstein 2000).

Furthermore since washing is a routine step in commercial egg processing, rewashing eggs prior to use is not necessary or recommended. When eggs are washed in warm water and then refrigerated, pressure changes in a cooling egg draw harmful exterior microbes in through the pores. Eggs should be stored cold at temperatures of approximately 40 °F (4.4 °C).

CULINARY ALERT! The common restaurant practice of "pooling" (commingling) eggs is not recommended. Pooled eggs are many eggs cracked together and stored ahead, ready to use such as for an omelet. Contamination likelihood is increased.

Prior to incorporation into recipes, either egg yolks or the egg white to be used for meringue may be heated over direct heat or water bath to raise temperature while controlling SE. If the egg is refrigerated, bacterial growth is extremely slow, and disease is not likely to result. Egg products are pasteurized and free of *Salmonella.*

Safety: Easter Egg Dyeing and Hunts—USDA

> Easter eggs are a fun tradition for many families but they can be a health hazard if not handled properly.
> **Dyeing eggs**: After hard cooking eggs, dye them in food-safe coloring and return them to the refrigerator within 2 h.
> **Blowing out eggshells**: Only use uncracked, refrigerated eggs. Wash the egg in hot water and then rinse in a solution of 1 tablespoon of unscented, liquid chlorine bleach in 1 gallon of water. This will destroy bacteria that may be present on the surface of the shell.
> **Hunting for eggs**: If you are preparing an Easter egg hunt with hard-boiled eggs, use extra caution when hiding the eggs to prevent the shell from cracking. If the shell cracks, bacteria and germs can contaminate the egg. Eggs should not be out of the refrigerator for more than 2 h. Once eggs have been found (within the 2 h time), re-refrigerate immediately and consume within 7 days. Do not eat if shell is cracked or if eggs were unrefrigerated for more than 2 h (or 1 h if over 90 °F).
> USDA

Hard cooked eggs reach a final cooking temperature that is sufficient to kill the natural bacteria of an egg, yet *recontamination* may occur. For example, recontamination may be due to the practice of "hiding Easter eggs" wherein the oil coating of the egg may be lost, and the pores of the egg may open. Subsequently, the egg may be contaminated with substances such as lawn chemicals, fertilizers, or droppings from household pets, birds, reptiles, and rodents (American Egg Board (AEB), Park Ridge, IL). Refrigerate after cooking—perhaps quick chill in an ice bath.

Accordingly, for consumers who follow the traditional practices of decorating and hiding Easter eggs, the USDA caution remains: Keep perishables at room temperature for no longer than 2 h. Decorated eggs are festive and may be very inexpensive and are thus included in many celebrations, but the recommendation is that separate eggs be used for eating and decorating or "hiding."

Egg White Resistance to Bacterial Growth

Egg whites have natural protection against microorganisms by several natural *chemical components*; however, they cannot be considered 100 % safe once the shell has been broken or cracked. These components include *avidin*, *lysozyme*, and *conalbumin*. *Avidin* in the raw egg white binds the vitamin biotin required for some microorganism growth. *Lysozyme* hydrolyzes cell walls of some bacteria and thus demonstrates antibacterial action, especially at lower pH levels. *Conalbumin* binds with the iron of the yolk preventing growth of the microorganisms that require iron for growth.

Pasteurized raw eggs in uncooked foods such as mayonnaise do not support the growth of bacteria as do unpasteurized shell eggs. Thus, only pasteurized egg products may be used in manufacturing or retail operations, where a food containing eggs is not subject to adequate heat treatment. Uncooked meringues prepared by shell egg whites that are not pasteurized are considered a "low-risk" food though, because they contain a large amount of sugar that ties up the water needed for bacterial growth. The water activity needs of the bacteria are not met, and the bacteria do not grow.

CULINARY ALERT! Some risk of bacterial presence exists in eggs and it is recommended that all raw egg parts, including egg whites, be cooked prior to use.

"In addition to containing antibacterial compounds such as lysozyme, layers of the white discourage bacterial growth because they are alkaline, bind nutrients bacteria need and/or don't provide nutrients in a form that bacteria can use. The thick white discourages the movement of bacteria. The last layer of white is composed of thick ropey strands which have little of the water that bacteria need but a high concentration of the white's protective materials. This layer holds the yolk centered in the egg where it receives the maximum protection from all the other layers." (American Egg Board (AEB), Park Ridge, IL)

USDA Sampling

The USDA-administered Egg Products Inspection Act of 1970 requires routine sampling and analysis, and routine inspection for wholesome, unadulterated eggs and egg products. Plants are inspected regardless of whether the shipment is intrastate, interstate, or out of the country. State standards, regulated by the state's Department of Agriculture, must be equivalent to federal standards.

Conclusion

Whole eggs, and their component parts, are important for their array of functional properties such as binding, emulsification, foaming, gelling, and thickening. These properties change with cooking as the egg protein denatures. The processes of grading and evaluation of quality, although not mandatory, are officially carried out by the USDA and their state counterparts.

Eggs are candled in order to evaluate and assign grades. In candling, the yolk, white, and air-cell size as well as the integrity of the shell are viewed prior to sale. Egg size is not a part of egg quality evaluation. The color of a shell is dependent on the breed of hen, and the yolk color is dependent on the feed.

Processing and preservation of eggs occurs with the use of mineral oil and the techniques of pasteurization, freezing, and dehydration, after which proper storage is important in maintaining safety and other aspects of food quality. The addition of other ingredients to an egg, such as salt and acid, promotes denaturation. Sugar exerts a protective effect, controlling the denaturation and coagulation processes. Some factors affecting the volume and stability of egg white foams include temperature, pH, salt, sugar, fat, and addition of liquid. The recommendation is that eggs not be old or cold.

Various forms of eggs, including pasteurized shelled eggs, are available, and egg substitutes may be purchased in the marketplace. Eggs have a biological value of 100, and are given the highest PDCAAS. The Food and Agriculture Organization of United Nations (FAO) has released a report that recommends that the Digestible Indispensable Amino Acid Score (DIAAS) replace the (PDCAAS) as the preferred method of measuring protein quality. Vigilance is necessary in the handling and use of eggs, a potentially hazardous food.

"Most healthy people can include eggs in their diet."(American Egg Board (AEB), Park Ridge, IL). "Egg ingredients are available in liquid, frozen or dried forms as whole eggs yolks and whites, or tailored to meet processing specifications." (American Egg Board 2013)

Notes

CULINARY ALERT!

Glossary

Air cell or air pocket A space between shell membranes where air is found within the shell, typically at the large end of an egg.

Biological value Eggs contain a score of 100 based on their efficiency in supporting the body's needs; reflects the amount of nitrogen retained in the body, due to the completeness of the protein. (An incomplete protein is deaminated and nitrogen is not retained.)

Beading Amber-colored syrup beads on top of baked meringue as a result of overcoagulation.

Binder Holds the ingredients of a mixture or its breading together.

Candling Viewing the inside and shell of an egg by holding it up to a bright light.

Clarify To remove foreign particles from a hot liquid.

Coagulation Extensive denaturation of protein molecules yielding a solid mass or gel.

Curdling The protein precipitates, shrinks, releases water, and becomes tough.

Denaturation Changes in the conformation of a protein caused by changes in temperature, an acidic pH, or by surface changes such as mechanical beating.

Digestible Indispensable Amino Acid Score (DIAAS) A measure of protein quality. The preferred method of measuring protein quality recommended to replace the PDCAAS.

Egg substitute Liquid or frozen egg white product with a "yolk" typically consisting of corn oil, NFMS, calcium caseinate, soy protein isolate, soybean oil, and other substances.

Emulsifier Material that allows two ordinarily immiscible substances to mix.

Flocculated Separation of overbeaten egg white foam into small masses.

Foam Increased volume of beaten egg white that holds shape as protein coagulates around air cells.

Gel A two-phase system where egg coagulates with liquid in a solid.

Pasteurization Heating for a specific time at a temperature that eliminates pathogens.

Protein-digestibility-corrected amino acid score (PDCAAS) A measure of protein quality that compares the amino acid balance with requirements of preschoolers and corrects for digestibility. Used by the FDA for labeling and by the WHO.

Digestible Indispensable Amino Acid Score (DIAAS) It has been proposed as a replacement for PDCAAS. More data needs to be developed to support full implementation.

***Salmonella enteritidis* (SE)** Pathogenic, infection-causing bacteria especially prevalent in poultry and eggs.

Syneresis "Weeping" or water leakage from coagulated egg.

Thickening agent Increases viscosity.

Ultrapasteurization High temperature, short time heat to kill pathogenic microorganisms.

Weeping Syneresis or release of water from undercoagulated or underbeaten egg whites.

References

American Egg Board (2013) The sunny side of egg protein. Food Product Design, p 88

Egg grading manual. Handbook no. 75. USDA, Washington, DC

Foster RJ (2013) Egg-stra egg-stra—read all about it! Food Product Design (March/April):18–21

Jordan R, Barr AT, Wilson MC Shell eggs: quality and properties as affected by temperature and length of storage. Purdue University Agricultural Experiment Station, West Lafayette, IN. Bulletin #612 March/April

Lockwood CM, Moon JR, Tobkin SE, Walter AW, Smith AE, Dalbo VJ, Cramer JT, Stout JR (2006) Minimal nutrition intervention with high-protein/low- carbohydrate and low-fat, nutrient-dense food supplement improves body composition and exercise benefits in overweight adults: a randomized controlled trial. Am J Clin Nutr 83:260–274

Mermelstein NH (2000) Cryogenic system rapidly cools eggs. Food Technology 54(6):100–102

Praxair Inc. (2000) Technologies target Salmonella in eggs. Food Eng 72:14

Bibliography

Centers for Disease Control and Prevention (CDC)

Functional Egg.Org

How to buy eggs. Home and garden bulletin no. 144. USDA, Washington, DC

http://www.fda.gov/Food/FoodborneIllnessContaminants/PeopleAtRisk/ucm082294.htm

http://www.naturalnews.com/029640_eggs_USDA.html#ixzz2VY3pMlpX

Model FDA Food Code

USDA ChooseMyPlate.gov

11 Milk and Milk Products

Introduction

Milk is the first food of young mammals produced by the mammary glands of female mammals. It is a mixture of fat and high-quality protein in water and contains some carbohydrate (lactose), vitamins, and minerals. Milk and milk products may be obtained from different species, such as goats and sheep, although the focus of this chapter is on *cow's* milk and milk products.

While fluid milk contains a very large percentage of water, it may be concentrated to form evaporated milk and cheeses. Throughout the world, it is used in a variety of ways, such as a beverage, cheese, yogurt, or in soups and sauces.

By law, milk and milk products must contain a designated percent of ***total milk solids*** (all of the components of milk except water), and also, the ***milk solids, not fat*** (***MSNF***) (all of the components of milk solids not including fat). The butterfat component of milk is the most expensive component of milk and its level determines if milk is offered for retail sale as whole milk or at some lesser percentage of fat, such as 2 % milk, 1 %, 1/2 %, or fat-free.

Milk may be cultured, dried, fortified, homogenized, or pasteurized and used to create products with different taste, texture, nutritive value, and shelf life. It may be processed into products such as buttermilk, cheese, cream, ice milk, ice cream, sour cream, and yogurt with different levels of fat content. Dried milk is added to a multitude of foods. It may be added to foods to increase the protein or calcium value.

The top eight milk-producing states produce 196.2 billions of pounds as follows:

California	21.1 %
Wisconsin	13.3
Idaho	6.8
New York	6.5
Pennsylvania	5.4
Texas	4.9
Minnesota	4.5
Michigan	4.3
All others	33.1 %
Source: USDA (2011)	

Total Milk Production has gone from 165.3 billions of pounds in 2001 to the 196.2 billion pounds in 2011 shown above.

High temperatures may curdle milk; therefore, care must be taken in the preparation of foods with milk. Milk requires safe handling and cold storage.

Milk is not well tolerated by a large portion of the population. The milk sugar, lactose, is not digested by persons lacking the enzyme lactase.

Definition of Milk

Milk means the lacteal secretion, practically free from colostrum, obtained by the complete milking of one or more healthy cows, which may be

V.A. Vaclavik and E.W. Christian, *Essentials of Food Science, 4th Edition*, Food Science Text Series,
DOI 10.1007/978-1-4614-9138-5_11,

> clarified and may be adjusted by separating part of the fat there from; concentrated milk, reconstituted milk, and dry whole milk. Water, in a sufficient quantity to reconstitute concentrated and dry forms may be added. (FDA)

Further useful Food and Drug Administration (FDA) definitions, such as that for cheese, appear later in this chapter.

Composition of Milk

Milk varies in physical and chemical composition depending on such factors as age and breed of the cow, activity level, stage of lactation, use of medication, and interval between milking. It consists mainly of water and contains some serum solids or milk solids, nonfat (MSNF) such as lactose, caseins, whey proteins, and minerals. Milk also naturally contains fat.

Water

Water is the largest component of milk and is present at a level of approximately 87–88 %. If that water is removed, the shelf life of milk products is greatly extended.

Carbohydrate

Carbohydrate is water-soluble and present in the aqueous phase of milk, at levels of slightly less than 5 %. The disaccharide *lactose* is the main carbohydrate. It exhibits low solubility and may precipitate out of solution as a grainy textured substance. It is converted to lactic acid (1) upon souring due to bacterial fermentation and (2) in the process of aging cheese. Therefore, aged cheese may be digestible by lactose-intolerant individuals even in the absence of the enzyme lactase. (The lactose content of milk and some milk products appears in Table 11.2. See section "Lactose Intolerance.")

Fat

Fat has a low density and may *easily* be centrifuged or *skimmed off* of the milk yielding low-fat or *skim milk*. The fat, or butterfat, exists at levels of approximately 3.5 % in whole milk, at lesser levels in reduced-fat or nonfat milks, and at *significantly* higher percentages in cream. Fat is the expensive component of milk and the basis on which dairy farmers are paid for milk. When fat and its carotenoids are removed, milk is bluish in color.

Fat globules are less dense than the water in the aqueous phase of milk and, therefore, rise to the top of the container in the ***creaming*** process. When emulsified during *homogenization*, there is an increase in the number of fat cells and greater viscosity because the fat is distributed throughout the fluid, and creaming does *not* occur. Membranes of lipids and protein, including lecithin, from each fat globule remain in the milk as it is processed.

Fat content in milk varies greatly in calories. The completeness of milking determines richness of the fat content. Milk either carries or may be fortified to contain the fat-soluble vitamins, and it contains the pigments carotene and xanthophyll. Fat contains the sterol cholesterol and phospholipids, although it is primarily triglyceride (95 %) with saturated, polyunsaturated, and monounsaturated fatty acid components. These have varying melting points and susceptibility to oxidation. The fatty acid chains contain many short-chain fatty acids such as the saturated butyric acid (4 °C) and caproic, caprylic, and capric acids.

A number of the more than 400 individual fatty acids have been identified in milk fat. It is approximately 15–20 *fatty acids* that make up 90 % of the milk fat. *Phospholipids* (such as phosphatidylcholine and sphingomyelin) in cell membranes compose approximately 1 % of the fat in milk.

Protein

Protein represents 3–4 % of the composition of milk and components may be fractioned out of milk by *ultracentrifugation*. ***Casein** is the primary protein of milk*, comprising approximately 80 % of the milk protein. The caseins are actually a group of similar proteins, which can be separated from the other milk proteins by acidification to a pH of 4.6 (the isoelectric point). At this pH, the caseins aggregate, since they are hydrophobic, are poorly hydrated, and carry no net charge. The other milk proteins, being more hydrophilic, remain dispersed in the aqueous phase.

There are three main casein fractions, known as alpha$_s$-, beta-, and kappa-casein (α_s-, β-, and κ-casein). Alpha$_s$-casein actually comprises two fractions: α_{s1}-casein and α_{s2}-casein. However, these two fractions are difficult to separate from each other. The four fractions α_s-, β-, κ-, and α_{s2}-casein occur in the weight ratio 3:3:1:0.8. All four fractions are phosphoproteins containing phosphate groups esterified to the amino acid serine. The α_s- and β-casein fractions contain several phosphate groups and as a result are "calcium-sensitive" and may be coagulated by addition of calcium. Kappa-casein contains only one phosphate group and is not calcium-sensitive. The α_s- and β-casein fractions are very hydrophobic. However, κ-casein is a glycoprotein containing an acidic (charged) carbohydrate section, and so it is much more hydrophilic.

In milk, the casein fractions associate with each other and with colloidal calcium phosphate to form stable spherical structures known as ***casein micelles***. The more hydrophobic α_s- and β-casein fractions exist mainly in the interior of the micelles, whereas the more hydrophilic κ-casein exists mainly on the micelle surface. It is the κ-casein that gives the micelles their stability in milk under normal handling conditions.

This is due to the negative charge and hydration of the κ-casein, coupled with the fact that the charged hydrophilic carbohydrate section of the molecule tends to protrude from the micelle surface in hairlike structures, which confer steric (or spatial) stability on the micelles. Also, since κ-casein is not sensitive to calcium, it protects the other caseins from the ionic calcium in milk, thereby increasing the stability of the micelles.

There have been several different views of the structure of the casein micelles, and their structure is still debated. Two major structural models have been postulated. The submicelle model was developed first and was prominent for many years. That model consists of aggregates of casein submicelles linked by calcium phosphate. It is suggested that there are κ-casein-rich and κ-casein-poor micelles, with the former being present at the surface of the micelles. However, there is not sufficient evidence for the existence of discrete submicellar particles.

The most current view of casein micelle structure is a nanocluster model, which is an open structure involving calcium phosphate nanoclusters surrounded by casein phosphopeptides. The caseins bind to more calcium phosphate or to other caseins, thus forming the casein micelles. This could be considered an inversion of the submicelle model. The calcium phosphate nanoclusters vary in density and provide for a porous structure that is able to hold a large amount of water. In both models, the κ-casein is mainly present at the surface and has a stabilizing effect on the micelles.

There are many reviews of casein micelle structure for those who would like to dig deeper on this subject. A good starting place would be the recent review by Dalgleish and Corredig (2012).

The casein micelles are *coagulated by addition of acid* at a pH of 4.6–5.2. As the micelles approach their isoelectric point, the charge and extent of hydration is reduced, and the κ-casein hairlike structures flatten, reducing steric hindrance. Hence, the micelles are no longer stable, and so they aggregate. This is the basis for the formation of cottage cheese, which is an acid cheese containing casein curds.

Acid also causes some calcium to be removed from the micelles, and so cottage cheese is relatively low in calcium compared with some other dairy products. The casein micelles may also be *coagulated by addition of the enzyme* ***rennin***, which may be added to milk to prepare rennet custard or cheese. Rennin cleaves a specific bond in κ-casein and causes the charged, hydrophilic hairlike structures to be removed from the micelles.

Accordingly, the micelle surface is uncharged, hydrophobic, and unstable, and so the micelles aggregate to form curds. The curds may be separated from the whey and processed to form cheese (see section on "Cheese"). Coagulation by rennin does not cause calcium to be removed from the micelles.

Casein micelles are relatively heat stable and are not denatured by heat (at neutral pH) unless temperatures are very high and heating is prolonged. This is not a problem under most cooking conditions. However, it is a potential problem in heated concentrated milk products such as evaporated milk. The problem is avoided by addition of carrageenan to protect the protein.

Caseins contain both hydrophobic and hydrophilic sections; in addition, they contain a high proportion of the amino acid proline, and so they are flexible proteins containing little regular, ordered secondary structure (see Chap. 8). As a result, they readily adsorb at an oil–water interface, forming a stable film that prevents coalescence of emulsion droplets (see Chap. 13), and so they make excellent emulsifiers.

A *second protein fraction* of milk is the ***whey*** or serum. It makes up approximately 20 % of milk protein and includes the *lactalbumins* and *lactoglobulins*. Whey proteins are more hydrated than casein and are denatured and precipitated by heat, rather than by acid. (More information is contained in this chapter in the section entitled "Whey.")

Additional significant protein components of milk include enzymes such as lipase, protease, and *alkaline phosphatase*, which hydrolyze triglycerides, proteins, and phosphate esters, respectively. The average measures of protein quality, including biological value, digestibility, net protein utilization, protein efficiency, and chemical score, for milk and milk products appear in Table 11.1.

Vitamins and Minerals

Vitamins in milk are both the water-soluble and fat-soluble varieties. The nonfat portion of milk is especially plentiful in the B vitamin B_2—riboflavin, a greenish fluorescent colored vitamin. It acts as a photosynthesizer and is readily destroyed upon exposure to sunlight.

Additional *water-soluble B vitamins* in milk beside riboflavin include thiamin (vitamin B_1), niacin (vitamin B_3), pantothenic acid (vitamin B_5), vitamin B_6 (pyridoxine), vitamin B_{12} (cobalamin), vitamin C, and folate.

The *fat-soluble vitamins A, D, E, and K* are dependent upon the fat content of the milk. Vitamin A is *naturally* in the fat component of *whole* milk, and more may be added prior to sale. If the milk is reduced fat (2 % fat), or low-fat (1 % fat), or even skim milk (fat skimmed off), fortification with vitamin A must occur in order to be made nutritionally equivalent to whole milk.

Whole milk is generally (98 %) *fortified* with *vitamin D* because it is naturally present only in *small* amounts. Vitamin D is present in milk to *some* extent due to the synthesis of vitamin D by the cow as it is exposed to sunlight and because vitamin D may be present in animal feed. *Low-fat* and *nonfat* milk, containing reduced levels or no fat, may be fortified with *both* of these fat-soluble A and D vitamins. Fortification with vitamin D is voluntary. Vitamins E and K are minor constituents of milk.

Minerals such as calcium and phosphorus are present at levels of approximately 1 % of milk, with a third of calcium in solution, and two-thirds of it colloidally dispersed. Calcium is combined with the protein casein as calcium caseinate, with phosphorus as calcium phosphate and as calcium citrate. Other minerals present in milk are chloride, magnesium, potassium, sodium, and sulfur.

Classification of Milk

Whole milk may be classified as a solution, dispersion, or emulsion as follows:

Table 11.1 Average measures of protein quality for milk and milk products

	BV	Digestibility	NPU	PER[a]	Chemical score
Milk	84.5	96.9	81.6	3.09	60
Casein	79.7	96.3	72.1	2.86	58
Lactalbumin	82	97	79.5[b]	3.43	[c]
Nonfat dry milk	–	–	–	3.11	–

Source: Adapted from the National Dairy Council

Note: *Biological* value (BV) is the proportion of absorbed protein that is retained. *Digestibility* (*D*) is the proportion of food protein that is absorbed. *Net protein utilization* (NPU) is the proportion of food protein intake that is retained (calculated as BV $\times$ *D*). *Protein efficiency ratio* (PER) is the gain in body weight divided by weight of protein consumed. *Chemical score* is the content of the most limiting amino acid expressed as a percentage of the content of the same amino acid in egg protein

[a]Often, PER values are adjusted relative to casein which may be given a value of 2.5

[b]Calculated

[c]Denotes no value compiled in Food and Agriculture Organization of the United Nations (FAO) report

Table 11.2 Composition of milks from different species (100-g portions)

Nutrient	Cow	Human	Buffalo	Goat	Sheep
Water (g)	87.99	87.50	83.39	87.03	80.70
Calories	61	70	97	69	108
Protein ($N \times 6.38$) (g)	3.29	1.03	3.75	3.56	5.98
Fat (g)	3.34	4.38	6.89	4.14	7.00
Carbohydrate (g)	4.66	6.89	5.18	4.45	5.36
Fiber (g)	0	0	0	0	0
Cholesterol (mg)	14	14	19	11	–
Minerals					
Calcium (mg)	119	32	169	134	193
Iron (mg)	0.05	0.03	0.12	0.05	0.10
Magnesium (mg)	13	3	31	14	18
Phosphorus (mg)	93	14	117	111	158
Potassium (mg)	152	51	178	204	136
Sodium (mg)	49	17	52	50	44
Zinc (mg)	0.38	0.17	0.22	0.30	–
Vitamins					
Ascorbic acid (mg)	0.94	5.00	2.25	1.29	4.16
Thiamin (mg)	0.038	0.014	0.052	0.048	0.065
Riboflavin (mg)	0.162	0.036	0.135	0.138	0.355
Niacin (mg)	0.084	0.177	0.091	0.277	0.417
Pantothenic acid (mg)	0.314	0.223	0.192	0.310	0.407
B_6 (mg)	0.042	0.011	0.023	0.046	–
Folate (μg)	5	5	6	1	–
B_{12} (μg)	0.357	0.045	0.363	0.065	0.711
Vitamin A (RE)	31	64	53	56	42
Vitamin A (IU)	126	241	178	185	147

Source: National Dairy Council

- *Solution*—contains the sugar lactose, the water-soluble vitamins thiamin and riboflavin, and many mineral salts such calcium phosphate, citrates, and the minerals chloride, magnesium, potassium, and sodium.
- Colloidal *dispersion* (sol)—casein and whey proteins, calcium phosphate, magnesium phosphate, and citrates.
- *Emulsion*—fat globules suspended in the aqueous phase (serum) of milk. The fat globules are surrounded by a complex membrane, the milk fat globule membrane, which contains mainly protein and phospholipids (and a few carbohydrate side chains at the outer surface). This membrane prevents coalescence of the fat droplets.

Grading of Milk

Grades are based on bacterial counts. Milk is a *potentially hazardous food* that must be kept *out* of the temperature danger zone. With its high water content and plentiful protein, vitamins, and minerals, milk is an ideal medium for supporting bacterial growth. Production, processing, and distribution of milk must ensure that products are kept *free* from pathogenic bacteria and *low* in nonpathogens. *Healthy cows* and *sanitary* conditions of handling lead to *low* bacterial counts. *Proper handling* also contributes to satisfactory shelf life, as well as appearance, flavor, and nutritive value.

The *temperature* of raw milk should reach 40 °F (4 °C) or less within 2 h of being milked from the cow. It should be kept well chilled, as it is highly perishable and susceptible to bacterial growth. The shelf life for properly refrigerated milk is 14 days or up to 45 days for *ultra-pasteurized* milk products including cream and lactose-reduced milk (see section "Pasteurization").

Numerous factors may lead to the spread of diseases by milk or milk products. A contaminated cow, *cross-contamination* at the farm or from workers hands, and unsanitary equipment or utensils may all become problematic. Traditionally, the diseases of *diphtheria*, *salmonellosis*, *typhoid fever*, *tuberculosis*, and *undulant fever* were spread by consumption of unsafe milk. Today, the incidence of these diseases is *rarely* attributed to milk transmission, as milk is pasteurized to destroy pathogens. The control of insects and rodents, as well as separation of animal waste products from the milking area, is also necessary for safe milk production.

The *US* Department of Agriculture (USDA) and *state* Departments of Agriculture regulate milk and milk products in interstate and intrastate commerce. Grades are based on bacterial counts. Grade "A" milk is available to the consumer for sale as fluid milk, although grades "B" and "C," with higher bacterial counts, are also safe and wholesome. The grades of US Extra and US Standard are given to dried milk. USDA official grades are given to all inspected milk on a voluntary fee-for-service basis.

Enzymes such as lipase, oxidation, and light may induce deterioration of the fat.

Flavor of Milk

The flavor of milk is mild and slightly sweet. The characteristic mouthfeel is due to the presence of emulsified fat, colloidally dispersed proteins, the carbohydrate lactose, and milk salts. Fresh milk contains acetone, acetaldehyde, methyl ketones, and short-chain fatty acids that provide aroma.

Less desirable, "barny" or rancid flavors, or other "off-flavors," may be due to the following:

- Slightly "cooked" *flavor* from excessive pasteurization temperatures.
- *Animal feed*, including ragweed and other weeds, or wild onion from the field.
- *Lipase activity* causes rancidity of the fat, unless destroyed by the heat of pasteurization. (Or, the short-chain butyric acid may produce an off-odor or

off-flavor due to bacteria, rather than lipase in the emulsified water of milk.)
- *Oxidation* of fat or phospholipids in the fat globule membrane, especially in emulsified, homogenized milk. Adequate pasteurization temperatures are necessary to destroy the enzyme which oxidizes fat.
- *Light-induced* flavor changes in the proteins and riboflavin because riboflavin acts as a photosynthesizer.
- *Stage of lactation* of the cow.

Flavor treatment to standardize the odor and flavor typically follows *pasteurization*. In this treatment process, milk is instantly heated to 195 °F (91 °C) with live steam (injected directly into the product) and subsequently subjected to a vacuum that removes volatile off-flavors and evaporates excess water produced from the steam.

Milk Processing

Pasteurization

"Pasteurized, when used to describe a dairy ingredient means that every particle of such ingredient shall have been heated in properly operated equipment to one of the temperatures (specified in the table) and held continuously at or above that temperature for the specified time (or other time/temperature relationship which has been demonstrated to be equivalent thereto in microbial destruction)" (FDA). Fluid milk is *not routinely sterilized* (see below); *rather*, *it is pasteurized*. This assures destruction of the pathogenic bacteria, yeasts, and molds, as well as 95–99 % of nonpathogenic bacteria. ***Pasteurization*** minimizes the likelihood of disease and extends the storage life of milk.

Pasteurization temperatures do *not* change milk components to any great extent (see section "Nutritive Value"). Vitamin destruction and protein denaturation are minimal, and the result is that milk is made safe for consumption. Several acceptable methods of pasteurization including thermal processing according to the International Dairy Food Association (IDFA) are shown in the following chart:

- 145 °F (63 °C) for 30 or more minutes—the batch or holding method is Vat Pasteurization and is considered low-temperature longer time (LTLT) pasteurization.
- 161 °F (72 °C) for 15 s—the flash method for this temperature is the high-temperature short-time (HTST) method of pasteurization, the most common method.
- 191 °F (88 °C) for 1 s. Higher heat shorter time (HHST) at this temp and above.
- 194 °F (90 °C) for 0.5 s. HHST.
- 201 °F (94 °C) for 0.1 s. HHST.
- 204 °F (96 °C) for 0.05 s. HHST.
- 212 °F (100 °C) for 0.01 s. HHST.

IDFA

"Another method, aseptic processing, which is also known as ultra high temperature (UHT), involves heating the milk using commercially sterile equipment and filling it under aseptic conditions into hermetically sealed packaging. The product is termed "shelf stable" and does not need refrigeration until opened. All aseptic operations are required to file their processes with the FDA's "Process Authority." There is no set time or temperature for aseptic processing; the Process Authority establishes and validates the proper time and temperature based on the equipment used and the products being processed."

"If the fat content of the milk product is 10 % or more, or if it contains added sweeteners, or if it is concentrated (condensed), the specified temperature shall be increased by 3 °C (5 °F). Provided that, eggnog shall be heated to at least the following temperature and time specifications:

Temperature Time Pasteurization Type
69 °C (155 °F) 30 min—vat pasteurization
80 °C (175 °F) 25 s—HTST pasteurization
83 °C (180 °F) 15 s—HTST"
(IDFA, http://www.idfa.org/files/249_Pasteurization Definition and Methods.pdf)

Pasteurization is required of all grade A fluid milk or milk products subject to interstate commerce for retail sale. Traditionally, prevention of tuberculosis (TB) was the primary concern in pasteurization; thus, temperatures of 143 °F (62 °C) were used to destroy *Mycobacterium tuberculosis*, the bacteria causing TB in humans. Actually, *Coxiella burnetii* that causes Q fever requires an even *higher* temperature for destruction; thus, the required 145 °F (63 °C) was set for pasteurization. The high pasteurization temperature, followed by rapid cooling, controls non-pathogenic growth.

A large US foodborne illness outbreak in recent years, where many thousands of people became ill, was attributed to raw milk that inadvertently entered the wrong pipeline (not effectively prevented from entering) before packaging and subsequently contaminated the already pasteurized milk.

Many foods use enzyme tests in determining adequate pasteurization. *Adequate* pasteurization is demonstrated by the absence of the enzyme *alkaline phosphatase*. The phosphatases are highly effective in an alkaline environment. The term may be used synonymously as basic phosphatase. This enzyme is *naturally* present in milk and is *destroyed* (thus no longer present) at temperatures similar to those required for adequate pasteurization. A simple test determines its presence in milk. For example, *inadequate* pasteurization of raw milk reveals the presence of a *high* alkaline phosphatase activity. Inversely, *adequate* pasteurization shows its absence.

"Except for Michigan, not a single state law expressly prohibits the sale of raw milk for animal consumption. The variables are the states' willingness to grant licenses to producers of raw milk for animal feed and how strictly state agencies would monitor licensees to make sure that raw milk sales did only go for animal consumption" (http://www.realmilk.com).

Sterilization (ultra-pasteurization [UP])—pasteurization occurs at *higher* temperatures with a different time:

- 280–302 °F (138–150 °C) for 2–6 s

"Ultrapasteurized when used to describe a dairy ingredient means that such ingredient shall have been thermally processed at or above 280 °F for at least 2 s" (FDA). This UP process still requires that the milk be refrigerated afterward.

The use of *sterilization temperatures* in combination with the use of *presterilized containers*, under *sterile conditions*, creates ultrahigh-temperature (UHT) processing. It *does not* allow spoilage or pathogenic bacteria to enter the milk. If *packaging too* is sterilized, the package is referred to as being "*aseptically* packaged." Thus, milk treated in this manner may be safely stored up to 3 months or longer. An example of this is milk in packaging similar to "juice boxes."

The typical HTST pasteurization of fluid milk does not significantly affect the vitamin content. However, the high heat treatment of ultrahigh-temperature (UHT) pasteurization does cause losses of some water-soluble vitamins.

Regarding minerals, calcium phosphate will travel both in and out of the casein micelle with changes in temperature. Yet this process is reversible at *moderate* temperatures, although at very high temperatures the calcium phosphate may precipitate out of solution and subsequently cause irreversible changes in the casein micelle structure.

Exposure to light will cause a decrease in the levels of riboflavin and vitamin A in milk. Therefore, milk is stored in opaque plastic or paperboard containers that provide barriers to light to maximize vitamin retention. ***See*** http://www.milkfacts.info/Milk%20Composition/Vitamins Minerals.htm.

Homogenization

The primary *function* of ***homogenization*** is to *prevent creaming*, or the rising of fat to the top of the container of milk (whole or reduced-fat milks). The *result* is that milk maintains a more uniform composition with improved body and texture, a whiter appearance, richer flavor, and more digestible curd.

Homogenization *mechanically* increases the number and *reduces* the size of the fat globules. The size is reduced from 18 m to less than 2 m, or 1/10 of their original size. The process of homogenization permanently *emulsifies* the fine fat globules by a method that pumps milk under high pressure [2,000–2,500 lb/in.^{2} (psi)] through small mesh orifices of a homogenizer.

Homogenization offers a *permanent* emulsification because as the surfaces of many new fat globules are formed, each fat globule becomes coated with a part of the lipoprotein membrane and additional proteins from casein and whey. Thus, these proteins adsorb onto the freshly created oil surface *preventing* globules from reuniting or coalescing, and the fat remains homogeneously distributed throughout milk.

Milk may be homogenized *prior to* or *subsequent to* pasteurization. The homogenization process is completed at a *fast* rate to ensure the control of bacteria and loss of quality.

Various characteristics of homogenized milk include the following:

- *No creaming* or separation of cream to the top of the container.
- *Whiter* milk due to finer dispersions of fat. There is an increase in the absorption and reflection of light due to the smaller fat particles.
- *More viscous* and creamy milk due to a greater number of fat particles.
- *More bland* due to smaller fat particles.
- *Decreased fat stability* as fat globule membranes are broken.
- *Less stable to light* and may exhibit light-induced favor deterioration by sunlight or fluorescent light. Thus, paperboard cartons and clouded plastic bottles are used for milk.

Fortification

Fortification is defined as the addition of nutrients at levels beyond/different from the original food. The addition of fat-soluble vitamins A and D to *whole milk* is optional fortification. Low-fat milk, nonfat milk, and low-fat chocolate milk *must* be ***fortified*** (usually before pasteurization) to carry 2,000 International Units (IU) or 140 retinol equivalents (RE) *vitamin A* per quart. It is *required* for milk subjected to *interstate* commerce. *Vitamin D* addition to reach levels of 400 IU per quart is *optional*; however, it is routinely practiced. Evaporated milks must be fortified.

CULINARY ALERT! Vitamin A and D are fat-soluble vitamins, thus are not naturally in milk without fat. Low-/nonfat milk is fortified to contain these vitamins.

In order to add to the viscosity and appearance, as well as the nutritive value of low-fat milk, *nonfat milk solids* (*NFMS*) may be *added* to milk. This addition allows milk to reach a 10 % NFMS (versus 8.25 % usually present), and it will state "protein fortified" or "fortified with protein" on the label.

Bleaching

Bleaching carotenoid or chlorophyll pigments in milk may be desirable. The FDA allows benzoyl peroxide (BP) or a blend of it with potassium alum, calcium sulfate, or magnesium carbonate to be used as a bleaching agent in milk. The weight of BP must not exceed 0.002 % of the weight of the milk, and the potassium alum, calcium sulfate, and magnesium carbonate, individually or combined, must not be more than six times the weight of the BP. Vitamin A or its precursors *may be destroyed* in the bleaching process; therefore, sufficient vitamin A is added into the milk or in the case of cheesemaking to the curd.

The use of whey proteins in food and beverage applications is chiefly derived from annatto-colored cheddar cheese. Since not all of the annatto is removed from the whey, bleaching occurs.

Types of Milk

Fluid Milk

Fluid milk may come from goats (Mediterranean countries), sheep (southern Europe), reindeer (northern Europe), and other animal sources throughout the world. It is *Holstein* cows that typically produce the greatest quantity of milk and are, therefore, the *primary milk cow* in the United States. The Guernsey and Jersey breeds produce milk with the highest percentage of fat—approximately 5 % fat.

Milk appears *white* due to the reflection of light from colloidally dispersed casein protein and calcium phosphate particles in the milk dispersion; however, an *off-white* color may be due to carotenoid pigment in the animal feed. A *bluish* color may be observed in milk skimmed of fat and thus devoid of carotenoid pigments.

Both the fat content and percent of MSNF of fluid milk are subject to FDA regulations and new technological developments. The butterfat and caloric content of milk are as follows:

Type of milk	Fat percent	Calories
Whole	3.25	150 calories/ 8 ounce
Reduced fat	2	120 calories/ 8 ounce
Low fat or light	0.5, 1.0	100 calories/ 8 ounce (1 %)
Nonfat, fat-free/"skim" (fat skimmed off)	<0.5	90 calories/ 8 ounce

Flavored milk contains fat, protein, vitamin, and mineral contents similar to the type of milk to which the flavoring was added—whole, reduced fat, and so forth. It will vary in caloric and carbohydrate values according to added ingredients.

CULINARY ALERT! Substitutes for 1 cup whole milk:

- 1/2 cup evaporated milk + ½ cup water—reconstituted
- 1/3 cup NFMS in measuring cup + water to reach 1 cup mark of cup—reconstituted
- 1 cup buttermilk + ½ tsp. baking soda

Evaporated and Concentrated Milks

Evaporated and *condensed*, or *sweetened condensed*, coupled with packaging in cans, extends the shelf life of the milk. *Cans* of evaporated milk may be adequately stored for extended time periods, although due to the Maillard reaction (more later), undesirable tan or brownish color or flavor changes may occur after 1 year's time. Rehydration may then be made difficult.

CULINARY ALERT! Discoloration is not indicative of possible foodborne illness. Once the can has been opened, it should be refrigerated and can be held for up to 1 week.

Evaporated milk is concentrated through the process of evaporation [at 122–131 °F (50–55 °C)] in a vacuum chamber. Either *whole* or *nonfat* milk with 60 % of the water removed is then homogenized, fortified with vitamins A and D, canned, and sterilized in the can [240–245 °F (115–118 °C)] in a pressure canner.

Whole evaporated milk must contain not less than 25 % total milk solids and not less than 7.5 % milk fat. Evaporated nonfat milk must contain not less than 20 % milk solids and no more than 0.5 % milk fat. It must be fortified with 125 and 25 IU of vitamins A and D, respectively.

Milk is increasingly *less* stable with the progression of concentration and heat and it may coagulate, so the stabilization of milk proteins is better assured by preheating (forewarming) milk prior to sterilization at temperatures of 203 °F (95 °C) for 10–20 min. This *forewarming* is designed to denature colloidally dispersed serum proteins and to shift salt balance of calcium chloride and phosphates that are in solution. Disodium phosphate or *carrageenan* may be added to stabilize casein against precipitation (Chap. 5).

As mentioned previously (see section "Safety/Quality of Milk"), an undesirable browning may occur in canned milk. The high temperature used in processing evaporated milk or a long storage of the product may produce a *light tan* color due to the early stages of the ***Maillard reaction*** between the milk protein and the milk sugar, lactose. This color change is *not* a microbial threat.

CULINARY ALERT! Evaporated milk is reconstituted (rehydrated) at a 1:1 ratio of evaporated milk and water, adding slightly less water than was removed in the 60 % evaporation.

Sweetened condensed milk is concentrated whole or nonfat milk with approximately 60 % of the water removed, and sugar levels of 40–45 % in the finished product. There is a calorie difference in this milk processing, as whole sweetened condensed milk contains no less than 8 % milk fat and 28 % total milk solids, and nonfat contains no more than 0.5 % milk fat and 24 % total milk solids.

Sweetened condensed milk is *pasteurized*, although *not sterilized*, because the *high* sugar content (usually at least 60 % in the water phase) plays a role in preventing bacterial growth. This is due to the *osmotic* effect of the sugar that competes with the bacteria for water and, thus, controls bacterial growth.

Dried Milk

Dried milk powder may be processed from either pasteurized *whole* or, more commonly, from *nonfat* milk. One method of drying involves *spray* drying. Milk is first condensed by removing two-thirds of the water and is typically sprayed into a heated vacuum chamber (spray drying) to dry to less than 5 % moisture levels. The drying process has *no* appreciable effect on the nutritive value of milk (National Dairy Council). Most nonfat dry milk is fortified with vitamins A and D.

Instant nonfat dry milk or *agglomerated* milk has some moisture added back to the spray-dried milk powder. As powder, it is easily pourable and dispersible in cold water. When reconstituted, the taste is best when the milk is prepared ahead and served well chilled.

CULINARY ALERT! Three and a half ounces (1–1/3 cups) of dried milk powder is needed to yield 1 quart of fluid milk. Nonfat dried milk (NFDM) may be added to foods to increase the protein or calcium content.

Dry whole milk is pasteurized whole milk with the water removed. It has limited retail distribution—mainly for use in infant feeding and for people without access to fresh milk, such as campers. Dry whole milk is usually sold to chocolate and candy manufacturers.

Tips on Dry Whole Milk: An opened package should be tightly sealed and stored in a cool, dry place. Dry whole milk develops off-flavors if not used soon after opening. (USDA)

In addition to whole or nonfat milk, *buttermilk* and *whey* may also be dried. Whey is of high biological value containing lactalbumins and lactoglobulins, with *one-half* of the protein and slightly *more* lactose than NFDM.

In particular, dried milk is an economical form of milk for shipping. It has an extended shelf life and is useful as an ingredient for addition to numerous other foods.

Cultured Milk/Fermentation

Cultured products are ***fermented*** by the addition of bacterial cultures, such as *Lactobacilli* and *Streptococci*, to fluid dairy products. These harmless bacteria (or bacterial enzymes) induce a chemical change in the organic substrates of milk solids. Lactose is fermented to lactic acid creating a low pH in the process, which

(1) controls both spoilage and pathogenic bacterial growth and (2) causes the casein to coagulate.

In earlier days, warm milk from various animals (cows, sheep, goats, camels) was preserved for several days or weeks, with no need for refrigeration. This was achieved by the addition of a small milk culture from a preceding batch.

Acidified products are produced by *souring* milk with an acid such as lactic, citric, phosphoric, or tartaric acid with or without microorganisms. The addition of lactic acid-producing bacteria is optional, and because cultured and acidified products contain different amounts of lactic acid, they differ in flavor.

The following milk products are examples of some commonly *cultured* milk products:

- ***Buttermilk***
  Traditionally, buttermilk was the liquid that *remained* when cream was churned to form butter. It was a by-product. *Today*, this is *not* the case commercially, because low-fat or skim milk, *not* cream, begins the process. Although its name (*buttermilk*) may mistakenly signify a high-fat content, the opposite is true! It is more correctly named "cultured low-fat milk" or "cultured nonfat milk."

  Buttermilk differs from nonfat milk in that it contains phospholipids and protein from the fat globule membrane, whereas nonfat milk does not. The texture is different as well.
- **Cultured buttermilk**
  Cultured buttermilk is the pasteurized low-fat, nonfat, or whole milk to which a starter culture of *Lactobacilli* and *Streptococci* (*S. lactis*) is added after the mix has been heated and then cooled. These bacteria ferment lactose, producing lactic acid, which clots the milk. Butter flakes or liquid butter, or low levels (0.01–0.15 %) of salt, may be added. *Leuconostoc citrovorum* and *L. destranicum* bacteria, 0.2 % citric acid, or sodium citrate may be added for flavor.
- **Sour cream**
  Traditionally, sour cream was made from heavy (whipping) cream that was soured. *Today*, it is made from pasteurized, homogenized, fresh, *light cream* (approximately 18 % fat) that is coagulated by a method similar to buttermilk (recall that while *buttermilk* starts *with low-fat or skim milk*, *sour cream* production begins with *18 % fat, or perhaps cream*). While inoculation and fermentation steps are *similar* to buttermilk production, fermentation is shortened.

  S. lactis and *Leuconostoc* bacteria may be added for flavor, and stabilizers such as gelatin or gums may be present. *Nonfat milk solids* may be *added* to thicken the cream. A bitter taste in sour cream that is stored more than 3–4 weeks may form due to proteolytic bacterial enzyme activity.
- **Yogurt**
  Yogurt is the food produced by culturing one or more of the pasteurized fluid dairy ingredients such as cream, milk, partially skimmed milk, or skim milk (used alone or in combination depending on the desired fat content) with a bacteria culture. In industrialized regions of the world, yogurt is made with cow's milk.

  Treatment of the milk is that it is both pasteurized and homogenized *before* the addition of a starter which contains the lactic acid-producing bacteria, *L. bulgaricus* and *S. thermophilus*. The process used to make yogurt is similar to buttermilk and sour cream, although the incubation temperature and types of bacteria are different.

  Denaturing proteins (unfolding the native chains or globular shape) becomes important for digestion and for yogurt production wherein the whey proteins bind water and provide a characteristic yogurt texture.

  Yogurt may be made using whole, low-fat, or skim milk. The formulation may include nonfat dry milk (NFDM) or condensed skim milk to boost its solids. It contains not less than 8.5 % MSNF and not less than 3.25 % milk fat. Or it may be prepared to be a reduced- or low-fat yogurt and have levels of 0.5–2.0 % milk fat or less. Other optional ingredients include buttermilk, whey, lactose, lactalbumins, lactoglobulins, or whey modified by partial or complete removal of lactose and/or minerals to increase the nonfat solid contents of the food. New research and development continues to explore additional optional ingredients.

Microorganisms in yogurt exist in a *friendly* form, known as *probiotic flora*. Such probiotic yogurt, with *Lactobacillus* and *Bifidobacterium*, is able to survive destruction during gastrointestinal (GI) passage and offer health benefits such as immune stimulation and positive balance to the GI microflora (Hollingsworth 2001). The Food and Agriculture Organization of the United Nations (FAO) defines probiotics as "live microorganisms administered in adequate amounts which confer a beneficial health effect on the host" (FAO). Most probiotics are bacteria, one is a yeast—*Saccharomyces boulardii* (Hollingsworth 2001).

The National Yogurt Association's "live and active cultures" seal indicates that the yogurt contains at least 100 million *L. acidophilus* bacteria per gram at the time it is manufactured, although this number diminishes with time and the microbial enzyme lactase.

Frozen yogurt may contain stabilizers for freezer stability, sugar, and added milk solids. The different types of yogurt, including sundae-style or, the blended, Swiss yogurt, are cultured and stored in different manners. Nutritive or nonnutritive sweeteners may be added, as well as flavoring agents, color additives, and stabilizers such as gelatin, gums, and pectin (Chap. 17).

- **Acidophilus milk**
  Acidophilus milk is a cultured product made from pasteurized low-fat, nonfat, or whole milk. *Lactobacillus acidophilus* is added and incubated at 99 °F (37 °C). Although *not* proven yet, a possible *benefit* of consumption is that ingestion can produce a number of B vitamins, thereby replacing what may have been destroyed during antibiotic treatment. A variation of this is sweet acidophilus milk. This sweet version has culture added, however is not incubated. It is thought to be therapeutic without the characteristic high acidity and flavor.

  Acidophilus produces the enzyme lactase and helps correct the symptoms of lactose intolerance. It is thought that lactase in combination with *L. acidophilus* is enabled to pass successfully through the stomach acids and reach the small intestine where it functions in lactose digestion, preventing the discomfort experienced by those individuals who are lactose intolerant and unable to digest lactose (National Dairy Council).
- **Kefir**
  Kefir is another, less well-known, fermented, probiotic milk product. It contains numerous bacteria including *Lactobacillus caucasicus* and the yeasts *Saccharomyces kefir* and *Torula kefir*. As well, it is slightly bubbly due to the fermentation process, and it may therefore contain a small amount—approximately 1 %—of alcohol.

Fermented dairy products have been and *are* used routinely throughout the world—it is a way of life and there is nothing novel about it. Yogurts, smoothies, and a plethora of flavors may be created using kefir. Each introduces live bacteria for good gut health. "… kefir in particular … adding excitement to the drinkable yogurt shelf." The lumps in kefir "grains" are not grains, but, rather, are "little clumps of bacteria, yeasts, sugars, proteins and lipids" (Decker 2012).

"Prebiotics are nondigestible carbohydrates that act as food for probiotics. When probiotics and prebiotics are combined, they form a synbiotic [the term 'synbiotic' should be used only if the net health benefit is synergistic—the United Nations Food & Agriculture Organization (FAO)]. Fermented dairy products, such as yogurt and kefir, are considered synbiotic because they contain live bacteria and the fuel they need to thrive.

Probiotics are found in foods such as yogurt, while prebiotics are found in whole grains, bananas, onions, garlic, honey and artichokes. In addition, probiotics and prebiotics are added to some foods and available as dietary supplements" (http://www.mayoclinic.com).

Additional specialty types of milk include low-sodium, lactose-reduced milk, calcium-fortified, as well as flavored milks, and shakes. Non-milks such as rice and soy "milk" are also consumed. The latter are especially useful to persons who are lactose intolerant.

Other Milk Products

Butter

Butter is a concentrated form of fluid milk, produced through churning of pasteurized cream. ***Churning*** involves agitation that breaks fat globule membranes so the emulsion breaks, fat coalesces, and water (buttermilk) escapes. Emulsions may be of two types. The original 20/80 *oil-in-water* type of emulsion of milk becomes a 20/80 *water-in-oil* emulsion. Milk is churned to form butter and the watery buttermilk. Butter may have a yellow color due to the fat-soluble animal pigment, carotene, or an additive.

> Butter is made by churning pasteurized cream. Federal law requires that it contain at least 80 percent milkfat. Salt and coloring may be added. Nutritionally, butter is a fat; one tablespoon contains 12 grams total fat, 7 grams saturated fatty acids, 31 milligrams cholesterol, and 100 calories. Whipped butter is regular butter whipped for easier spreading. Whipping increases the amount of air in butter and increases the volume of butter per pound. The USDA grade shield on butter packages means that butter has been tested and graded by experienced government graders. In addition to checking the quality of the butter, the graders also test its keeping ability. (USDA)

Today, there are various blends of butter and margarine in the market. The fat composition and taste differ from the original. Margarine, or oleomargarine, is the food in plastic form or liquid emulsion containing not less than 80 % fat. It may be produced from water and/or milk and/or milk product, unsalted, or lactose-free. It contains vitamin A and may contain vitamin D.

Sweet cream butter is made by the addition of *S. diacetylactis*, which ferments the citrate in milk to acetaldehyde, acetic acid, and diacetyl, the last being the major flavor compound of butter. Commercially, it may contain salt, yet is known as "sweet cream" butter, because today, the butter is prepared from sweet, not the traditional soured cream. The USDA grade AA is of superior quality, USDA grade A is very good, and grade B is standard.

Spreads contain a higher percentage of water and may not be suitable for some baking and cooking applications.

Cream

> Cream means cream, reconstituted cream, dry cream, and plastic cream. Water, in a sufficient quantity to reconstitute concentrated and dry forms, may be added. (FDA)

Cream is the high-fat component separated from whole milk as a result of the creaming process. It has a higher proportion of fat droplets to milk than regular fluid milk, and according to federal standards of identity, cream must contain *18 % milk fat* or more. Due to this high-fat content of cream compared to milk, some yellow, fat-soluble pigments may be apparent. Some fats are naturally small and do not coalesce.

Various liquid creams available for use in foods include the following:

- Light (coffee) cream—18–30 % butterfat
- Light whipping cream—30–36 % butterfat
- Heavy cream—36 % butterfat, minimum
- "Half-and-Half" cream diluted with nonfat milk—10.5 % butterfat
- Whipping cream packaged under pressure in aerosol cans—may be nonfat or contain various levels of fat, sugar, flavoring, emulsifiers, and a stabilizer

Ice Cream

Ice cream is sometimes referred to as an "indulgent" food, meaning that while fat is reduced elsewhere in the diet, ice cream consumption may not decrease! While ice mixes were enjoyed for centuries prior to this, the first commercial, wholesale ice cream was manufactured in 1851, in Baltimore, MD.

Ice cream is a food produced by freezing, while stirring a pasteurized mix containing *dairy* product. The mix consists of one or more dairy ingredients such as cream, milk, skim milk, sweet cream buttermilk or sweetened condensed milk, and optional caseinates. *In addition* to the dairy ingredient, sherbet, low-fat ice cream, and ice creams contain other ingredients. Typically, sugar (sucrose, dextrose, which flavors and depresses the freezing point), cookies, eggs, fruit, nuts, and other ingredients such as coloring or flavoring agents, emulsifiers [such as egg yolks, polysorbate 80 (a sorbitol ester consisting of a glucose molecule bound to the fatty acid, oleic acid), or mono- and diglycerides], stabilizers (gelatin, vegetable gum), and water are added.

The ice cream mix is subject to pasteurization, homogenization, holding (for aging), and quick freezing. Slow freezing creates larger ice crystals. Air is naturally incorporated into an ice cream mixture by agitation, although excessive air may *not* be whipped into a mix as specified by federal and state standards. The increase in volume due to air is ***overrun*** and is calculated as

$$\begin{aligned}\%\ \text{Overrun} &= (\text{Volume of ice cream} \\ &\quad - \text{Volume of mix}) \times 100 \\ &= \text{Volume of mix}\end{aligned}$$

For instance, if a 1-gallon container of ice cream contains an equal measure of ice cream mix and air, it has 100 % overrun. Overrun in ice creams may range from 60 % to greater than 100 %.

Ice cream contains not less than 10 % milk fat, nor less than 10 % MSNF, except when it contains milk fat at 1 % increments above the 10 % minimum.

Low-fat ice cream (formerly ice milk) contains less fat and more MSNF, and deluxe ice cream contains more milk fat and less MSNF. Other frozen desserts may include milk and varying percentages of milk fat or perhaps a fat substitute.

Blended milk products are fruit juices and milk, which may contain added lactic acid, or caffeine, plus other ingredients, and may be prepared using herb teas and additional sugars.

Sherbet contains 1–2 % milk fat and 2–5 % total milk solids. A greater amount of sugar and less air (hence 30–40 % overrun) than ice cream are standard.

Percent milk fat	Minimum percent MSNF
10	10
11	9
12	8
13	7
14	6

Whey

Whey has previously been discussed as the aqueous (serum) protein in milk, yet it warrants further discussion due to its *increasing* use in consumer products. Research is ongoing to target separating milk serum proteins from liquid milk prior to cheesemaking. Some cheese such as ricotta cheese may be made *partially* of whey.

Whey comprises approximately *20 %* of the protein in milk. It contains the albumins and globulins, the majority of *lactose*, and the water-soluble nutrients, such as riboflavin. Whey is the *by-product* of cheesemaking, the liquid that remains after curds are formed and drained (recall the nursery rhyme Little Miss Muffet—eating her curds and whey!). A tremendous quantity of cheese is manufactured, and, currently, more satisfactory ways of using whey are being explored.

Whey is a nutritious product. It may also be used in beverages, frozen dairy desserts, and baked goods. In a dried form, it may have useful applications as an emulsifier and in providing extra protein to foods. Whey also has foaming and gelling applications. Yet, because it contains lactose, which the majority of the world cannot digest (see section "Lactose Intolerance"), it *cannot* be used in worldwide feeding.

Whey begins to precipitate at temperatures below the coagulation temperature of casein, yet is not precipitated at a pH of 4.6 or by rennin, as is casein. Evidence of whey precipitation is seen when the lactalbumin coagulum (as well as calcium phosphate) sticks to the bottom of the pan and scorches.

In addition to some uses previously listed, whey is concentrated by ultrafiltration to yield *whey protein concentrates* (WPCs). WPC/whey protein isolate (WPI) is also used in sports supplements and bars due to their high nutritional value; WPCs are frequently added to yogurt and dried for use in such items as coffee whitener, whipped toppings, meringue, fruit beverages, chocolate drinks, and processed meats.

Further purification steps may be added to yield WPIs. For example, WPIs are used in infant formulas, and whey refinery may yield proteins used to fortify clear bottled drinks, including sodas. Fractionation in the whey refining process could lead to products without phenylalanine and thus to products with useful ingredients to people with phenylketonuria (PKU) (Food Eng 2000a).

Cooking Applications

Cooking applications subsequent to the mild denaturation or change in molecular structure of proteins may form cross-links and coagulate milk. ***Coagulation*** and precipitation of clumps or aggregates may occur with heat or when acid, enzymes, or salts are in a formulation. In more severe heat, acid, enzyme, or salt treatment, unwelcome curdling may be expected to occur. Some of these effects are as follows:

- **Heat**: Heat, especially direct or high heat, may *denature*, *coagulate*, *or curdle* milk. Slow, *low*, or moderate heat such as indirect heating over a water bath should be used for milk-based products. Increasing temperatures and length of heating may break the fat emulsion if the protein film around the fat globules breaks. Thus, the fat will coalesce. *High* heat also forms greater amounts of coagulum at the bottom of the pan than low heat. The *same* calcium phosphate compound that forms at the bottom of the pan by scorching also forms a skin (scum or film) at the surface of the food as water evaporates. This surface skin may "hold in" heat and lead to a boilover of the milk product subjected to heat. Prevention includes use of a pan lid or surface application of an agent such as fat.

CULINARY ALERT! Cooking with a cover is recommended in order to prevent skin formation. Stirring constantly also helps avoid protein precipitation on the sides of the cooking vessel.

- **Acid**: Acid may come from a variety of sources. It be *added* to food or be a *part* of a food, or it may be produced by bacteria. It coagulates milk mixtures by forming unstable *casein* proteins. Casein precipitates at a pH of approximately 4.6 (recall that *whey* proteins are not precipitated by acid). Use of a white sauce may control precipitation.
- **Enzyme coagulation**: As will be discussed in the chapter section on "Cheese," *several* sources of enzymes are responsible for coagulation and curd formation—animal, plant, or microbial enzymes. However, the *primary* enzyme used to coagulate milk in cheese or ice cream is rennin (commercially known as rennet).

  Rennin requires a slightly acidic environment and functions best at temperatures of 104–108 °F (40–42 °C), rather than high temperatures. Calcium is retained if the coagulation of milk is achieved by rennin rather than acid (e.g., some custard-like desserts and cottage cheese).
- **Polyphenolic compound coagulation**: *Phenolic* compounds (formerly called tannins) are in some plant materials including fruits and vegetables (e.g., potatoes, tomatoes), tea, and coffee, and they coagulate milk. Although baking soda (alkali) may be added to milk

combinations to shift the pH and control curdling, it is not recommended, as it destroys vitamin C in the product. Low heat and a gelatinized starch buffer (white sauce) may be used for controlling this undesirable coagulation.

- **Salt coagulation**: *Calcium* and *phosphorus salts* present in milk are less soluble with heat and may coagulate milk protein. Salty foods such as ham as well as some vegetables and salt flavorings that are added to milk frequently may cause the milk to curdle. As with acid-cause coagulation, a gelatinized starch buffer is used to prevent undesirable precipitation.

Fig. 11.1 Cheeses (courtesy of SYSCO® Incorporated)

Cheese

Cheese as defined by the FDA is "a product made from curd obtained from the whole, partly skimmed, or skimmed milk of cows, or from milk of other animals, with or without added cream, by coagulating with rennet, lactic acid, or other suitable enzyme or acid, and with or without further treatment of the separated curd by heat or pressure, or by means of ripening ferments, special molds, or seasoning" (FDA).

With a look at an amount of US cheese consumption over 25 years, beginning in 1985, the per capita cheese rose steadily from 22.5 lb in 1985 to a projected 34.9 lb (Wisconsin Milk Marketing Board).

Cheese (Fig. 11.1) is a concentrated form of milk that contains casein; various percentages of fat, primarily saturated fat; mineral salts; and a small portion of milk serum (whey proteins, lactose, and water-soluble vitamins). It is the curd that forms as a result of casein coagulation by the enzyme rennin (also known as chymosin) or lactic acid. It requires approximately 10 lb of milk to make a pound of cheese.

"Chymosin, known also as rennin, is a proteolytic enzyme synthesized by chief cells in the stomach. Its role in digestion is to curdle or coagulate milk in the stomach, a process of considerable importance in the very young animal. If milk were not coagulated, it would rapidly flow through the stomach and miss the opportunity for initial digestion of its proteins.

"Chymosin efficiently converts liquid milk to a semisolid like cottage cheese, allowing it to be retained for longer periods in the stomach. Chymosin secretion is maximal during the first few days after birth, and declines thereafter, replaced in effect by secretion of pepsin as the major gastric protease.

In days gone by, chymosin was extracted from dried calf stomachs for this purpose. Presently, the cheesemaking industry has expanded beyond the supply of available young calves. Many proteases are able to coagulate milk by converting casein to paracasein; and alternatives to chymosin are readily available. "Rennet" is the name given to any enzymatic preparation that clots milk."

Animal (calf, bovine pepsin), plant (papain), and microbial protease *enzymes* clot milk to form curds. Genetic engineering of bacteria has produced new options. Cheesemaking typically uses rennin and pepsin. Rennin produces clots that are rich in calcium (although slightly tougher curds form with rennin than lactic acid).

- ***Rennin*** is from the stomach of milk-fed *calves*. Although rennin is active at neutral pH, the enzyme clots milk much faster in acidic conditions, such as when lactic acid is used.

Biotechnology has enabled the *specific gene* that produces rennin to be reproduced in *bacteria without* extracts from the calves' stomach. Rennet (the commercial name for rennin) is then produced through fermentation. In fact, half of rennet in cheese production is produced through fermentation (IFIC).

- *Pepsin* is from the stomach of *pigs* (swine).
- *Proteases* from fungi.
- *Plant enzymes* such as *papain* (from papaya) and *ficin* (from figs) may be used by industry to clot milk casein and form some cheeses.

In general, cheese is classified according to (1) the moisture content, producing either very hard, hard, semisoft, or soft cheeses, and (2) the kind and extent of ripening. A brief explanation appears below.

Moisture content
- *Very hard cheese*—e.g., Parmesan and Romano.
- *Hard cheese contains 30–40 % water. It has very tiny fat globules and is a near-perfect emulsion.* For example, cheddar, Colby, Gouda, and Swiss cheese.
- *Semisoft cheese*—blue, feta, Monterey Jack, mozzarella, Muenster cheese, and provolone cheese.
- *Soft cheese* contains 40–75 % water and has large fat globules. It is only slightly emulsified. For example, Brie, Camembert, ricotta, and cottage cheese.

Ripening

Ripening may require 2–12 months. In that time, the changes involve the following:

- *Carbohydrate* lactose is fermented by lactase to lactic acid.
- *Fat* is hydrolyzed by lipase.
- *Protein* undergoes mild proteolysis to amino acids by rennin.

Ripening refers to the *chemical and physical changes* that occur in the cheese in the time between curd precipitation and satisfactory completion of texture, flavor, aroma, and color development. Ripening modifies the characteristics mentioned, as well as continuing to ferment residual lactose.

First, milk proteins are coagulated with enzymes (rennet) and acids. Then, aging or ripening by bacteria or mold occurs. It may be due to bacteria, bacterial enzymes (chiefly rennin), or the fungus mold and yeast. Some example follows:

- Cheeses, such as cottage cheese or cream cheese, are *not* ripened. Other popular unripened cheeses include feta cheese and ricotta.
- Cheeses may be *ripened with bacteria*. Examples include cheddar cheese, Colby, Parmesan, and Swiss cheese. For example, the holes or eye formation in Swiss cheese is evidence of gas-producing bacteria that exist throughout the interior of the cheese.
- Camembert and Brie, for example, are *ripened by mold* that is sprayed onto the surface of the cheese, or mold may be introduced internally as in the *ripening* of blue cheese that is inoculated with *Penicillium roqueforti*.

According to USDA preliminary 2011 statistics, the average personal consumption rose from 29.8 lb per capita in 2000 to 34.9 lb in 2011 (USDA). In descending order, American, cheddar, and mozzarella cheese are by far the leaders in sales, followed by a distant Monterey Jack, Swiss, and Colby cheese. Many American kitchens also contain Parmesan and perhaps blue cheese.

Cheese production and markets have emerged as important elements of the dairy industry over the past three decades. Supply-and-use analysis shows an upward trend in total cheese consumption over the past three decades. Nielsen 2005 retail Homescan data were used to analyze cheese consumption by location as well as by income, age, and racial/ethnic groups. ...To the extent that increases in consumers' food expenditure translate into more cheese purchases, it is expected that total cheese consumption will continue to rise. However, changes in the demographic profile of the U.S. population may somewhat slow future growth. USDA Outlook No. (LDPM-193-01) 19 pp, August 2010 (Davis et al. 2010)

In the United States, the FDA has requirements for specific standardized cheese that must be followed by manufacturers, packers, and distributors. For some cheese varieties, a starter culture is used.

Curd development begins as a starter culture is added to milk. Once a curd has formed, it is cut, cooked to shrink the curd, and drained of any remaining whey (syneresis). Next, it is salted to provide flavor, draw whey from the curd, and retard microbial growth. The curd is then pressed and fermented with various microorganisms at 40–55 °F (4–13 °C).

Cottage cheese is an example of a cheese that may be made without bacteria or yeast, however with lactic acid. The origins were not in industry, yet rather individual "cottages." Thus, the name! Cottage cheese is a no/low-fat, soft, acid cheese formed by coagulation of casein with lactic acid. It is made from pasteurized skim milk, to which is added *either* lactic acid or a bacterial culture that produces lactic acid to reduce the pH to 4.6.

Cheese is cut and packaged under hundreds of names worldwide. Despite the abundance of names given to various cheeses throughout the world, there are only approximately 18 types that differ in flavor and texture (Potter and Hotchkiss 1998). These types are listed as follows:

- Brick (USA)—semisoft, ripened primarily by bacteria
- Camembert (France)—soft, mold externally applied (*Oidium lactis* and then *P. camemberti*); thin edible crust
- Cheddar (England)—hard, bacteria ripened (*S. lactis and S. cremoris*); most common cheese used for cooking in the United States, colored by annatto (a seed pod extract)
- Cottage cheese—soft, unripened; creamed, low-fat, nonfat, or dry curd
- Cream cheese (USA)—soft, unripened; may be flavored
- Edam (the Netherlands)—hard, ripened; ball shaped with a red paraffin coating
- Gouda (the Netherlands)—semisoft to hard, ripened; similar to Edam
- Hand—soft
- Limburger (Belgium)—soft, surface bacteria ripened (Bacterium linens)
- Neufchatel (France)—soft, unripened in the United States; ripened in France
- Parmesan (Italy)—hard, bacteria ripened
- Provolone (Italy)—hard, ripened
- Romano (Italy)—very hard, ripened
- Roquefort (France)—semisoft, internally mold ripened (*P. roqueforti*)
- Sapsago (Switzerland)—very hard, ripened by bacteria
- Swiss, Emmentaler (Switzerland)—hard, ripened by gas-forming bacteria (*S. lactis* or *S. cremoris* and *S. thermophilus*, *S. bulgaricus*, and *P. shermani*)
- Trappist—semisoft, ripened by bacteria and surface microorganisms
- Whey cheeses, such as ricotta (Italy), which may be a combination of whole and low-fat milk or whey; coagulated by heating, not rennin (Potter and Hotchkiss 1998; some definition in fourth and fifth edn.)

More details of the types of cheese available to the consumer are listed and explained below:

- *Natural cheese* is the curd of precipitated casein—either ripened or unripened. It may be overcoagulated and allow water to be squeezed out, or the fat emulsion to break when exposed to *high* heat, in which case it shows a separated appearance and stringy texture. Therefore, *low* heat should be used when cooking with natural cheese.

- *Pasteurized process(ed) cheese* is the most common cheese produced in the United States. By FDA ruling, it is "prepared by comminuting and mixing, with the aid of heat, one or more cheeses of the same, or two or more varieties, except cream cheese, neufchatel cheese, cottage cheese, low-fat cottage cheese, cottage cheese dry curd, cook cheese, hard grating cheese, semisoft, part-skim cheese, part-skim spiced cheese, and skim milk cheese for manufacturing with an emulsifying agent . . . into a homogeneous plastic mass" (CFR 21).

  The mixture is pasteurized (which halts ripening and its flavor development) for 3 min at 150 °F (66 °C), and salt is added. An emulsifier such as disodium phosphate or sodium citrate is incorporated to bind the calcium and produce a more soluble, homogeneous, and smooth cheese that can withstand higher heat than natural cheese, without coagulating. The melted cheese is placed in jars or molds such as foil-lined cardboard boxes or single-slice plastic wrap.

  This cheese may also contain an optional mold-inhibiting ingredient consisting of not more than 0.2 % by weight of sorbic acid, potassium sorbate, sodium sorbate, or any combination of two or more of these or consisting of not more than 0.3 % by weight of sodium propionate, calcium propionate, or a combination of sodium and calcium propionate. It may contain pimentos, fruits, vegetables, or meats.

  The moisture content of a process cheese made from a *single variety* of cheese is not more than 1 % greater than the maximum moisture content prescribed by the definition and standard of identity, for the variety of cheese used, if there is one. In no case is the moisture more than 43 % (except 40 % for process washed curd and process Colby cheese and 44 % process Swiss and Gruyere).

  The moisture content of a process cheese made from *two or more* varieties of cheese, as opposed to the aforementioned one is not more than 1 % greater than the arithmetical average of the maximum moisture contents prescribed by the definitions and standards of identity, if there is one, for the cheeses used. In no case is the moisture content more than 43 % (40 % cheddar, Colby, 44 % Swiss and Gruyere).

  The fat content of process cheese made from a single variety of cheese is not less than the minimum prescribed by the definition and standard of identity for the variety of cheese used, and in no case is less than 47 % (except process Swiss 43 % and process Gruyere 45 %). The fat content of process cheese made from two or more varieties of cheese is not less than the arithmetical average of the two cheeses, as described above, and in no case is less than 47 % (except the mixture of Swiss and Gruyere 45 %).

- *Pasteurized process(ed) cheese food* is comminuted and mixed and contains not less than 51 % cheese by weight. The moisture is not more than 44 %, and the fat content is not less than 23 %. Thus, it contains less cheese and more moisture than process cheese. It may contain cream, milk, nonfat milk, NDM, whey, and other color or flavoring agents. It has a soft texture and melts easily.

  An emulsifying agent may be added in such quantity that the weight of the solids of such an emulsifying agent is not more than 3 % of the weight of the pasteurized process cheese food (FDA).

- *Pasteurized process cheese spread* is comminuted and mixed. It has a moisture content of 44–60 % and a milk fat level of not less than 20 %. Therefore, it has more moisture and less fat than processed cheese food and can be spread. Gelatin and gums such as carob bean, cellulose gum (carboxymethylcellulose), guar, tragacanth, and xanthan, as well as carrageenan, may be added if such substances are not more than 0.8 % of the weight of the finished product

(FDA). Sodium may be added to retain moisture, and sugar or corn syrup may be added for sweetness.

- *Cold-pack cheese* preparation involves grinding and mixing natural cheese without heat. The moisture content of a cold-pack cheese made from a single variety of cheese is not more than the maximum prescribed for the variety of cheese used (if there is a standard of identity), and the fat content is not less than the minimum prescribed for that cheese, yet is not less than 47 % (except 43 % cold-pack Swiss and 45 % Gruyere).

Although cold-pack cheeses may contain various flavor combinations, manufacturers have/have used the technology to create custom-colored and custom-flavored specialty cheeses as needed (Food Eng 2000b). When made from two or more varieties of cheese, the moisture content should be the arithmetical average of the maximum of the two cheeses, as prescribed by the definition or standard of identity, yet in no case more than 42 %. The fat content is not less than the arithmetical average of the minimum percent of fat prescribed for the cheeses, if there is a standard of identity or definition, but in no case less than 47 % (cold-pack Swiss and Gruyere 45 %).

The lactose content of ripened cheese *decreases* during ripening and is virtually *absent* in several weeks. It is the whey that contains lactose, which some individuals cannot consume (lactose intolerance). The majority of vitamins and minerals remain after ripening, some protein is hydrolyzed by rennin or proteases, and some fat is digested. Grades of US grade AA and A are assigned to some commonly consumed cheeses such as cheddar and Swiss cheese.

CULINARY ALERT! Tips for lengthening shelf life of cheese involve cold storage and lowering the pH. This is achieved by refrigeration and wrapping in a vinegar-soaked cheesecloth.

If mold forms on the cheese, it may not be acceptable to the would-be consumer. Yet, the rule of thumb is not necessarily to discard the entire piece of cheese. Rather, any apparent mold should be cut off deeper than what is seen in order to cut out the roots. The mold may produce a toxin. (Keep in mind that mold is acceptable in certain cheeses such as blue cheese.)

CULINARY ALERT! Blue cheese is made from cow's milk; Roquefort cheese is made from sheep milk. If other cheeses show mold, they can be consumed if mold is cut away deeper than what can be seen. It is recommended to cut off ¼ to 1 in. of this moldy product.

According to a research study by Oregon State University, "Imitation cheese is made from vegetable oil: it is less expensive, but also has less flavor and doesn't melt well. For the record, *Velveeta*® is pasteurized process cheese spread and *Velveeta Light*® is pasteurized process cheese product. *Cheez Whiz*® is labeled as pasteurized process cheese sauce, although that type isn't noted in the *Code of Federal Regulations*" (OSU).

Milk Substitutes Imitation Milk Products

Milk substitute and imitation milk products were officially defined in 1973. At that time, the FDA differentiated between *substitute* and *imitation* products by establishing regulations regarding the use of the two names. More details follow each introduction:

A milk ***substitute*** product is one that resembles the traditional product and is nutritionally equal. A substitute is pasteurized, homogenized, and packaged like milk. It is more economical than real dairy products because it does not contain the costly butterfat.

Filled Milk

Filled milk is an example of milk ***substitute*** and *does not contain milk fat*. It consists of a *vegetable* fat or oil, and nonfat milk solids, so it is a not a substitute for persons with milk allergies. The vegetable fat has *traditionally* been coconut oil, although it may be partially hydrogenated corn, cottonseed, palm, or soy oil. Oil, water, an emulsifier such as monoglycerides or diglycerides, color such as carotene, and flavoring may be added. Filled milk contains no cholesterol.

An ***imitation milk*** product may look and taste like the traditional product, yet is nutritionally inferior. Specifying the term "imitation" on labels is no longer a legal requirement.

Imitation Milk

Imitation milk usually contains *no* milk products at all—*no* milk fat or milk solids. It is composed of water, vegetable oil, corn syrup, sugar, sodium caseinate, or soy, and stabilizers and emulsifiers. Vitamins and minerals may be added to the product to improve the nutritional value. Again, the term "imitation" on labels is no longer a legal requirement.

Food items that are available in the marketplace, including nondairy dry and liquid creamer, may fit into these above categories.

CULINARY ALERT! Milk and milk products with the "Real" symbol on the package indicate that the product is made from real dairy products, not substitutes or imitations.

CULINARY ALERT! Flavored "milk," "butter," "cream cheese," whipped "cream," and other imitation products are readily available in the marketplace. Nondairy "creamers" or whiteners are prevalent in fluid and dehydrated form.

Nutritive Value of Milk and Milk Products

The 1996 FDA ruling for nutritive value of milk and milk products *revoked* the "standard of identity" (prescribed formulation or recipe that the manufacturer needed to follow). The nutrient claims such as "fat-free" and others similar to those carried by other products became the *rule* for dairy labels.

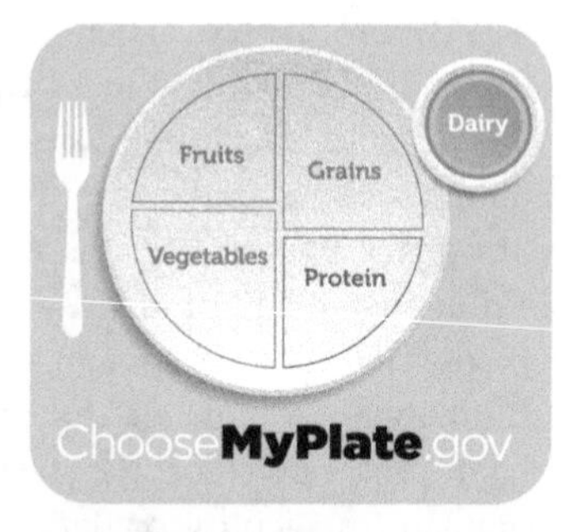

Open to some discussion and further research on fats, the current American Heart Association recommendation states that "We recommend that adults and children age 2 and older use milk that's *low* in dairy fats. This includes fortified fat-free (skim or nonfat) milk, fortified nonfat milk powder, and 1/2 and 1 % low-fat milk. The label on the container should show that the milk has been fortified with *vitamins A and D*. We also recommend buttermilk made from skim milk and canned evaporated skim milk.

10 tips
Nutrition Education Series

got your dairy today?

10 tips to help you eat and drink more fat-free or low-fat dairy foods

The Dairy Group includes milk, yogurt, cheese, and fortified soymilk. They provide calcium, vitamin D, potassium, protein, and other nutrients needed for good health throughout life. Choices should be low-fat or fat-free—to cut calories and saturated fat. How much is needed? Older children, teens, and adults need 3 cups* a day, while children 4 to 8 years old need 2½ cups, and children 2 to 3 years old need 2 cups.

1 "skim" the fat
Drink fat-free (skim) or low-fat (1%) milk. If you currently drink whole milk, gradually switch to lower fat versions. This change cuts calories but doesn't reduce calcium or other essential nutrients.

2 boost potassium and vitamin D, and cut sodium

Choose fat-free or low-fat milk or yogurt more often than cheese. Milk and yogurt have more potassium and less sodium than most cheeses. Also, almost all milk and many yogurts are fortified with vitamin D.

3 top off your meals

Use fat-free or low-fat milk on cereal and oatmeal. Top fruit salads and baked potatoes with low-fat yogurt instead of higher fat toppings such as sour cream.

4 choose cheeses with less fat
Many cheeses are high in saturated fat. Look for "reduced-fat" or "low-fat" on the label. Try different brands or types to find the one that you like.

5 what about cream cheese?
Regular cream cheese, cream, and butter ***are not*** part of the dairy food group. They are high in saturated fat and have little or no calcium.

* What counts as a cup in the Dairy Group? 1 cup of milk or yogurt, 1½ ounces of natural cheese, or 2 ounces of processed cheese.

6 ingredient switches
When recipes such as dips call for sour cream, substitute plain yogurt. Use fat-free evaporated milk instead of cream, and try ricotta cheese as a substitute for cream cheese.

7 choose sweet dairy foods with care
Flavored milks, fruit yogurts, frozen yogurt, and puddings can contain a lot of added sugars. These added sugars are empty calories. You need the nutrients in dairy foods—not these empty calories.

8 caffeinating?
If so, get your calcium along with your morning caffeine boost. Make or order coffee, a latte, or cappuccino with fat-free or low-fat milk.

9 can't drink milk?
If you are lactose intolerant, try lactose-free milk, drink smaller amounts of milk at a time, or try soymilk (soy beverage). Check the Nutrition Facts label to be sure your soymilk has about 300 mg of calcium. Calcium in some leafy greens is well absorbed, but eating several cups each day to meet calcium needs may be unrealistic.

10 take care of yourself and your family

Parents who drink milk and eat dairy foods show their kids that it is important. Dairy foods are especially important to build the growing bones of kids and teens. Routinely include low-fat or fat-free dairy foods with meals and snacks—for everyone's benefit.

Go to www.ChooseMyPlate.gov for more information.

DG TipSheet No. 5
June 2011
USDA is an equal opportunity provider and employer.

Avoid substitutes that contain *coconut oil*, *palm oil or palm kernel oil*. These oils are very high in *saturated* fats. Saturated fats tend to raise the level of cholesterol in the blood. High blood cholesterol is one of the six major risk factors for heart disease that can be changed, treated or modified. It can also lead to developing other heart and blood vessel diseases (AHA)."

Proteins

High-quality proteins are in milk—casein and whey. According to the American Diabetes Association (ADA) Exchange List, an 8-ounce serving of fluid milk contains 8 g of protein, regardless of fat content.

Several milklike substitutes are available on the market—rice milk, soy milk, and others may be found at the grocery store and specialty stores. Such products meet special allergy and nutrition needs.

Fats and Cholesterol

Labeling changes have served both to benefit processors' creativity, such as in developing "light" milk, and to better assist consumers in lowering their fat and saturated fat intake. As shown above (Types of Milk), a label may state whole milk, reduced fat, or fat-free. The calorie levels differ according to the fat content. For example, whole milk contains 150 calories per 8 ounce and skim milk contains 90 calories per 8 ounce. Cholesterol levels range from 4 to 33 mg per cup.

According to the USDA, milk sales have indicated an increase in the sales of reduced-fat and skim milk, while there has been a decrease in sales of full-fat, whole milk.

Carbohydrates

The carbohydrate content of 8 ounces of milk is 12 g regardless of the level of fat. A discussion of lactose intolerance follows.

Vitamins and Minerals

The fat-soluble vitamins A, D, E, and K are present in whole and some reduced-fat milk. Fortification beyond vitamin A and D is not allowed in current standards of identity. Milk is a major source of riboflavin (B_2) in the diet of many populations. Losses of B_2 may occur due to exposure to sunlight as riboflavin is a photosynthesizer. Milk also contains the amino acid tryptophan, a precursor to niacin. Milk is a good source of the mineral calcium.

No apparent undesirable effect on protein, fat, carbohydrates, minerals, and vitamins B_6, A, D, and E is observed with pasteurization. Vitamin K is slightly diminished, and there is less than 10 % loss of thiamin and vitamin B_{12}.

One 8-ounce cup (240 mL) of whole fluid cow's milk contains the following minerals: potassium, calcium, chlorine, phosphorus, sodium, sulfur, and magnesium. Milk does not contain iron. The composition of milks from different species appears in Table 11.2.

Low-sodium milk may be included in diets with sodium restrictions. Sodium may be reduced from a normal amount of 49 mg to about 2.5 mg/100 g of milk by replacing the sodium with potassium in an ion exchange.

Flavored milks are an alternative to such beverages and may assist in consuming calcium (Johnson et al. 2002).

CULINARY ALERT! There is no appreciable effect on the availability of calcium or protein to humans when normal quantities of *chocolate* are added to milk (National Dairy Council).

FDA Report Summary

The Food and Drug Administration (FDA) is announcing that the International Dairy Foods Association (IDFA) and the National Milk Producers Federation (NMPF) have filed a petition requesting that the Agency amend the standard of identity for milk and 17 other dairy products to provide for the use of any safe and suitable sweetener as an optional ingredient. FDA is issuing this notice to request comments, data, and information about the issues presented in the petition. (FDA 2013)

Last modified on December 26, 2012 is An Overview of U.S. State Milk Laws—http://www.realmilk.com.

Lactose Intolerance

Lactose intolerance is an inability to digest the principal milk sugar, lactose. Many individuals demonstrate a permanent loss of the enzyme used to digest lactose. It may be due to the absence of, or insufficient amount of, lactase, a birth deficit, or physical impairment. Caucasians are among the *few* population groups who can digest lactose.

A reminder: Lactose is a disaccharide of glucose and galactose and represents the slightly less than 5 % carbohydrate that is in milk.

If lactose remains undigested by lactase in the intestine, it is fermented by microflora to form short-chain fatty acids and gases such as carbon dioxide, hydrogen, and, in some individuals, methane. Symptoms of lactose intolerance include flatulence, abdominal pain, and diarrhea due to the high solute concentration of undigested lactose. A correct understanding of tolerable doses may be more liberal than expected. Both the lactose-intolerant individual and the food industry may benefit. Also, acidophilus milk contains the needed enzyme lactase and is readily available at many grocery markets.

Lactose assists in the absorption of calcium, phosphorus, magnesium, zinc, and other minerals from the small intestine brush border. Nondairy "milk" such as rice or soy milk or other imitation milk contains no lactose and may be consumed by individuals with milk allergies and by those who would otherwise not drink milk.

The loss of lactase activity in the intestine affects, to some extent, approximately 75 % of the world's population. Individuals with lactose intolerance may compensate by consuming lactase-treated milk (which reduces lactose by 70 %) or purchase the lactase enzyme and administer it directly to milk prior to consumption. It has been shown that small servings (120 mL = 6 g of lactose) of milk and hard cheeses (less than 2 g of lactose) may be consumed without an increase in intolerance symptoms. Hard cheeses contain less lactose than soft cheeses. Up to 12 g of lactose are tolerated, especially if the individual consumes other foods with the source of lactose.

A quantity of fermented products, such as cheese, is tolerated if lactose has sufficiently been converted to lactic acid. Aged cheese is an example of such food. The lactose content of some milk and milk products is given in Table 11.3.

Safety/Quality of Milk

Safe handling was previously discussed in this chapter.

Milk is a highly perishable substance, high in water, with significant amount of protein and a near-neutral pH (6.6)—the qualities that *support bacteria growth*. Details of sanitation are previously mentioned, but it is important to know about the care and safety of milk. Depending on the ingredients, even nondairy imitation "milks" *may* require refrigeration or freezing comparable to the dairy product that they resemble.

Packaging contains a date on the carton that should be followed for a retail sale. Milk may remain fresh and usable for several days past this "sell-by date" if the following directions, suggested by the Dairy Council, are observed:

Table 11.3 Lactose content of milk and milk products

Type of milk	Weight 1 cup (g)	Average percentage	Grams/cup
Whole milk	244	4.7	11.5
Reduced-fat milk (2 %)	245	4.7	11.5
Low-fat milk (1 %)	245	5	12.3
Nonfat milk	245	5	12.5
Chocolate milk	250	4.5	11.3
Evaporated milk	252	10.3	26.0
Sweetened condensed milk	306	12.9	39.5
Nonfat dry milk (unreconstituted)	120	51.3	61.6
Whole dry milk (unreconstituted)	128	37.5	47.9
Acidophilus milk (nonfat)	245	4.4	10.8
Buttermilk	244	4.3	10.5
Sour cream	230	3.9	8.9
Yogurt (plain)	277	4.4	10.0
Half-and-Half	242	4.2	10.0
Light cream	240	3.9	9.3
Whipping cream	239	2.9	6.9

Source: National Dairy Council

- Use proper containers to protect milk from exposure to sunlight, bright daylight, and strong fluorescent light to prevent the development of off-flavor and a reduction in riboflavin, ascorbic acid, and vitamin B_6 content.
- Store milk at refrigerated temperatures [45 °F (7 °C)] or below as soon as possible after purchase.
- Keep milk containers closed to prevent absorption of other food flavors in the refrigerator. An absorbed flavor alters the taste but the milk is still safe.
- Use milk in the order purchased.
- Serve milk cold.
- Return milk container to the refrigerator immediately to prevent bacterial growth. Temperatures above 45 °F (7 °C) for fluid and cultured milk products for even a few minutes reduce shelf life. Never return unused milk to the original container.
- Keep canned milk in a cool, dry place. Once opened, it should be transferred to a clean opaque container and refrigerated.
- Store dry milk in a cool, dry place and reseal the container after opening. Humidity causes dry milk to lump and may affect flavor and color changes. If such changes occur, the milk should not be consumed. Once reconstituted, dry milk should be treated like any other fluid milk: covered and stored in the refrigerator.
- Serve UHT milk cold and store in the refrigerator after opening.

"In 1924, the United States Public Health Service (USPHS), a branch of the FDA, developed the Standard Milk Ordinance, known today as the Pasteurized Milk Ordinance (PMO). This is a model regulation helping states and municipalities have an effective program to prevent milk borne disease. The PMO contains provisions governing the production, processing, packaging and sale of Grade "A" milk and milk products. It is the basic standard used in the, a program all 50 states, the District of Columbia and U.S. Territories participate in.

Forty-six of the 50 have adopted most or all of the PMO for their own milk safety laws with those states not adopting it passing laws that are similar. California, Pennsylvania, New York and Maryland have not adopted the PMO.

Section 9 of the PMO states in part that, "only Grade 'A' pasteurized, ultra-pasteurized or aseptically processed milk and milk products shall be sold to the final consumer, to restaurants, soda fountains, grocery stores or similar establishments" (http://www.realmilk.com).

How USDA's Dairy Grading Program Works

The **US grade AA or grade A** shield is most commonly found on butter and sometimes on cheddar cheese.

US Extra Grade is the grade name for instant nonfat dry milk of high quality. Processors who use USDA's grading and inspection service may use the official grade name or shield on the package.

The "**Quality Approved**" shield may be used on other dairy products (e.g., cottage cheese) or other cheeses for which no official US grade standards exist if the products have been inspected for quality under USDA's grading and inspection program. USDA

. . . the composition or milkfat content given for each product (except for butter) is required under FDA regulations. State laws or regulations may differ somewhat from FDA's. The milkfat content of butter is set by a Federal law. FDA has established a regulation that allows a product to deviate from the standard composition in order to qualify for a nutrient content claim. Products such as nonfat sour cream, light eggnog, reduced fat butter, and nonfat cottage cheese fall into this category. (USDA)

Carbohydrate browning reactions with their color and flavor changes are observed in canned or dry milk that has been subject to either long or high-temperature storage. It should be mentioned here that the browning does not indicate contamination or spoilage. Rather, it is the nonenzymatic ***Maillard browning*** or "carbonyl-amine browning" reaction between the free carbonyl group of a reducing sugar and the free amino group of protein.

Marketing Milk

Marketing milk has made use of the National Milk Mustache "got milk?" Campaign. See the Milk Processor Education Program at http://www.milknewsroom.com/index.htm. This site is designed to be a resource for the media to access information about milk research, milk programs, and the National Milk Mustache "got milk?" Campaign.

Conclusion

Milk is the first food of mammals. It contains major nutrients, carbohydrate, fat, and protein, with water being predominant (88 %). The two major proteins in milk are casein and whey, with additional protein found in enzymes. The fat content of milk is designated by law according to the specific product and jurisdiction.

Milk is pasteurized to destroy pathogens and is homogenized to emulsify fat and prevent creaming. Grade A milk must be treated in this manner if subjected to interstate commerce. Milk may be fluid, evaporated, condensed, dried, or cultured and made into butter, cheese, cream, ice cream, or a variety of other products. It is a potentially hazardous food due to its high protein, water activity, and neutral pH and must be kept cold.

Notes

CULINARY ALERT!

Glossary

Buttermilk, cultured Pasteurized low-fat or nonfat milk to which bacteria are added to ferment lactose to the more acidic lactic acid that clots the casein in milk.

Casein Primary protein of milk, colloidally dispersed.

Casein micelles Stable spherical particles in milk containing α_s-, β-, and κ-casein, and also colloidal calcium phosphate. The micelles are stabilized by κ-casein, which exists mainly at the surface; the α_s- and β-casein fractions are located mainly in the interior of the micelles.

Cheese Coagulated product formed from the coagulation of casein by lactic acid or rennin; may be unripened or bacteria ripened; made from concentrated milk.

Churning Agitation breaks fat globule membranes so the emulsion breaks, fat coalesces, and water escapes.

Coagulate The formation of new cross-links subsequent to the denaturation of a protein. This forms a clot, gel, or semisolid material as macromolecules of protein aggregate.

Creaming Fat globules coalesce (less dense than the aqueous phase of milk) and rise to the surface of unhomogenized, whole, and some low-fat milk.

Cultured See fermented.

Evaporated milk Concentrated to remove 60 % of the water of ordinary fluid milk; canned.

Fermented (Cultured) enzymes from microorganisms or acid that reduce the pH and clot milk by breaking down the organic substrates to smaller molecules.

Fortified Increasing the vitamin content of fresh milk to contain vitamins A and D to levels not ordinarily found in milk.

Homogenization Dispersion of an increased number and smaller fat globules to prevent creaming.

Imitation milk Resembles (looks, tastes like) the traditional product but is nutritionally inferior—contains no butterfat or milk products.

Lactose intolerance Inability to digest lactose due to the absence or insufficient level of intestinal lactase enzyme.

Maillard reaction The first step of browning that occurs due to a reaction between the free amino group of an amino acid and a reducing sugar; nonenzymatic browning.

Milk solids nonfat (MSNF) All of the components of milk solids except fat.

Milk substitute Resembles (looks, tastes like) traditional product and is nutritionally equal; contains no butterfat (e.g., filled milk).

Overrun The increase in volume of ice cream over the volume of ice cream mix due to the incorporation of air.

Pasteurization Heat treatment to destroy pathogenic bacteria, fungi (mold and yeast), and most nonpathogenic bacteria.

Rennin Enzyme from the stomach of milk-fed calves used to clot milk and form many cheeses.

Ripening The time between curd precipitation and completion of texture, flavor, and color development in cheese. Lactose is fermented, fat is hydrolyzed, and protein goes through some hydrolysis to amino acids.

Sterilization Temperature higher than that required for pasteurization, which leaves the product free from all bacteria.

Sweetened, condensed milk Concentrated to remove 60 % of the water, contains 40–45 % sugar.

Total milk solids All of the components of milk except for water.

Whey Secondary protein of milk, contained in serum or aqueous solution; contains lactalbumins and lactoglobulins.

References

A new way to separate whey proteins. Food Eng 2000a; 72(December):13

Davis CG, Blayney DP, Dong D, Stefanova S, Johnson A (2010) Long-term growth in U.S. cheese consumption may slow. United States Department of Agriculture. A report from the Economic Research Service

Decker KJ (2012) Culture splash: fermented dairy beverages. Food Prod Des November:44–53

Hollingsworth P (2001) Food technology special report. Yogurt reinvents itself. Food Technol 55(3):43–49

Johnson RK, Frary C, Wang MQ (2002) The nutritional consequences of flavored-milk consumption by school-aged children and adolescents in the United States. J Am Diet Assoc 102:853–855

Potter N, Hotchkiss J (1998) Food science, 5th edn. Springer, New York

Research yields new reasons to say cheese. Food Eng 2000b; 72(November):16

Bibliography

American Dairy Products. Chicago, IL

American Whey. Paramus, NJ

Associated Milk Producers (AMPI). New Ulm, NM

Centers for Disease Control and Prevention (CDC)

Cheese varieties and descriptions. Handbook, vol 54. USDA, Washington, DC

Dairy and Food Industries Supply Association, Inc. McLean, VA

Dalgleish DG, Corredig M (2012) The structure of the casein micelle of milk and its changes during processing. Annu Rev Food Sci Technol 3:449–467

How to buy cheese. Home and garden bulletin no. 193. USDA, Washington, DC

How to buy dairy products. Home and garden bulletin no. 201. USDA, Washington, DC

http://www.mayoclinic.com/health/probiotics/AN00389—Is it important to include probiotics and prebiotics in a healthy diet?

Model FDA Food Code

National Dairy Council. Rosemont, IL

Standards of identity for dairy products—http://milkfacts.info/MilkProcessing/StandardsofIdentity.htm, Part 131—milk and cream—http://www.access.gpo.gov/nara/cfr/waisidx_06/21cfr131_06.html, Part 133—cheeses and related cheese products—http://www.access.gpo.gov/nara/cfr/waisidx_06/21cfr133_06.html, Part 135—frozen desserts—http://www.access.gpo.gov/nara/cfr/waisidx_06/21cfr135_06.html

USDA ChooseMyPlate.gov

Part IV

Fats in Food

12 Fat and Oil Products

Introduction

Fat is a principal component of the diet. It is enjoyed in the diet due to such characteristics as its *flavor/mouthfeel*, *palatability*, *texture*, and *aroma*. Fats also carry *the fat-soluble vitamins* A, D, E, and K. Sources of fats and oils may be animal, vegetable, or marine that may be manufactured in some combination in industrial processing. *Fats* appear solid at room temperature, whereas *oils* are liquid at room temperature.

Several fats are essential, such as linolenic and linoleic fatty acids, indicating that the body *can either not* make them or make *enough*. Fats and oils are *insoluble* in water and have a greasy feel that the consumer may feel or see evidence of on a napkin or dinner plate. Fats may be *processed* into monoglycerides and diglycerides—glycerol units that have one or two fatty acid chains, respectively—and they may be *added* to many food products functioning as emulsifiers and more.

Some of the functions of fat in food preparation are as follows:

- Add or modify flavor, texture
- Aerate (leaven) batters and doughs
- Contribute flakiness
- Contribute tenderness
- Emulsify (see Chap. 13)
- Transfer heat, such as in frying
- Prevent sticking
- Provide satiety

Edible oils are used in margarines, spreads, and dressings, as retail bottled oils, as frying oils, and more. Soybean oil is currently the highest volume vegetable oil used in the United States. It is incorporated into a variety of products.

Various fat replacements attempt to mimic fat in mouthfeel and perception so that it is good tasting and low-fat. With the use of fat replacements, the caloric and cholesterol level may be made significantly less than a fat. Fats and oils are in many food groups, yet, they are not part of the composition of fruits and many vegetables.

Most current health recommendations state that, as a group, fats and oils should be used sparingly in the diet. Fats and oils are triglycerides, the major constituent of lipids. Overall, lipid is the umbrella term that includes the triglycerides, phospholipids, and sterols.

Structure and Composition of Fats

Glycerides

Glycerides include *mono*glycerides (Fig. 12.1), *di*glycerides, and *tri*glycerides. The first two act

V.A. Vaclavik and E.W. Christian, *Essentials of Food Science, 4th Edition*, Food Science Text Series,
DOI 10.1007/978-1-4614-9138-5_12, © Springer Science+Business Media New York 2014

Fig. 12.1 Formation of a monoglyceride

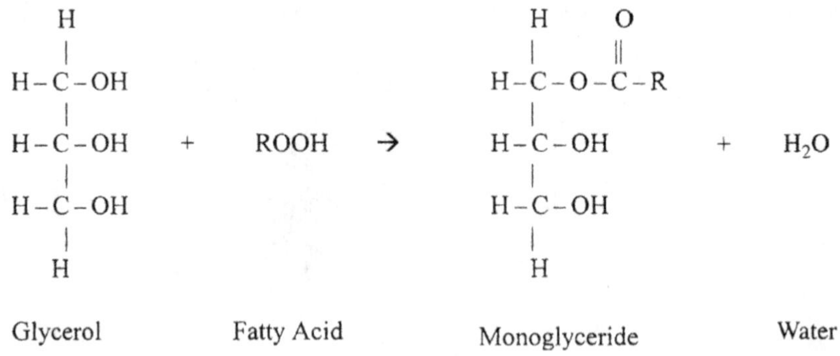

```
   H                                  H     O
   |                                  |     ||
H–C–OH                             H–C–O–C–R
   |                                  |
H–C–OH     +     ROOH     →        H–C–OH        +     H2O
   |                                  |
H–C–OH                             H–C–OH
   |                                  |
   H                                  H

Glycerol       Fatty Acid        Monoglyceride         Water
```

as emulsifiers in foods, while the most abundant fatty substance in food—more than 95 %—is the latter, triglycerides. Triglycerides are insoluble in water and may be either liquid or solid at room temperature, with liquid forms generally referred to as oils and solid forms as fats.

If two fatty acids are esterified to glycerol, a diglyceride is formed, and three fatty acids undergoing the same reaction make a triglyceride. If a triglyceride contains three identical fatty acids, it is called a *simple* triglyceride; if it contains two or three different fatty acids, it is called a *mixed* triglyceride. Spatially, there is no room for all three fatty acids to exist on the same side of the glycerol molecule; thus, triglycerides are thought to exist in either a stair-step (chair) or a tuning-fork arrangement (Fig. 12.2). The arrangement and specific type of fatty acids on the glycerol determine the chemical and physical properties of a fat.

Minor Components of Fats and Oils

In addition to glycerides and free fatty acids, a lipid may contain small amounts of phospholipids, sterols, tocopherols, fat-soluble vitamins, and some pigments. Each is discussed, if only briefly, in this section of the text.

Phospholipids are similar to triglycerides but contain only *two* fatty acids esterified to glycerol. In place of the third fatty acid, there is a polar group containing phosphoric acid and a nitrogen-containing group; the most common phospholipid is ***lecithin*** (Fig. 12.3, and for more, see the chart at the closing of the chapter). Lecithin is found in nearly every living cell. The word is derived from the Greek *lekithos* that means "yolk of an egg," and lecithin is in egg yolk. However, the primary *commercial* source of lecithin is the soybean (Central Soya Company, Inc., Ft. Wayne, IN). Sunflower lecithin is also commercially available.

The two fatty acids of a phospholipid are attracted to fat, whereas the phosphorus and nitrogen portions are attracted to water. Therefore, a phospholipid forms a bridge between fat and water, two ordinarily *immiscible* substances, and thus, *emulsification* is observed (see section "Emulsification," Chap. 13). "Refined" lecithins are modified to provide important surface-active properties to a variety of foods such as instant drink mixes, infant formulas, meat sauces and gravies, dispersible oleoresins, pan releases, chewing gum, and fat-replacer systems (Central Soya Company, Inc., Ft. Wayne, IN).

Lecithins are significant in the food industry, and they are available in numerous forms—the standard fluid, a modified chemical lecithin, a modified enzymatic lecithin, and a deoiled or powdered form. There exist two lecithin properties of significance—acetone insolubles (AI) and hydrophilic/lipophilic balance (HLB). The AI for a standard fluid lecithin is 62–64 %; deoiled lecithin has a minimum of 97 % AI. The HLB value for a standard fluid lecithin is 2–4; deoiled lecithin has a 7–10 HLB. HLB values are indicative of the size and strength of the groups on the lecithin emulsifier. See Table 12.1.

The presence of lecithin promotes a more stable formation of oil-in-water and water-in-oil emulsions (see more in Seabolt 2013).

Fig. 12.2 Fatty acid tuning-fork (*left*) and stair-step or chair arrangements (*right*)

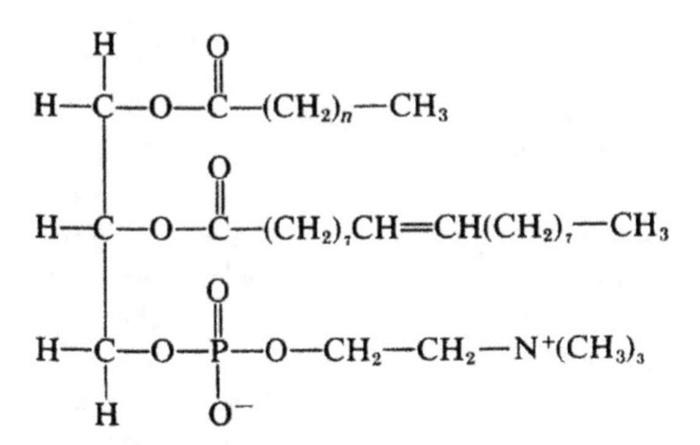

Fig. 12.3 Lecithin (phosphatidylcholine)

Sterols contain a common steroid nucleus, an 8–10 carbon side chain and an alcohol group. The chemists' view of sterols is *unlike* triglycerides or phospholipids—sterols are round in shape. Cholesterol is the primary *animal sterol* (Fig. 12.4) although *plant sterols* or stanols also exist; the most common ones are sitosterol and stigmasterol. Other plant sterols are found in "margarine"-type products, including those marketed under the trade name Benecol®.

Tocopherols are important minor constituents of most *vegetable oils*; *animal fats* contain *little or no* tocopherols. Tocopherols are antioxidants, helping to prevent oxidative rancidity, and are also sources of vitamin E. They are *partially removed* by the heat of processing and may be *added* after processing to improve oxidative stability of oils. If vitamin E is added to oil, for example, the oil is frequently marketed as a source of vitamin E, or as an antioxidant-containing oil.

Vitamins soluble in fat can be carried by fat. The fat-soluble vitamins A, D, E, and K, and if not in a food naturally, or at significant levels, may be added to foods—such as margarine and milk or a wide variety of other foods—in order to increase nutritive value. Fats in the diet promote the absorption of these fat-soluble vitamins.

Pigments such as carotenoids and chlorophylls may be present in fats, and these may impart a distinct color to a fat. Such colors may be removed by bleaching during processing (e.g., milk).

Structure of Fatty Acids

Fatty acids are long hydrocarbon chains, with a methyl group (CH_3) at one end of the chain and a carboxylic acid group (COOH) at the other. Most natural fatty acids contain from 4 to 24 carbon atoms, and most contain an *even* number of carbon atoms in the chain. For example, butyric acid is the smallest fatty acid, having four carbon atoms, and it is found in butter; lard and tallow contain fatty acids with longer hydrocarbon chains.

Fatty acids may be *saturated*, in which case they contain single carbon-to-carbon bonds and have the general formula $CH_3(CH_2)_nCOOH$. They have a linear shape, as shown in Fig. 12.5, and appear solid at room temperature with a high melting point. Fatty acids may be *unsaturated*, containing one or more carbon-to-carbon double bonds. *Monounsaturated* fatty acids, such as oleic acid, contain only one double bond, whereas *polyunsaturated* fatty acids (PUFAs), such as linoleic and linolenic acids, contain two or more double bonds. Generally, *unsaturated* fats are *liquid* at room temperature and have low melting points.

The double bonds in fatty acids occur in either the *cis* or the *trans* configuration (Fig. 12.6), representing different isomeric structures. In the ***cis*** form, the hydrogen atoms attached to the carbon atoms of the double bond are located on the *same* side of the double bond. In the ***trans*** configuration of the isomer, the hydrogen atoms are located on *opposite* sides of the double bond, across from one another.

This *configuration* of the double bonds affects both melting point and shape of a fatty acid molecule. The *trans* double bonds have a *higher* melting point than the *cis* configurations, and *trans* configurations do *not* significantly change

Table 12.1 Lecithin

Lecithin properties
Emulsification; capacity to prevent sticking; improve wettability and dispersibility of powders; in every cell, not strictly plants
Benefits
Provides a "clean" label; dough improvements; anti-staling
Composition—amounts and ratios vary with the plant
Phospholipids—acetone insoluble, glycolipids, neutral lipids, and sugar

PC phosphatidylcholine, *PE* phosphatidylethanolamine, *PI* phosphatidylinositol, *PA* phosphatidic acid

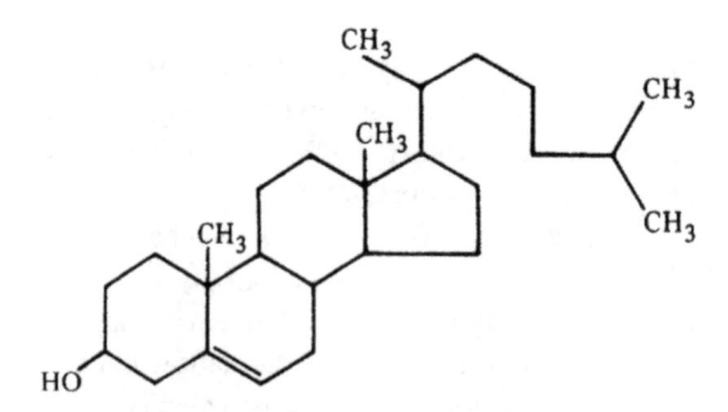

Fig. 12.4 Cholesterol, phytosterols

the linear shape of the molecule. However, a *cis* double bond causes a *kink* in the chain. (A *cis* double bond introduces a bend of about 42° into the linear hydrocarbon chain.) Such kinks affect some of the properties of fatty acids, including their melting points as was mentioned.

Almost all naturally occurring fats and oils that are used in food exist in the *cis* configuration. (Vaccenic acid [11-octadecenoic acid] is a naturally occurring *trans*-fatty acid found in small amounts in the fat of ruminants and in dairy products such as milk, butter, and yogurt. In fact, the name is derived from the word vacca, which is the Latin word for cow. Vaccenic acid comprises about 2.7 % of the fatty acids of milk (MacGibbon and Taylor 2006). *Trans* isomers of conjugated linoleic acid may also occur in trace amounts in these sources; they are synthesized from vaccenic acid by bacteria in the gut.)

In the process of *hydrogenation* of oils, a conversion of some double bonds to the *trans* configuration may be the result in foods (see *trans* fat). The National Cholesterol Education Program (NCEP) has stated that "*trans* fatty acids are another LDL-raising fat that should be kept to a low intake." Specific labeling that includes *trans*-fatty acid content had been desired by some nutrition activists (Huffman 2001, Federation of American Societies for Experimental Biology (FASEB), Bethesda, MD). Effective January 2006, it became law that Nutrition Facts food labels and advertisements must *include* data on *trans*-fatty acids in foods.

As a result of this legislation, some food manufacturing companies made an early decision to simply not use *trans* fats in their products. Food manufacturers may only be required to list *trans* fats if they total more than 0.49 g per serving. Thus, some food content modification may have been necessary for better labeling. The last 15 years have seen a lot of developments in the industry in terms of *trans*-fat-free oils and fats for multiple uses: frying, fillings, and so forth.

Isomerism

Fatty acids may have geometric or positional ***isomers***, which may be *similar* in number of C, H, and O, but which form *different* arrangements, thus offering different chemical and physical properties. Oleic and elaidic acids are examples of *geometric* isomers, existing in the *cis* and *trans* forms, respectively. *Positional* isomers have the same chemical formula; however, the *position* of the double bonds varies. Examples include alpha-linolenic acid, which has double bonds at carbons 9, 12, and 15, counting from the acid end of the chain, and the rare isomer gamma-linolenic acid, which has double bonds at positions 6, 9, and 12.

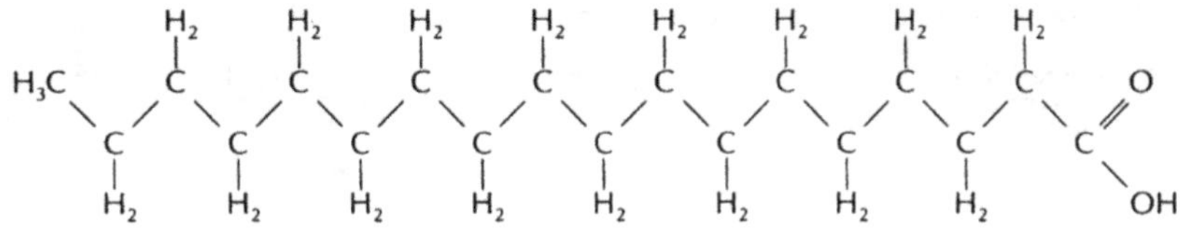

Fig. 12.5 Example of a fatty acid

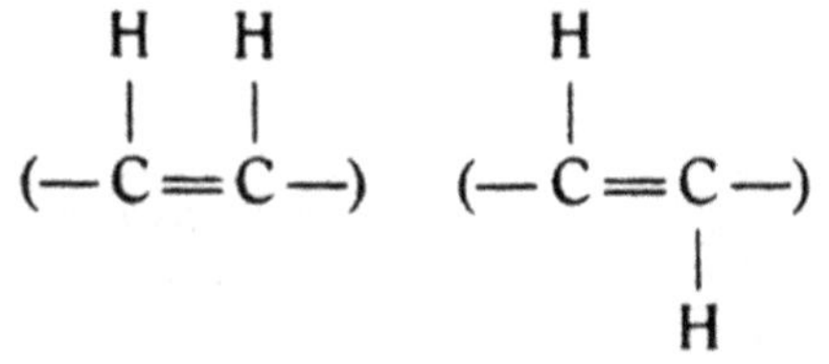

Fig. 12.6 *Cis* (*left*) and *trans* (*right*) configurations representing isomeric structures of fatty acids

Commercial *modification* of fats may produce *either* geometric or positional isomers. Geometric isomers tend to be produced during hydrogenation of fats, and positional isomers may be formed during interesterification or rearrangement of fats.

Nomenclature of Fatty Acids

Fatty acids are named in three ways: (1) each has a *common* or *trivial* name, which has been used for many years, and they also have (2) a *systematic* or *Geneva* name, which is more recent and has the advantage of describing the structure of the fatty acid to which it belongs. In addition, there is (3) the *omega* system, which classifies fatty acids according to the position of the first double bond, counting from the methyl end of the molecule. This system was developed to classify families of fatty acids that can be synthesized from each other in the body. Examples of all three names for some of the most common fatty acids are given in Table 12.2.

Fatty acids are also denoted by two numbers, the first signifying the number of carbon atoms in the chain and the second indicating the number of double bonds present. For example, oleic acid, which contains 18 carbon atoms and one double bond, could be written as 18:1 (Table 12.2).

Geneva or Systematic Nomenclature

The Geneva naming system is a systematic method of naming the fatty acids, and each name completely describes the structure of the fatty acid to which it belongs. Each unsaturated fatty acid is named according to the number of carbon atoms in the chain, as shown in Table 12.1. For example, stearic acid, which has 18 carbon atoms in its chain, has the name octadecanoic acid; **octadec** means 18. The *oic* ending signifies that there is an acid group (COOH) present, and **anoic** signifies that there are no double bonds in the chain. Palmitic acid, which contains 16 carbon atoms, is named hexadecanoic acid. **Hexadec** means 16, and the **anoic** ending again shows that there are no double bonds in this fatty acid chain (the **oic** equals presence of an acid group).

Fatty acids that contain double bonds are also named according to the number of carbon atoms they contain. Therefore, oleic acid (18:1), linoleic acid (18:2), and linolenic acid (18:3) all have **octadec** as part of their name, signifying that they each contain 18 carbon atoms. The rest of the name differs, however, because they contain one, two, or three double bonds, respectively. The number of double bonds and their position in the fatty acid chain are both specified in the name.

It is important to note that the position of each double bond is specified counting from the functional group or **acid end** of the molecule, *not* from the methyl end. Thus, oleic acid has the name 9-octadecenoic acid. The number 9 refers to the position of the double bond between carbon-9 and carbon-10, counting from the acid end. Note that the name ends with **enoic** acid, the **en** signifying that there is a double bond present.

Linoleic acid is named 9,12-*octadecadienoic* acid. Again, the position of double bonds is specified, counting from the acid end. **Octadeca** means that there are 18 carbon atoms in the

Table 12.2 Nomenclature of some common fatty acids

Systematic name	Common name	Carbons: double bonds	Melting point °F (°C)
Ethanoic	Acetic	2	
Butanoic	Butyric	4	18 (−7.9)
Hexanoic	Caproic	6	26 (−3.4)
Octanoic	Caprylic	8	62 (16.7)
Decanoic	Capric	10	89 (31.6)
Dodecanoic	Lauric	12	112 (44.2)
Tetradecanoic	Myristic	14	130 (54.4)
Hexadecanoic	Palmitic	16	145 (62.9)
Octadecanoic	Stearic	18	157 (69.6)
Eicosanoic	Arachidic	20	168 (75.4)
Docosanoic	Behenic	22	176 (80.0)
9-Octadecenoic	Oleic	18:1	61 (16.3)
9-Octadecenoic[a]	Elaidic	18:1	110.7 (43.7)
11-Octadecenoic[a]	Vaccenic	18:1	111.2 (44)
9,12-Octadecadienoic	Linoleic/omega-6	18:2	20 (−6.5)
9,12,15-Octadecatrienoic	Linolenic/omega-3	18:3	9 (−12.8)

Source: Adapted from Institute of Shortening and Edible Oils (Decker 2012)
[a]All double bonds are in the *cis* configuration except for elaidic acid and vaccenic acid, which are in the *trans* configuration. Vaccenic occurs naturally; elaidic is produced by hydrogenation

chain, and **dien** signifies that there are two double bonds in the chain. Similarly, linolenic acid, which contains three double bonds, is named 9,12,15-*octadecatrienoic* acid. The letters **trien** indicate that there are three double bonds in the chain, and again their positions are specified counting from the acid end of the molecule.

The configuration of the double bonds may also be specified in the name. For example, oleic acid and elaidic acid are geometric isomers, because the double bond in oleic acid exists in the *cis* configuration, whereas elaidic acid contains a double bond in the *trans* configuration. The complete name for oleic acid is ***cis*, 9-octadecenoic acid**, and elaidic acid is named ***trans*, 9-octadecenoic acid**.

By looking at a systematic name for a fatty acid, it is possible to tell how many carbon atoms it contains, and how many double bonds and where they are located. Each name gives important information about the fatty acid that is not available just by looking at the trivial or omega name of the acid.

The Omega Naming System

The omega naming system is used for unsaturated fatty acids and denotes the position of the first double bond in the molecule, counting from the **methyl** (CH_3) end, not the acid (as in the Geneva system). This is because the body lengthens fatty acid chains by adding carbons at the acid end of the chain. Using the omega system, a family of fatty acids can be developed which can be made from each other in the body. For example, an omega-6 fatty acid contains its first double bond between carbon-6 and carbon-7, counting from the methyl end. Linoleic acid is an example of an omega-6 fatty acid, and it is the primary member of the omega-6 family. Given linoleic acid, the body can add two carbon atoms to make arachidonic acid (20:4), which is also an omega-6 fatty acid.

The primary omega-3 fatty acid is linolenic acid, which contains three double bonds. The first double bond is located on carbon-3, counting from the methyl end. The body can synthesize both eicosapentaenoic acid (EPA: 20:5) and docosahexaenoic acid (DHA: 22:6) from linolenic acid. Both EPA and DHA are omega-3 fatty acids, because their first double bond is located at carbon-3 (again, counting from the methyl end of the molecule).

Properties of Fats and Oils

Crystal formation: When liquid fat is cooled, the molecular movement slows down as energy is removed, and the molecules are attracted to each other by van der Waals forces. These forces are weak and of minor significance in small molecules. However, their effect is cumulative, and in large or long-chain molecules, the total attractive force is appreciable. Consequently, fat molecules can align and bond to form crystals.

Symmetrical molecules and molecules with fatty acids that are similar in chain length align most easily to form crystals. Fats containing *asymmetrical* molecules and molecules containing kinks due to double bonds align less easily, because they cannot pack together closely in space. Molecules that align easily need less energy to be removed before they will crystallize, and so they have high melting points. They also tend to form *large* crystals. Molecules that do not align easily have low melting points, because more energy must be removed before they crystallize and they tend to form *small* crystals.

Polymorphism

Fats can exist in different crystalline forms, and this phenomenon is known as ***polymorphism***. A fat may crystallize in one of four different crystal forms, depending on the conditions during crystallization and on the composition of the fat. The smallest and least stable crystals are called **alpha** (α) **crystals**. These are formed if fats are chilled rapidly. The alpha crystals of most fats are unstable and change readily to **beta prime** (β′) **crystals**. These are small needlelike crystals, approximately 1 μm long. Fats that can form stable β-crystals are good for use as shortenings, as they can be creamed easily, and give a smooth texture. Unstable β′-crystals change to the **intermediate crystal** form, about 3–5 μm in size, and finally convert to **coarse beta** (β) **crystals**, which can range from 25 to 100 μm in length. Beta crystals have the highest melting point.

Formation of small crystals is favored by rapid cooling with agitation. This allows formation of many small crystals, instead of slow growth of fewer large crystals. (Smaller crystals are desirable if a fat contributes aeration to a food.) Growth of large crystals occurs if cooling is slow. (The reader may want to read more about fat polymorphism and its effects on chocolate bloom.)

The more heterogeneous the fat, the more likely that the molecules form small stable crystals. Homogeneous fats readily form large crystals. Lard is an example of a homogeneous fat; more than 25 % of the molecules contain stearic acid, palmitic acid, and one unsaturated fatty acid molecule (usually oleic acid). Therefore, lard exists in the coarse *beta* crystalline form. However, lard can be modified by interesterification, which causes the fatty acids to migrate and recombine with glycerol in a more random manner.

Rearranged lard forms stable β′-crystals, because it is more heterogeneous. Acetoglycerides are able to form stable α-crystals, because they contain acetic acid esterified to glycerol, in place of one or two fatty acids. This increases the heterogeneity of the fatty acid composition of each individual triglyceride, which hinders the formation of large crystals.

All other things being equal, a fat with small crystals contains many more crystals and a much greater total crystal surface area than does a fat containing large crystals. Fats with small crystals are harder fats, have a smooth, fine texture, and appear to be less oily because the oil is present as a fine film surrounding the crystals, whereas the reverse is true of fats with large crystals.

The food industry uses controlled polymorphism to obtain fats with crystal sizes that improve their functional properties in foods.

For example, fats used for creaming must contain small, stable crystals in the β′ form; thus, crystallization is controlled during the manufacturing process.

Melting Points

The melting point of a fat or oil is an index of the force of attraction between molecules. The greater the attractive forces between molecules, the more easily they will associate to form a solid, and the harder it is to separate them when they are in the crystalline form and convert them to a liquid. A lot of energy in the form of heat must be put in to convert a solid to a liquid; thus, the melting point will be high. In other words, a high melting point indicates a strong attractive force between molecules. A strong attractive force indicates a good degree of fit between the molecules. Molecules that do not fit together well do not have strong attractive forces holding them together, and so they have lower melting points.

A fat or oil, which is a mixture of several triglycerides, has a lower melting point and a broader melting range than would be expected based on the melting points of the individual components. However, the *melting range* is dependent on the fatty acids of the component triglycerides. Fats may also be plastic at room temperature, containing some triglycerides that are liquid and some that are solid.

Generally speaking, oils, which are liquid at room temperature, tend to be more unsaturated, have shorter chains, and have lower melting points than fats, which are plastic or solid, with long chains and high melting points at room temperature. (See Table 12.1 for melting points of several fatty acids.) However, this is not the case always, as illustrated by coconut oil (see Tropical Oils), which has a high level of saturates (90 %), with a low *melting range* [75–80 °F (24–27 °C)]. It is liquid at room temperature because it contains an appreciable number of relatively short-chain (12 carbons) fatty acids, as is the case with palm and palm kernel oils. Lard, on the other hand, contains only about 37 % saturates, with mostly long-chain fatty acids, and so it is semisolid at 80 °F (27 °C).

As mentioned, the melting point of a fat or oil is actually a range, not a sharply defined temperature. The melting range depends on the composition of the fat. Each fat or oil contains triglycerides that melt at different temperatures, depending on their component fatty acids. Some fats have a wide melting range, whereas others, such as butter or chocolate, have a narrow melting range. Chocolate has a narrow melting range that is close to body temperature, and this accounts for its characteristic melt-in-your-mouth property.

The melting points of individual fatty acids depend on such factors as *chain length*, *number of double bonds* (degree of saturation), and *isomeric configuration*, because all these factors affect the degree of fit and the force of attraction between fatty acid molecules.

Chain length: *Long-chain* fatty acids have a higher melting point than *short-chain* fatty acids, because there is more potential for attraction between long chains than there is between short chains. The attractive forces are cumulative and can be appreciable if the chain is long enough. (In other words, you can think of them as having a zipper effect. A long zipper is much stronger than a short one, because more teeth are intersecting with each other.) For example, butyric acid (4:0) has a melting point of 18 °F (−7.9 °C), whereas stearic acid (18:0) has a higher melting point of 157 °F (69.6 °C). Stearic acid is a crystalline solid at room temperature, whereas butyric acid is a liquid unless the temperature drops below the freezing point of water.

Number of double bonds: A second factor that determines melting point is the *number of double bonds*. As the number of double bonds *increases*, the melting point *decreases*. Double bonds introduce kinks into the chain, and it is harder for molecules to fit together to form crystals; thus, the attractive forces between the molecules are weaker. This is demonstrated by comparing the melting points of stearic, oleic, linoleic, and linolenic acids, as shown in Table 12.1.

Isomeric configuration: A third influence on melting point is *isomeric configuration*. Geometric isomers have different melting points, because the *cis* double bond configuration introduces a much bigger kink into the molecule than does the *trans* configuration. Consequently, the *cis* isomer has a lower melting point than the *trans* isomer, because molecules in the *cis* configuration do not fit together as well as molecules in the *trans* configuration. This can be seen by comparing the melting points of oleic and elaidic acids. Oleic acid (*cis*) has a lower melting point than elaidic acid (*trans*—*see* Table 12.1). Low-*trans* liquid shortening such as the high oleic, monounsaturated sunflower oil requires no *trans* or "hydrogenated" reporting on labels, because it has a level of less than 2 % *trans*-fatty acids. A standard shortening may contain more than 30 % *trans*-fat levels.

The melting point of a triglyceride depends on the melting point of the component *fatty acids* as discussed above. *Simple* triglycerides can fit together easily, because the three fatty acid chains are identical and therefore allow for close packing of the molecules and high melting points. In general, the more *heterogeneous* triglycerides will not fit together as well, and so will have lower melting points. The melting point of a fat increases with each shift in polymorphic form, from alpha to coarse beta crystals.

Plastic Fats

Fats may be either liquid, solid, or plastic at room temperature. A ***plastic fat*** is moldable because it contains both *liquid* oil and *solid* crystals of triglycerides. Its consistency depends on the *ratio* of solid to liquid triglycerides: the more *liquid* the triglycerides, the *softer* the fat will be, and the more *solid* the triglycerides, the *harder* it will be. A plastic fat is a two-phase system, containing solid fat crystals surrounded by liquid oil. The liquid phase acts as a lubricant, enabling the solid crystals to slide past one another, and thus conferring moldability to the fat. A fat that contains only solid triglycerides is hard and brittle and cannot be molded, because the crystals cannot move past each other.

CULINARY ALERT! Fats that are "creamed" as per a recipe set of instructions (for some cookies or shortened cakes) must be plastic, so that they are easily workable and incorporate air into a mixture without breaking.

Ideally, plastic fats should be semisolid or plastic over a *wide* temperature range, so that creaming can be carried out at different (high or low) temperatures. Fats with a wide plastic range contain some triglycerides that are *solid* at high temperatures and some triglycerides that are *liquid* at low temperatures.

Fats with a *wide* plastic range are obtained by commercial modification, including the processes of hydrogenation and interesterification. Examples of such fats include *partially hydrogenated* soybean oil (found in margarine) and *interesterified lard*. Shortenings that are to be creamed must also contain small crystals, preferably in the β' form. Rearranged lard forms stable β'-crystals, and so has a fine-grained texture that is suitable for creamed fats.

Butter has a *narrow* plastic range and is, therefore, *not* a good choice for a fat that needs to be creamed. It cannot be creamed if taken straight out of the refrigerator, because it is too hard; neither can it be creamed if it sits on the counter on a warm day, because it will be too liquid.

Composition of Dietary Fats and Oils

A table showing fatty acid composition of various fats and oils frequently used by the consumer in food preparation is shown in Fig. 12.7. Time has shown variability as to which oil is best! The pendulum has swung from one product to another!

Polyunsaturated fats are liquid at room temperature and found primarily in *plants*. Safflower oil is 76 % polyunsaturated, sunflower oil is 71 %, soybean oil 54 %, and corn oil 57 % ("partially hydrogenated" oils are hydrogenated to have a *greater* degree of saturation).

Monounsaturated fats are liquid at room temperature and found chiefly in *plants*. Olive oil is 75 % monounsaturated, and canola (rapeseed oil) is 61 % monounsaturated. These fats are associated with a *decrease* in serum cholesterol and a decreased risk of coronary heart disease (CHD). There is not uniformity among researchers in suggesting that one of these fats is the best of all fats/oils to consume.

DIETARY FAT	Fatty acid content normalized to 100%			
	Saturated Fat	Polyunsaturated	Alpha Linolenic	Monounsaturated Fat
Canola oil	7%	21%	11%	61%
Safflower oil	10%	76%	Trace	14%
Sunflower oil	12%	71%	1%	16%
Corn oil	13%	57%	1%	29%
Olive oil	15%	9%	1%	75%
Soybean oil	15%	54%	8%	23%
Peanut oil	19%	33%	Trace	48%
Cottonseed oil	27%	54%	Trace	19%
Lard*	43%	9%	1%	47%
Beef tallow*	48%	2%	1%	49%
Palm oil	51%	10%	Trace	39%
Butterfat*	68%	3%	1%	28%
Coconut oil	91%	--	2%	7%

*Cholesterol content (mg/Tbsp): Lard 12; beef tallow 14; butterfat 33.
(No cholesterol in any vegetable-based oil.)

Alpha-Linolenic Acid (an Omega-3 Fatty Acid)

Source: POS Pilot Plant Corporation, Saskatoon, Saskatchewan, Canada June 1994

Canola Council of Canada, 400·167 Lombard Avenue, Winnipeg Manitoba Canada R3B OT6

Fig. 12.7 Comparison of composition of dietary fat (*Source*: Canola Council of Canada)

Saturated fats are solid at room temperature and found primarily in *animals*, although they are found in some tropical oils (see the listing below). These saturated fats are implicated in a *greater* rise in serum cholesterol than that produced by intake of dietary cholesterol!

Animal Fats

Animal fats typically have 18 carbons in the fatty acid chain. These long chains are made of various fatty acids and are chiefly *saturated*. Such fats may be *rendered* for use in baking and cooking applications (see section "Rendered Fat"). Animal fats derived from hogs and cattle include the following:

- **Lard**. Rendered from hogs, 43 % saturated fatty acids
- **Tallow (suet)**. Rendered from cattle, 48 % saturated fatty acids

Tropical Oils

Oils derived from plants grown in tropical areas of the world are referred to as *tropical oil*. Unlike most plants, these in particular are high in saturated fat content and contain an appreciable amount of short-chain fatty acids. Examples of tropical oils include the following:

- **Cocoa butter**. Extracted from cocoa beans, typically used in candies and chocolate confections
- **Coconut oil**. Highest saturated fat vegetable oil—over 90 % saturated; very stable against oxidation and, to a lesser degree, stable against hydrolysis
- **Palm oil**. 50 % saturated fatty acids; stable against oxidation
- **Palm kernel oil**. 84 % saturated fatty acids; derived from the kernel of the palm tree; stable against oxidation

CULINARY ALERT! In part due to the fact that animal fats contain cholesterol, saturated fat, and a pronounced flavor, the use of animal fat such as lard and tallow in foods has declined in favor of vegetable oils.

Production and Processing Methods

Crops are bred to increase the grower's yield while offering health benefits to *consumers* who want desirable health features in fats and oils. Both groups—growers and consumers—desire shelf stability. A brief discussion of the *conventional* as well as *nonconventional* approaches to breeding appears in the following text. Techniques are provided by the molecular geneticist and are available to growers and oilseed processors so that suppliers of edible oils can make both shelf stability *and* consumer health their priorities.

For example, *ordinary* soybean oil is *not* shelf stable because it contains 7.6 % linolenic acid, an unstable, 18-3, PUFA. To improve on this, *conventional* cross-breeding and selection has developed a low-linolenic soybean oil (LLSO), containing 2.5–3 % linolenic content This lower-linolenic soybean oil, derived from selected soybeans, is more stable than ordinary soybean oil and does not require hydrogenation for protection against rancidity. Consumers who want *less* saturated fat may make this oil their choice.

Unconventional approaches to breeding include gene modification, which produces a more stable oil that does not require hydrogenation. Then, stability as well as lower saturated fat may be achieved in a product. So, either conventional cross-breeding or unconventional genetic modification offers increased shelf stability without loss of health benefits, and this may be desirable.

Deodorized Oils

Deodorized oils are those that have undergone the process of removing odors by heat and vacuum or by adsorption onto charcoal. For example, olive oil may be deodorized to provide broader use in *baking* applications, without imparting its characteristic odor and flavor to food.

Rendered Fat

Rendered fat is the solid, usable fat derived from animal fat after it is heated and freed from connective tissue and then cooled. Food manufacturers render *hog* fat and process it to become *lard*, or *cattle* fat to become *tallow*. On a small scale, the consumer renders fat by (1) cutting animal fat into small pieces and gently boiling the pieces to extract liquid fat and then (2) cooling, until it becomes solid. The leftover rind, devoid of usable fat, has uses outside the scope of this discussion of fats.

In structure, the large crystalline structure of lard is composed of many similar triglycerides that are used to produce a highly desirable, flaky piecrust. Today, lard may be processed to contain smaller crystals, and then it functions more like a hydrogenated shortening. The addition of antioxidants such as butylated hydroxyanisole (BHA) and butylated hydroxytoluene (BHT) protects it against rancidity.

As previously mentioned, lard and tallow are not as commonly used in cooking as they were in the past, partially because of the pronounced flavor, saturated fat, and cholesterol content. As well, animals are now bred to be leaner, so lard is less available. Today, there are many convenient, commercially prepared shortenings on the market that replace lard in cooking.

Modification of Fats

Hydrogenation

Hydrogenation is the process of adding hydrogen to *un*saturated fatty acids to reduce the number of double bonds. The purpose of hydrogenation is twofold:

- To convert liquid oils to semisolid or plastic fats
- To increase the thermal and oxidative stability of the fat, and thus the shelf life

Hydrogenation of unsaturated fatty acid occurs when *hydrogen gas* is reacted with *oil* under controlled conditions of *temperature* and *pressure* and in the presence of a nickel, copper, or other *catalyst*. The reaction is carefully controlled and stopped when the desired extent of hydrogenation has been reached. As the reaction progresses, there is a gradual production of *trans*-fatty acids which increases the melting point of the fat or oil and creates a more solid product. *Solid* shortening is created out of a hydrogenated *oil*.

The extent of the hydrogenation process is carefully controlled to achieve stability and/or the physical properties required in the finished food product. If the reaction is taken to completion, a *saturated* fat is obtained, and the product is hard and brittle at room temperature. However, this is not usually the aim of hydrogenation, as *partial* hydrogenation is normally desired for foods, providing an *intermediate* degree of solidification, reducing the number, yet, while not eliminating all double bonds. In fact, approximately 50 % of the total fatty acids present in *partially hydrogenated vegetable shortening* products are *mono*unsaturated and about 25 % are *poly*unsaturated.

Polyunsaturated fats are subject to oxidative rancidity. Thus, reducing the number of double bonds by hydrogenation serves to increase their stability. Once saturated though, consumption of the fat contributes more toward the elevation of serum cholesterol than does dietary cholesterol intake. The process of hydrogenation causes conversion of some *cis* double bonds to the *trans* configuration. Most of the *trans*-fatty acids formed are monounsaturated. Tub margarines, for examples, typically contain *trans*-fatty acid at levels of 13–20 %.

A previous *Federation of American Societies for Experimental Biology (FASEB)* report published in 1985 (Huffman 2001) concluded that there was little cause for concern with the safety of dietary *trans*-fatty acids, both at present and expected levels of consumption. However, this was challenged by later research.

"A small amount of *trans* fat is found naturally, primarily in dairy products, some meat, and other animal-based foods" (FDA). The majority is formed when manufacturers add hydrogen to turn liquid oils into partially/hydrogenated oils. Thus, *trans* fat *can* be found in hydrogenated vegetable shortenings, some margarines (not butter), crackers, snack items, and convenience fast food. The advice is to read labels.

Plastic fats have useful functional properties for use in margarines or shortenings that are to be creamed. Hydrogenated fats are frequently specified in batter and dough recipes that depend on the creaming ability of solid fats for aeration (Chap. 14). Creaming increases volume by incorporating air and results in numerous air cells. As a result, the grain of the crumb in baked products is small and even.

Interesterification

Interesterification, or ***rearrangement***, causes the fatty acids to migrate and recombine with glycerol in a more random manner. This causes new glycerides to form and *increases* the *heterogeneity* of the fat. However, it does *not* change the degree of unsaturation or the isomeric state of the fatty acids.

Lard is an example of a fat that is modified in this way to improve its functional properties. In its natural state, lard is a relatively *homogeneous* fat, as has already been mentioned. Therefore, it has a *narrow* plastic range and is *too firm* to be used straight from the refrigerator and *too soft* at temperatures above normal room temperature. Lard also contains coarse β-crystals. Rearrangement increases the *heterogeneity* of lard, enabling it to form stable β′-crystals and *increasing* the temperature range over which it is plastic or workable. This significantly enhances its use as a shortening product.

Hydrogenation may be used in conjunction with interesterification and may either precede or follow it. This gives a shortening manufacturer the ability to produce fats with a *wide* range of properties.

Acetylation

Acetoglycerides or ***acetin fats*** are formed when one or two fatty acids in a triglyceride are replaced by acetic acid (CH_3COOH). Acetin fats may be liquid or plastic at room temperature depending on the component fatty acids. However, the presence of acetic acid lowers the melting point of the fat, because the molecules do not pack together as readily. It also enables the fat to form stable α-crystals.

Acetin fats are used as edible lubricants; they also form flexible films and are used as coating agents for selected foods such as dried raisins and produce to prevent moisture loss.

Winterization

Winterized oil is oil that has been *pretreated* to control undesirable cloudiness. The large, high-melting-point triglyceride crystals in oil are subject to crystallization (forming solids) at refrigeration temperatures. Therefore, in the process of winterization, oil is refrigerated and subsequently filtered to remove those large, undesirable crystals, which could readily disrupt a salad dressing emulsion. The treated oil is called *salad oil*, which is specially used in salad dressing.

CULINARY ALERT! *Salad oils* are clear and are bleached, deodorized, and refined, in addition to undergoing winterization. Salad oils differ from *cooking oils*, the latter of which do not undergo winterization.

Deterioration of Fats

Fats deteriorate either by absorbing odors or by becoming *rancid*. Both of these are described below. For example, deterioration by *absorbing odors* becomes evident when *chocolate fat* absorbs the odor of (1) smoke in a candy store environment or (2) soap packaged in the same grocery bag at the supermarket. Butter may also deteriorate by readily absorbing *refrigerator* odors. When *rancidity* causes deterioration, it produces a disagreeable odor and flavor in fatty substances.

CULINARY ALERT! Processing does not remove *all* chance of fat and oil deterioration and rancidity, but it prolongs the life of a fat or oil.

Deterioration by rancidity may occur in two ways (details below) making fats undesirable for use in foods. One way is ***hydrolytic rancidity*** which involves reaction of fats with water and liberation of free fatty acids. The other, ***oxidative rancidity***, is a more complex and potentially more damaging reaction. In this second case, the fat is oxidized and decomposes into compounds with shorter carbon chains such as fatty acids, aldehydes, and ketones all of which are volatile and contribute to the unpleasant odor of rancid fats.

Hydrolytic Rancidity

Fats may become rancid by hydrolytic rancidity when the triglycerides react with water and free their fatty acids from glycerol. The reaction is shown in Fig. 12.8. If one molecule of water reacts with a triglyceride, *one* fatty acid is liberated and a *di*glyceride remains. To liberate glycerol, *all three* fatty acids must be removed from the molecule. The reaction is catalyzed by heat and by enzymes known as lipases. Butter contains lipase, and if left on the kitchen counter on a warm day, a characteristic rancid smell frequently develops due to liberation of the short-chain butyric acid. (Unlike long-chain fatty acids, these short-chain fatty acids may form an unpleasant odor and flavor.)

Hydrolytic rancidity is also a problem with deep-fat frying, where the temperature is high

$$\begin{array}{lcl} CH_2-OOC-R & & CH_2-OH \\ | & & | \\ CH-OOC-R + 3H_2O & \longrightarrow & CH-OH + 3RCOOH \\ | & & | \\ CH_2-OOC-R & & CH_2-OH \end{array}$$

Fig. 12.8 Hydrolytic rancidity

and wet foods are often introduced into the hot fat. The continued use of rancid oil results in additional breakdown of the oil. To avoid this type of rancidity, fats should be stored in a cool place and, if possible, lipases should be inactivated.

CULINARY ALERT! Fats should be kept away from water, and foods to be fried should be as dry as possible before they are added to hot fat. The kind of fat used for frying should be selected based on stability.

Oxidative Rancidity or Autoxidation

Oxidative rancidity is the *predominant* type of rancidity. In this process, the unsaturated fatty acids are subjected to oxidative rancidity or autoxidation, and the *more* double bonds there are, the *greater* the opportunity for addition of oxygen to double bonds, increasing risk that the fat or oil will become rancid. ***Autoxidation*** is complex and is promoted by heat, light, certain metals (iron and copper), and enzymes known as lipoxygenases. The reaction can be separated into three stages: initiation, propagation, and termination.

The **initiation stage** of the reaction involves formation of a free radical. A hydrogen on a carbon atom adjacent to one carrying a double bond is displaced to give a free radical, as shown in Fig. 12.9. There is chemical activity around and in the double bonds. (The bold type indicates the atoms or groups of atoms involved in the reactions.) As previously mentioned, this reaction is catalyzed by heat, light, certain metals such as copper and iron, and lipoxygenases. The free radicals that form are unstable and very reactive.

The **propagation stage** follows the initiation stage and involves oxidation of the free radical to yield activated peroxide. This, in turn, displaces hydrogen from another unsaturated fatty acid, forming another free radical. The liberated hydrogen unites with the peroxide to form a hydroperoxide, and the free radical can be oxidized as just described. Thus, the reaction repeats, or propagates, itself. Formation of one free radical, therefore, leads to the oxidation of many unsaturated fatty acids.

Hydroperoxides are very unstable and decompose into compounds with shorter carbon chains, such as volatile fatty acids, aldehydes, and ketones. These are responsible for the characteristic odor of rancid fats and oils. The two reactions of the propagation stage of autoxidation are shown in Fig. 12.10.

The **termination stage** of the reaction involves the reaction of free radicals to form nonradical products. Elimination of all free radicals is the only way to halt the oxidation reaction.

Prevention of Autoxidation

Oxidation can be prevented or delayed by *avoiding* situations that would serve as catalysts for the reaction. For example, fats and oils *must be stored* in a cool dark environment (offering temperature and light change controls) and in a closed container (to minimize oxygen availability). *Vacuum packaging* of fat-containing products controls oxygen exposure, and *colored glass* or wraps control fluctuations in light intensity. Fats must also be stored *away from metals* that could catalyze the reaction, and any cooking utensils used must be free of copper or iron. Lipoxygenases should be inactivated.

CULINARY ALERT! Store fats and oils in a cool dark environment and in a closed container. Colored glass jars or wraps control rancidity.

In addition, *sequestering agents* and *antioxidants* can be added to fats to prevent autoxidation, increasing keeping quality and shelf life of fats.

Sequestering agents bind metals, thus preventing them from catalyzing autoxidation. Examples of sequestering agents include EDTA (ethylenediaminetetraacetic acid) and citric acid.

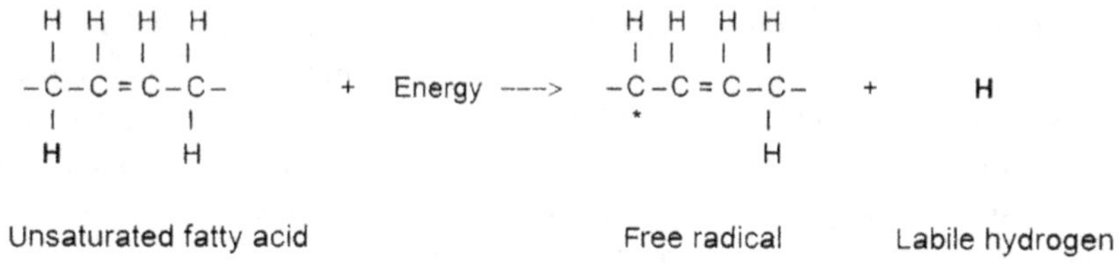

Fig. 12.9 The initiation stage of autoxidation

```
 H  H  H  H                             H  H  H  H
 |  |  |  |                             |  |  |  |
-C -C = C -C-    +  Energy --->  -C -C = C -C-    +    H
 |           |                    *         |
 H           H                              H

Unsaturated fatty acid             Free radical      Labile hydrogen
```

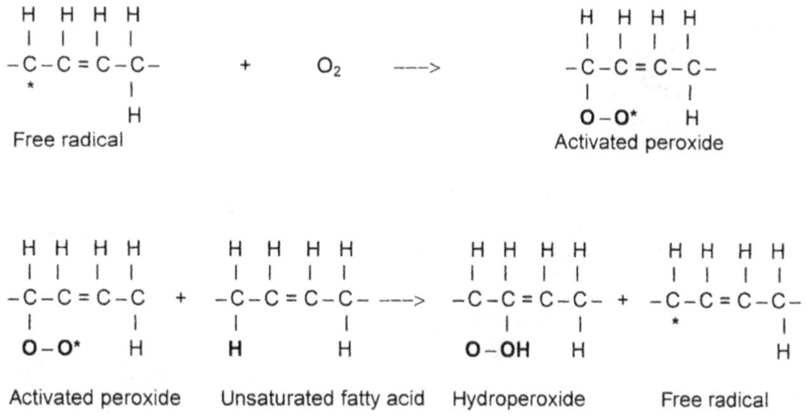

Fig. 12.10 The two reactions of the propagation stage of autoxidation

```
 H  H  H  H                                   H  H  H  H
 |  |  |  |                                   |  |  |  |
-C -C = C -C-      +      O2    --->         -C -C = C -C-
 *           |                                |           |
             H                                O - O*      H
Free radical                                 Activated peroxide
```

```
 H  H  H  H         H  H  H  H            H  H  H  H         H  H  H  H
 |  |  |  |         |  |  |  |            |  |  |  |         |  |  |  |
-C -C = C -C   +  -C -C = C -C- --->    -C -C = C -C-  +   -C -C = C -C-
 |           |      |           |              |    |          *          |
 O - O*      H      H           H              O-OH H                     H

Activated peroxide  Unsaturated fatty acid  Hydroperoxide     Free radical
```

Antioxidants help prevent autoxidation with its formation of fatty acid free radicals. Antioxidants prevent rancidity by donating a hydrogen atom to the double bond in a fatty acid and preventing the oxidation of any unsaturated bond. They halt the chain reaction along the fatty acid, which leads to rancidity.

Most antioxidants are phenolic compounds. Those approved for use in foods include ***BHA***, ***BHT***, ***TBHQ*** (tertiary butylhydroquinone), and propyl gallate. These are *all synthetic* antioxidants. The effectiveness of antioxidants may be increased if they are used together. For example, propyl gallate and BHA are more effective when *combined* than if used separately.

BHA is a waxy white solid that survives processing to create a stable product. It is effective in preventing oxidation of *animal* fats yet *not vegetable* oils. BHT is a white crystalline solid that may be combined with BHA. It is effective in preventing oxidation of animal fats. TBHQ is a white-to-tan-colored powder that functions best in frying processes rather than baking applications.

Tocopherols are naturally occurring antioxidants that are present in vegetable oils. They can be added to both animal and vegetable oils to prevent oxidation. The tocopherols are also sources of essential nutrient vitamin E.

Use of antioxidants in foods containing fat increases their keeping quality and shelf life. Examination of food labels reveals that antioxidants are widely used in many food products, from potato chips to cereals. Without them, the quality of fat-containing foods would not be as good, and off-flavors and odors due to oxidative rancidity would be commonplace.

Shortening and Shortening Power of Various Fats and Oils

Plant, animal, or numerous plant–animal blends of fats and oils may be used for shortening, and, typically, the blend is creamy. The shortening potential of a fat or oil is influenced by its fatty acid composition (see Fig. 12.5), and various fats and oils may function as shortenings. "Shortenings" may include many types, from *pourable liquids to stiff solids*, with the latter being most commonly considered shortening. A shortening is hydrogenated oil and it functions to physically shorten platelets of protein–starch structure developed in manipulated wheat flour mixtures.

Shortening power of some fats and oils appears below.

Lard has a large fatty acid crystal structure, unless it is interesterified. It forms a desirable flaky product. This solid fat when cut into pea-sized chunks or smaller melts within the gluten structure of flour, creating many layers or *flakes* in baked piecrusts or biscuits (more later).

Butter and margarine contain water and milk (20 %) in addition to a variety of fat or oils (80 %). One stick is derived from 2.5 quarts of milk. Due to this water, butter and margarine have less shortening potential than lard, hydrogenated shortening, or oil that contains 100 % fat. When butter or margarine is incorporated into flour-based formulations, they toughen the mixture, as its water component hydrates the starch.

CULINARY ALERT! A recipe substituting butter or margarine for lard or hydrogenated shortening adds water; thus, the recipe requires less additional water and yields a *less flaky* piecrust.

A replacement for butter originated in 1869 when margarine was formulated by a French pharmacist. Today, margarine may contain part cultured skim milk or whey, optional fat ingredient(s), emulsifier, and color (annatto or carotene) and may include added salt, flavoring, and vitamins A and D. The margarine is likely to be high in PUFAs, if oil is listed as the first ingredient on a margarine label. If *partially hydrogenated oil* is listed on a label as the first ingredient, there is less PUFA. A product must be labeled "spread" if it does not meet the Standard of Identity for margarine. Also, today, margarine *substitutes* may be milk-free, sodium-free, or even fat-free.

Hydrogenated fats are saturated and easily workable. When creamed, they incorporate air into a mixture. They are processed to be *without* a pronounced flavor and have a *wide* plastic range. Hydrogenated fats contain 100 % fat and have greater shortening power than butter or margarine. Finished food products may be *flaky*, however not as flaky as if lard is used.

Oils contain a *high* liquid-to-fat crystal ratio and are *un*saturated. They shorten strands of protein *mechanically* by coating the platelets. Oil controls gluten development and subsequent toughness because *less* water contacts the gluten proteins. Oil helps produce a *tender* product, but in pastries, *flakiness* may be sacrificed. Flakes are *not* readily obtained because there are no large chunks of fat to melt between layers of dough.

Tenderization Versus Flakiness Provided by Fats and Oils

Lipids provide either ***tenderization*** or ***flakiness***, as discussed, and impart distinct characteristics of a food product. The differences are especially evident in finished piecrusts and can also be observed in biscuits. *Tender* products are easily crushed or chewed; they are soft and fragile—i.e., oil piecrusts. *Flaky* products contain many thin pieces or layers of cooked dough, i.e., puff pastry and lard piecrusts.

Some factors that affect these two distinct attributes are presented in Table 12.3. The *type of fat or oil* chosen to be incorporated into food, its *concentration*, *degree of manipulation*, and *temperature* each affect the flakiness and tenderness of a product. Fats and oils should be selected and used with knowledge of these factors. Yet, health attributes of a fat or oil may supersede other quality attributes creating products that do not meet traditional product standards. For example, for health reasons, a piecrust may not incorporate solid fat but may be prepared using oil. If that is the case, the finished piecrust will sacrifice flakiness but will be tender and crumbly.

CULINARY ALERT! In order to control formation of an undesirable crumbly food product, some gluten formation may be needed *prior* to the addition of the fat or oil. This may be achieved by adding fat to a recipe, *after* some hydration and manipulation has formed gluten.

Table 12.3 Factors affecting the tenderness and flakiness of a product

The type of fat or oil—Chunks of *solid fat* create layers or flakes in the gluten starch mixture as they melt, whereas *oil* coats flour particles more thoroughly, *creating less layers* and a mealy product. Substituting one fat or oil for another may not produce acceptable or expected results
Fat concentration—Fat may be reduced or omitted in a formulation, or the fat that *is* used may not be 100 % fat; it may be a butter, margarine, or "spread." Adequate levels of fat or oil must be present in foods if they are to meet acceptable standards. For example, sufficient fat in flour-based mixtures is needed to control gluten development and generate a tender crumb. Imitation "butters" or "spreads" have a high water content and may not have the high percentage of fat needed to perform satisfactorily in all baking, sautéing, or "buttering" processes
Degree of manipulation—An insufficient degree of manipulation may result in poor distribution of fat throughout the food mixture. Inversely, excess manipulation may cause the fat to spread or be softened, thus minimizing the possibility of flakes. For example, a flaky piecrust is produced when solid fat is incorporated in the formulation as pea-sized chunks
Temperature—Depending on the type of fat, *cold* shortenings (solid or liquid) provide less covering potential than *room temperature* shortenings and produce more flaky biscuits and piecrusts. Food items prepared with cold shortenings also remain slightly more solid in the hot oven while the item bakes. When a shortening is *melted*, it displays a greater shortening potential than an *unmelted* solid shortening; it coats better than the same amount of unmelted solid fat. Melted shortening produces a more tender, less flaky product

Emulsification (See Chap. 13)

Fats and oils are *not* emulsifiers; however, in addition to providing flavor, aerating batters and doughs, and shortening, fats and oils are important *constituents* of emulsions. An emulsion consists of a three-phase system composed of (1) a ***continuous phase***, the phase or medium in which the dispersed phase is suspended; (2) a ***dispersed phase***, the phase which is disrupted or finely divided within the emulsion; and (3) an ***emulsifier*, which** is present at the interface between the dispersed phase and the continuous phase and keeps them apart. An emulsifier acts in the following ways:

- It adsorbs at the interface between two immiscible liquids such as oil and water.
- It reduces the interfacial tension between two liquids, enabling one liquid to spread more easily around the other.
- It forms a stable, coherent, viscoelastic interfacial film, which prevents or delays coalescence of the dispersed emulsion droplets.

Molecules that can act as emulsifiers contain both a polar, ***hydrophilic*** (water-loving) section, which is attracted to water, and a ***hydrophobic*** (or water-hating) section, which is attracted to hydrophobic solvents such as oil. In order for the hydrophilic section to be dispersed in the water phase and for the hydrophobic section to be dispersed in the oil phase, the molecule must adsorb at the *interface* between the two phases, instead of being dispersed in either bulk phase.

Good emulsifiers are able to interact at the interface to form a *coherent* film that does not break easily. Therefore, when two droplets collide, the emulsifier film remains intact, and the droplets do *not* coalesce to form one big droplet. Instead, they drift away from each other.

The *best* emulsifiers are proteins, such as egg yolk (lipoproteins) or milk proteins, because they are able to interact at the interface to form stable films and hence to form stable emulsions. However, many *other* types of molecules are used as emulsifiers.

Mono- and diglycerides are examples of emulsifiers that are added to products in order to provide ease of mixing. They adsorb at the interface, reducing ***interfacial tension*** and increasing the spreadability of the continuous phase or the wettability of the dispersed phase.

In some cases, finely divided *powders* such as dry mustard or spices are used to act as emulsifiers in oil-in-water mixture. The mustard and spices adsorb at the interface and reduce interfacial tension. However, they *cannot* form a stable film around oil droplets, and so they are *unable* to form a stable emulsion. Therefore, they should *not* really be considered as emulsifiers.

Emulsions may be temporary or permanent. A *temporary emulsion* separates upon standing. The emulsion is not permanent because the hydrophobic oil and hydrophilic water components separate upon standing. This is because the emulsifiers used are unable to form a stable interfacial film to prevent coalescence of the droplets of the dispersed phase. As coalescence occurs, the droplets combine to form bigger ones, and eventually the two phases separate out completely. An example of a temporary emulsion would be French dressing, which separates out a few seconds after it has been shaken.

A *permanent emulsion* is formed when two ordinarily nonmiscible phases, such as water and oil, are combined with an emulsifier. One phase (usually the oil phase) is dispersed within the other as small droplets. These remain dispersed in the continuous phase (usually water), because they are surrounded by a stable film of emulsifier that resists coalescence, and so prevents separation of the two phases.

Thus, the time of separation of oil and water is dependent upon the effectiveness of an emulsifier and the degree of agitation. As mentioned, more detail on emulsification is provided in Chap. 13.

Various examples of emulsified mixes are cake mixes, mayonnaise, and salad dressings, discussed below.

Cake mixes contain an emulsifier that aids in incorporation of air upon stirring or beating. The emulsifiers are usually monoglycerides and diglycerides, which act by dispersing shortening in smaller particles. This creates a maximum number of air cells that *increase* cake volume and creates a more even grain in the baked product (Chap. 14).

Mayonnaise is an emulsified product. A real mayonnaise as opposed to salad dressing (mayo type) is described in the 1952 Standard of Identity. Mayonnaise is an emulsified semisolid, with not less than 65 % by weight, edible vegetable oil.

Salad dressings are typically emulsified, containing oil, vinegar, water, salts, and so forth. Oil coats the salad contents and disperses herbs, spices, and other substances. Early application may *wilt* the salad due to salt in the dressing. Winterized oils are used. Some dressings are available in no-fat formulations. Except for bacon dressing, which uses bacon fat, *solid* fats are generally *not* acceptable for use in a dressing.

Hydrocolloids (see section "Fat Replacements" and Chap. 5) such as gelatin, gums, pectin, and starch pastes may be added in the preparation of salad dressings, but they contain only a hydrophilic section and are *not* considered *emulsifiers*. Rather, they act as stabilizers in emulsions, and help to prevent or lessen coalescence, because they increase the viscosity of the continuous phase.

Frying

Frying with melted fat or oil is a common cooking technique because frying is a *rapid* heat transfer method that achieves a *higher* temperature than boiling or dry heat temperature. The characteristics of fats for frying include that the fat must be colorless, odorless, and bland and have a high smoke point.

Smoke Point

The ***smoke point*** is the temperature at which fat may be heated before continuous puffs of blue smoke come from the surface of the fat under controlled conditions. The presence of smoke indicates that free glycerol has been further hydrolyzed to yield *acrolein*, a mucous membrane irritant. Monoglycerides, in hydrogenated shortenings, and diglycerides are hydrolyzed more easily than triglycerides and they tend to have a low smoke point. Therefore, they are not recommended in frying oils.

When fat *exceeds* the smoke point, it may reach *flash point*, when small flames of fire begin in the oil. Subsequently, it reaches the *fire point* where a fire is sustained in the oil. Oils such as cottonseed or peanut oil have a high smoke point of 444 or 446 °F (229 or 230 °C), respectively. Other oils with a lower smoking point may not perform satisfactorily when exposed, for example, to the high heat of a wok.

CULINARY ALERT! Lard, butter, margarine, and animal fats have a low smoke point and less tolerance of heat when compared to hydrogenated fat and oils.

Changes During Frying

Frying exposes the food product to high temperatures, removes internal water, and allows a level of oil absorption. The duration of frying, composition of the food, surface treatment, and other factors determine levels of oil uptake.

The subsequent thermal decomposition of oil occurs in fat as air, water, and prolonged high temperature lead to fat oxidation and hydrolysis. Oil may become an unwanted orange or brown color or it may become more viscous and foam. The smoke point decreases as oil is repeatedly used for frying. And the quality is reduced.

Numerous factors are reported to affect oil uptake during frying, and a better understanding of how oil is absorbed during frying can lead to improved food quality of fried foods. For example, porosity requires more study in order to determine its effect on oil uptake. Some of these factors that affect oil uptake during frying are addressed in Table 12.4.

Low-Fat and No-Fat Foods

Consumer interest in eating reduced-fat or fat-free foods has increased, as is evidenced by the trend for more healthy foods. Yet, the per capita consumption of fats and oils has *not* decreased to meet the Surgeon General's recommendation (<30 % of a day's calories from fat) in the Report on Nutrition and Health. This may be in part due to the fact that the function, flavor, and mouthfeel of fat have *not* been duplicated by any nonfat component in the diet.

Overcoming flavor challenges in low-fat frozen desserts may involve the removal of fat in ice cream products that affects flavor and aroma, texture, and mouthfeel. Overcoming flavors is challenging.

The USDA reports one attempt at meeting flavor challenges. Utilizing a starch–lipid ratio varying from 10:1 to 2:1, oil droplets are suspended in cooked starch dispersions and then added as an ingredient to embellish flavor, texture, and mouthfeel (USDA).

> The fats and oils in dressings and sauces play several roles and provide a number of attributes" "When you consider a full-fat salad dressing may contain as much as 30 % to 50 % oil, and mayonnaise or sauces from it, fully 80 %—you gain a better understanding of why the low-fat, fat-reduced, "lite", or fat-free versions fall so short of expectations. (Decker 2013)

Fat Replacements

Fat replacements in a formulation may be protein-, carbohydrate-, or fat-based. Of course, the noncaloric water and air may be added if it works! Replacements are "useful when they help with calorie control and when their use encourages the consumption of foods delivering important nutrients" (The Academy of Nutrition and Dietetics, Eatright.org).

The use of a particular fat replacement may be determined by answering the question: What properties of fat are fat replacers attempting to simulate?

Today, there are many materials designed to replace fat; they are derived from several different categories of substances. Some replacers that attempt to simulate fat include protein-, carbohydrate-, and fat-derived fat replacements described below.

Using "the systems approach" in problem-solving, the Calorie Control Council reports that "…a variety of synergistic components are used to achieve the functional and sensory characteristics of the full-fat product. Combinations of ingredients are used to compensate for specific functions of the fat being replaced. These combinations may

Table 12.4 Selected factors that affect oil uptake during deep-fat frying

Frying temperature, duration, and product shape—Increases in temperature decrease oil uptake due to short frying duration
Pressure frying decreases duration and oil uptake
A high surface-to-mass ratio or surface roughness increases oil absorption
Composition—The addition of soy protein, egg protein, or powdered cellulose decreases oil uptake. High sugar, soft flour, or developed gluten increase oil uptake
Prefrying treatments—Blanching, prewashing with oil containing emulsifiers, freezing, and steam pretreatment have been shown to decrease oil uptake
Surface treatment—Hydrocolloids (see section "Fat Replacements") and amylose coatings may function as barriers to fat uptake

include proteins, starches, dextrins, maltodextrins, fiber, emulsifiers and flavoring agents. Some fat replacers are now available that are themselves a combination or blend of ingredients (for example, one ingredient currently in use is a combination of whey, emulsifiers, modified food starch, fiber and gum)" (Calorie Control Council, CalorieControl.org).

The Academy of Nutrition and Dietetics (formerly the American Dietetic Association [ADA]) states "Fat replacements provide an opportunity for individuals to reduce intake of high-fat foods and enjoy reduced-fat formulations of familiar foods while preserving basic food selection patterns." It is the position of the ADA that "the majority of fat replacers, when used in moderation by adults, can be safe and useful adjuncts to lowering the fat content of foods and may play a role in decreasing total dietary energy and fat intake. Moderate use of low-calorie, reduced-fat foods, combined with low total energy intake, could potentially promote dietary intake consistent with the objectives of *Healthy People 2010* and the 2005 *Dietary Guidelines for Americans*" (The Academy of Nutrition and Dietetics, Eatright.org).

Reported by the Institute of Food Technologists is that nutrition, a healthy lifestyle, regular exercise, and a reduction of total dietary fat are significant in lifestyles that incorporate fat.

http://www.caloriecontrol.org

Featured in the following text are a discussion, examples, and label designation for each group of derived fat replacers.

Carbohydrate-Derived Fat Replacements

Fat replacements may be derived from carbohydrates with 0–4 kcal/g instead of 9 kcal/g. *Starches* work well as fat replacements in high moisture systems to absorb water and form gels that mimic fat. They have been utilized in the bakery industry for many years.

Fruit purees or dried puree powder is also used to replace fats, as are *cellulose*, *gums*, *fiber*, *dextrins*, *maltodextrins*, *modified food starch*, *modified dietary fibers*, *and polydextrose*. Starch hydrolysis derivatives known as ***maltodextrins*** (classified as hydrocolloids) are bland in flavor and have a smooth mouthfeel. They are fat-replacing ingredients of commercial cakes and also assist in maintaining product moisture. Gelling, thickening, and stabilizing are desirable functional properties.

The plant root, tapioca, and the tuber, potato, as well as the cereal starches corn and rice, are also used as fat replacers. An oat-based fat replacement is made by partial hydrolysis of oat starch using a food-grade enzyme, and barley is being investigated for use as a possible fat substitute.

Fat replacers may be basically hydrocolloid materials or contain hydrocolloids as an important part of their ingredient composition (see below).

Hydrocolloids are long-chain polymers, principally carbohydrate, that thicken or gel in aqueous systems, creating the creamy viscosity that mimics fat. Some are listed below. They include the starch derivatives, hemicelluloses, β-glucans, soluble bulking agents, microparticulates, composite materials [i.e., carboxymethyl cellulose (CMC) and microcrystalline cellulose or xanthan gum and whey], and functional blends (gums, modified starches, nonfat milk solids, and vegetable protein).

Polydextrose may be used as a 1 kcal/g substitute for either fat or sucrose. Polydextrose is a bulking agent created by the random polymerization of glucose, sorbitol, and citric acid, 89:10:1. It may be used in a variety of products such as baked goods, chewing gum, salad dressings, and gelatins, puddings, or frozen desserts.

Several dried-fruit-based substances are available for replacement of fat in recipes. Raisin, plum, and other fruit mixtures are available for consumer use at this time. Applesauce is also used to partially replace fat in formulations. Many additional fat replacers are being explored, including the use of encapsulated technologies (USDA).

Examples of carbohydrate-derived Food and Drug Administration (FDA)-approved *or* currently researched fat replacers:

Examples of Carbohydrate-Based Fat Replacers

Cellulose (Avicel® cellulose gel, Methocel™, Solka-Floc®)
Various forms are used. One is a non-caloric purified form of cellulose ground to microparticles which, when dispersed, form a network of particles with mouthfeel and flow properties similar to fat. Cellulose can replace some or all of the fat in dairy-type products, sauces, frozen desserts and salad dressings.

Dextrins (Amylum, N-Oil®)
Four calories per gram fat replacers which can replace all or some of the fat in a variety of products. Food sources for dextrins include tapioca. Applications include salad dressings, puddings, spreads, dairy-type products and frozen desserts.

Fiber (Opta™, Oat Fiber, Snowite, Ultracel™, Z-Trim)
Fiber can provide structural integrity, volume, moisture holding capacity, adhesiveness and shelf stability in reduced-fat products. Applications include baked goods, meats, spreads and extruded products.

Gums (KELCOGEL®, KELTROL®, Slendid™)
Also called hydrophilic colloids or hydrocolloids. Examples include guar gum, gum arabic, locust bean gum, xanthan gum, carrageenan and pectin. Virtually non-caloric; provide thickening, sometimes gelling effect; can promote creamy texture. Used in reduced-calorie, fat-free salad dressings and to reduce fat content in other formulated foods, including desserts and processed meats.

Inulin (Raftiline®, Fruitafit®, Fibruline®)
Reduced-calorie (1–1.2 calories/g) fat and sugar replacer, fiber and bulking agent extracted from chicory root. Used in yogurt, cheese, frozen desserts, baked goods, icings, fillings, whipped cream, dairy products, fiber supplements and processed meats.

Maltodextrins (CrystaLean®, Lorelite, Lycadex®, MALTRIN®, Paselli®D-LITE, Paselli®EXCEL, Paselli®SA2, STAR-DRI®)
Four calorie per gram gel or powder derived from carbohydrate sources such as corn, potato, wheat and tapioca. Used as fat replacer, texture modifier or bulking agent. Applications include baked goods, dairy products, salad dressings, spreads, sauces, frostings, fillings, processed meat, frozen desserts, extruded products and beverages.

Nu-Trim
A beta-glucan rich fat replacer made from oat and barley using an extraction process that removes coarse fiber components. The resulting product can be used in foods and beverages such as baked goods, milk, cheese and ice cream, yielding products that are both reduced fat and high in beta-glucan. (The soluble fiber beta-glucan has been cited as the primary component in oats and barley responsible for beneficial reduction in cardiovascular risk factors.)

Oatrim [Hydrolyzed oat flour] (Beta-Trim™, TrimChoice)
A water-soluble form of enzyme treated oat flour containing beta-glucan soluble fiber and used as a fat replacer, bodying and texturizing ingredient. Reduced calorie (1–4 calories/g) as used in baked goods, fillings and frostings, frozen desserts, dairy beverages, cheese, salad dressings, processed meats and confections.

Polydextrose (Litesse®, Sta-Lite™)
Reduced-calorie (1 calorie/g) fat replacer and bulking agent. Water-soluble polymer of dextrose containing minor amounts of sorbitol and citric acid. Approved for use in a variety of products including baked goods, chewing gums, confections, salad dressings, frozen dairy desserts, gelatins and puddings.

Polyols (many brands available)
A group of sweeteners that provide the bulk of sugar, without as many calories as sugar (1.6–3.0 calories/g, depending on the polyol). Due to their plasticizing and humectant properties, polyols also may be used to replace the bulk of fat in reduced-fat and fat-free products.

Starch and modified food starch (Amalean®I & II, Fairnex™VA15, & VA20, Instant Stellar™, N-Lite, OptaGrade®#, Perfectamyl™AC, AX-1, & AX-2, PURE-GEL®, STA-SLIM™)
Reduced-calorie (1–4 calories/g as used) fat replacers, bodying agents, texture modifiers. Can be derived from potato, corn, oat, rice, wheat or tapioca starches. Can be used together with emulsifiers, proteins, gums and other modified food starches. Applications include processed meats, salad dressings, baked goods, fillings and frostings, sauces, condiments, frozen desserts and dairy products.

Z-Trim
A calorie free fat replacer made from insoluble fiber from oat, soybean, pea and rice hulls or corn or wheat bran. It is heat stable and may be used in baked goods (where it can also replace part of the flour), burgers, hot dogs, cheese, ice cream and yogurt.
#Appears as corn starch on the ingredient statement, others appear as food starch modified.
(Calorie Control Council—http://www.caloriecontrol.org)

Some Carbohydrate-Based Fat Replacers on Food Labels

Carrageenan, cellulose, gelatin, gellan gum, gels, guar gum, maltodextrins, polydextrose, starches, xanthan gum, modified dietary fibers.

The ingredient may be used for reasons OTHER than fat replacement.

Fat-Derived Fat Replacements

Fat-derived fat replacements, such as Olestra, offer 0 calorie/g. Other replacements offer less than 9 kcal/g of fat. The majority are *emulsifiers, emulsions with little fat, or analogs*—triglycerides or similar, with a changed configuration (see the underlined items listed below as examples). It is reported by the International Food Information Council (IFIC) that "Some fat-based ingredients, such as Caprenin and Salatrim, are actually fats tailored to contribute fewer calories and less available fat to foods. Others such as olestra, are structurally modified to provide no calories or fat" (Calorie Control Council, CalorieControl.org).

Olestra, marketed under the brand name Olean®, differs from fats and oils in its chemical composition and properties. Olestra is a sucrose polyester (SPE), predominantly sucrose octaester, which is synthesized by reacting six to eight fatty acids with the eight free hydroxyl groups of sucrose. (Recall that fats are a glycerol backbone with three fatty acids attached.) Each fatty acids may be 12–20 or more carbons in length and may be either saturated or unsaturated. Fatty acids may be derived from corn, coconut, palm, or soybean sources.

Olestra became the latest of several food ingredients approved without generally recognized as safe (GRAS) status [others are TBHQ (1972), aspartame (1981), polydextrose (1981), and acesulfame K (1988) (Chap. 17)]. Its chemical makeup and configuration make olestra indigestible and it is not absorbed. Its numerous fatty acids are attached to the sucrose in a manner that cannot be easily penetrated by digestive enzymes in the length of time it is in the digestive tract. As a result, olestra provides *no* calories.

Unlike *protein-derived* fat replacements, which by their nature *cannot* be exposed to high heat, olestra is used for frying applications. It was first patented in 1971 and sought FDA approval as a cholesterol-lowering drug. Approval was denied, because such use was not shown.

A subsequent petition in 1987 requested use of olestra as a direct food additive. It was to be used as a fat replacement for (1) up to 35 % of the fat in home-use cooking oils and shortenings and (2) up to 75 % of the fat in commercial deep-fat frying of snack foods. The petition was amended in 1990 and approved in 1996 to allow the Procter & Gamble Olean® to be used as a 100 % replacement for fats in savory snacks (salty, piquant, but not sweet, such as potato chips, cheese puffs, and crackers), including the frying oil and any fat sources in the dough (conditioners, flavors, etc.). All other uses of olestra require separate petitions.

The FDA conclusions regarding the major chemical changes in frying and baking applications of olestra are that changes are similar to triglycerides. The fatty acid chains oxidize in both cases. In baking, there is slower degrading of the fatty acids, but the same by-products are produced. Olestra has baking and frying applications and may be used in dairy-based or oil-based foods.

A distinctive label statement is required for all Olean®-containing products. Labels must state "This Product Contains Olestra. Olestra may cause abdominal cramping and loose stools. Olestra inhibits the absorption of some vitamins and other nutrients. Vitamins A, D, E, and K have been added." In three small test markets, the major user of Olean® has not observed nor has there been evidence of severe abdominal cramps and loose stools resulting from the consumption of products containing Olean® (Frito-Lay).

Health concerns regarding the use of olestra have been addressed in part by over 150 Procter & Gamble studies.

Caprenin® (Procter & Gamble) is another fat replacement that contains 5 calories/g. It contains a glycerol backbone with three fatty acids. Two of the fatty acids are medium chain, caprylic and capric, which are metabolized similarly to carbohydrates, and the other chain consists of a long fatty acid—behenic acid—that is incompletely absorbed. These fatty acids are selected on the basis of specific, desired properties.

Nabisco Foods has developed a proprietary family of low-calorie **salatrim** fats—named for **s**hort **a**nd **l**ong **a**cyl**tri**glyceride **m**olecule. Salatrim is a patented ingredient of conventional glycerol backbones to which long-chain fatty acids and short-chain fatty acids are added. The long-chain stearic acid is combined with the short-chain acetic, propionic, and butyric acids on a glycerol molecule.

Nabisco states that salatrim is different from other fat replacers because it is made from real fat, whereas other fat substitutes are made from protein and carbohydrates.

Salatrim received GRAS status by the FDA in 1994. It was approved for use in baked products, chocolates and confections, dairy products, and snacks, but it cannot be used successfully in frying applications.

A nutritional advantage of using these fat replacers is that they contain 5 kcal/g, instead of the normal fat amount of 9 kcal/g. This calorie reduction may be due to hydrolysis of short-chain fatty acids that are rapidly hydrolyzed to carbon dioxide and long-chain fatty acids that are incompletely absorbed.

Examples of fat-derived FDA-approved *or* currently researched fat replacers:

Examples of Fat-Based Fat Replacers

Emulsifiers (Dur-Lo®, EC^{T}-25)

Examples include vegetable oil mono- and diglyceride emulsifiers which can, with water, replace all or part of the shortening content in cake mixes, cookies, icings, and numerous vegetable dairy products. Same caloric value as fat (9 calories/g) but less is used, resulting in fat and calorie reduction.

Sucrose fatty acid esters also can be used for emulsification in products such as those listed above. Additionally, emulsion systems using soybean oil or milk fat can significantly reduce fat and calories by replacing fat on a one-to-one basis.

Salatrim (Benefat^{T})

Short and long-chain acid triglyceride molecules. A 5 calories/g family of fats that can be adapted for use in confections, baked goods, dairy and other applications.

Lipid (fat/oil) analogs

- **Esterified propoxylated glycerol (EPG)**
  Reduced-calorie fat replacer. May partially or fully replace fats and oils in all typical consumer and commercial applications, including formulated products, baking and frying.
- **Olestra** (Olean®)
  Calorie-free ingredient made from sucrose and edible fats and oils. Not metabolized and unabsorbed by the body. Approved by the FDA for use in replacing the fat used to make salty snacks and crackers. Stable under high heat food applications such as frying. Has the potential for numerous other food applications. For more information on olestra, check out the new olestra brochure.
- **Sorbestrin****
  Low-calorie, heat stable, liquid fat substitute composed of fatty acid esters of sorbitol and sorbitol anhydrides. Has approximately 1.5 calories/g and is suitable for use in all vegetable oil applications including fried foods, salad dressing, mayonnaise and baked goods.

  **Brand names are shown in parentheses as examples.*

  ***May require FDA approval.*

(Calorie Control Council, CalorieControl.org) www.caloriecontrol.org/fatreprint.html

Protein-Derived Fat Replacements

Proteins may be used in place of fat. They contribute 1–4 kcal/g, instead of 9. An easily recognized type is gelatin; however, there are others. The International Food Council states that "Some protein-based ingredients, such as Simplesse®, are made through a process that gives fat-like textural properties to protein. Other proteins are heated and blended at high speed to produce tiny protein particles that feel creamy to the tongue. ... protein-based fat reducers cannot be used as substitutes for oils and other fats in frying" (International Food Information Council, IFIC.org).

Simplesse® is a natural fat substitute developed by the NutraSweet Company and approved by the FDA in 1990. It is a *microparticulated protein* (*MPP*). Simplesse® uses a patented process that heats and intensely blends naturally occurring food proteins such as *egg white and milk* proteins, along with water, pectin, and citric acid. The protein remains chemically unchanged, yet aggregates under controlled conditions that allow formation of small aggregates or microparticles.

The *blending* process produces small, round uniformly shaped protein particles—about 50 billion per teaspoon—that create the creamy mouthfeel of full fat. The microparticulated particle size is near the lower range of *MPPs* that naturally occur in milk, egg white, grains, and legumes. For example, casein (milk protein) micelles range in size from 0.1 to 3.0 mm in diameter and are perceived as creamy to the tongue. In comparison, a larger particle size, 10–30 mm in diameter, is found in powdered (confectionery) sugar, which is perceived as more powdery and gritty.

Initially, Simplesse® was an ingredient approved by the FDA for use in dairy-based frozen desserts. Today, it has many more food applications in products such as butter spreads, cheese (creamed, natural, processed, baked cheesecakes), creamers, dips, ice cream, and sour cream. It is also successfully incorporated into oil-based products such as margarine spreads, mayonnaise, and salad dressings. Many are Kosher approved, and with proper storage, they have a shelf life of 9 months (The Academy of Nutrition and Dietetics, Eatright.org).

Due to its milk and egg protein composition, individuals *allergic* to milk or eggs *cannot* eat this fat substitute. It contains 1.2 calories/g (*not* a 0-calorie food), approximately one-third the calories of protein, and significantly lowers fat intake. Simplesse® is a GRAS substance.

Whey protein concentrates (WPCs), and *isolates* (WPIs), and isolated soy protein (legumes) are proteins that can be used to provide some of the functional properties of fat without the same number of fat calories. Dairy-Lo® is an example of a WPC and uses include dairy products, baked goods, frostings, mayonnaise-type products, and salad dressings.

Soy may be used for emulsification or gelling and is approved for addition of up to 2 % in cooked sausage and cured pork. It may be used at higher levels in ground meat and poultry.

Examples of protein-derived FDA-approved *or* currently researched fat replacers:

Examples of Protein-Based Fat Replacers

Microparticulated protein (Simplesse®)
Reduced-calorie (1–2 calories/g) ingredient made from whey protein or milk and egg protein. Digested as a protein. Many applications, including: dairy products (e.g., ice cream, butter, sour cream, cheese, yogurt), salad dressing, margarine- and mayonnaise-type products, as well as baked goods, coffee creamer, soups and sauces.

Modified whey protein concentrate (Dairy-Lo®)
Controlled thermal denaturation results in a functional protein with fat-like properties. Applications include: milk/dairy products (cheese, yogurt, sour cream, ice cream), baked goods, frostings, as well as salad dressing and mayonnaise-type products.

Other (K-Blazer®, ULTRA-BAKE™, ULTRA-FREEZE™, Lita®)
One example is a reduced-calorie fat substitute based on egg white and milk proteins. Similar to MPP yet made by a different process. Another example is a reduced-calorie fat replacer derived from a corn protein. Some blends of protein and carbohydrate can be used in frozen desserts and baked goods.

(Calorie Control Council—caloriecontrol.org. http://www.caloriecontrol.org/articles-and-video/feature-articles/glossary-of-fat-replacers)

Beyond the scope of this discussion is more information that defines fat replacers and extenders. For example, fat substitutes, fat analogs, fat mimic, fat extender, and fat barriers are terms better defined elsewhere in the literature.

Nutritive Value of Fats and Oils

Most health authorities in the United States take the stance that fat should be limited—yet, not *all* currently agree with this recommendation. Fats are needed for numerous functions in the human body, and two PUFAs are essential—linoleic and linolenic acid are required for human growth. In addition to the many roles fat plays in functionality of foods, fats are a very concentrated energy source—providing 9 calories/g. This is 2¼ times as many calories per gram as either carbohydrates or protein.

The health-conscious consumer may make choices of reducing certain foods that are major contributors of less desirable fatty acids, and, as well, substitute foods, possibly *increasing* fats that are major contributors of the fatty acids that are desired (Pszczola 2000). The food industry may have a major impact in reducing heart disease, as they have *changed* formulations. See health and nutrition article "Feeling better about fat" (Decker 2012).

Similar to the role of cholesterol in animal cell membranes, *phytosterols* and *phytostanols* perform the same role in *plants*. Phytostanols are the saturated form of plant sterols. The structures are similar to cholesterol, differing only in the side chain (Fig. 12.4). *Plant sterols* are commercially available in margarines (such as Benecol®, which contains stanols, and Take Control ®, which contains sterols) and salad dressings, and although there are several theories suggested, and the precise mechanism is unknown, these phytonutrients have been shown for many decades to significantly reduce low-density lipoprotein (LDL) or "bad" cholesterol. They inhibit the uptake of endogenous and dietary cholesterol

(ISEO Technical Committee 2006). More recently, it has been shown that *dietary* cholesterol does not bear a direct negative influence on *serum* cholesterol in the healthy individual.

The cost factor continues to be a challenge, as is the marketing of any "healthy food" that incorporates new ingredients. Benecol® and Take Control® are much more expensive than other types of margarines or spreads.

Research is ongoing regarding the type and amount of fats as a part of optimal nutrition.

OILS (ChooseMyPlate.gov)

What Are "Oils?"

Oils are fats that are liquid at room temperature, like the vegetable oils used in cooking. Oils come from many different plants and from fish. Oils are NOT a food group, but they provide essential nutrients. Therefore, oils are included in USDA food patterns. Some oils are used mainly as flavorings, such as walnut oil and sesame oil. A number of foods are naturally high in oils.

Foods that are mainly oil include mayonnaise, certain salad dressings, and soft (tub or squeeze) margarine with no *trans* fats. Check the Nutrition Facts label to find margarines with 0 g of *trans* fat. Amounts of *trans* fat are required to be listed on labels.

Most oils are high in monounsaturated or polyunsaturated fats, and low in saturated fats. Oils from plant sources (vegetable and nut oils) do not contain any cholesterol. In fact, no plant foods contain cholesterol.

A few plant oils, however, including coconut oil, palm oil, and palm kernel oil, are high in saturated fats and for nutritional purposes should be considered to be solid fats.

Solid fats are fats that are solid at room temperature, like butter and shortening. Solid fats come from many animal foods and can be made from vegetable oils through a process called hydrogenation.

Safety of Fats and Oils

Safety of the original fats and oils may be compromised. For example, rancidity due to lengthy or improper storage conditions, including temperature, may destroy fats and oils. Of course, the presence of harmful, foreign substances in any food material, as well as skin burns from hot oil products, poses severe dangers in the workplace.

Conclusion

Fats and oils add or modify flavor, aerate batters and doughs, contribute flakiness and tenderness, emulsify, transfer heat, and provide satiety. They are composed of a glycerol molecule with one, two, or three fatty acids attached creating mono-, di-, or triglycerides, respectively. Minor components of fats and oils include phospholipids, sterols, tocopherols, and pigments. Fatty acid chains of even number may exist as geometric or positional isomers. Nomenclature may be according to a common name, systemic or Geneva name, or omega system.

Fats and oils exist in several crystalline forms and have different melting points. Solid fats have higher melting points than oils. Fats and oils may be processed by being deodorized or rendered. They are modified by hydrogenation, interesterification, acetylation, or winterization.

The deterioration of fats and oils occurs as they absorb odors or become rancid. Hydrolytic rancidity releases free fatty acids, and oxidative rancidity produces shorter, off-odor free radicals catalyzed by heat, light, metals, or enzymes. Prevention of oxidation by avoiding catalysts in the environment or by the addition of sequestering agents or antioxidants may be useful in extending shelf life.

Monoglycerides and diglycerides have uses as emulsifiers, permitting fats and liquids to mix. Fats and oils are useful as shorteners; they tenderize and produce flakes in baked products. They may also be used in the preparation of salad dressings and for frying applications. Foods may contain reduced-fat, low-fat, or no-fat formulations using a variety of fat replacers derived from carbohydrates, proteins, or fats.

The cost factor continues to be a challenge, as is the marketing and healthy value of any "healthy food" that incorporates new ingredients.

Plant breeders are researching the development of healthier fats. A variety of vegetable oils continue to be available to food processors and, to a lesser extent, to the consumer. Stability without increased saturation is the goal of processors. Advanced hybridization of vegetable sources of oil may reduce saturated fatty acids, and thus improve nutritional value. Fats and oils should be used sparingly in the daily diet.

Notes

CULINARY ALERT!

Glossary

Acetin fat A triglyceride with one or two fatty acids on a triglyceride replaced by acetic acid; this decreases the melting point.

Acetoglyceride Acetin fat.

Antioxidant Prevents, delays, or minimizes the oxidation of unsaturated bonds by donating an H atom to the double bond in a fatty acid.

Autoxidation Progressive oxidative rancidity in an unsaturated fatty acid promoted by heat, light, the metals iron and copper, and lipoxygenases.

BHA Butylated hydroxyanisole; an antioxidant.

BHT Butylated hydroxytoluene; an antioxidant.

***Cis* configuration** A double-bond formation when H atoms attach to the C atoms of the double bond on the same side of the double bond.

Continuous phase The phase or medium in which the dispersed phase is suspended in an emulsion.

Deodorized oils Oils that have undergone the process of removing odors by heat and vacuum or by adsorption onto charcoal.

Dispersed phase A phase that is disrupted or finely divided in the continuous phase of an emulsion.

Emulsifier Bipolar substance with a hydrophilic and hydrophobic end, which reduces surface tension and allows the ordinarily immiscible phases of a mixture to combine.

Fat replacement A substance used to replace fat in a formulation; these may be protein-, carbohydrate-, or fat-based.

Flakiness Thin, flat layers formed in some dough products desirable in biscuits or piecrusts.

Hydrocolloid Long-chain polymers; colloidal material that binds and holds water.

Hydrogenation Process of adding H to unsaturated fatty acids to reduce the number of double bonds; an oil becomes more solid and more stable in storage.

Hydrolytic rancidity Reaction of fats with water to liberate free fatty acids.

Hydrophilic Water-loving substance attracted to water.

Hydrophobic Water-fearing substance attracted to fat.

Interesterification Rearrangement as fatty acids migrate and recombine with glycerol in a more random manner.

Interfacial tension See surface tension.

Isomer Fatty acids have the same number of carbons, hydrogens, and oxygens, but form different arrangements that create different chemical and physical properties.

Lecithin Phospholipid of two fatty acids esterified to glycerol and a third group of phosphoric acid and choline as the N group; useful as an emulsifier.

Maltodextrin Hydrocolloid; starch derivative of tapioca, potato, corn, rice, oats, or barley that may be used to replace fat in a formulation.

Oxidative rancidity Fat is oxidized and decomposes into off-odor compounds with shorter-chain fatty acids, aldehydes, or ketones.

Phospholipid A lipid containing two fatty acids and a phosphoric acid group esterified to glycerol.

Phytosterols and phytostanols Natural substances obtained from plants, which are related to cholesterol, but are able to reduce blood cholesterol levels. Stanols are the saturated form of plant sterols. These substances are contained in margarines such as Benecol (contains stanols) and Take Control (contains sterols).

Plastic fat Able to be molded and hold shape; contains both liquid and solid triglycerides in various ratios.

Polymorphism Fats existing in different crystalline forms: α, β' intermediate, and β.

Rearrangement Interesterification of fatty acids on glycerol, i.e., modified lard.

Rendered Fat freed from connective tissue and reduced, converted, or melted down by heating; for example, lard is rendered hog fat.

Sequestering agent Binds metals, thus preventing them from catalyzing autoxidation; for example, EDTA and citric acid.

Smoke point The temperature at which fat may be heated before continuous puffs of blue smoke come from the surface of the fat.

Sterols A lipid containing a steroid nucleus with an 8–10 C side chain and an alcohol group; cholesterol is the most well known.

Surface tension (Interfacial tension) force that tends to pull molecules at the surface into the bulk of a liquid and prevents a liquid from spreading. Reduction of surface tension enables a liquid to spread more easily.

TBHQ Tertiary butylhydroquinone; an antioxidant.

Tenderization Easily crushed or chewed, soft, fragile, baked dough.

Tocopherols Minor component of most vegetable fats; antioxidant; source of vitamin E.

***Trans* configuration** A double-bond formation in fatty acids where the H atoms attach to the C atoms of the double bond on opposite sides of the double bond.

Winterized Salad oil that is pretreated prior to holding, to control undesirable cloudiness from large, high-melting-point triglyceride crystals.

References

Decker KJ (2012) Feeling better about fat. Food Prod Des April:58–68

Decker KJ (2013) Pouring it on thin. Food Prod Des May/June:70–86

Huffman M (2001) "Trans fat" labeling? J Am Diet Assoc 101:28

ISEO Technical Committee (2006) Food fats and oils, 9th edn. Institute of Shortening and Edible Oils, Washington, DC

MacGibbon AHK, Taylor MW (2006) Composition and structure of bovine milk lipids. In: Fox PF, McSweeney PLH (eds) Advanced dairy chemistry. Springer, New York, pp 1–42

Pszczola DE (2000) Putting fat back into foods. Food Technol 54(12):58–60

Seabolt KA (2013) Learning about lecithin. Food Prod Des May/June:22–25

Bibliography

American Soybean Association, St. Louis, MO

Code of Federal Regulations (CFR), Title 21 Section 101.25(c)(2)(ii)(a & b)

Coultate T (2009) Food the chemistry of its components, 5th edn. RSC Publishing, Cambridge, UK

Gurr MI (1992) Role of fats in food and nutrition. Chapman & Hall, New York

Hicks KB, Moreau RA (2001) Phytosterols and phytostanols: functional food cholesterol busters. Food Technol 55(1):63–67

USDA—Choosemyplate.gov

Food Emulsions and Foams

13

Introduction

Many convenience foods, such as frozen desserts, meat products, margarine, and some natural foods, such as milk and butter, are emulsions. That is, they contain either water dispersed in oil or oil dispersed in water. These water and oil liquids do not normally mix, and so when present together, they exist as two separate layers. However, when an emulsion is formed, the liquids are mixed in such a way that a single layer is formed with droplets of one liquid dispersed within another. Food emulsions need to be stable; if they are not, the oil and water will separate out. Stability is usually achieved by adding a suitable emulsifier. In some cases, a stabilizing agent is also required.

Food foams, such as beaten egg white, are similar to emulsions except that instead of containing two liquids, they contain a gas (usually air or carbon dioxide) dispersed within a liquid. The factors affecting stability of emulsions also apply to foams. Some foods, such as ice cream and whipped cream, are highly complex being both an emulsion and a foam.

Understanding of food emulsions and foams is complex, yet is important if progress is to be made in maintaining and improving the stability and hence the quality of these types of foods. This chapter will discuss the principles of formation and stability of emulsions and foams and the characteristics of the ingredients necessary to stabilize them.

Emulsions

Definition

An ***emulsion*** is a ***colloidal system*** containing droplets of one liquid dispersed in another, the two liquids being immiscible. The droplets are termed the ***dispersed phase***, and the liquid that contains them is termed the ***continuous phase***. In food emulsions, the two liquids are oil and water. If water is the continuous phase, the emulsion is said to be an ***oil-in-water*** or ***o/w emulsion***, whereas if oil is the continuous phase, the emulsion is termed a ***water-in-oil*** or ***w/o emulsion***. Oil-in-water emulsions are more common and include salad dressings, mayonnaise, cake batter, and frozen desserts. Butter, margarine, and some icings are examples of water-in-oil emulsions.

An emulsion must also contain an ***emulsifier***, which coats the emulsion droplets and prevents them from ***coalescing*** or recombining with each other. Emulsions are colloidal systems because of the size and surface area of the droplets (in general, around 1 μm, although droplet size varies considerably, and some droplets may be a lot larger than this). Emulsions are similar to colloidal dispersions or sols, except that the dispersed phase is liquid and not solid. Colloidal dispersions are mentioned in Chap. 2.

V.A. Vaclavik and E.W. Christian, *Essentials of Food Science, 4th Edition*, Food Science Text Series,
DOI 10.1007/978-1-4614-9138-5_13,

Surface Tension

To form an emulsion, two liquids that do not normally mix must be forced to do so. To understand how this is achieved, we must first consider the forces between the molecules of a liquid. Imagine a beaker of water placed on a desk (Fig. 13.1).

The water molecules are attracted to one another by hydrogen bonds as described in Chap. 2. A molecule in the center of the beaker has forces acting on it in all directions, because water molecules surround it. The net force on this molecule due to attraction by other water molecules is zero, because these forces are acting in all directions. However, this is not the case for a water molecule on the surface. Since there are no water molecules above it, there is a net downward pull on the molecule. This results in the molecule being pulled in toward the bulk of the liquid.

This downward pull can be seen when one fills a narrow tube such as a pipette or a burette with water. The surface of the liquid curves downward at the center, and the curve is called the meniscus. The greater the attractive forces between the liquid molecules, the greater the depth of the meniscus. Water molecules have strong attractive forces among them, and so it is relatively hard to penetrate the surface, or to get the water to spread. Try placing a needle gently on the surface of clean or distilled water. It will float, because the attractive forces between the water molecules keep it on the surface. (To make it sink, see below.)

If there are strong attractive forces among the molecules of a liquid, the force required to pull the molecules apart, to expand the surface, or to spread the liquid will be high. This force is known as ***surface tension***. A liquid such as water, with strong attractive forces between the molecules, has a high surface tension. This makes it hard to spread. You can see this if you put water on a clean surface. It will tend to form droplets rather than spreading evenly as a thin film across the surface. (A droplet has minimal surface area and maximal internal volume, and so it is the most energetically favorable shape for liquids with a high surface tension, where the molecules are being pulled into the interior.)

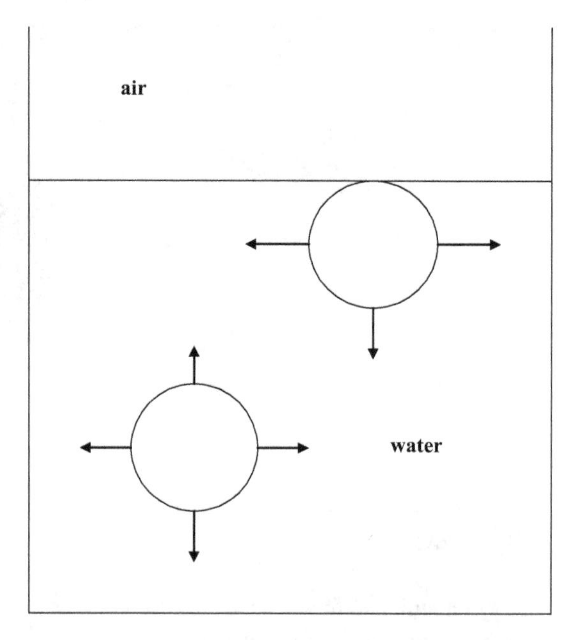

Fig. 13.1 Schematic diagram of the forces acting on water molecules in the bulk and at the surface of the liquid

The term surface tension is normally used when a *gas* (usually air) surrounds the *liquid* surface. When the surface is between two *liquids*, such as water and oil, the term ***interfacial tension*** is used.

A high surface or interfacial tension makes it hard to mix the liquid either with another liquid or with a gas. This is a drawback when making an emulsion or foam and needs to be overcome. So how can surface tension be reduced?

Surface-Active Molecules

To reduce the surface or interfacial tension, something must be done to decrease the attractive forces between the liquid molecules, so that it is easier to spread them. This can be achieved by adding a ***surface-active*** molecule, or a ***surfactant***. As their name suggests, surface-active molecules are active at the *surface* of a liquid, rather than at the bulk of it. Surfactant molecules prefer to exist at the surface of a liquid rather than at the bulk because of their structure. In all cases, a section of

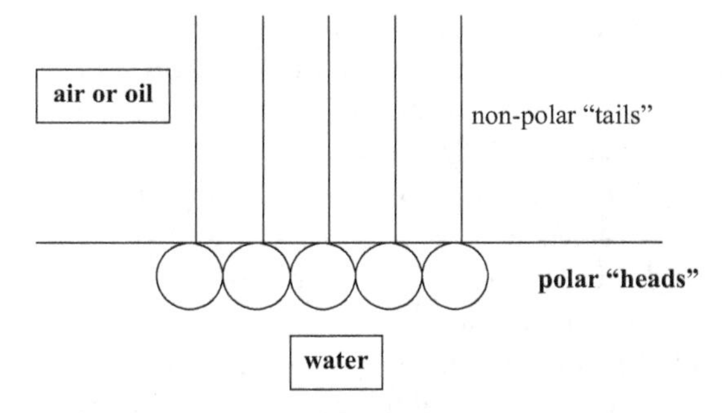

Fig. 13.2 Orientation of amphiphilic molecules at an interface

the molecule is water-loving or ***hydrophilic*** because it is polar or charged, and a section is water-hating or ***hydrophobic*** because it is apolar. In other words, the molecules are ***amphiphilic***.

The apolar section has little or no affinity for water, and so it is energetically favorable for this section to be as far away from the water as possible. However, the polar section is attracted to the water and has little or no affinity for the oil. Therefore, the molecule orients at the surface with the polar section in the water, with the apolar section either in the air or in the oil (see Fig. 13.2).

Due to the fact that the molecule ***adsorbs*** at the surface, it reduces the attractive forces of the water molecules for themselves and makes it easier to expand or spread the surface. In other words, it reduces the surface or interfacial tension.

Detergent is an example of a surfactant. When detergent is added to water, it enables the water molecules to spread much more easily, so that they wet a surface more readily. After adding detergent, water will flow over a surface, forming a thin sheet, instead of tending to gather into droplets. Going back to the example of the needle floating on water (see above), if a small drop of detergent is added, the needle will sink. The surface tension is reduced, allowing the water molecules to spread more easily, and so the needle no longer stays on the surface.

Obviously, detergents are not used as food ingredients! (However, they are used when washing dishes, because they enable the water to spread across the surface and remove food particles more easily.) There are many food ingredients that are surfactants. Polar lipids such as lecithin, which has a polar "head" and an apolar "tail," are surfactants and may be used as food additives to increase the wettability and aid in mixing of products like hot chocolate mix.

Proteins are surface-active because they contain both hydrophilic and hydrophobic sections. The nature and extent of these sections depend on the specific amino acid sequence of each protein, and some proteins orient at the surface more readily than others do (Proteins are discussed in Chap. 8).

Some spices, such as dry mustard and paprika, are also used as surface-active ingredients. These finely divided powders tend to gather at the surface rather than the bulk of the liquid.

Molecules that are either hydrophilic or hydrophobic do not orient at an interface. The molecules remain in the bulk of the liquid. For example, sugars, which are hydrophilic, or salt, which dissociates into ions, will be located in the bulk water phase. These types of molecules are not surface-active and will not decrease the interfacial tension. In fact, they may increase it, depending on their ability to bind the water molecules, hence increasing molecular attraction.

Emulsion Formation

An emulsion is formed when oil, water, and an emulsifier are mixed together. Although there are different food emulsions, they *all* contain these three components. To form an emulsion, it is necessary to break up either the oil or the water phase into small droplets that remain dispersed throughout the other liquid. This requires energy and is usually carried out using a mixer or a homogenizer. As the oil and water are mixed, droplets are formed. (They may be oil or water, yet are usually oil droplets.) An emulsifier is adsorbed at the surface of new droplets, decreasing the interfacial tension and allowing formation of more and smaller droplets. The lower the surface or interfacial tension of the oil and water, the more easily one liquid can be disrupted to form droplets and the more easily the other liquid will flow around the droplets.

The liquid with the higher interfacial tension will tend to form droplets, and the other liquid will flow around the droplets to form the continuous phase. The emulsifier generally determines the liquid that would form the continuous phase. Emulsifiers that are more easily dispersed in water (and therefore are more hydrophilic overall) tend to reduce the interfacial tension of the water more than that of the oil, promoting formation of o/w emulsions. Emulsifiers that disperse more readily in the oil phase tend to form w/o emulsions. The emulsifier is usually dispersed in the preferred phase before the oil and water are mixed together.

Principles of Formation of a Stable Oil-in-Water Emulsion

- Emulsifier is dispersed in the aqueous phase.
- Oil is added and the interfacial tension of each liquid is reduced by the emulsifier.
- Energy is supplied by beating or homogenizing the mixture.
- The oil phase is broken up into droplets, surrounded by water.
- Emulsifier adsorbs at the freshly created oil droplet surfaces.
- Small droplets are formed, protected by an interfacial layer of emulsifier.
- The interfacial area of the oil becomes very large.
- The aqueous phase spreads to surround each oil droplet.
- The emulsion may become thick due to many small oil droplets surrounded by a thin continuous phase.
- If the interfacial film is strong, the emulsion will be stable.

An emulsifier does not simply reduce interfacial tension. It must also form a stable film that protects the emulsion droplets and prevents separation of the emulsion. The droplets are continually moving through the continuous phase, and so they constantly encounter or collide with each other. When two droplets collide, one of three things happens, as shown in Fig. 13.3:

(a) The emulsifier film stretches or breaks, and the droplets combine to form one larger droplet (or in other words, they ***coalesce***). This ultimately leads to separation of the emulsion.
(b) The two emulsifier layers surrounding the droplets interact and an aggregate is formed. This occurs when a cream layer develops on top of fresh milk.
(c) The droplets move apart again.

Which of these three events occurs depends on the nature of the emulsifier molecules and on their ability to completely coat all the emulsion droplets with a stable, cohesive, ***viscoelastic*** film. A viscoelastic film tends to flow to coat any temporarily bare sections of the surface and is also able to stretch instead of breaking when, for example, another droplet bumps into it. Therefore, it is less likely to break when droplet collisions occur. As the droplets are formed, their surface or interfacial area increases dramatically, and sufficient emulsifier must be present to completely coat all the droplet surfaces. Incompletely coated droplets will coalesce resulting in larger droplets and ultimately in separation of the emulsion.

Emulsifiers

Emulsifiers must be able to:

- Adsorb at the interface between two liquids such as oil and water
- Reduce the interfacial tension of each liquid, enabling one liquid to spread more easily around the other
- Form a stable, coherent, viscoelastic interfacial film
- Prevent or delay coalescence of the emulsion droplets

Reduction of the interfacial tension facilitates emulsion formation, because it reduces the amount of energy needed to break up one liquid into droplets and to spread the other liquid around them. Formation of a film that prevents coalescence promotes emulsion stability.

All emulsifiers are surfactants, because all emulsifiers adsorb at the surface and reduce

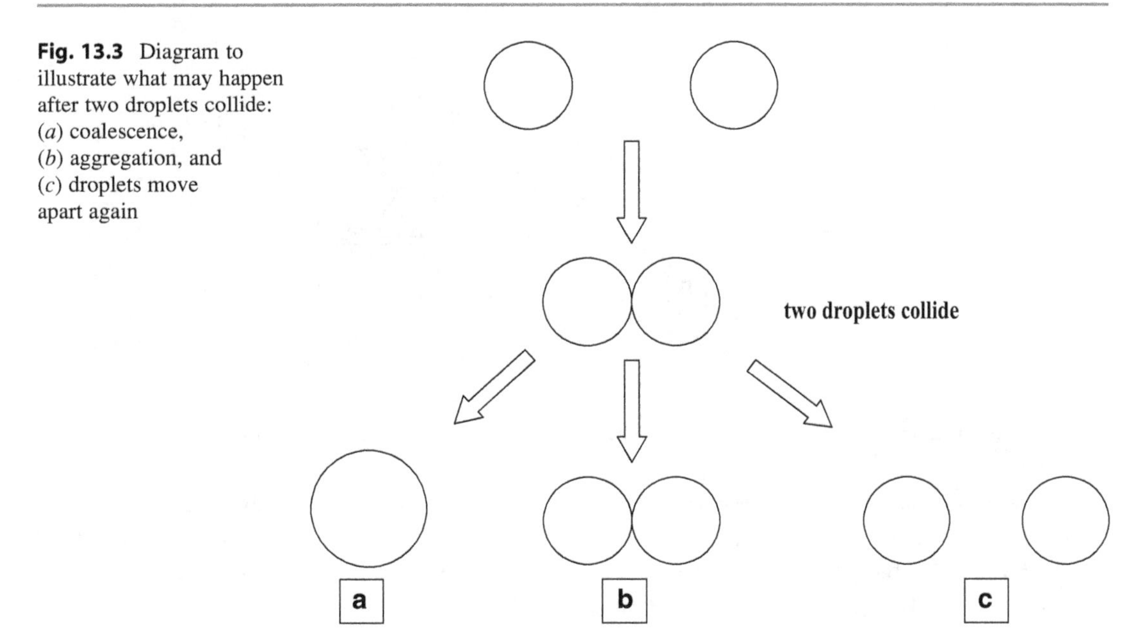

Fig. 13.3 Diagram to illustrate what may happen after two droplets collide: (*a*) coalescence, (*b*) aggregation, and (*c*) droplets move apart again

interfacial tension. However, all surfactants do *not* make good emulsifiers, because not all surfactants are able to form a stable film at the interface and prevent coalescence. The stability of the film is important in determining the stability and shelf life of the emulsion. Some emulsifiers work better than others do, in terms of forming a stable emulsion.

In general, large macromolecules such as proteins form stronger surface films than smaller surfactant molecules such as lecithin because of their greater ability to extend over the droplet surface. They also have a greater ability to interact with other groups within the same molecule or on different molecules and are able to form viscoelastic surface films.

Small molecules are not usually able to form stable interfacial films by themselves, and their role is normally that of a surfactant rather than an emulsifier, in that they lower interfacial or surface tension and promote spreading or wettability. Although they do not make good emulsifiers, they are often called emulsifiers. Many food scientists do not differentiate between surfactants and emulsifiers, and so the words may be used interchangeably in some cases. However, in the world of a colloid scientist, there is a clear distinction between the two!

Characteristics of an Emulsifier

- Contains hydrophilic and hydrophobic sections (amphiphilic)

Functions of an Emulsifier

- Adsorbs at the oil/water interface
- Reduces interfacial tension

} facilitates formation emulsion

- Forms a stable interfacial film
- Prevents coalescence

} promotes emulsion stability

Natural Emulsifiers

The best emulsifiers are proteins, which uncoil or denature and adsorb at the interface, and interact

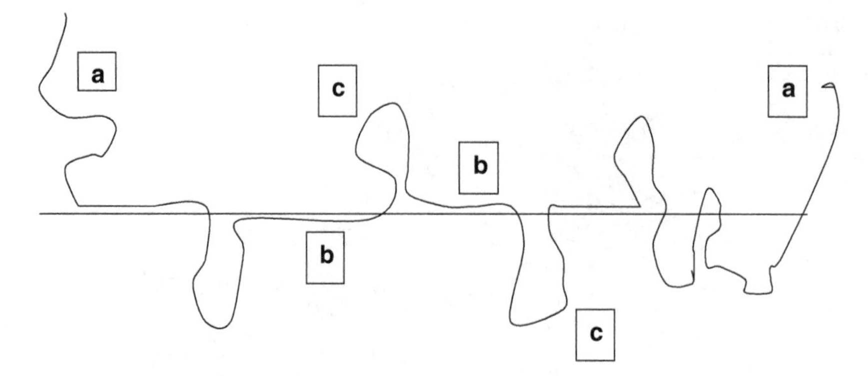

Fig. 13.4 Schematic diagram of a protein adsorbed at an interface: (*a*) tails, (*b*) trains, and (*c*) loops

to form a stable interfacial film. Proteins tend to uncoil such that their hydrophobic sections are oriented in oil, and their hydrophilic sections are oriented in water. Hence, a series of loops, trains, and tails may be envisioned at the interface, as shown in Fig. 13.4.

The loops and tails are able to interact with each other, thus forming a stable film that resists rupture. The proteins of *egg yolk* tend to be the best emulsifiers, as exemplified by their use in mayonnaise. These proteins are lipoproteins and are associated with each other and with phospholipids such as lecithin, in structures known as micelles. These micellar structures appear to be responsible for the excellent emulsifying properties of egg yolk proteins.

The *caseins* of milk are also excellent emulsifying agents. They are important emulsifiers in homogenized milk and in dairy desserts. In fresh (unhomogenized) milk, the caseins are associated with each other in structures known as casein micelles. Electron micrographs have shown that after homogenization, intact micelles are present at the fat globule surfaces, as well as individual protein molecules. It is thought that the micelles are responsible for the stability of homogenized milk, rather than the individual protein molecules.

Other food proteins used as emulsifiers include *meat* proteins and *soy* proteins. Lecithin is often considered to be an emulsifier. Lecithin is a surfactant and is useful for promoting wettability and aiding mixing of products such as hot drink mixes. It is also an essential ingredient in chocolate, where it aids in dispersion of the sugar and fat. Lecithin does not usually form strong interfacial films by itself and so would not be the emulsifier of choice unless other emulsifiers or stabilizers were added.

However, proteins are usually present in food emulsions, which may allow for formation of a strong interfacial film involving lecithin. Soy lecithin may be added to emulsions containing egg yolk, in order to reduce the amount of egg yolk needed, since soy lecithin is cheaper than egg yolk.

Synthetic Emulsifiers or Surfactants

Most synthetic emulsifiers would more correctly be termed *surfactants*, because they are relatively small molecules compared with proteins, and they are used mainly to aid in dispersion of fat, rather than to stabilize emulsions.

Surfactants such as *mono-* and *diglycerides* are added to shortening and to cake mixes, to aid in dispersion of the shortening. Cakes are complex, in that they contain fat droplets and air bubbles, and so are both emulsions and foams. (Foams are discussed later in this chapter.) The mono- and diglycerides enable the shortening to be dispersed into smaller particles, and this promotes incorporation of a large number of air cells, which increases cake volume and promotes a more even grain in baked products (Chap. 15).

Glycerol monostearate is an example of a monoglyceride that is commonly used in foods. Acids may be esterified with monoglycerides to give another group of surfactants, including sodium stearoyl-2-lactylate, which is often used in baked products. Two other groups of manufactured surfactants include the *SPANS*, which are fatty acid esters of sorbitan, and the *TWEENS*, which are fatty acid esters of polyoxyethylene sorbitan. Although all surfactants are amphiphilic, they have different degrees of hydrophobic (***lipophilic***) and hydrophilic character. This can be expressed as the ***hydrophilic/lipophilic balance***, or ***HLB***.

An HLB scale has been developed, which goes from 1 to 20. Surfactants with a low HLB (3–6) have more hydrophobic or lipophilic character. These would be used to form a w/o emulsion. Examples include glycerol monostearate and sorbitan monostearate (SPANS 60). Surfactants with a high HLB (8–18) have more hydrophilic character and form w/o emulsions. Examples would be polyoxyethylene sorbitan monostearate (TWEENS 60) or sodium stearoyl-2-lactylate. SPANS usually have a low HLB and form w/o emulsions, whereas TWEENS have a high HLB and form o/w emulsions. Use of the HLB scale may be going out of favor, yet is useful to food scientists to help them in determining which emulsifier is most suited to their needs.

Examples of Emulsions

French dressing is an example of a ***temporary emulsion***, or in other words, an unstable emulsion that separates fairly soon after formation. The basic ingredients of French dressing are oil (the dispersed phase), vinegar (the continuous phase), dry mustard, and paprika. Other ingredients may be added for flavor. The "emulsifiers" used here are the mustard and paprika. Combining the ingredients and shaking them vigorously forms the emulsion. The mustard and paprika adsorb at the interface and reduce interfacial tension as the dressing is shaken, thus facilitating formation of an emulsion, yet they do not interact at the interface to form a stable film. Hence, when shaking is stopped, the oil droplets are not protected, and so they soon coalesce, and the oil and vinegar layers separate.

Mayonnaise is an example of a ***permanent emulsion***, since it is stable and does not separate under normal handling conditions. The main ingredients of mayonnaise are oil (the dispersed phase), vinegar (the continuous phase), and egg yolk. The egg yolk proteins, being excellent emulsifiers, protect the oil droplets against coalescence. Mayonnaise usually contains about 75 % oil, which exists as stable droplets surrounded by a thin aqueous film. It is unusual in that it contains so much more dispersed phase than continuous phase. Generally, the continuous phase of an emulsion is present in greater quantity.

Mayonnaise is made by slowly pouring small amounts of oil at a time into the vinegar and egg yolk mixture and continuing to beat to break up the oil into droplets and form the emulsion. As more oil is added, more droplets are formed, and the surface area increases dramatically. The continuous phase spreads out to surround the oil droplets and becomes a thin film. It is hard for the droplets to move around, since they are packed tightly together, and separated only by a thin film of aqueous phase, and so the mayonnaise becomes very thick and may even be stiff enough to cut. Some salad dressings may be similar to mayonnaise, except that they contain less oil and have a thinner consistency. Adding stabilizers such as gums or starches often enhances the stability of the emulsion.

Milk is an example of an emulsion that occurs in nature (Chap. 11). Milk contains about 3.5 % fat in emulsified form. In fresh (unhomogenized) milk, the fat droplets are stabilized by a complex protein membrane known as the milk fat globule membrane. Fresh milk is a stable emulsion; however, it will cream fairly quickly if left to stand. The fat droplets vary in size from about 0.1 to 10 μm. There are many more small droplets than large ones; however, because of their size, the larger ones account for most of the fat. Because of the density difference between the milk fat and the aqueous phase, the fat droplets tend to rise though the milk. This is especially true for the larger droplets.

Milk fat globules are unique in that as they rise, they tend to cluster together. This results in larger fat particles, which rise even faster. Hence, after a few hours, a cream layer can be seen at the top of the milk. This is not a true separation of oil and water, since the cream layer is still an emulsion and the interfacial film is still intact. The milk has separated into a concentrated emulsion and a dilute one. The cream can be removed and either used as cream or made into butter.

Homogenizing the milk, which breaks up the fat globules into much smaller ones, prevents this creaming effect. By Stokes law, the smaller particles would take almost infinite time to coalesce and aggregate, thus remaining as small droplets.

Factors Affecting Emulsion Stability

Obviously, the main factor affecting emulsion stability is the *emulsifier* itself. As has been discussed, emulsifiers that form stable interfacial films produce stable emulsions. There must also be sufficient emulsifier to completely coat the surface of all the droplets in order to ensure stability. *Droplet size* is also important because larger droplets are more likely to coalesce. Also, because of the density difference between oil and water, large oil droplets will tend to rise through the emulsion more quickly, creating a more concentrated emulsion closer to the surface, as is seen in milk. This may cause the emulsion to break.

Changing the *pH* by adding acid or changing the *ionic strength* by adding salts may reduce the stability of the interfacial film, especially if it is made of protein. Such changes may denature the protein, as explained in Chap. 8, and cause the emulsion to separate.

Another factor affecting emulsion stability is the *viscosity* of the emulsion. The thicker the emulsion, the slower the movement of the molecules within the system and the longer it will take for the two phases to separate. Emulsions can be made thicker by adding ingredients such as gums, pectin, or gelatin. If gums are added to French dressing, a permanent emulsion may be formed without the need for egg yolk as the emulsifier.

Gums are often added to emulsions as stabilizers. They are not emulsifiers themselves, and they do not normally adsorb at an interface, because they are hydrophilic. However, they act by increasing the viscosity of the system, which slows movement, and hence reduces the number of collisions between droplets. This slows down and may even prevent separation of emulsions.

Storage and *handling* affect emulsion stability. Although some emulsions are termed permanent, it should be noted that all emulsions are delicate systems that are inherently unstable, because they contain two immiscible liquids, and the wrong handling conditions can cause emulsion breakage.

Temperature also affects emulsion stability. When emulsions are warmed, the oil droplets become more fluid and coalescence is more likely. On the other hand, cooling an emulsion to refrigeration temperatures may cause some solidification of the oil droplets, depending on the composition of the oil. This may enhance stability. Most emulsions do not survive freezing conditions. This is usually because the proteins at the interface become denatured, or because the interfacial film is physically disrupted by the formation of ice crystals. *Gums* are often added to emulsions that are to be frozen to enhance their stability.

Heat and *violent shaking* are also likely to disrupt emulsions. For example, cream is converted to butter by churning the warm emulsion. The emulsion breaks, the aqueous phase is drained off, and a water-in-oil emulsion is formed, with water droplets (approximately 18 %) dispersed throughout the butterfat.

Factors Affecting Emulsion Stability

- Type of emulsifier
- Concentration of emulsifier
- Droplet size
- Changing pH or ionic strength
- Viscosity
- Addition of stabilizers
- Heating, cooling, freezing, and/or shaking

Foams

Foams make a vital contribution to the volume and texture of many common food products. They give volume and a distinctive mouthfeel to products such as whipped cream and ice cream and they give a light, airy texture to baked goods. Improperly formed or unstable foams result in dense products with a low volume, which are unacceptable to consumers. Foams are inherently unstable, and it is imperative that food scientists increase their knowledge of the factors affecting foam stability, in order to enhance the quality and shelf life of these products.

A foam contains gas bubbles dispersed in a liquid continuous phase. The liquid phase may be a simple dispersion, as in egg white, which is a dilute protein dispersion, or it may be complex, containing emulsified fat droplets, ice crystals, and/or solid matter. Examples of complex food foams include ice cream, angel food cake, marshmallows, and yeast-leavened breads. Foams such as meringue and baked goods are heat-set, which denatures the protein and converts the liquid phase to a solid phase. This gives permanence to the foam structure.

Comparison Between Foams and Emulsions

Foams are similar to emulsions, in that the gas bubbles must be protected by a stable interfacial film otherwise they will burst. Therefore, the factors affecting emulsion formation and stability also apply to foams, and, in general, good emulsifying agents also make good foaming agents. However, there are some important differences between emulsions and foams. The bubbles in foams are generally much bigger than the droplets in emulsions, and the continuous phase surrounding the gas bubbles is very thin.

In fact, it is the continuous phase that has colloidal dimensions, rather than the dispersed phase. The density difference between the two phases is much greater in a foam, and there is a tendency for the liquid continuous phase to drain due to gravity, and for the gas bubbles to escape. The factors affecting formation are similar for both emulsions and foams. However, there are additional factors involved in foam stability.

Foam Formation

In order to produce a foam, energy must be supplied (by whipping) to incorporate gas into the liquid, to break up large bubbles into smaller ones, and to spread the liquid phase around the gas bubbles as they form. The foaming agent, which is contained in the liquid phase, adsorbs at the surface of the liquid, reducing surface tension and also forming a film around the gas bubbles. It is important that the surface tension is low, so that the liquid will spread rapidly around the gas bubbles during whipping. If newly formed gas bubbles are not immediately coated with foaming agent, they will burst or coalesce and be lost.

The amount of energy supplied during whipping is also important; the higher the energy, the smaller the bubbles and the greater the foam volume, provided that sufficient foaming agent is present to completely coat and stabilize the bubbles.

Foam Stability

The stability of a foam may be measured in terms of loss of foam volume over a period of time. When a liquid is whipped to form a foam, the volume of the liquid increases due to incorporation of air. If the foam is stable, the volume does not change very much. However, loss of air from an unstable foam may cause a considerable reduction in volume.

Foam stability may be *reduced* due to the following factors:

- The tendency of the liquid film to drain due to gravity. As it drains, a pool of liquid gathers at the bottom of the container, and the film surrounding the gas bubbles becomes very thin. This may allow the gas to escape and the volume of the foam to shrink.

- The tendency for the film to rupture and allow coalescence or escape of gas bubbles.
- Diffusion of gas from small bubbles to larger ones. This results in fewer bubbles and the foam shrinks.
- Evaporation of the continuous phase also affects foam stability, but to a lesser extent. If the liquid evaporates, gas bubbles burst and foam volume is reduced.

If gas bubbles are lost due to any of these factors, a more dense, low-volume foam is produced, which is not usually desirable, especially in foods such as angel food cake or ice cream.

To produce a *stable* foam with a high volume, film rupture, liquid drainage, and evaporation must be prevented or at least minimized. As with emulsions, the gas bubbles must be stabilized by the presence of a stable interfacial layer, which resists rupture. However, the composition of the continuous phase is also very important in determining foam stability. The liquid phase must have a low vapor pressure, so that it does not evaporate readily at storage and handling temperatures.

More importantly, drainage of the continuous phase must be minimized. Thick liquids drain more slowly than thin ones, and so increasing the viscosity of the continuous phase will reduce drainage. A high viscosity is essential if a stable foam is to be produced.

Foaming Agents

The two most important characteristics of a foam are foam *volume* and foam *stability*. Foam volume depends on the ability of the foaming agent to adsorb at the interface and rapidly reduce interfacial tension and on the level of energy input during whipping. Foam stability depends on the ability of the foaming agent to produce a stable interfacial film and a viscous continuous phase. Although all surfactants are able to reduce surface tension and produce foams, not all are able to form stable foams. In fact, some may act as foam suppressants!

A good foaming agent has the same characteristics as an emulsifier, in that it is able to adsorb at the interface, reduce interfacial tension, and form a stable interfacial film that resists rupture. As might be expected, the best foaming agents used in foods are proteins. Although many proteins are able to produce foams, egg white proteins are superior foaming agents and are used in food foams such as meringues, angel cake, and other baked goods. Other proteins used as good foaming agents include gelatin and milk proteins.

When egg white is whipped (Chap. 10), the proteins denature at the interface and interact with one another to form a stable, viscoelastic, interfacial film. Some of the egg white proteins are glycoproteins containing carbohydrate. When these proteins adsorb at the interface, the carbohydrate sections orient toward the aqueous phase. Being hydrophilic, they bind water and increase the viscosity of the liquid. This helps to reduce drainage, thereby contributing to foam stability.

Gelatin is a good foaming agent, and a warm gelatin sol can be whipped to three times its original volume. When cooled, the gelatin solidifies or forms a gel, which traps the air bubbles and stabilizes the foam; marshmallows are gelatin foams.

The Effect of Added Ingredients on Foam Stability

Many food foams have additional ingredients added to *enhance stability*. For example, egg white foams, such as meringue or angel food cake, also have sugar added. The *sugar* increases the viscosity of the liquid, aiding stability. It also protects the proteins from excessive denaturation and aggregation at the interface. Too much interaction results in an inelastic film which is not resistant to rupture, and in reduced foam volume. Therefore it is important to guard against this when making egg white foams.

Factors Affecting Foam Stability
- Drainage of the liquid film between gas bubbles
- Rupture of the interfacial film around gas bubbles
- Diffusion of gas from small to large bubbles
- Evaporation of the continuous phase

Factors Promoting Foam Stability
- Stable viscoelastic surface film
- Very viscous continuous phase
- Low vapor pressure liquid

Effects of Added Ingredients

Foam stabilizers	Foam destabilizers
Gums	Lipids
Thickeners	Phospholipids
Sugar	Small molecule surfactants
Acid	Salts
Solid particles	

Acid, such as cream of tartar or lemon juice, may also be used to increase foam stability. Addition of acid reduces the pH, which reduces the charge on the protein molecules and usually brings them closer to their isoelectric point. This generally results in a stronger, more stable interfacial film. When added to egg whites, acid prevents excessive aggregation at the interface. However, acid delays foam formation. It may therefore be added toward the end of the whipping process. In the case of egg whites, it is often added at the "foamy" stage. Whipping is not complete until the egg whites have formed stiff peaks. (Egg white foams are discussed in more detail in Chap. 10.)

Other ways to increase viscosity of the continuous phase include addition of *gums* and other *thickening agents*. Also, addition of *solid matter* may promote stability. Whipped cream, for example, is stabilized by solidified fat globules that are oriented in the continuous liquid film. The emulsified fat increases viscosity and is responsible for the stability of whipped cream. To form a stable foam, cream to be whipped must contain at least 30 % fat. Creams with lower fat contents may be whipped successfully if thickening agents such as carrageenan are added. If the cream is warm and too much of the butterfat is liquid, then whipping will not produce a stable foam. Instead, the emulsion will break and the cream will be converted to butter.

Ice cream is another example of a complex foam, which is stabilized by emulsified fat droplets and small ice crystals oriented within the continuous phase. Angel food cake contains solid particles, in the form of flour, which are folded into the egg white/sugar foam. The flour contributes to stability by increasing the viscosity of the liquid, which minimizes film drainage. The increased viscosity and presence of solid particles also reduces breakage of the interfacial film, hence minimizing loss of foam volume.

Anti-foaming Agents and Foam Suppressants

As all cooks know, egg whites will not whip to a stable foam if there is any egg yolk present (Chap. 10). This is because the phospholipids and lipoproteins in the yolk adsorb at the surface, in competition with the egg white proteins, and interfere with formation of a stable egg white protein film. Unlike the egg white glycoproteins, which are hydrophilic, the phospholipids and lipoproteins are unable to increase the viscosity of the continuous phase, because they are hydrophobic, and so orient away from the water. This prevents formation of a stable foam.

Such molecules are termed ***foam suppressants***. They suppress foam volume because they adsorb at the interface, thus suppressing adsorption of the desired foaming agent and preventing it from forming a stable foam. They do not have the properties required to form a stable film or to sufficiently increase the viscosity of the continuous phase. Hence, their presence makes formation of a stable foam impossible. Typical foam suppressants include fats, phospholipids, and other small amphiphilic molecules.

Salts also tend to act as foam suppressants, because they weaken interactions between the protein molecules at the surface, thus weakening the interfacial film around the gas bubbles. However, their effect is not as important as surfactant molecules, because they do not adsorb at the interface.

Anti-foaming agents are able to break up foams or prevent them from forming. Anti-foaming agents are added to fats and oils used in frying, to prevent foaming during the frying process. Like foam suppressants, they act by adsorbing at the air/liquid interface in place of the foaming agents, and because they do not have the characteristics of a foaming agent, they prevent foam formation.

Other Colloidal Systems

Although this chapter covers emulsions and foams, gels should be mentioned, since they are also colloidal systems. A ***gel*** consists of a liquid dispersed phase held within a solid continuous phase. Gels are formed when conditions allow the solid dispersed phase of a colloidal dispersion or sol to bond at strategic points, forming a three-dimensional network that traps liquid within itself. Conditions that are likely to cause formation of such a network include heating, cooling, addition of calcium or other divalent ions, and/or change of pH. Important food gels include starch gels (discussed in Chap. 4), pectin gels (Chap. 5), and gelatin, egg white, and other protein gels (Chap. 8).

Conclusion

Food emulsions and foams are complex colloidal systems, and understanding of their formation and stability is important if the quality and shelf life of these products is to be improved.

Emulsions contain liquid droplets stabilized by an interfacial layer of emulsifier and dispersed throughout a liquid continuous phase. Foams are similar, although the dispersed phase consists of large gas bubbles surrounded by a very thin, continuous, liquid film. The nature of the emulsifier or foaming agent is crucial in determining stability. It must adsorb at the interface, reduce surface tension, and form a stable, viscoelastic interfacial layer that resists rupture, so that coalescence of liquid droplets or loss of gas bubbles is avoided. Additional factors are important in foam stability; it is important that the liquid film between the gas bubbles is very viscous, so that drainage due to gravity is minimized. Evaporation of the liquid must also be prevented during normal storage and handling conditions.

Both natural and synthetic emulsifying agents are available to food companies. The best emulsifiers and foaming agents are proteins. Egg yolk proteins are known as the best emulsifiers, whereas egg white proteins are considered to be the best foaming agents used in food products.

Notes

CULINARY ALERT!

Glossary

Adsorb To bind to a surface.

Amphiphilic A molecule containing both hydrophobic and hydrophilic sections.

Coalescence (coalescing) Two liquid (or gas) droplets merge (merging) to form one larger droplet.

Colloidal system Emulsions, foams, dispersions (or sols), and gels are all colloidal systems. A colloidal system contains one phase (usually the dispersed phase) with dimensions ranging mainly from 0.1 to 10 μm. The dispersed phase contains large numbers of small droplets or particles, and so the surface or interfacial area of this phase is very large. This is an important characteristic of colloidal systems.

Continuous phase The phase or substance that surrounds the liquid droplets or gas bubbles in an emulsion or foam.

Dispersed phase The discrete bubbles (air, carbon dioxide, or liquid) that are surrounded by liquid in an emulsion or foam.

Emulsifier A substance that enables two normally immiscible liquids to be mixed together without separating on standing.

Emulsion An emulsion contains liquid droplets stabilized by a layer of emulsifier and dispersed throughout a liquid continuous phase.

Foam A foam contains gas bubbles coated with a stable interfacial layer and surrounded by a thin, viscous liquid continuous phase. In food foams, the gas is usually air or carbon dioxide.

Foaming agent A molecule that is able to promote foam formation. Useful foaming agents in foods are also able to promote foam stability by forming a stable interfacial layer and also by increasing the viscosity of the continuous phase.

Foam suppressant A molecule that prevents or hinders foaming, generally by adsorbing to the interface in place of the desired foaming agent and interfering with the action of the foaming agent.

Gel A two-phase system consisting of a solid continuous phase and a liquid dispersed phase. A gel may be considered to be a three-dimensional network with liquid trapped within its spaces.

Hydrophilic Water-loving. Hydrophilic molecules are either charged or polar and have an affinity for water.

Hydrophilic/lipophilic balance or HLB A scale that goes from 1 to 20 and indicates the ratio of hydrophilic and hydrophobic groups on a molecule. It is used to determine the suitability of emulsifiers when formulating an emulsion. A high HLB indicates a molecule with more hydrophilic groups, which is suitable for o/w emulsions. A low HLB indicates that there are more lipophilic groups, and the molecule has a greater affinity for oil and is more suited for w/o emulsions.

Hydrophobic Water-hating. Hydrophobic molecules are nonpolar and have an affinity for apolar solvents.

Interfacial tension The force required to increase the interfacial area of a liquid, or to spread it over a surface such as oil. See also, surface tension.

Lipophilic Fat-loving, or water-hating. Lipophilic molecules are nonpolar and have an affinity for lipids and other apolar solvents.

Oil-in-water or o/w emulsion An emulsion containing oil droplets dispersed in water. Oil is the dispersed phase and water is the continuous phase.

Permanent emulsion A stable emulsion that does not separate over time.

Surface tension The force required to increase the surface area of a liquid, or to spread it over a surface. Surface and interfacial tension are often used interchangeably. Generally, surface tension applies at the surface of a liquid (i.e., when it is in contact with air), whereas interfacial tension applies when two liquids are in contact with each other.

Surface-active A molecule that adsorbs at the surface of a liquid. Surface-active molecules contain both hydrophobic and hydrophilic sections, and it is energetically favorable for them to exist at the interface rather than in the bulk phase of a liquid.

Surfactant A surface active molecule (see above).

Temporary emulsion An unstable emulsion, which separates into two layers on standing.

Viscoelastic Exhibits both viscous (liquid) and elastic (solid) properties. In other words, the material will flow if force is applied, but it will also stretch. When the force is removed, the material does not return completely to its original position. It is important for an emulsifier film to flow around droplets to cover temporary bare patches, and also to be able to stretch, so that when disrupted, it does not break.

Water-in-oil or w/o emulsion An emulsion containing water droplets dispersed in oil. Water is the dispersed phase and oil is the continuous phase.

Bibliography

Charley H, Weaver C (1998) Foods: a scientific approach, 3rd edn. Prentice-Hall, Upper Saddle River, NJ

Coultate T (2009) Food: the chemistry of its components, 5th edn. RSC, Cambridge

McWilliams M (2012) Foods: experimental perspectives, 4th edn. Prentice-Hall, Upper Saddle River, NJ

Ritzoulis C (2013) Introduction to the physical chemistry of foods. CRC, Boca Raton, FL

Setser CS (1992) Water and food dispersions. In: Bowers J (ed) Food theory and applications, 2nd edn. Macmillan, New York, pp 7–68

Walstra P, van Vliet T (2007) Dispersed systems—basic considerations. In: Fennema O (ed) Food chemistry, 4th edn. CRC, Boca Raton, FL

Part V

Sugars, Sweeteners

14 Sugars, Sweeteners, and Confections

Introduction

Sugars are simple carbohydrates classified as *monosaccharides* or *disaccharides* (see Chap. 3). The common granulated or table sugar is the disaccharide sucrose, made of glucose and fructose. This chapter on sugars, sweeteners, and confections examines the sources, roles, and properties of sugars, the various types of nutritive sweeteners, and sugar substitutes added to foods. As well, confections and factors influencing candy types are addressed. Sugar should be used sparingly in the diet, and depending on serum glucose and lipid goals, nutritive and nonnutritive sweetener intake should be individualized by consumers.

Sources of Sugar

Table sugar comes from *two* sources. It naturally exists as syrup in the *sugar cane* or in *sugar beet*, both of which are *identical* in chemical composition. Sugar cane has been used for centuries. It is washed, shredded, pressed, and heated, and the extracted juice is centrifuged to create raw sugar with its slightly brown color. As the juice is centrifuged, molasses separates from the crystals and become a by-product of sugarcane sugar production. The crystals are then further refined for uses in various forms.

Roots of the beet are less frequently used to produce sugar and were first extracted in the 1790s. They too are washed, shredded, and so forth. Then, roots are treated with lime to remove impurities and further refined to yield usable sugar.

Roles of Sugar in Food Systems

The roles of sugars are diverse (some are listed below). Sugar may be utilized in *trace* amounts or it may be the *primary* ingredient of a formulation. Sugar imparts sweetness, tenderness, and browning, is hygroscopic (water retaining), and functions in various other ways in food systems, as may be seen in the following examples of sugar function.

Sweetness

Sugar provides flavor appeal to foods and is therefore incorporated into many foods. It is a significant ingredient of candy, many baked goods and frostings as well as some beverages, and may be used in a less significant manner or not at all in other foods. Around the world, there is an *innate* desire for sweetness. Some individuals consume fruit "picked off of a tree," as a piece of fruit, while others consume snacks they "pick out from the office vending machine"!

V.A. Vaclavik and E.W. Christian, *Essentials of Food Science, 4th Edition*, Food Science Text Series, DOI 10.1007/978-1-4614-9138-5_14, © Springer Science+Business Media New York 2014

Tenderness

A batter/dough formula *with* sugar is *more* tender than one *without* sugar, because sugar binds with each of the two proteins gliadin and glutenin and absorbs water so they do *not* form gluten.

Browning

Browning in some varieties of *fruits and vegetables* is due to *enzymatic, oxidative browning*. Yet it is *sugar* that browns and imparts color to foods by two types of *nonenzymatic* browning including (1) the low-temperature Maillard browning reaction and (2) high-temperature caramelization.

Maillard browning involves the reaction of the carbonyl group of a reducing sugar with the amine group of an amino acid and occurs with *low*-temperature heat, a high pH, and low moisture. Maillard browning is responsible for the color changes that occur in many baked breads, cakes, and pie crusts, canned milks, meats, as well as caramel candies (which, although the name is used, is *not* caramelization).

Caramelization is a *nonenzymatic* browning process that occurs in sugars heated to *high* temperatures. As sugar is heated to temperatures above its melting point [338 °F (170 °C)], it dehydrates and decomposes. The sugar ring (either pyranose or furanose) opens and loses water. The sugar becomes brown, more concentrated, and develops a caramel flavor as it continues to increase in temperature. The dessert flan is an example.

CULINARY ALERT! Caramelization in culinary terms refers to any sugar in food that is broken down to produce enhanced color and flavor upon reaching a high temperature. Most notable is caramelized, dark brown, onions.

Additional roles of sugar in food systems (not all inclusive!):

- Functions as a *separating agent* to prevent lump formation in starch-thickened sauces (Chap. 4)
- *Reduces starch gelatinization* (Chap. 4)
- *Dehydrates* pectin and permits gel formation in jelly-making (Chap. 5)
- Stabilizes *egg white foams* (Chap. 10)
- *Raises the coagulation temperature* of protein mixtures (Chap. 10)
- Adds *bulk* and body to foods such as yogurt (Chap. 11)
- Helps *aerate* batters and dough (Chap. 15)
- Reduces gluten structure by competing with gliadin and glutenin (Chap. 15) for water, thus increasing *tenderness* (Chap. 15)
- Acts as the substrate that *ferments* to yield CO_2 and alcohol (Chap. 15)
- Adds *moisture retention* properties to baked products (Chap. 15)
- Slows/prevents crystallization in candies if invert sugar is used (Chap. 14)

Types of Sugars and Sugar Syrups

Types of sugars "-ose," those used in food preparation, and syrups are discussed below. Sugar substitutes will be discussed later in this chapter.

Sucrose. Sucrose is a *disaccharide* consisting of the monosaccharides glucose and fructose. It is commonly referred to as "sugar," "white sugar," or "granulated sugar."

Fructose. Fructose is a *monosaccharide* that combines with glucose to form the *disaccharide* sucrose. It is known as fruit sugar, since it is contained in many fruits. Fructose is 1.2–1.8 times as sweet as sucrose.

Glucose. Glucose is a *monosaccharide* that combines with fructose to form sucrose, with galactose to form lactose, and with another glucose to form the *disaccharide* maltose.

Galactose. Galactose is a *monosaccharide* that combines with glucose to form the disaccharide lactose.

Lactose. Lactose is a *disaccharide* (a glucose and galactose molecule) known as milk sugar. It is less sweet than sucrose.

Maltose. Maltose is a *disaccharide* (two glucose molecules) formed by the hydrolysis of starch.

*All monosaccharides contain a free carbonyl group and are reducing sugars, as is the disaccharide maltose.

Specific recipe preparation may require use of the following sugars (the use of artificial sweeteners and sugar alcohols will be discussed in a later section of this chapter):

Brown sugar. Brown sugar has a molasses film on the sugar crystals, which imparts the brown color and characteristic flavor of this sugar. It contains approximately 2 % moisture and requires storage protection against moisture loss.

Confectioners' sugar. Confectioners' sugar (confectionery sugar) is also known as *powdered sugar* and is derived from either sugar cane or the sugar beet. Sugar grains are pulverized by machine to change the sugar grain to a powdered substance and form such sugars as "6× sugar" (pulverized 6 times to create "very fine"), "10× sugar" (pulverized 10 times to form "ultrafine"), and so on. Confectioners' sugar typically contains 3 % cornstarch to prevent caking.

Invert sugar. Invert sugar is created when *sucrose* is treated by acid or enzyme to form an equal amount of *fructose and glucose*. It is more soluble and sweeter than sucrose and is commonly used in confections, including that which will become the liquid center of chocolate-covered cherries.

Raw sugar. Raw sugar has a larger grain than ordinary granulated sugar. It is 97–98 % sucrose. It is *not* approved by the FDA for sale in the United States since impurities and contaminants remain in the granule. (It is not the same as "Sugar In The Raw®.")

Turbinado sugar. Turbinado sugar is raw sugar with 99 % of the impurities and most of the by-product of sugar crystal formation (molasses) removed.

Syrups (Liquids)

The conversion of starch yields dextrose (glucose). Syrups are then measured as dextrose equivalents (D.E.). Syrups may have a D.E. of 36–55. More pure glucose yield is 96–99 D.E.

Corn syrup: Corn syrup is a mixture of carbohydrates (glucose, maltose, and other oligosaccharides) formed from the hydrolysis of cornstarch by the use of acid or enzymes (HCl, or α and β amylases). Following hydrolysis, it is subsequently refined and concentrated. The sugar solution contains approximately 25 % water and is viscous.

$$\text{Starch} + \text{Water} \xrightarrow[\text{enzymes}]{\text{acid+heat}} \text{Dextrins} + \text{Maltose} + \text{Glucose}$$

Corn syrup, due to its high glucose content, more readily participates in Maillard reactions. As a reducing sugar, corn syrup (its glucose) is a major browning enhancer. Adding just 1 tablespoon (15 mL) of corn syrup to otherwise paler cookie dough significantly increases browning.

Read more from cooking pro Shirley Corriher at http://www.tipsonhomeandstyle.com/food/the-cure-for-common-cookie-problems#ixzz2IqVGlh2o

High-fructose corn syrup (HFCS): HFCS is a specialty syrup prepared by the same three steps as corn syrups—it is hydrolyzed, refined, and concentrated. In addition, isomerization occurs whereby the principle sugar, glucose, is made into a more soluble fructose by the enzyme action of another enzyme, isomerase.

The HFCS contains approximately 42 % fructose and may undergo a fractionation process to further remove glucose and create syrup that is 55 or 90 % fructose. HFCS containing 42 and 55 % fructose are *generally recognized as safe* (GRAS). Many beverages contain HFCS, and although sugar consumption in the United States may show a downward trend, HFCS is increasingly ingested.

See *What the Science Says about HFCS* at www.sweetsurprise.com/what science-says-about-hfcs (SweetSurprise.com).

Honey: Honey is made from the nectar of various flowers and therefore differs in color, flavor, and composition. It contains approximately 20 % water and a mixture that is glucose and fructose (predominantly the latter), with no more than 8 % sucrose. Due to the hygroscopic property of fructose, the addition of honey to a formulation favorably increases its level of moisture.

Darker colored honey is *more* acidic and *more* strongly flavored than light colored honey. The strains of alfalfa and clover honey, commonly sold in the United States, are mild-flavored honey. "Strained honey" is honey from a crushed honeycomb that is strained.

Maple syrup: Maple syrup is obtained from the sap of the maple trees. The sap is boiled and evaporated, and the final product contains no more than 35 % water (40 parts sap = 1 part maple syrup).

Molasses: Molasses is the syrup (plant juice) separated from raw sugar beet or sugar cane during its processing into sucrose, and it is thus a by-product of sugar making. The predominant sugar is sucrose, which becomes more invert sugar with further processing. Molasses provides very low levels of the minerals, calcium, and iron, although blackstrap molasses is the product of further sugar crystallization and contains a *slightly* higher mineral content.

The Sugar Association states that almost 60 % of sweetener intake is from corn sweeteners, especially those used in sodas and sweetened drinks. The other 40 % is from table sugar or sucrose (The Sugar Association, Washington, DC).

Properties of Sucrose

Properties of sucrose, in addition to supplying sweetness, are important in food systems, such as confections. These are discussed in the following subsections.

Solubility

Solubility of sugars varies with sugar type. For example, sucrose is more soluble than glucose and less soluble than fructose. This influences candy type and product success (see later chapter section on Formation of Invert Sugar).

In its dried, granular form, sugar becomes *increasingly* soluble in water with an increase in temperature. At *room temperature*, water is capable of dissolving sucrose in a ratio of 2:1 (67 % sucrose and 33 % water). If that same water is *heated*, more sugar is dissolved, and as the sugar–water is further heated and brought to a boil, water evaporates and the sugar syrup becomes increasingly concentrated. This is seen in the amount of sugar held by equal amounts of iced tea and hot tea beverages. Hot tea holds more dissolved sugar.

Sugar may precipitate from solution, forming an undesirable grainy, crystalline product. Therefore, to increase the solubility of sucrose and reduce possible undesirable crystallization, sucrose may be treated by inversion to become *invert sugar*.

Types of Solutions

Solutions are the homogeneous mixtures of ***solute***, dissolved in ***solvent***. Depending on the amount of dissolved solute that the water is holding at any specific temperature (see Sugar Concentration),

Table 14.1 Boiling point of sucrose–water syrups of different concentrations[a]

Percent of sucrose in syrup	Percent water	Boiling point in °F (°C)
0 (All water, no sugar)	100	212 (100)
20	80	213.1 (100.6)
40	60	214.7 (101.5)
60	40	217.4 (103)
80	20	233.6 (112)
90	10	253.1 (123)
95	5	284 (140)
99.5	0.5	330.8 (166)

[a]At sea level

solutions may be dilute (unsaturated), saturated, or supersaturated.

Elevation of Boiling Point

The boiling point of sugar *elevates* with *increasing concentrations of sucrose* in solution as shown in Table 14.1. The boiling point also rises as the liquid evaporates from a boiling solution and causes there to be a greater concentration. This more concentrated solution now has a reduced vapor pressure that elevates boiling point because more heat is needed to raise the reduced vapor pressure found in a more concentrated sugar solution.

The addition of sugars other than sucrose, as well as the addition of interfering agents (see Interfering Agents), may also *elevate the boiling point*. At sea level, water boils at 212 F (100 °C). For every gram molecular weight of *sucrose* that is dissolved in water, there is a 0.94 °F (0.52 °C) increase in boiling point. This is why sugar solutions reach a very high temperature and cause more severe burns than boiling water.

CULINARY ALERT! High elevation lowers the point at which water boils. For each 500 ft in elevation above sea level, there is progressively less atmospheric pressure and the boiling point *decreases* 1 °F. [Therefore, at an elevation of 5,000 ft, the boiling point is *lowered* by 10 °F, to 202 °F (94 °C)]. Also, the boiling point is *lowered* above sea level.

Formation of Invert Sugar

Another property of sugar is that *invert sugar* is formed by sucrose hydrolysis in the process of ***inversion***. The inversion process yields equal amounts of the monosaccharides *glucose* and *fructose*, and the latter is more soluble than sucrose (solubility of fructose > sucrose > glucose).

CULINARY ALERT! Due to its increased solubility, the use of invert sugar in confections is desirable in candies. It is used to slow crystallization and help keep crystals small. Invert sugar is combined in a ratio of 1:1 with sucrose in many product formulations.

As seen in the formula, sucrose may be hydrolyzed into invert sugar by either *weak acids*, such as in cream of tartar (the acid salt of weak tartaric acid), or by *enzymes* such as invertase. Each is described below.

$$C_{12}H_{22}O_{11} + H_2O \xrightarrow[enzyme]{acid+\ heat} C_6H_{12}O_6 + C_6H_{12}O_6$$

Sucrose Water Glucose Fructose

Concerning *acid hydrolysis*, it is both (1) the amount of acid and (2) the rate and length of heating that determine the quantity of invert sugar that forms. This is addressed below:

- **Amount of acid**:
  Too much acid, such as cream of tartar, may cause *too much* hydrolysis, which forms a soft or runny sugar product.
- **Rate and length of heating**:
  A *slow* rate and slow attainment (long length of heating) of the boiling point increases inversion opportunity, whereas a *rapid* rate provides less inversion opportunity.

In *enzyme hydrolysis*, sucrose is treated with the enzyme invertase (also known as sucrase) to form glucose and fructose.

CULINARY ALERT! Enzyme hydrolysis may take several days, as is the case with invert sugar that is responsible for forming the liquid in chocolate-covered cherries.

The *glucose* that forms from inversion is *less* sweet than sucrose, and the *fructose more* sweet, with the overall reaction producing a sweeter, more soluble sugar than sucrose. Invert sugar is combined in a ratio of 1:1 with untreated sucrose in many formulations to control crystal formation and achieve small crystals.

Hygroscopicity

Hygroscopicity, or the ability to readily absorb water, is a property of *sucrose*. However, other sugars that are high in *fructose*, such as *invert sugar*, *HFCS*, *honey*, *or molasses*, are *more* hygroscopic than sucrose. It is therefore important to control the degree of inversion that these more hygroscopic sugars undergo, or they may exhibit *runny* characteristics in storage. Sugar alcohols such as mannitol are nonhygroscopic.

Sugar that is stored in a *humid storeroom* location and candy that is prepared on a *humid day* are *both* situations that demonstrate this property of hygroscopicity in that the sugar becomes lumpy and the finished candy is soft. (This hygroscopicity property of sugar carried caution for the preparation of meringues in the Egg chapter.)

Due to this hygroscopic property of sucrose, product developers may *encapsulate* or coat sugars so that sugars are time released.

Fermentable

One more property of sugar is that it is fermentable. It undergoes ***fermentation*** by the biological process in which bacteria, mold, yeast, and enzymes anaerobically convert complex organic substances, such as sucrose, glucose, fructose, or maltose, into *carbon dioxide* and *alcohol*.

The next discussion encompasses substances with very different properties than the aforementioned organic substances.

Sugar Substitutes

Sugar substitutes include two categories: (1) *artificial (or high-intensity) sweeteners* (noncaloric, nonnutritive) and (2) *sugar alcohols* (caloric, nutritive). Each of these sugar substitutes may be utilized with varying degrees of success in food products including cereals, cakes, pies, ice cream, soda, and candies. Americans regularly consume low- or no-calorie or sugar-free sugar substitutes in order to cut back on calories or sugars.

Artificial or High-Intensity Sweeteners

[You may prefer to differentiate between artificial sweeteners and high-intensity sweeteners because natural high-intensity ones are being discovered/developed such as steviosides and Reb A (covered on in this chapter).]

Artificial sweeteners or high-intensity sweeteners are one category of sugar substitute. They are noncaloric, nonnutritive, intense sugar substitutes, whose use has grown in response to increased consumer demand. They must be FDA approved before use. As with foods discussed throughout other book chapters, an individual aversion, medically or otherwise, may exist.

Various examples of artificial sweeteners are included as follows.

Acesulfame K. *Acesulfame potassium* is a noncaloric, synthetic derivative of acetoacetic acid. It received FDA approval in 1988. Acesulfame K is an organic salt consisting of carbon, hydrogen, nitrogen, oxygen, potassium, and sulfur and is *not* metabolized by the body; however, it is rather excreted *unchanged*. It is 200 times (thus high intensity) sweeter than sucrose and is heat stable, able to successfully be used for baking and cooking purposes in addition to use as a tabletop sweetener.

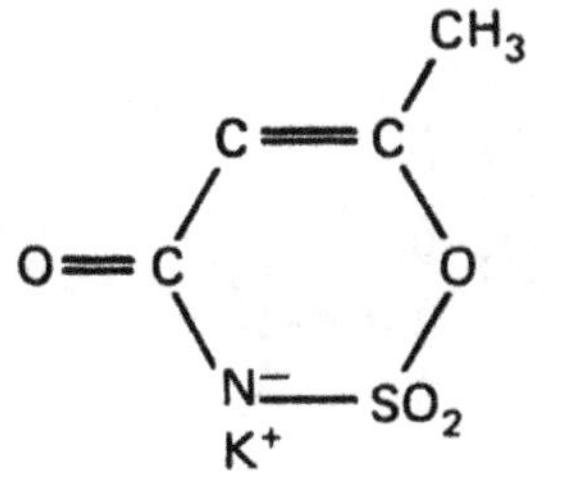

Fig. 14.1 Chemical structure of acesulfame K

Acesulfame K has no bitter aftertaste and may be used *alone or in combination* with the other sweeteners saccharin or aspartame (Fig. 14.1). Some brand name examples of acesulfame K are Sunett®, Sweet One®, Swiss Sweet®, and Nutra Taste®.

CULINARY ALERT! *Sweet One® 12 packets = 1 cup sugar.*

Advantame—developed by Ajinomoto—is made from aspartame and vanillin and is 20,000 times as sweet as sugar and 100 times as sweet as aspartame.

Aspartame. *Aspartame* is a *nutritive* sweetener that contains the *same* number of calories per gram as sugar (4 cal/g). However, due to the fact that it is much sweeter and used in minute amounts, it is *not* a significant source of either calories or carbohydrates and is often put in the category of *nonnutritive*, *noncaloric* sweeteners. Aspartame is a methyl ester comprising two amino acids: aspartic acid and phenylalanine. Thus, because of the latter, phenylalanine, it should *not* be consumed by those with the genetic disease phenylketonuria (PKU) because the phenylalanine is *not* metabolized (Fig. 14.2).

Aspartame is one of the most thoroughly tested and studied food additive the FDA has ever approved (FDA). It gained FDA approval in 1981 and is 180–200 times sweeter than sucrose. It is marketed under the trade names NutraSweet® and Equal®. (Equal® is the tabletop low-calorie sweetener with NutraSweet®.) Aspartame was *not* originally intended for use in heated products; however, it may be encapsulated in hydrogenated cottonseed oil with a time–temperature release, which makes its inclusion in baked products acceptable.

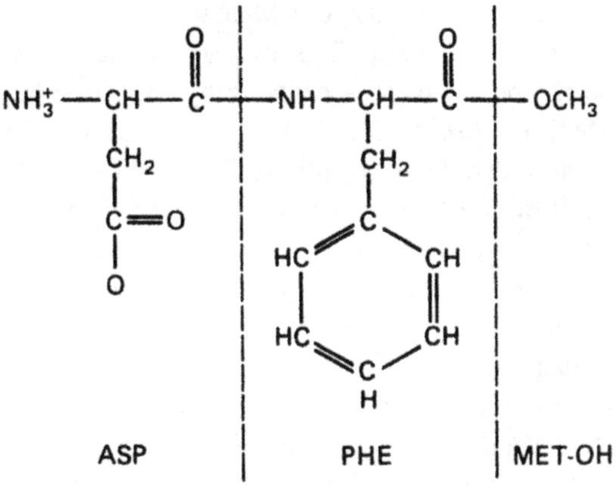

Fig. 14.2 Chemical structure of aspartame. *ASP* aspartic acid, *PHE* phenylalanine, *MET-OH* methyl alcohol. ASP ¼ aspartic acid; PHE ¼ phenylalanine; MET-OH ¼ methyl alcohol

CULINARY ALERT! *Equal® 24 packets = 1 cup sugar.*

Commonly, aspartame and acesulfame K are used together at a ratio of 50:50 or so, "Their synergy together covers the entire sweetness curve" (Hazen 2012a).

Neotame. Neotame is chemically similar to the artificial sweetener aspartame. It is between 7,000 and 13,000 times as sweet as sucrose. It was granted FDA approval in 2002.

Saccharin. *Saccharin* is a noncaloric substance produced from methyl anthranilate, a substance naturally found in grapes. It has been used as a noncaloric sweetener since 1901 in the United States and is 300–700 times sweeter than sucrose.

The use of saccharin was periodically reviewed as specified by US Congress in the *Saccharin Study and Labeling Act*. The ruling required that foods containing saccharin must be labeled to read as follows: "Use of this product may be hazardous to your health. This product contains saccharin which has been determined to cause cancer in *laboratory animals*."

However, following a moratorium on banning saccharin, which was extended by Congress several times, pending further safety studies, it was shown that saccharin has *not* demonstrated any carcinogenicity applicable to *humans*. Therefore, after several decades the safety of saccharin has been shown, and the use of a warning label is *no* longer required. The use of saccharin has been reported to be acceptable by the American Medical Association, the American Cancer Society, and the Association of Nutrition and Dietetics (formerly American Dietetic Association).

In December 2000, Congress passed H.R. 5668—the **S**accharin **W**arning **E**limination via **E**nvironmental **T**esting **E**mploying **S**cience and **T**echnology (**SWEETEST**) Act. It is approved for use in more than 100 countries.

Calcium or sodium saccharin, combined with dextrose (nutritive, glucose) and an anticaking agent, may be used in tabletop sweeteners. Saccharin may also be used in combination with aspartame. Brand name examples include Sweet'N Low®, Sugar Twin®, Necta Sweet®, and Sweet-10®.

CULINARY ALERT! *Sweet'N Low® 12 packets = 1 cup sugar.*

Sucralose. *Sucralose* gained FDA approval in 1998 for use in 15 specific food and beverage categories, including baked goods and baking mixes; beverages and beverage mixes; chewing gum; coffee and tea; confections and frostings; dairy product analogs; fats and oils (salad dressings); frozen dairy desserts and mixes; fruit and water ices; gelatins, puddings, and fillings; jams and jellies; milk products; processed from it and fruit juices; sweet sauces, toppings, and syrups; and sugar substitutes.

Sucralose is a *noncaloric* trichloro derivative of sucrose [three hydroxyl (hydrogen–oxygen) groups on a sugar molecule are selectively replaced by three atoms of chlorine], plus maltodextrin, which gives it bulk and allows it to be measured cup for cup, like table sugar. It is 400–800 times sweeter than sucrose.

Several advantages to its approval are that (1) it is the only *noncaloric* sweetener made from sugar; (2) it is *stable* under a wide range of pH, processing, and temperature scenarios, for example, it is water- and ethanol-soluble and heat stable in baking and cooking; and (3) it carries *no health warnings*. Splenda® is the brand name under which sucralose is marketed (Fig. 14.3).

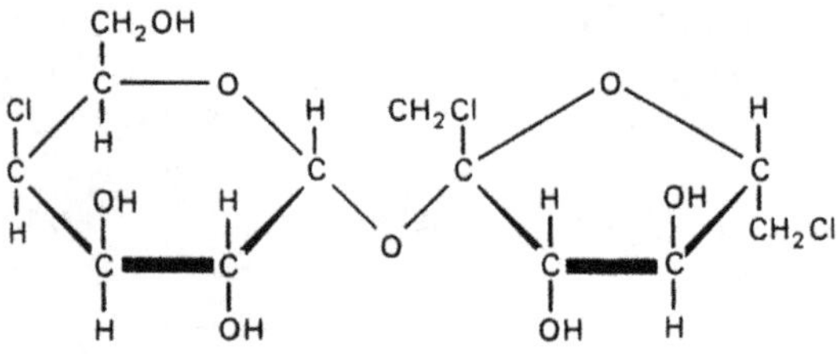

Fig. 14.3 Chemical structure of sucralose

Cyclamate. Cyclamate does not have FDA approval although it is still used as a sweetener in many parts of the world, including Europe. It was discovered "accidently" in a US university research lab in 1937 and was used through the 1960s, although banned in the United States in 1970 and suspect for bladder cancer, liver damage, and other health issues. Currently, the FDA is considering a petition for reapproval, as evidence of its connection with bladder cancer is not verified. It is noncaloric and 30 times sweeter than sucrose.

Cyclamate: Calorie Control Council

Substantial scientific evidence supports cyclamate's safe use by the millions of consumers who seek to control their intake of carbohydrate-based sweeteners and calories…

No low-calorie sweetener is perfect for all uses. However, with several low-calorie sweeteners available, each can be used in the applications for which it is best suited. Also, when used in combination (as would most often be the case with cyclamate), the strengths of one sweetener can compensate for the limitations of another, providing for increased stability, improved taste, lower production costs and more product choices for the consumer. (The Calorie Control Council, Atlanta, GA)

Sugar Alcohols (Polyols)

A category of sugar substitute with a *distinct* classification from *artificial sweeteners* is *sugar alcohol.* Sugar alcohols are *caloric*, chemically reduced carbohydrates (slightly less calories than sugar) that provide sweetness to foods. Examples of sugar alcohols include erythritol, HSH, isomalt, mannitol, sorbitol, xylitol as well as lactitol, and maltitol. **Polyols** is another term for the sugar alcohols which, although they are carbohydrates, are neither sugars nor alcohols.

Q and A: What Other Names Are Used for Polyols?

Since "polyols" is not a consumer friendly term, many nutritionists and health educators refer to polyols as "sugar replacers" when communicating with consumers. Scientists call them sugar alcohols because part of their structure chemically resembles sugar and part is similar to alcohols. However, these sugar-free sweeteners are neither sugars nor alcohols, as these words are commonly used. Other terms used primarily by scientists are polyhydric alcohols and polyalcohols. (The Sugar Association, Washington, DC)

The sugar alcohols are similar in chemical structure to glucose, yet with an *alcohol* group that replaces the aldehyde group of glucose. The sugar alcohol classification includes:

Erythritol. NECTRESSE® is 150 times sweeter than sucrose. A monk fruit extract is blended with other natural sweeteners to create NECTRESSE™ Sweetener. The result is 0 cal per serving and the sweet taste of sugar.

The monk fruit extract is combined with small amounts of sugar, molasses, and erythritol. Erythritol is a sugar alcohol that is found in many fruits and vegetables. According to McNeil Nutritionals (McNeil Nutritionals Fort Washington, PA) in 2012:

"Erythritol is an all-natural, sugar alcohol that is naturally fermented from sugars and is found in many vegetables and fruits. Erythritol contributes zero calories per serving of NECTRESSE™ Sweetener. Consuming erythritol from NECTRESSE™ Sweetener is not expected to result in laxative or other gastrointestinal effects that are known to sometimes occur with other sugar alcohols.

Monk Fruit Extract is about 150× sweeter than sugar and contributes zero calories per serving to NECTRESSE™ Natural No Calorie Sweetener. Like other no-calorie sweeteners, NECTRESSE™ Sweetener contains a small amount of carbohydrate (1–2 g per serving) from other food ingredients to provide needed volume and texture. These food ingredients, which include small amounts of erythritol, sugar, and molasses, contribute so few calories per serving that NECTRESSE™ Natural No Calorie Sweetener Products meet the FDA's criteria for no-calorie foods (<5 cal/serving) (McNeil Nutritionals Fort Washington, PA)."

CULINARY ALERT! One packet of NECTRESSE™ is equal to 2 teaspoons of sugar.

Hydrogenated starch hydrolysate (HSH) and hydrogenated glucose syrup (HGS) are other sugar alcohols. According to the Calorie Control Council, polyols that do not contain a specific polyol as the majority component continue to be referred to by the general term "hydrogenated starch hydrolysate" (The Calorie Control Council, Atlanta, GA).

Isomalt. Isomalt is 45–65 % as sweet as sucrose; 2 cal/g is in HSH and HGS has 3 cal/g. Isomalt is a disaccharide comprised of two glucose molecules sharing a 1–6 link

Mannitol provides half the sweetness of sucrose and provides 1.6 cal/g. Mannitol has a low glycemic index and therefore does not stimulate an increase in blood glucose; thus, it may be used as a sweetener for people with diabetes and in chewing gums.

Mannitol and sorbitol sugar alcohols provide half the sweetness of sucrose and may be used in various foods. They are isomers. Mannitol is in a wide variety of natural products, including almost all plants, including seaweed.

Sorbitol is commercially produced from glucose and contains 2.6 cal/g.

It provides half the sweetness of sucrose. In *combination* with aspartame and saccharin, it provides the volume, texture, and thick consistency of sugar. It is also used as a bulking agent.

Xylitol. Xylitol is approximately as sweet as sucrose with 33 % fewer calories.

Sugar alcohols may be *sugar-free*; however, they are *not calorie-free*! The body does not metabolize sugar alcohols, so persons with diabetes may use them without a rise in their blood sugar. Large amounts of sugar alcohols may cause intestinal diarrhea; therefore, they are not recommended for use in significant amounts.

Novel Sweeteners

"A few sweeteners are considered novel sweeteners because of their chemical structure" (Mayo Foundation for Medical Education and Research (MFMER)).

Stevia—Stevia is from the leaves of the stevia plant with 300× the sweetness of sugar and 0 cal. The FDA once labeled stevia as an "unsafe food additive" and restricted its import. The FDA's stated reason was "toxicological information on stevia is inadequate to demonstrate its safety." Further studies have shown it to be safe, and it began to be used in the United States in 2008.

CULINARY ALERT! *24 packets = 1 cup of sugar.*

Stevia is, according to Webster's definition:

1. Any of a genus (*Stevia*) of composite herbs and shrubs of tropical and subtropical America; *especially*: a white-flowered tender perennial (*S.rebaudiana*) native to Paraguay
2. A white powder composed of one or more intensely sweet glycosides derived from the leaves of a stevia (*S. rebaudiana*) and used as noncaloric sweetener

Its first known use was in 1806. The many types of stevias on the market are listed in the general "stevia" category.

"Extracts from the stevia plant glycosides or steviosides vary in sweetness and flavor profiles. The combinations and percentages of these glycosides differ from manufacturer to manufacturer."

> They don't all taste the same, so it is important for food scientists to try out the different types that are available. If one doesn't work for their needs there are plenty of others to choose from. Stevia extracts also come in a variety of percentages (e.g. 95%) but the numbers don't really say anything about the taste profile. It's still very much a formulator's world where art meets science. (Hazen 2012b)

Three newer sweeteners are becoming more frequently publicized as sugar replacers. These are fructo-oligosaccharide, tagatose and trehalose. Each is made from different carbohydrate sources, and each bestows slightly different functional properties. (These are categorized by the FDA as GRAS substances.)

What is a fructo-oligosaccharide?

Like many of the starch-based sugar replacers, the term "fructo-oligosaccharide" represents a family of ingredients, not a single product. Fructo-oligosaccharides (FOS) are manufactured by fragmenting a large molecule. In the case of FOS, that molecule (polysaccharide) is inulin. Inulin is a polysaccharide in which a single glucose unit ends a chain of up to sixty fructose units linked together.

Inulin occurs naturally in chicory, Jerusalem artichokes, wheat, onions, and bananas. Chicory and Jerusalem artichoke are the commercial sources of FOS products. Since commercial FOS products can have various numbers of fructose units linked to the ending glucose unit, the Food and Drug Administration has ruled that "fructooligosaccharide" is the term approved for an ingredient list.

FDA has agreed with manufacturers' conclusions that FOS products are safe food ingredients. FOS may be used in hard and soft candies, baked goods like biscuits, cakes, cookies and crackers, frozen dairy desserts, cereals, jams and jellies, flavored and unflavored milks, and soups. Additionally, FOS has been approved for use a binder and stabilizer in a variety of meat and poultry products.

What is tagatose?

Tagatose occurs naturally in dairy products, but the commercial product is manufactured from lactose (milk sugar) by a patented process. It is very similar to fructose in structure.

Tagatose has the bulk of sugar, and is almost as sweet. However, it has only 1.5 cal per gram since less than 20 % of ingested tagatose is absorbed in the small intestine. Although tagatose is digested the same as fructose, its limited absorption means that it is metabolized mainly in the large intestine. The short chain fatty acids promote the growth of the two bacteria recognized to improve colon health. Consequently, the prebiotic potential of tagatose is often stressed for the foods using this sugar replacer.

Tagatose was launched in the U.S. in 2003 after the Food and Drug Administration issued a letter agreeing with the manufacturer's determination that it is a safe food ingredient. Tagatose may be used in foods like soft and hard candies, frozen dairy desserts, cereals, frostings and fillings, and chewing gum.

What is trehalose?

Trehalose is found naturally in such diverse foods as honey, mushrooms, shrimp and lobster, and in foods produced with baker's or brewer's yeast. It is found naturally in such diverse foods as honey, mushrooms, shrimp and lobster, and in foods produced with baker's or brewer's yeast.

Commercially, trehalose is manufactured from cornstarch. Although trehalose is a disaccharide of two glucose units, its molecular bonding makes it different than maltose, the other glucose disaccharide made from cornstarch. Trehalose has four calories per gram—same as sugar—but is only half as sweet (The Sugar Association, Washington, DC).

Confections

The word *confections* has several uses and meanings. For example, chocolates may be known as *chocolate confectionery*, cakes and pastries may be referred to as *flour confectionery*, and the term *sugar confectionery* may signify any sugar-based products. Sweet food products may utilize the terms "confections" or candy. However, in the United States, both chocolates and the various sugar-based confections are simply referred to as "candy."

CULINARY ALERT! In the manufacture of confectionery products, sugar syrups achieve a very high temperature and can cause severe skin burns.

Candy-making is primarily dependent on the *concentration* of sugar in boiled sugar syrups and *controlling or preventing crystal formation*. Various ingredients, such as gelatin, fruit, nuts, milk, and acids, in addition to sugar, may be added to sugars to produce specific candies.

Sugar substitutes are not generally used for candy-making although there exist "chocolates" and other confections for consumption by those with diabetes mellitus. Since they are used in small quantities, and cannot add bulk to candy formulation/recipes, and due to the fact that they do not crystallize, sugar substitutes do *not* produce satisfactory results in all candies. Real sugar may be necessary as a major recipe ingredient in successful candy-making.

During the preparation of candies, the sugar solution must be ***saturated***—holding the *maximum* amount of dissolved sugar it is capable of holding at the given temperature needed for the specific candy type. Upon cooling, the solution becomes ***supersaturated***—holding *more* dissolved sugar that it can theoretically hold at a given temperature.

CULINARY ALERT! Sugar is hygroscopic. Therefore, high humidity during candy preparation results in excess moisture retention of the sugar and less than desirable results.

Major Candy Types: Crystalline and Amorphous Candies

Two major types of candies are *crystalline* and *amorphous* candies. Each will be discussed in this chapter section. ***Crystalline*** candies are formed in the process of ***crystallization*** as heat is given off—***heat of crystallization***. This type of candy has crystals suspended in a saturated sugar solution. Crystals may be *large and glasslike*, as in rock candy, or they may be *small and smooth textured*, breaking easily in the mouth, as in fondant or fudge candies.

Crystalline candies have a highly structured crystalline pattern of molecules that forms around a nuclei or seed, and therefore, it is required that the sugar mixture for crystalline candies must be left undisturbed (more later) to cool. Again, examples of crystalline candies include:

- Rock candy
- Fondant
- Fudge candies

Amorphous or noncrystalline candies are those without a crystalline pattern and include several types as follows:

- Caramel and taffies are chewy amorphous candies.
- Brittles are hard amorphous candies.
- Marshmallows and gumdrops are aerated, gummy amorphous candies.

In general, amorphous candies contain a:
- *High* sucrose concentration (Table 14.2)
- Large amount of agents that interfere with (see Interfering Agents) or prevent crystal formation
- Low moisture level (the molecular ring opened and water is lost)
- More viscous as syrup than crystalline candies

Factors Influencing Degree of Crystallization and Candy Type

Crystals are closely packed molecules that form definite patterns around ***nuclei*** as a sugar solution is heated and subsequently cooled. Crystal development (crystalline candies) or lack of it (amorphous candies) is dependent on factors discussed in the following text. Such factors include the *temperature*, *type and concentration of sugar*, *cooling method*, and *use of added substances* that interfere with crystal development.

Crystalline formation in a sugar solution occurs due to ***seeding***. It is desirable. Yet seeding may occur prematurely. For example, stray sugar crystals remaining on the side of the pan after stirring may later fall into the mixture in the pan. To prevent this unwelcome addition, use of a *pan lid* is recommended for initial cooking so that all crystals dissolve.

CULINARY ALERT! It is recommended that the pan lid remain on the sugar mix for a few minutes initially so that steam can dissolve stray sugars and prevent seeding.

Within this upcoming chapter section, various factors that influence the degree of crystallization and, consequently, candy type are presented.

Temperature. *Temperature* of a sucrose solution is an indication of its concentration. Specific temperature requirements must be met for cooking each type of candy (Table 14.2). If the designated temperature has been *exceeded*, water may be added to the sugar solution in order to *dilute* its concentration and *lower* the temperature. This helpful addition of water is possible

Table 14.2 Major candy types

Candy type	Final temperature in °F (°C)	Percent sucrose
Crystalline		
Fudge	234 (112)	80
Fondant	237 (114)	81
Amorphous		
Caramel	248 (118)	83
Taffy	265 (127)	89
Peanut brittle	289 (143)	93

only as long as the sugar solution has not yet reached the caramelization stage.

A *slow rate* of achieving the boiling point of a sucrose and water solution is desirable. A slow rate increases the time available for inversion of sucrose, allows increases in the solubility of the sugars, and produces a softer final product compared to *rapid* heating.

CULINARY ALERT! Candy-making temperatures exceed the boiling point of water, and as water evaporates, the sugar syrup becomes viscous, causing more severe burns than boiling water if it contacts the skin.

Sugar Type. *Sucrose* molecules are able to align and form large lattice arrangements of crystals. Other sugars, such as the monosaccharides *glucose and fructose* (or invert sugar), possess different shapes that interfere with aggregation and crystal development (thus, a candy with too much invert sugar will fail to harden and is deemed unsatisfactory). HFCS, *honey, and invert sugar* are examples of sugars that are added to syrup in candy-making in order to *prevent* the formation of large crystals.

Sugar Concentration. As previously mentioned, candy-making is dependent upon the sugar *concentration*. A sugar solution is *dilute* (unsaturated) if the concentration of a solute is less than maximum at a given temperature. Initially, this is true in candy-making. Then, as the sugar solution boils, water evaporates, and the solution becomes *saturated*. When the saturated solution is cooled, it becomes *supersaturated* and easily precipitates sugar.

Amorphous candies have a *higher* sugar concentration (Table 14.2) than crystalline candy because more sugar is incorporated and more water has evaporated at the higher temperature. The candy mixture is so viscous that crystals *cannot* form.

Cooling Method and Timing of Agitation/Beating. The cooling method and timing of agitation determine adequacy of crystalline candy. *Crystalline* candy must *not* be disturbed by *premature* agitation/beating during cooling. *Crystalline* candy is best formed by *slowly* cooling the sugar solution to approximately 100–104 °F (38–40 °C) *before* stirring or beating. (In reading temperature, stray crystals/seeds are prevented from entering the mixture if the thermometer is free from sugar residue.)

Once cooled to the desirable cooling temperature, the timing is correct and agitation becomes necessary because timely agitation produces/keeps *many small* **nuclei** in the supersaturated solution. Then, with agitation, excess sugar molecules in the solution are prevented from attaching to already developed crystals. The crystal size remains small.

Amorphous candy is formed from a very supersaturated solution (Table 14.2), and an undisturbed cooling method is *not* crucial for success. The solution is too viscous to allow aggregation of solute molecules and crystal formation.

Interfering Agents. These influence the degree of crystallization and, consequently, candy type. There are two types of interfering agents—chemical and mechanical:

*Chemica*l interfering agents include *corn syrup or cream of tartar*. Both reduce the quantity of excess sucrose (the solute) available for formation of the crystalline lattice (see Sugar Type, above). Corn syrup contains *glucose*, and the acid cream of tartar inverts sucrose to glucose and fructose. These *non-sucrose* molecules (*glucose and fructose*) do not fit properly (are not able to join) onto existing sucrose lattice structures, thereby keeping crystals *small*. Both *small crystals* and the resultant smooth-textured candy are produced by the addition of cream of tartar or corn syrup to the solution.

Mechanical interfering agents used in candy-making *adsorb* to the crystal surface and physically *prevent* additional sucrose from attaching to the crystalline mass; thus, crystals are many and small. Some examples of mechanical interfering agents are *fat*, the *fat in milk or cream*, and the *proteins* in milk and egg whites.

In crystalline products, ***interfering agents*** *reduce the speed of crystallization* and help to *prevent* undesirable growth of crystal structures that result in the formation of large, crystalline, gritty candies.

Again, examples of crystalline candies include:

- Rock candy
- Fondant
- Fudge candies

In *amorphous* products, interfering agents *prevent crystallization* and add flavor.

To repeat from *Major Candy Types* above, amorphous or noncrystalline candies are those without a crystalline pattern and include several types as follows:

- Caramel and taffies are chewy amorphous candies.
- Brittles are hard amorphous candies.
- Marshmallows and gumdrops are aerated, gummy amorphous candies.

Factors Affecting Candy Hardness. Candies vary in their moisture content. Moisture in the air and other added ingredients affect candy hardness or softness. A hard candy has 2 % moisture, while gummy candy, such as gumdrops, contains 15–22 % moisture.

Ripening

Crystalline candies must ripen in order to produce an acceptable candy. Ripening occurs in the initial period of *storage*, *following* the cooking, cooling, and crystallization of a sugar solution, as the moisture level (sugar is hygroscopic) increases, and small crystals are redissolved in the syrup, preventing unwanted crystallization. Smoothness of the finished candy is desired.

Nutritive Value of Sugars and Sweeteners

Sucrose is a carbohydrate that contains 4 cal/g. It supplies *energy*, although *no* nutrients to the body. Use of nutritive sweeteners should be based on a patient's eating habits along with serum glucose levels and lipid goals. For example, the diabetic must manage *blood serum glucose* levels, and others watch their levels of *serum lipids* that are adversely affected by large amounts of fructose. "There is nothing unusual about craving sweets ... Humans have an appetite for sugary things. But in excess, sugary foods can take a toll. Large quantities add up to surplus calories, which can contribute to weight gain" (FDA).

Sugar substitutes, including (1) the nonnutritive, artificial sweeteners and (2) caloric sugar alcohols, may pose adverse health effects for some individuals. If that is the case, intake of that product should be limited or eliminated from the diet. For example, aspartame contains phenylalanine, a substance phenylketonurics are unable to properly rid from their bodies, and excessive levels of sugar alcohols may cause diarrhea.

A healthful diet uses sugars sparingly, as high consumption equates to a diet with low nutrient density. "Sugars" that appear on the Nutrition Facts label include (1) the total sugars found *naturally* in foods and (2) *added sugars*. Labeling criteria require that *all monosaccharides* and *disaccharides* be listed as "sugars" on the

Nutrition Facts label, regardless of whether they are a natural part of the food or added to the product.

Clarification of natural and added sugars may be determined by reading the foods' *ingredients* list.

HFCS has been receiving much attention. See more at *Myth* vs. *Facts* (above in HFCS) or in reference (SweetSurprise.com).

CULINARY ALERT! An example of "sugars" on labels is seen in orange juice which reports "sugars" on the label although it may not contain added sugar.

Sugars have a recommended intake of 10 % of calories, yet no % Daily Value. "Sample dietary patterns recommend limiting total added sweeteners, on a carbohydrate-content basis to no more than 6 tsp/day at 1,600 kcal, 12 tsp/day at 2,200 kcal, and 18 tsp/day at 2,800 kcal . . . 6–10 % of energy" (USDA). Reducing added sugar is often recommended (Hazen 2012b).

The designation "sugar-free" signifies that there is less than 0.5 g of sugar per serving. "Reduced sugar" indicates that the food contains 25 % less sugar per serving than the regular product. "No added sugar" signifies that the product has no sugar added. Product labels may state that the product is a *reduced-* or *low-calorie* food, if the food meets the necessary requirements of those definitions.

The Academy of Nutrition and Dietetics Position Statement with regard to sweeteners is as follows:

> . . . consumers can safely enjoy a range of nutritive and non-nutritive sweeteners when consumed in moderation and within the context of a diet consistent with the Dietary Guidelines for Americans.

Peruse more http://www.livestrong.com/article/319513-the-calories-in-sugar-alcohol/#ixzz2PJuR6xie.

> There are eight commonly used sugar alcohols and each provides a different amount of calories. Per gram each sugar alcohol contains the following amounts of calories: erythritol provides 0.2, polyglycols have up to 3, isomalt has 2, lactitol provides 2, maltitol has 2.1, mannitol contains 1.6, sorbitol offers 2.6, and xylitol gives you 2.4 cal per gram.

In February 2013 the FDA released standards (proposed) to limit individual vending machine foods to 200 cal. High intensity sweeteners may be called on to come to the rescue for ingredient innovation! (Decker 2013).

Safety

Safety of foods is always important. Although *foodborne illness* could certainly occur in sugary products, bacterial contamination and multiplication in sugary products is deterred by competition for life-sustaining water between many microbes and sugar. There *is* a safety and health concern for persons with illnesses such as *diabetes mellitus* though, because sugars may not be properly utilized.

Adverse personal health effects for some individuals may be the result of either consuming too many calories and therefore gaining weight or using *sugar substitutes* including (1) the non-nutritive, artificial sweeteners and (2) caloric sugar alcohols. In either case, intake of too much food or various "diet" products should be limited or eliminated from the diet.

A healthful diet uses sugars sparingly, due to the fact that high sugar consumption equates to a diet with low nutrient density. For some individuals though, *no* sugar consumption is their way of life. In general, wise intake of sugar is recommended to be 10 % or less of the total daily calories.

Conclusion

Sugar comes from sugar cane or sugar beets, both of which have the same chemical structure. The roles of sugar are many and include providing flavor, color, and tenderness. Real sugars elevate boiling point and are soluble in water, hygroscopic, and fermentable. A variety of sweeteners, including sugar substitutes, and syrups are incorporated into food systems to provide sweetness at a lesser amount of calories.

In order to control the rate of crystallization and the formation of small crystals and to ensure a smooth texture, interfering agents are incorporated into a sugar formulation. Chemical interfering agents produce invert sugar (glucose and fructose), thereby slowing crystallization and increasing the solubility of solute. Mechanical interfering agents such as fat and protein help to keep crystals small by preventing the adherence of additional sugar crystals onto the nuclei. According to the USDA, a healthful diet uses sugars sparingly.

Notes

CULINARY ALERT!

Glossary

Amorphous Noncrystalline candies without a crystalline pattern; may be hard candies and brittles, chewy caramel and taffies, gummy **marshmallows**, and gumdrops.

Artificial sweetener Noncaloric, nonnutritive sugar substitute; examples are acesulfame K, aspartame, and saccharin.

Caramelization Sucrose dehydrates and decomposes when the temperature exceeds the melting point; it becomes brown and develops a caramel flavor, nonenzymatic browning.

Crystalline A repeating crystal structure; solute forms a highly structured pattern of molecules around a nuclei or seed; includes large crystal, glasslike rock candy, or small crystal fondant and fudge.

Crystallization Process whereby a solute comes out of solution and forms a definite lattice or crystalline structure.

Fermentation The anaerobic conversion of carbohydrates (complex organic substances), such as sucrose, glucose, fructose, or maltose, to carbon dioxide and alcohol by bacteria, mold, yeast, or enzymes.

Heat of crystallization The heat given off by a sugar solution during crystallization.

Hygroscopicity The ability of sugar to readily absorb water; sugars high in fructose such as invert sugar, HFCS, honey, or molasses retain moisture more than sucrose.

Interfering agent Used in crystalline products to reduce the speed of crystallization and help prevent undesirable growth of large crystal structures; interference is by mechanical or chemical means.

Inversion The formation of equal amounts of glucose and fructose from sucrose, by acid and heat or enzymes; invert sugar is more soluble than sucrose.

Maillard browning Browning is a result of reaction between the amino group of an amino acid and a reducing sugar.

Nuclei An atomic arrangement of a seed needed for crystalline formation; fat is a barrier to seeding of the nuclei.

Saturated A sugar solution holding the maximum amount of dissolved sugar it is capable of holding at the given temperature.

Seeding To precipitate sugar from a supersaturated solution by adding new sugar crystals (the seed may originate from sugar adhering to the sides of the cooking utensil).

Solute That which is dissolved in solution; the amount of solute held in solution depends on its solubility and the temperature.

Solution Homogeneous mixture of solute and solvent; it may be dilute, saturated, or supersaturated.

Solvent Medium for dissolving solute; i.e., water dissolves sugar.

Sugar alcohol Caloric sugar substitute; chemically reduced carbohydrates that provide sweetening; examples are mannitol and sorbitol.

Supersaturated Solution contains more solute than a solution can hold at a specified temperature; formed by heating and slow, undisturbed cooling.

References

Decker KJ (2013) Finding the sweet spot: confections for a slimmer society. Food Product Design:39–48

Hazen C (2012) Optimizing flavors and sweeteners. Food Product Design (November):30–42

Hazen C (2012) Reducing added sugars. Food Product Design (May):40–52

Bibliography

http://www.polyol.org/fap/fap_HSH.html

Pfizer Food Science Group, New York, NY

Part VI

Baked Products

Baked Products: Batters and Dough 15

Introduction

This baked products chapter builds on knowledge of the functional properties of carbohydrates, fats, and proteins discussed in previous chapters. Specific batter and dough ingredients that are discussed in this chapter include previously studied commodities, such as flour, eggs, milk, fats and oils, and sweeteners. Among other important points, this chapter will view the functions of various ingredients in a *general* manner and the role of those ingredients in *specific* baked products.

A majority of baked products contain flour (of course not flourless cake!), especially wheat flour, as the primary ingredient. Baked products vary significantly in their fat and sugar content. Pastries and some cakes are *high* in fat, whereas other cakes such as angel food cake and breads may be *low-fat* or *fat-free*. Many baked products such as breads, cakes, and cookies are increasingly available as gluten-free, accommodating a growing segment of the population.

Batters and *dough* each contain different proportions of liquid and flour and therefore are manipulated differently—by stirring, kneading, and so forth. Some batters and dough contain a well-developed gluten protein network, while others do not have this characteristic and, as mentioned above, are gluten-free. In some food products, the network may hold many additional substances, such as starch, sugar, a leavening agent to produce CO_2, liquid, flavoring agents, and perhaps eggs and fats or oils. Other items including salts or acids are also found in baked products.

The gas cell size and shape as well its surrounding ingredients create the "grain" and texture of a baked product. Most batters and dough are "foams" of coagulated proteins around air cells. For example, angel food cakes and sponge cakes form definite foam structures.

By way of introduction, a *quick bread* is one that is relatively quick to mix before baking and is leavened primarily by added *chemical* agents, such as baking powder or baking soda, not by yeast. It may be leavened by steam or air. Pancakes and waffles, biscuits, and muffins are examples. *Yeast breads*, on the other hand, are leavened *biologically* by yeast and are therefore not quick, rather more time-consuming to prepare. More detailed discussion on leavening will follow.

Ready-to-eat (r.t.e.) and ready-for-baking products continue to replace some baking "from scratch." Low-fat products are popular. Proper storage extends shelf life.

Imagination is the limit to creative baked products!

Classes of Batters and Dough

Batters and dough are classified according to their ratio of liquid to flour (Table 15.1), and

V.A. Vaclavik and E.W. Christian, *Essentials of Food Science, 4th Edition*, Food Science Text Series,
DOI 10.1007/978-1-4614-9138-5_15,

Table 15.1 Batters and doughs: ratio of liquid to flour

Type	Liquid	Flour
Batter		
Pour batter	1 Part	1 Part
Drop batter	1 Part	2 Parts
Dough		
Soft dough	1 Part	3 Parts
Stiff dough	1 Part	6–8 Parts

they each utilize various mixing methods. While exact ingredient proportions of both batters and dough vary by recipe, for use as a planning guide or in recipe analysis, the ratios in Table 15.1 provide useful guidelines.

Batters are flour–liquid mixtures that are *beaten or stirred*, and as their formulations indicate, these incorporate a considerable amount of *liquid* as the *continuous medium*. Batters are classified as either (1) pour batters or (2) drop batters. *Pour* batters, such as those batters used in the preparation of items such as pancakes and popovers, are thin and have a 1:1 ratio of liquid to flour. *Drop* batters contain *more* flour than a pour batter with a ratio of 1:2 of liquid to flour. Muffins and some cookies are examples of products prepared with drop batter.

Dough is distinguished from batter by being thicker than batter. Dough does not contain a lot of liquid and is *kneaded*, not beaten or stirred. The *flour/gluten matrix*, not liquid (as batters), is the *continuous medium*. The flour mixtures are classified as soft or stiff dough. For example, *soft* dough, such as that used in biscuit preparation, or yeast bread has a liquid-to-flour ratio of 1:3. *Stiff* dough may have a ratio of 1:6 or higher and might be used for cookies or pastry dough, such as piecrust.

Gluten

Gluten, or the gluten matrix, is noted for its *strong*, three-dimensional, viscoelastic structure that is created by specific proteins. Specifically, it is the hydrophobic, *insoluble gliadin proteins* that contribute *sticky*, fluid properties to the dough and the *insoluble glutenins* that contribute ***elastic*** properties to the dough. Not all flour and therefore not all dough form gluten. Non-gluten flours contain *starch* that provides some structure; however, it is *gluten* protein that provides the major *framework* for many batters and dough.

Upon hydration and manipulation, the two proteins aggregate and form disulfide bridges, producing a gluten protein matrix that is subsequently *coagulated* upon baking. This is a three-dimensional structure capable of stretching without breaking, although it may break with overextension if dough is kneaded too much. The gluten determines the texture and volume of the finished product. Oftentimes directions will state, "rest the dough," and to the extent that the dough contains gluten, resting serves to relax the gluten structure (Fig. 15.1).

Many baked products contain flour that is derived from wheat and especially *hard wheat, rye, or barley* (see Chap. 6). These flours have ***gluten-forming potential***, while oat (more below), corn, rice, and soy do *not* have gluten-forming potential due to inherent differences in protein composition. Oats may be cross-contaminated with gluten during shipping or processing and are therefore often avoided by persons following a gluten-free diet. (According to many researchers, including those at the University of Chicago Celiac Disease Center, "Regular, commercially available oats are frequently contaminated with wheat or barley. However.... pure, uncontaminated oats can be consumed safely in [limited] quantities....It is important that you talk to your physician and your registered dietitian prior to starting oats.")

Yeast breads made with wheat flour are kneaded to create an extensible structure. The dough requires extensive ***gluten development*** to be able to expand. *Without* gluten, the *latter* types of flour listed above are incapable of any structure expansion when CO_2 is generated from yeast.

The gluten structure in a batter/dough mixture is embedded with numerous recipe ingredients. This includes the starch in the flour which itself contributes to dough rigidity, added fat or sugar,

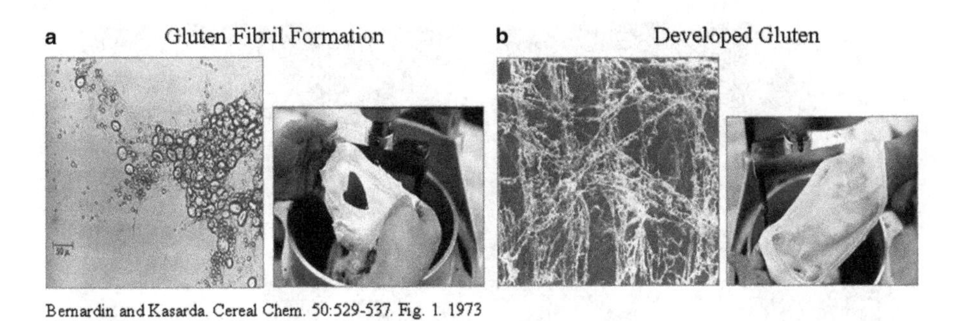

Fig. 15.1 Gluten Fibril Formation. Bernardin and Kasarda. Cereal Chemistry. 50:529–537. Figures 15.1 and 15.2 (1973)

liquid, or leavening. These added ingredients (see Functions of Various Ingredients in Batters and Dough) influence the development of the gluten structure, the dough strength, and the finished baked product. For example, dough does not reach its maximum strength when the recipe includes high levels of (1) sugar, which competes with gliadins and glutenins for available water, or (2) fat, which *covers* flour particles and prevents water absorption needed for gluten development.

Dough such as biscuit dough has a liquid-to-flour ratio that makes it more likely than *batters* to become tough due to the large proportion of flour. This is true especially if the biscuits are overstirred or overkneaded resulting in extensive development of gluten.

Of the *batter* types, *pour* batters do not exhibit a significant difference in gluten development between *adequately mixed* and *overmixed* batter. *Drop* batter, such as a muffin batter, has more flour than a pour batter and consequently has a greater chance of developing gluten. If gluten is *overdeveloped*, batters and dough may exhibit obvious internal holes in a tunnel formation (see Mixing Methods for Various Batters and Dough).

With the use of *less* flour, *less* gluten is likely to be produced. It follows that *sifted* flour also incorporates *less* flour into a recipe, and so there is *less* gluten-forming potential compared to an equal measure of *unsifted* flour. The sifting process also incorporates air that provides leavening.

CULINARY ALERT! Specifying "flour, sifted" or "sifted flour" as directions in a product formulation/recipe are two different instructions. Measure first, then sift is the former; sift first, then measure is the latter!

With the aim of physically seeing the gluten in flour, manipulated dough may be washed in *cold water* (not hot water as heat will gelatinize starch). This washing removes the nonprotein components of the flour. Then, only the gummy gluten (remember—a protein) component of flour remains. It resembles already chewed chewing gum! When this gluten ball is subsequently baked, the entrapped water becomes steam and leavens the now hollow structure. Figure 15.2 shows the size of raw and baked gluten balls, which indicates the relative amount of gluten in the various types of flour. Of course, some flours contain *no* gluten-forming proteins. In that case, there is no gummy material created or retained and therefore no dough to show in a picture.

Gluten in a dried form may be added to other flours, providing extra strength and several times the gluten-forming potential of that flour. Extracted gluten is used to fortify protein content of some breakfast cereal, for binding breading on meat, poultry or fish, and as an extender for fish and meat products. As well, nonfood uses of gluten may be as a constituent of mascara and pharmaceutical tablets.

Fig. 15.2 Unbaked and baked gluten balls. *Left to right*: gluten balls prepared from cake flour, all-purpose flour, and bread flour (*Source*: Wheat Flour Institute)

A view of the Codex Standard for gluten-free food, daily gliadin consumption, and studies on the safety of wheat starch-based gluten-free foods is found in other literature (Thompson 2000, 2001).

Function of Various Ingredients in Batters and Dough

Certainly baked products do not necessarily need all of these ingredients that follow. The watery mixture of substances that these ingredients create bakes *around gas cells* and subsequently determines the texture, flavor, and appearance of baked products.

Flour Function

Flour provides structure to baked goods because of its *protein* and, to a lesser degree, its *starch* components. For example, to the extent that the *gluten-forming proteins* are present in flours, there is dough elasticity and structure (see Gluten) due to formation of a gluten matrix. *Starch* contributes structure to a batter or dough, as it gelatinizes and makes the *crumb more rigid*. Additionally, flour is a source of fermentable sugar that is acted upon by yeast in producing CO_2 for leavening.

Many types of flour (Chap. 6) are used in the preparation of baked goods.

Wheat flour is derived from the endosperm of *milled* wheat and is the most common flour used in the preparation of baked goods in the United States. Specifically, ***all-purpose flour*** is chosen for use. It is produced by blending *hard and soft* wheat during milling and has applications in many baked products. Consumers refer to it simply as "flour."

- *Hard* wheat flour, such as bread flour, has a *high* gluten potential that is important for structure and expansion of yeast dough. It absorbs *more water* than an equal amount of soft wheat flour.
- *Soft* wheat flour, such as cake flour, contains *less* gluten-forming proteins and is effectively used in the preparation of the more tender (due to less gluten) cakes and pastries. An equal amount of soft wheat flour absorbs *less water* than hard wheat flour, and it becomes important to know that the two flours are not an even exchange.

CULINARY ALERT! All "flour" used in a recipe is not created equal! High-protein "hard" flour absorbs more water than low-protein "soft" flour, and thus flour cannot be interchanged in all cases.

Whole wheat flour differs from *wheat* flour, as it contains *all* of the three kernel parts, including the endosperm, germ, and bran (Chap. 6). *Bran* has sharp edges that *cut* through the developing protein structure and results in a *lower*-volume baked product, especially when a recipe replaces *all* of the flour with *whole wheat* flour.

Improved food results are seen when the whole wheat flour is *finely ground*. Finely ground flour introduces *less* sharp edges that cut and can reduce volume. Due to the presence of whole wheat's *bran*, the percentage of protein is lower in whole wheat flour than refined wheat flour.

Whole wheat flour also contains the *germ* that may cause rancidity over time. Yet, baked products may not remain uneaten for too long though!

CULINARY ALERT! Generally, when a whole grain flour is desired in a baked product, the recipe may replace the flour with no more than *half* whole grain flour used in combination with *half* bread flour.

While wheat flours are the most common types of flour used in baked products, *non-wheat* flours such as corn, rice, and soy are also popular in bread making. When the formulation combines use of these flours with wheat flour, there are more desirable baking results.

Despite the type of flour used, it is typically *sifted* prior to measurement, as sifting standardizes the amount of flour added to a formulation and assures consistency in product preparation. Consistency is *also* more likely when ingredients are *weighed*, not *measured.*

Flour shows variance in the *same* brand of flour milled in *different milling locations* throughout the country. Due to these variations the same recipe may yield a slightly different finished product in different locations.

CULINARY ALERT! Adhere to appropriate measuring techniques, as well as local standardized recipes and flour type, especially when cooking/baking or in various parts of the country or world!

Liquids Function

Liquids are crucial in hydrating the *proteins* required for *gluten* formation and the *starch* element that undergoes gelatinization. These proteins and starch form the texture of the baked crumb. Additionally, liquids are the solvent for dissolving many recipe ingredients such as the leavening agent, baking powder, and baking soda, as well as salt and sugar. Liquids produce steam that leavens and expands air cells during baking.

According to federal regulations, the water level of a finished commercially prepared bread loaf may not exceed 38 %. *Liquids* though may contribute *more* than water. For example, while *milk* contains a very high percentage of water, it also *contains protein*, *milk salts*, and the *milk sugar lactose*. Juices, sugar syrups, eggs, and so forth may also be part of the liquid in a recipe.

In general, its lactose milk produces a softer crumb, holds moisture in the product, and contributes both flavor and color from the Maillard browning. The *near-neutral pH* of milk causes it to act as a buffer, preventing an acid environment that would be unacceptable to the growth of baker's yeast.

The practice of *scalding* milk is thought to be *unnecessary*. However, milk that is *not* scalded may contain whey protein that results in *diminished volume and poor quality*. This negative effect is especially true with the use of reconstituted, scalded nonfat milk solids (NFMS). *Unreconstituted* NFMS powder may also be added to recipes to increase *nutritive value*.

Leavening Agents Function

***Leavening** agents* are presented in more detail in a later section of this chapter. Overall, leavening agents raise dough or "make light and porous." They include *air*, *carbon dioxide* (CO_2), and *steam* discussed below. Virtually, *all* baked products are leavened to some extent, if not solely, by *air*. The amount of air depends on the mixing method, sifting flour prior to addition, beating, creaming, and so forth. Consequently, there may be great variance in the *amount of air* that is incorporated into a batter or dough mixture.

Carbon dioxide gas is a leavening agent produced *chemically* by baking soda and baking powder. It is produced *biologically* by yeast. These agents fill existing air cells and the gluten structures that then expand with the CO_2 they produce.

A third leavening agent is steam. *Steam* works in combination to further expand cell size, making batters and dough light and porous. Leavening agents make foam out of batters as they fill air pockets or cells, contributing to the ***grain*** of the product. Holes in the crumb may be large or small; they may be intact or exploded.

Eggs Function

Eggs function in various manners in the batter or dough. They are *binders*, holding ingredients together. The *whole eggs* and *yolks* contain emulsifiers that distribute fat in the batter (a greater percentage of egg is necessary in a high-fat formulation compared to a low-fat or high-liquid formulation). Eggs leaven, provide coagulated structure, nutritive value, color, flavor, and more.

Egg whites contribute aeration and *leavening* when beaten due to the presence of air cells that are filled with CO_2 or expanded by steam. Egg whites produce a lighter, drier finished product. Eggs contribute elasticity to products such as popovers and cream puffs; thus when *omitted* from a formulation, the baked product is significantly (and unacceptably) lower in volume.

Structure provided by *flour* proteins forming a gluten matrix has been discussed previously. *Egg* proteins contribute to the structure as well, as they *coagulate* by heat, beating, or a change in pH. *Egg whites* incorporate air and may play an important part in *nutrition* as they function as a substitute for a portion of the whole egg in a formulation, thus reducing cholesterol levels.

The color and flavor imparted by eggs is especially significant in specialty ethnic and holiday breads and cookies. (More information on eggs is contained in Chap. 10.)

CULINARY ALERT! ***Large***-size eggs are generally used in a formulation that requires the addition of eggs.

Fat Function

Fats and oils are discussed in Chap. 12, and the reader is referred to that chapter for more specific information. Fat functions in *various* ways in batters and dough as is seen in Table 15.2 that illustrates effects of fats and oils on baked products. Fats and oils *tenderize* baked products by *coating* flour proteins in the batter or dough and physically interfering with the development of the gluten protein. They *shorten* by controlling gluten strand length; they create the *flakes* or *dough layers* seen in biscuits or piecrust. Fats *leaven* by incorporating air (creaming solid fats with sugar). Fats and oils help *prevent the staling* process of baked products.

Plastic fats such as hydrogenated shortening or some other solid fats may be spread or perhaps molded to shape; they do not pour. Hydrogenated vegetable shortenings and lard may contain emulsifiers (monoglycerides or diglycerides). These emulsifiers increase fat distribution and promote greater volume of the developed protein matrix, allowing it to stretch more easily without breaking.

Polyunsaturated oils yield a more ***tender***, mealy, and crumbly product than *saturated* fats. This is because the oil covers a larger surface area of flour particles than *saturated* fat, and it helps control/limit absorption of water (Chap. 12). Saturated fat such as lard covers less and produces a less tender, yet ***flaky***, piecrust with many layers in the dough. As discussed in Chap. 12, these two attributes cannot exist together.

Clearly when milk, especially whole milk, is used in a recipe, it contains *more* fat than juices or

Table 15.2 Effects of fats and oils on baked products

Coating and mechanical tenderizing effect—Fats and oils shield gluten protein from water, thus physically interfering with the hydration needed for gluten development. Both fats and oils tenderize baked products by coating, although oil (liquid at room temperature) coats more completely and yields a more tender product than solid fats; if coating is extreme, the texture of the product will be mealy, and the dough will show reduced gluten formation Fats containing emulsifiers help water and fat to mix and may promote the stretching of gluten strands, yielding a higher volume of the baked product
Shortening—Fats and oils minimize the length of developing gluten protein platelets; that is, they keep them short
Flakiness—Plastic fat that is cut into pea-sized chunks in piecrust doughs (or smaller in biscuits) contributes the characteristic of flakiness to baked products as it melts in the dough, forming layers in the dough. Fats contribute *flakiness*, and oil provides *tenderness*
Leavening—Plastic fats may be creamed in order to incorporate air and aerate batters and doughs
Less staling—Fats with monoglyceride addition, such as hydrogenated shortenings and commercially available lard, soften the crumb and function to retain moisture. It is primarily the amylopectin component of starch that forms a dry crumb

water, and therefore milk creates a *more* tender finished product than either juices or water. *Chilled* oils or fats exhibit slightly more *flakiness* in the baked product compared to room temperature versions, as the *covering potential* is reduced.

CULINARY ALERT! In order to reduce saturated fat intake and for culinary success, *well-chilled oils* may be utilized instead of room temperature oils.

Cup for cup, the various fats and oils *cannot* be substituted for one another and produce the same quality of baked product.

- Oils, hydrogenated vegetable oils, or animal fat (such as lard) are *100 %* fat.
- Margarine and butter contain approximately 20 % water.
- Reduced-fat "spreads" have an even *higher* percentage of water than margarine.

Fats containing water in the mix are *less* effective in their shortening ability than 100 % fats. Often, specially modified recipes are required to assure success in baking with reduced-fat replacements.

Baked products such as angel food cake do *not* contain added fat in the formulation, whereas other products such as shortened cake and pastries are high in fat content. With a low-fat modification, products may be missing some of the flavor, tenderness, or flakiness that fat provided in the original version.

CULINARY ALERT! As appropriate, 1 cup of margarine or butter may be substituted with 7/8 cup of oil.

Salt Function

Salt is a *necessary* component of yeast breads because it dehydrates yeast cells and controls the growth of yeast with its CO_2 production. In typical yeast dough the salt exerts an *osmotic effect*, competing with other substances for water absorption. Specifically there is *less* water for gluten development and *less* for starch gelatinization in salted yeast dough compared to unsalted dough. Salt contributes *flavor* to baked products.

The *absence* of salt in yeast bread dough allows rapid yeast development and rapid rising. This produces a collapsible, extremely porous structure, as gluten is overstretched and strands break.

CULINARY ALERT! Salt is a necessary component of yeast breads—its use controls overproduction of yeast.

Sugar Function

In addition to contributing flavor, sugar functions in many additional ways in batters and dough. The presence of sugar makes a product tender. This is because as the sugar in a recipe *competitively absorbs water* (instead of flour proteins and starch), there is *less* water available for gluten formation and less for starch gelatinization. Sugar also *elevates the temperature* at which the protein coagulates and starch gelatinizes, thus extending the time for CO_2 to expand the baking dough.

Sugar is a *substrate for the yeast* organism to act upon, producing CO_2, acids, alcohols, and a number of other compounds. Granulated white sugar, brown sugar, corn syrup, honey, and molasses are substrates for yeast, whereas artificial sweeteners cannot be fermented. Sugar exhibits *hygroscopic* (water-retaining) tendencies. Therefore baked products may become overly moist, gummy, or runny, especially if the formulation is high in fructose (i.e., honey). Reducing sugars, such as the lactose in milk, provide browning due to the Maillard browning reaction, and sugars also caramelize.

The amount of sugar usage varies. A *small* amount of sugar is helpful to include in yeast bread formulations because it is fermented by yeast to produce CO_2. *A large* amount (more than 10 % by weight) *dehydrates* yeast cells and *reduces* dough volume. Thus, a sweetened dough requires more kneading and rising time due to this osmotic effect of sugar. High levels of sugar are more easily tolerated in breads and cakes leavened by *baking soda or baking powder* than by *yeast*, since yeast cells are dehydrated by sugar. (As shown later, there may be occasions for using both leavens.)

Further types of sweeteners include the following:

Honey may be used in baked products. It imparts varied flavors. When honey is used as a baking ingredient, it makes a sweeter and moister baked product because it contains *fructose*, which is *sweeter* and *more hygroscopic* than sucrose.

CULINARY ALERT! One cup of sugar in a recipe may be replaced by 3/4 cup of honey plus 1 tablespoon of sugar; liquid is reduced by 2 tablespoonfuls.

Molasses imparts its own characteristic flavor that may be very strong. It may be used as the sweetener in baked products, yet, because it is more acidic than sugar, it should *not* be used to replace *more than half* of the total amount of sugar in a recipe. In order to control acidity, it may be necessary to add a small quantity of baking soda. As is the case with honey, when molasses is substituted for sugar, there needs to be a reduction in the amount of liquid in the recipe.

Sugar substitutes provide sweetness; however, they do not provide the functional properties of sugar, including browning, fermenting, tenderizing, and hygroscopic properties of sugars (see Chap. 14). Among sugar substitutes, an equal replacement of one sugar substitute for another, by weight, is *not* possible due to inherent differences in bulk and sweetness. *Acesulfame K*, *aspartame* (if encapsulated), and *saccharin* are examples of heat-stable sugar substitutes successfully incorporated to some degree into baked products (more in Chap. 14).

The Leavening Process of Baked Products

Leavening of quick breads and yeast breads occurs when the air spaces or gluten structure is filled with a leavening agent. For example, after gluten has formed in dough and the dough has subsequently been fermented, the gluten structure becomes extensible for the leaven inside. As previously discussed, ***leavening** agents* include air, steam, or CO_2, which become incorporated into the structure. The latter is produced either biologically or chemically.

As dough *is proofed* or rises in its final rising (usually yeast dough is raised two times), the gluten structure expands, and dough increases in volume and makes a product light and porous.

Air as a Leavening Agent

Air, which is incorporated to some extent into *almost every* batter and dough, expands upon heating and increases the volume of the product. It may be the only leavening agent in "unleavened" baked products such as some breads, crackers, or piecrusts. Air may be incorporated by *creaming* fat and sugar for a cake, by *beating* egg whites/whole eggs for angel food or sponge cake, by sifting ingredients, or by *folding* (lifting and turning) the airy egg into the mixture. After its introduction into the food, air cells expand with heat in baking, and another leaven, such as steam or CO_2, diffuses into the air space, enlarging it.

Steam as a Leaven

Steam too partially leavens *almost everything*. One part of water creates 1,600 parts of steam vapor. Steam is produced from liquid ingredients, including water, juices, milk, or eggs. Products such as cream puffs or popovers are dependent on steam formation for leavening and a hollow interior. They obtain their characteristic high volume and hollow interior as dough protein expands due to steam development and as the egg protein denatures and coagulates. A *high* liquid-to-flour ratio and a *high* oven temperature are needed for water vaporization and dough expansion in products leavened mainly by steam.

Carbon Dioxide as a Leaven

Carbon dioxide is a *major* leavening agent in batters and dough. The amount required in a formulation is proportional to the amount of flour. For example, a formulation that is high in flour (dough) requires more CO_2 production for leavening than does a high-liquid (batter) product; therefore, the recipe must contain *more* of the ingredient responsible for forming CO_2.

Chemical Production of CO_2

CO_2 may be produced *chemically* by the reaction of sodium bicarbonate with an acid (wet or dry), or it may be produced *biologically*, through bacteria or yeast fermenting sugar. CO_2 is easily *released* into a batter and it may also easily *escape*, becoming unavailable for leavening. This occurs if a batter or dough is left unbaked for an extended time period, or if the gluten structure is not sufficiently developed to allow extension with the CO_2.

Baking Soda

One means of chemical leavening is by *baking soda* or sodium bicarbonate. It chemically produces CO_2 as follows:

$NaCO_3$	+	heat → Na_2CO_3	+	CO_2	+	H_2O
Sodiumbicarbonate		Sodium carbonate		Carbon dioxide		Water

When incorporated *alone*, baking soda reacts *quickly* with heat and CO_2. It may *escape* from the raw batter before it is able to leaven. Therefore, baking soda must be *combined* with another substance to make it useful. The choices are either (1) *liquid acid* or (2) *dry acid*, plus liquid, in order to *delay* production of CO_2 and prevent escape from the mixture. Examples of liquid and dry acids are as follows:

- Liquid acids—applesauce, buttermilk, citrus juices, honey, molasses, and vinegar
- Dry acid—cream of tartar (potassium acid tartrate, a weak acid), shown below

$HKC_4H_4O_6$ +	$NaHCO_3$ →	$NaKC_4H_4O_6$
Cream of tartar	Sodium bicarbonate	Sodium Potassium tartrate
+ CO_2	+ H_2O	
Carbon dioxide	Water	

If a batter or dough is made *too alkaline* with the addition of baking soda, sodium carbonate is produced in the food product, and it forms a *soapy flavor, spotty brown color, and yellowing* of the flavonoid pigment. This may occur in buttermilk (soda–acid) biscuits, if soda is present in greater amounts than the acid with which it reacts. Soda–acid biscuits exhibit *more* tenderness than baking powder biscuits because the *soda* softens the gluten.

In contrast to alkalinity, the pH may be *too acidic*. If it is too acidic, baked products such as biscuits exhibit whitening in color.

CULINARY ALERT!

- Baking soda is added to the recipe ingredient mixture along with the *dry* ingredients. If it is added with the *liquid* ingredients, CO_2 may be prematurely released into the liquid and escape from the mixture during manipulation.
- Baking soda may be used to neutralize mildly acidic juice.

Baking Powder

A *second* means of supplying CO_2 gas *chemically* is via the use of baking *powder*. It was first produced in the United States in the early 1850s and quickly provided consumers with the convenience of a premixed leaven. Baking powder contains three substances: sodium bicarbonate (baking soda), a dry acid, and inert cornstarch filler. The starch filler keeps the soda and acid from reacting with each other prematurely and standardizes the weight in the baking powder canister.

Commercial baking powder must yield at least 12 % available CO_2 gas by weight (each 100 g of baking powder must yield 12 g of CO_2), and home-use powders yield 14 % CO_2.

Baking powders are classified in several manners. One method is according to the *type of acid* component. The acids differ in strength, and thus each determines the rate of CO_2 release. While in the past, tartrate and phosphate were used as the dry acid, now consumers use the more common *SAS* phosphate (sodium aluminum sulfate phosphate).

Baking powders are also classified according to their *action rate*, or how quickly they react with water and heat to form CO_2. A *fast-acting* baking powder, such as the monocalcium phosphate, is a *single-acting* powder whose soluble acids release CO_2 almost *immediately* upon moistening/mixing with liquid at room temperature. The SAS phosphate is *slow*-acting and a *double-acting* powder that releases CO_2 *two* times. The first release of CO_2 occurs as the mixture is *moistened*; the second occurs as the mixture is *heated*.

If an *excessive* quantity of baking powder is added to a formulation, cell walls may be stretched and break. This breakage results in a coarse-textured, low-volume product—due to an overstretched, collapsed structure and release of CO_2 bubbles. The use of excessive baking powder also results in a *soapy* flavor, *yellow* crumb, and overly *browned* exterior.

Cracks may form in some SAS phosphate baking powder used in biscuits due to attempted leavening with insufficient dough stretching. This might be due to inadequate manipulation of dough required for gluten development.

Inversely, if *too little* baking powder is used, the product is *not* sufficiently leavened. The finished baked product is soggy with a compact grain of small air cells in the batter or dough.

A distinction between the use of baking *soda* and baking *powder* is seen when *baking* various biscuits. For example:

- *Baking soda* + buttermilk (liquid acid) for buttermilk biscuits (baking soda requires an acid)
- *Baking powder* for baking powder biscuits

The occasional inclusion of *both* baking powder and baking soda may be necessary in a recipe if the amount of soda and liquid acid would not amply supply the CO_2 that is needed to leaven the mixture.

CULINARY ALERT! One teaspoon of baking powder is replaced by 1/4 teaspoon baking soda and ½ teaspoon of cream of tartar.

Biological Production of CO_2

Leavening may occur by the abovementioned nonfermentation methods, using air, steam, or chemical CO_2 production. Leavening may be the result of *fermentation*, the *biological* process in which the microorganisms, *bacteria or yeast*, function to metabolize fermentable organic substances.

Bacteria

An example is *Lactobacillus sanfrancisco*. This is the bacteria responsible (along with a non-baker's yeast *Saccharomyces exiguus*) for forming sourdough bread. The bacteria function to degrade maltose, yielding acetic and lactic acid and producing CO_2. It is common that *starters* or *sponges* of dough containing the bacteria, along with yeast, may be saved from one baking and used in a subsequent baking (see below).

CULINARY ALERT! Commonly shared among friends in their home kitchens, a "starter" culture or "sponge" may be used in bread making. The starter is retained from a previous baking, and therefore fresh yeast is not required each time bread is prepared.

Yeast

The most common strain of yeast used in bread making is *Saccharomyces cerevisiae*. It is a microscopic, one-celled fungi, a plant without stems or chlorophyll that grows by a process known as budding—a new cell grows and comes from an existing cell. It releases zymase, which metabolizes *fermentable sugars*—***fermentation***—in an anaerobic process, yielding ethanol and CO_2 (with more yeast cells, more CO_2 is produced). Most of the alcohol is then volatized in baking and the CO_2 provides leavening. The three main forms of yeast used in food include those listed in Table 15.3.

Table 15.3 Forms of yeast

Forms of yeast
Active dry yeast (ADY)
1 teaspoon ADY = 1 cake of compressed yeast (CY)
Contains approximately 2–1/4 teaspoons per envelope
Leavens 6–8 cups of flour
Has a longer shelf life than CY
Less moisture than CY
Cake or compressed yeast (CY)
Moist yeast with starch filler
Short shelf life—must be refrigerated or yeast cells die
***Quick*-rising dry yeast**
Is rapidly rehydrated
Raises a mixture rapidly
Is formed by protoplast fusion of cells

Yeast is a fungus that leavens. In the leavening process it is developed by warm water and fed by the substrate sugar, fermenting it to yield carbon dioxide. In the presence of liquid and temperatures of 105–115 °F (41–46 °C), each yeast cell rehydrates and *buds*, producing new cells (see below). Reaching temperatures *greater than* 130 °F (54 °C) has a negative effect (thermal death) on yeast development, and *colder temperatures* are ineffective.

Due to the osmotic pressure that sugar exerts, more time is necessary to leaven *sweetened* yeast dough. It is possible that leavening may utilize baking soda or baking powder (chemical leavens) along *with* the yeast (biological leaven), especially if the recipe uses a high level of sugar that inhibits gluten development and the subsequent rise. Either may be added to the dough at the second rise, to provide extra leavening.

CULINARY ALERT! Directions for using the more recently developed "quick-rise yeast" may differ from what the experienced bread maker has learned about how to use yeast. Best advice: read the label!

There is a noticeable effect of spices on yeast activity. Spices such as cardamom, cinnamon ginger, and nutmeg greatly *increase* yeast activity, as does the more savory addition of thyme. The use of dry mustard has the opposite effect and *decreases* yeast activity (Spice Science and Technology).

CULINARY ALERT! Think of the festive and holiday breads you create and the wonderful effect of the added spice.

Ingredients in Specific Baked Products

Through an application of previously presented *general* concepts, the role of ingredients in some *specific* baked products will be examined.

Yeast Bread Ingredients

Yeast breads are prepared from *soft dough* (1:3–1:4 ratio of liquid to flour) using *hard wheat* flours to form a gluten structure that is strong and elastic. The structure may contain starch and sugar or other ingredients such as eggs and fat. The yeast is responsible for the production of CO_2 within the gluten structure, and in turn, the CO_2 is responsible for the reduction of pH from 6.0 to 5.0.

Four mandatory yeast bread ingredients in the United States are *flour, liquid, yeast, and salt.* Additionally, sugar and the commercial enzyme α-amylase may be added during the commercial preparation of bread loaves. (The enzyme α-amylase is naturally present in flour and may cause *unwanted* hydrolysis of the starch; however, it may be added in order to form *desirable* structure and texture in bread making and in creating food for yeast. Starch-hydrolyzing enzymes in flour such as *amylase* are important dough ingredients in commercial bread making because they produce such fermentable sugars upon which yeast acts [*α-amylase* breaks off one glucose unit, at a time, immediately yielding glucose, and *β-amylase* breaks off two glucose units—yielding maltose].)

Yeast will be addressed later in this chapter; however, the following are yeast bread ingredients.

Flour. Yeast breads are made with *hard flour.* Adequate gluten development and viscoelasticity are required for the entrapment of the CO_2 evolved from yeast fermentation. Some flours are *not* suitable for bread making, as they do *not* have gluten potential. Isolated gluten may also be added to flour to yield high-gluten flour. The *starch* component of flour also contributes to structure as it is gelatinized. In part it is converted to sugar, which provides food for yeast.

Liquid. Liquid is necessary to hydrate flour proteins, starch, and yeast cells. Milk or water, warmed to approximately 105–115 °F (41–46 °C) allows yeast cells to begin development (to bud). Higher or lower temperatures do not activate yeast or may destroy it.

Salt. Salt is a *required* ingredient in yeast formulas. It is added for flavor and to control gluten development so that the gluten stretches sufficiently yet not too much, causing breaking. If salt is omitted from a formulation, a collapsible structure would result from weak, overstretched gluten.

Sugar. The initial incorporation of a *small* amount of sugar with yeast promotes the yeast growth. Sugar also functions to brown the crust of yeast breads by the Maillard browning reaction, and it tenderizes dough if added in large amounts. *High* amounts of sugar *inhibit* yeast development. With *high* amounts of sugar, *less salt* or *more yeast* may be added.

Optional Ingredients Used in Yeast Breads. Optional ingredients in yeast bread making are *many*, determined in part by cultural or family preference. Yeast breads may include *sugar, fat,* and *eggs*. Fat may be added for flavor and tenderness; eggs may be added to provide emulsification, for nutritive value, flavor, or color. The incorporation of various *spices* including ginger, cinnamon, cardamom, and thyme increases gas production in dough by chemically enhancing yeast fermentation.

CULINARY ALERT! Many ingredients from A to Z are added to batters and dough. This assortment includes apples, carrots, cheese, dried beans, citrus fruit zest, dill, herbs, nuts, olives, sun-dried tomatoes, and zucchini, to name a few!

Quick Bread Ingredients

Quick breads, as their name implies, are relatively *quick* to mix before baking and are baked immediately without the lengthy waiting period as required of yeast breads. The leaven is typically *chemically* produced by baking powder, baking soda, or steam and/or air and is *not biologically* created such as yeast. Quick breads include biscuits, loaf breads, muffins, pancakes, popovers, and waffles among the variety of other baked products.

Flour. *All-purpose flour* is used to provide an adequate gluten structure for quick breads. The high liquid-to-flour proportion in a quick bread formulation limits gluten development and yields a tender product. Too much flour may produce excessive gluten, tunnels, and a dry crumb.

Liquid. Water, juice, or milk may be used as the dispersing medium for sugar, salt, and the leavening agent. As the liquid is heated, it forms steam, which leavens, gelatinizes starch, and contributes rigidity to the crumb.

Eggs. Eggs provide *structure* as they coagulate. They emulsify quick bread batters, allowing the *lipid* part to combine with *liquid* due to the presence of phospholipids in the yolk. Eggs also impart nutritive value and color.

Fat. Various fats and oils are used in quick bread production. *High* levels of fat limit the development of gluten. *Oil* coats flour granules, *covering* them to prevent water absorption. For example, oil is used in pancakes and muffins, and *pea-sized* chunks of *solid fat* are used in the preparation of biscuits to form flakes.

When formulations are modified for health purposes, such as may occur with the substitution of *oil for fat*, there is a noticeable change of quality. For example, the absence of flakes in biscuits becomes apparent. When a formulation is reduced fat or fat-free, it produces a *less* tender crumb due to the *increased* development of gluten.

Leaven. Typically, quick breads are chemically leavened quickly, baking powder (e.g., baking powder biscuits) or baking soda and a liquid acid (e.g., buttermilk biscuits) [a substitution: 2 tsp. baking powder + 1 cup milk = 1/2 tsp. soda + 1 cup buttermilk].

Sugar. Sugar provides sweetness and tenderization. It also assists in the Maillard browning reaction. *High* levels inhibit gluten development.

Pastry Ingredients

Depending upon the specific product desired, the quantity and type of fat/oil, flour, liquid, and so forth will vary. Piecrust is made from a *high-fat* stiff dough that has one of two distinct features—either *tenderness* or *flakiness*. Pastries may also be made from layered puff pastry dough or from a thick paste such as *choux paste*—for cream puffs and éclairs. These latter two forms of pastry dough may rise to several times the raw dough size when baked.

The function of various ingredients in pastry is identified in the following chapter subsections.

Flour. Pastry flour is best to use because it is *soft wheat flour* that is low in protein. If unavailable, it may be created using a blend *of hard and instantized flour*. Pastry flour or the blend of flour produces *less* gluten than either all-purpose or hard wheat flour and yields a *more tender* product.

CULINARY ALERT! If all-purpose or hard flour is used when the recipe specifies pastry flour, recall that hard flour absorbs more water and therefore less of it must be used than soft types to yield a similar consistency.

Liquid. The liquid in pastry dough is chiefly water. Water hydrates flour, promotes gelatinization, and forms cohesiveness. A pastry may depend upon steam from an egg to leaven. As the liquid of egg changes to steam, it leavens the mixture, i.e., cream puffs, and it contributes a gel-like interior to the hollow wall. (Although not made from a thick paste or high-fat pastry, another hollow baked product is the popover, a quick bread made from a high-liquid pour batter.)

Fat. Solid pea-sized chunks of *fat* in pastries melt to form many *flaky* layers in a crust, such as a piecrust. The use of *oil* in a recipe coats flour particles and permits *less* hydration of the flour. With the use of oil, piecrusts will exhibit a crumbly mealy nature and produce a pastry crust that is *not flaky*, but *tender*.

Lard and *hydrogenated shortenings* are solid shortenings that produce *very flaky* pastries, while *butter* and *margarine* are solid at room temperature and contain 80 % fat and 20 % liquid, thus *reducing* flakiness. Reduced-fat and fat-free margarines do not contain sufficient levels of fat to function well in pastry.

Pastries that are typically *high in fat* will not be as tender if the recipe is subject to fat reduction (Chap. 12).

Other Pie Pastry Ingredients. A piecrust may contain other ingredients depending upon the pie type. For example, *savory* crusts for a quiche may contain cheese or herbs. *Sweet* dessert crust may contain other spices, chocolate, or sugar for color or flavor. If made with sugar, crusts easily brown.

Cake Ingredients

Cakes commonly contain *fat* and *sugar*. Obviously there are many types and varieties of cake! However, this discussion is applicable to typical layer cakes. Many of the ingredients affect the cake volume and texture. Some functions of cake ingredients are presented in the following subsections.

Flour. *Soft* wheat (7–8.5 % protein) cake flour is desirable for cakes. The soft flour particles are small in size, and the cake is more lofty and tender with a *finer grain* than hard flour with its higher gluten-forming ability. Thinner walls, increased volume, and a less coarse cake result from using soft cake flour.

If the flour is *bleached*, as is often the case with cake flour, there are two advantages: (1) the pigment is whiter and (2) the baking performance is improved because, among other characteristics, bleaching oxidizes the surfaces of the flour grains and weakens the protein limiting gluten development. Higher loaf volume and finer grain result.

CULINARY ALERT! At the household level, 1 cup of all-purpose flour minus 2 tablespoons is used to replace 1 cup of cake flour in a formulation.

Liquid. Liquid gelatinizes starch and develops *minimal* gluten. *Fluid* milk hydrates protein and starch, providing structure and crumb texture. The milk sugar, lactose, and protein are valuable in determining the color of a finished cake. Milk proteins combine with sugars in nonenzymatic, Maillard browning.

Eggs. The protein of whole eggs or egg whites provides *structure* and may toughen the mixture as the protein coagulates. Egg *whites* leaven because they are beaten to incorporate air and they provide liquid, which leavens as it becomes steam. *Sponge cakes* incorporate whole egg, and *angel food cakes* are prepared with beaten egg whites to create volume. Egg *yolks*, due to their lipoprotein content, function as emulsifiers. The addition of extra fat and sugar offsets the toughness of egg in a formulation.

Fat. Fat functions in tenderizing cakes, since it shortens protein–starch strands. It provides increased volume, especially if creamed in a recipe or if monoglycerides and diglycerides are used as emulsifiers in the fat. *Butter* in a formulation may require more creaming than hydrogenated shortening because it is not as aerated and it has a narrow plastic range (Chap. 12). *Lard* has a large crystal size and therefore creams less well than most plastic fats. *Oils* produce tenderization.

Fat in a recipe also functions to retain moisture in the mixture and it softens the crumb. Shortened cakes differ from sponge cakes in that the latter have no fat besides egg.

CULINARY ALERT! Extra fat in the form of sour cream or eggs provides tenderness and flavor.

Sugar. Sugar imparts a sweet flavor to cakes and is often added to cake batters in large amounts. It competes with the protein and starch for water and inhibits both *gluten development and starch gelatinization*. Sugar also functions to incorporate air when plastic fats are creamed with sugar prior to inclusion in a batter. Even if not creamed, its addition increases the number of air cells in the batter, contributing to the *tenderness* of the grain.

Leaven. Leavening is created in several ways. The grain shows evidence of numerous air cells that hold expanding gases released by the leaven.

The process of creaming fat and sugar incorporates air to leaven. Baking soda reacts with an acid ingredient to leaven. The use of chemical leavening by both baking soda and baking powder is common, as is steam and air.

Since early times (and applicable today), there were ingredients added to baked products to lengthen shelf life. Some of these ingredients included spices such as cinnamon, ginger, clove, and garlic as well as honey. Effective cleanliness including hand washing is also recommended for extending product shelf life.

Mixing Methods for Various Batters and Doughs

The *function* of batter and dough ingredients and *ingredients* in specific baked products has been addressed in former sections. This section covers *specific mixing methods* for various batters and dough.

The purpose of *mixing* is to distribute ingredients, including leavening agents, and to equalize the temperature throughout a mixture. *Dough* such as that in biscuits and pastries are manipulated by *kneading*; cakes, muffins, and pour *batters* are *stirred.*

CULINARY ALERT! Depending on the mixing method utilized, two baked products with the exact same ingredients and proportions may yield two different end results! Due to various mixing methods, the volume, texture, and grain size may differ.

Biscuits

Biscuits are quick breads made of *soft* dough. The recommended mixing method is to cut a *solid fat*, pea size or smaller, into the sifted, dry mixture. Next, *all* of the liquid is added, a ball is formed, and the dough is kneaded. Kneading (see Yeast Dough, Kneading) 10–20 times develops gluten and orients the direction of gluten strands, necessary to create flakes. It mixes all ingredients, such as the *baking powder or soda* and *acid*, which leaven.

Underkneading produces a biscuit that fails to rise sufficiently.

Overkneading or rerolling overproduces gluten and results in a smaller volume, *tougher* biscuit, which will not rise evenly because CO_2 escapes through a weak location in the gluten structure.

Cakes

Cake batters may be prepared by several different methods. *Conventionally*, they are mixed by first creaming a plastic fat with sugar, which provides aeration of the cake batter. Next, the egg is added, and the dry and wet ingredients are added alternately. A second method or *dump* method mixes all of the items and then adds the leaven at the end.

CULINARY ALERT! With a lack of creaming, product loftiness is sacrificed because the number of air cells that can be filled with CO_2 is reduced.

Muffins

Muffins are a quick bread prepared from *drop* batter. The optimal mixing method for muffins is to pour *all* of the liquid ingredients into *all* of the sifted, dry ingredients and *mix minimally*. Overmixing a high-gluten-potential batter develops long strands of gluten and results in the formation of tunnels or peaks in the muffin (see below).

Tunnels, or hollow internal pathways, form long strands of gluten, allowing gases to escape from the interior. Muffins may also take on a peaked appearance if the oven temperature is too high, allowing a top crust to form while the interior is still fluid and maximum expansion of the muffin has not occurred. A center tunnel forms for gases' escape, creating a ***peak***.

Of particular interest are:

Bran muffins: Pieces of the bran physically *cut through* the developing gluten strands during mixing, and thus, bran muffins do not rise as much as non-bran-containing muffins.

Corn muffins: Since corn, a non-gluten flour, is used in a formulation, it is best to mix it with an *equal* amount of wheat flour in order to obtain a desirable structure.

Pastries

The mixing method for pastry is similar to biscuit preparation. It involves cutting the large amount of solid fat into the sifted, dry ingredients, then adding all of the liquid.

The mixture may be is stirred, then kneaded, and cut to desired shape. Croissant pastry dough must be repeatedly *folded*, not stirred or kneaded, numerous times over the course of several hours. This folding produces layers in the dough.

If oil is incorporated, *well-chilled* oil restricts the covering potential of room temperature oil and produces a slightly flaky product.

Pour Batters

Food items such as pancakes, popovers, and waffles contain a high proportion of liquid to flour and do *not* require a definite manner of mixing. Overmixing a pour batter is *unlikely* to affect the shape or texture of the finished product due to the high level of water and low level of gluten development.

Dough, Yeast Dough

Preparation of yeast dough includes kneading, fermenting, punching down, resting, shaping, and proofing the dough.

Subsequent to combining all of the ingredients into a ball, *kneading* must occur to stretch and develop the elastic-like gluten. This is done by pressing dough down, folding it in half, and giving a half-turn to the dough in between each pressing and folding. ***Kneading*** incorporates and subdivides air cells, promotes evenness of temperatures throughout the dough (75–80 °F [27 °C]), removes the excess CO_2 (which may overstretch the gluten structure), and distributes the leavening agent.

Kneading may be accomplished by utilizing a heavy-duty mixer, bread machine, or food processor, perhaps requiring 10, 5, and 1–2 min, respectively. *Underkneading*, or use of non-gluten-forming flour, will produce less/no gluten strands and thus breads with less volume. *Overkneading* is also possible, especially with the use of machine kneading. If this is the case, the gluten strands may break, resulting in a less elastic mass of dough that fails to rise satisfactorily.

Subsequent to kneading, yeast dough is left to rise as yeast cells undergo *fermentation* where fermentable sugar is converted into ethanol and CO_2. The dough has doubled in size when the rise is complete. Then, the dough is *punched down*. Punching down is beneficial in that it allows the heat of fermentation and CO_2 to escape, introduces more oxygen, controls the size of air cells, and prevents overstretching and collapse of gluten. If the dough is allowed to rise *too much*, gluten is overstretched, causing the dough to be inelastic and inextensible.

This step of punching down the dough provides yeast contact with a fresh supply of food (the sugar) and oxygen. The dough is punched down and left to rest for 15–20 so that gluten strands *rest or relax*, and the starch absorbs water in the dough to make it less sticky. In this time period, fermentation continues and the gluten network becomes easier to manipulate.

Next is the rest period where gluten is relaxed. The dough is *shaped* and allowed to rise a second time—it is ***proofed***. In this second rise, the dough will *double* in volume as many more yeast cells have budded and produced additional CO_2. It is ready to bake when a slight indentation mark remains in the dough when it is pressed lightly with the fingers.

In the case of over-risen dough, it should be punched down again and allowed to rise a third time so that it is not baked in a condition where the overstretched structure will collapse. If the stretched gluten structure collapses, volume decreases due to CO_2 loss, and the texture is thus noticeably coarse, open, and dense instead of fine and even.

CULINARY ALERT! Knead enough however not too much!

Baking Batters and Doughs

Unbaked batters and dough are foams of watery substance surrounding air cells. This surrounding mixture forms the *grain* of the finished product as it "sets" or coagulates around air cells. Major product changes that occur during baking involve protein, starch, gases, browning, and importantly a release of aroma!

- *Proteins* in the flour, or added protein ingredients, harden or coagulate by heat.
- *Starch* granules lose their birefringence, swell, and gelatinize as they imbibe moisture.
- *Gases* expand and produce leavening.
- *Water* evaporates and a browning of the crust becomes evident due to the Maillard browning reaction.
- The *alcohol* by-product of yeast fermentation evaporates, albeit not completely.

The qualities of a finished baked product may be determined by the degree of manipulation (stirring, kneading) and oven temperature. The type of flour, the amount of liquid, and an almost unlimited list of possible added ingredients affect quality.

CULINARY ALERT! A curved and split top seen in a rectangular shaped baked cake is due to the setting of structure on the outer surface while the interior is still fluid.

In the oven for a few minutes, yeast breads will exhibit an initial rising of the loaf known as ***oven spring***. Then, the rise is due to expansion with heat, yeast's CO_2, and the steam from water.

Gases expand the gluten strands until they form a rigid structure. However, *over*-fermentation and *over*-proofing result in the ballooning of the loaf of bread, followed by a likely collapse in structure, as previously discussed. Flavor develops as the crust browns with water loss and aroma is released.

Altitude-Adjusted Baking

As a rule, cooking and baking differ at elevations other than at sea level—both above and below sea level. Around 1/3 of Americans live at high altitudes (which is anything over 3,000 ft).

Water boils at 212° at sea level and at a *lower* temperature at higher elevations—a few °C or approximately 10 °F less. For example, every 1,000-ft change in elevation up or down changes the boiling point by approximately 2°. Therefore water boils at a temperature of 202 ° F (94 °C) at 5,000 ft and at less than 200 °F at 7,500 ft. This equates to the fact that at higher elevations, foods cooked in water have to be cooked substantially longer to get them done. Even when water comes to a rapid boil, it is not as hot at high altitudes as rapidly boiling at sea level! It follows that baking requires an increase in time too.

When a product is baked at *high* altitude, there is less atmospheric resistance and it takes longer to bake. The lowered air pressure also tends to cause the air bubbles in baked goods to rise faster, producing increased dough expansion. Then these air bubbles escape to the atmosphere causing the cake to fall. The inverse is true regarding low altitudes and high atmospheric pressure where water boils at a higher temperature. Therefore, *local* instructions specific to the altitude must be followed in manufacturing, foodservice, or home recipes.

At high elevations (5,000 ft or more above sea level), a *reduction in sugar* and *less leaven* is needed. A reduction in sugar provides *less* competition for the water, and therefore water is available to develop a strong gluten structure. *Less* leaven prevents the overexpansion of dough that may so easily occur with the lower atmospheric pressure.

"With less air pressure weighing them down, leavening agents tend to work too quickly at higher altitudes, so by the time the food is cooked, most of the gasses have escaped, producing a flat tire. For cakes leavened by egg whites, beat only to a soft-peak consistency to keep them from deflating as they bake. Also, decrease the amount of baking powder or soda in your recipes by 15 % to 25 % (one-eighth to one quarter teaspoon per teaspoon specified in the recipe) at 5,000 ft, and by 25 % or more at 7,000. For both cakes and cookies, raise the oven temperature by 20° or so to set the batter before the cells formed by the leavening gas expand too much, causing the cake or cookies to fall, and slightly shorten the cooking time.

> "Flour tends to be drier at high elevation, so increase the amount of liquid in the recipe by 2 to 3 tablespoons for each cup called for at 5,000 feet, and by 3 to 4 tablespoons at 7,000 ft. Often you will want to decrease the amount of sugar in a recipe by 1 to 3 tablespoons for each cup of sugar called for in the recipe" (Food News Service, Brunswick, ME)."

"There are some standard adjustments you can make. At 7,000 ft., for each cup of liquid called for in the recipe, increase it by 3 to 4 tablespoons. For each teaspoon of baking powder called for, decrease the amount by 1/4 teaspoon. For each cup of sugar in the recipe, decrease the amount by 1 to 3 tablespoons."

"For cakes leavened by egg whites, beat only to a soft peak consistency to keep them from expanding too much as they bake. For both cakes and cookies, lower the oven temperature by 20 degrees or so and slightly shorten the cooking time. You will want to keep the changes on the small side the first time you prepare a recipe, and adjust as needed subsequently" (Food News Service, Brunswick, ME).

Death Valley in California is the lowest land point in the United States, 282 ft below sea level. Inversely compared to high altitudes, cooking and baking below sea level will require a decreased cooking time of 5–10 %. Baked goods will rise more slowly and retain more moisture at lower altitudes. Increasing the amount of baking powder or baking soda typically used at higher elevations may be necessary.

Storage of Baked Products

Proper storage of baked products extends shelf life and maintains the best flavor and texture. Covering and elimination of external air is a step normally taken to protect baked products; thus, a good wrap or airtight storage is recommended. Such storage may also deter staleness. For long-term storage, use of a freezer wrap prior to freezer storage minimizes dryness or freezer burn.

Nutritive Value of Baked Products

The nutritive value of baked products varies according to the type and amount of ingredients used in the formulation. The primary ingredient of many baked products is flour. However, there may be a significant amount of fat or sugar. Generally, food choices that provide less sugar and fat in the diet should be selected, as fats and sweets should be used sparingly.

Whole grains, fruits, grated vegetables such as carrots or zucchini, nuts, and NFMS may be used in recipes, providing appearance, texture, and flavor benefits and boosting nutritive value. Individuals following a gluten-free dietary regimen may avoid specific flours such as wheat and instead choose flour such as rice flour to bake.

Reduced-Fat and No-Fat Baked Products

Some baked products may be successfully prepared with a reduction in the fat content, and this modification may fit into many fat- or calorie-restricted diets. However, the product will be *less* tender and flavorful than the unmodified original counterpart. The result of reducing or eliminating fat is altered flavor, more gluten development, and less tenderness than a product with the standard amount of fat.

CULINARY ALERT! Reduced and low-fat baked products may not result in quality that is acceptable to all individuals.

Safety Issues in Batters and Doughs

Microbial Hazards

"Rope" is a condition attributed to bacilli *bacteria* in flour. It may be present in the field from which a crop was obtained to produce flour. Its presence causes a syrupy ropelike interior of bread—it stretches and appears as a rope! An acid environment (pH 5 or 4.5) prevents this growth of bacteria.

Mold spoilage is also possible. Therefore, mold inhibitors such as *sodium or calcium propionate* or *sodium diacetate* are commonly added to commercially prepared bread to inhibit mold and bacteria.

Nonmicrobial Hazards

Nonmicrobial deterioration may occur due to rancidity or staling (Chap. 5). Both terms have previously been discussed in earlier chapters. A little about staling is justified herein. Staling is defined as all those changes occurring *after* batters and dough are baked. It is thought that deterioration primarily involves recrystallization of amylopectin, and it includes a change in flavor, a harder, less elastic crumb, and less water-absorbing ability. In order to partially restore flavor, brief reheating is recommended. If heat is prolonged or too high, a dry crumb is evident.

Foreign substances also pose hazards if found in foods. Controls must be established and enforced to protect against deterioration and hazards (Chap. 16).

Conclusion

Batters and dough are made with different types and proportions of liquids, flour, and other ingredients such as leavening agents, fat, eggs, sugar, and salt. Depending on the amount of flour, batter may be a pour-type or drop batter, and dough may be soft or stiff. A formulation that includes wheat flour forms a protein network known as gluten, and liquid gelatinizes starch as the batter or dough bakes. Both gluten and gelatinized starch contribute to the structure of baked products. A quick bread is quick to prepare, whereas yeast breads require more lengthy time periods for the yeast to raise bread prior to baking.

Sugar and salt contribute flavor and exert an osmotic effect on dough as they compete with other added substances for water absorption. A small amount of sugar serves as the substrate for yeast in fermentation, whereas a large amount of sugar interferes with CO_2 development by dehydrating yeast cells. Salt is needed for control of yeast growth.

Baked products may be leavened with air, steam, or CO_2 that enlarges air cells and raises dough. Carbon dioxide may be produced biologically by yeast or chemically by baking powder or baking soda. Leavening is also accomplished by air or steam.

Fat is considered optional in some batters and dough and mandatory in other baked products. Liquid oil coats flour particles more thoroughly than solid fat, limiting gluten development and contributing tenderness. Solid fat, cut into pea-sized chunks or less, melts, forming layers in piecrusts and biscuits, respectively. Eggs may be added to batter and dough formulation.

Egg whites may be beaten to incorporate air; whole eggs or yolks contribute nutritive value, color, flavor, and emulsification. The nutritive value of baked products is dependent on the individual recipe ingredients.

Imagination is the limit to creative baked products!

Notes

CULINARY ALERT!

Glossary

All-purpose flour The flour created by a blend of hard and soft wheat milling streams.

Batters Thin flour mixtures that are beaten or stirred, with a 1:1 or 1:2 ratio of liquid to flour, for pour batters and drop batters, respectively.

Dough Thick flour mixtures that are kneaded, with a 1:3 or 1:6–8 ratio of liquid to flour for soft and stiff dough, respectively.

Elastic Flexible, stretchable gluten structure of dough.

Fermentation A biological process where yeast or bacteria, as well as mold and enzymes, metabolize complex organic substances such as sucrose, glucose, fructose, or maltose into relatively simple substances; the anaerobic conversion of sugar to carbon dioxide and alcohol by yeast or bacteria.

Flaky Thin, flat layers of dough formed in some dough such as biscuits or piecrusts; a property of some pastries that is inverse to tenderness.

Gluten Three-dimensional viscoelastic structure of dough, formed as gliadin and glutenin in some flour are hydrated and manipulated.

Gluten-forming potential Presence of the proteins gliadin and glutenin that may potentially form the elastic gluten structure.

Gluten development The hydration and manipulation of flour that has gluten potential.

Grain The cell size, orientation, and overall structure formed by a pattern or structure of gelatinized starch and coagulated protein of flour particles appearing among air cells in batters and dough.

Kneading To mix dough into a uniform mass by folding, pressing, and stretching.

Leavening To raise, make light and porous by fermentation or nonfermentation methods.

Oven spring The initial rise of batters and doughs subject to oven heat.

Peak A center tunnel where gases escape from a muffin.

Plastic fat Solid fat able to be molded to shape, but does not pour.

Proofed The second rise of shaped yeast dough.

Tender Having a delicate, crumbly texture, a property of some pastries that is inverse to flakiness.

Tunnels Elongated air pathway formed along gluten strands in batters and doughs, especially seen in over manipulated muffins.

Wheat flour Flour derived from the endosperm of milled wheat.

Whole wheat flour Flour derived from the whole kernel of wheat—contains bran, endosperm, and germ of wheat.

References

Thompson T (2000) Questionable foods and the gluten-free diet: survey of current recommendations. J Am Diet Assoc 100:463–465

Thompson T (2001) Wheat starch, gliadin, and the gluten-free diet. J Am Diet Assoc 101:1456–1459

Part VII

Aspects of Food Processing

Food Preservation

16

Introduction

This chapter is in the newly named *Aspects of Food Processing* section of the text. The chapters covering *food additives* and *food packaging* components of the food processing section appear in Chaps. 17 and 18, respectively.

The objective of food preservation is to *slow down* or *stop* (kill) bacterial spoilage activity that would otherwise exhibit loss of taste, textural quality, or nutritive value of food. Techniques of food preservation include heating, cold refrigeration, freezing, freeze drying, dehydration, concentrating, microwave heating, or other means as discussed in this chapter. For added clarification, and a succinct explanation, the following is utilized. See:

> Food *processing* and *preservation* are two techniques that are used to maintain the quality and freshness of foods. In terms of how they are performed, food processing and preservation are different; food preservation is just part of the entire procedure of processing foods. Food *processing* mostly involves both packaging and *preservation*, while food preservation is concerned with the control and elimination of the agents of food spoilage. Additionally, food processing is performed to turn food into something that is more palatable and convenient to eat. There are various methods of *food preservation, which include the addition of chemicals, dehydration, and heat processing*. . . .
>
> . . . Food processing and preservation are interrelated, as food is preserved to ensure quality before being packed for processing . . . (http://www.wisegeek.com/what-is-the-difference-between-food-processing-and-preservation.htm) (italics added)

Food Preservation

Techniques of *food preservation* may occur by heating (e.g., cooking, mild heat treatment methods such as blanching or pasteurization, severe heat treatment such as canning or bottling), cold refrigeration, freezing, freeze drying, dehydration, concentrating, microwave heating, or other means. These topics appear in this chapter.

Use of *additives* (Chap. 17) may be employed in preservation, namely, preservatives including those used in fermentation, chemical preservation, irradiation (FDA labels this an additive) salt (e.g., salting), sugar, and vinegar (e.g., pickling). As well, preservation may entail *packaging* (Chap. 18), including modifying or removing oxygen as a preservation technique.

In food processing, raw food ingredients are turned into foods that we might more readily eat. Food processing may be achieved in industry or in the home and may be simply referred to as cooking or baking.

V.A. Vaclavik and E.W. Christian, *Essentials of Food Science, 4th Edition*, Food Science Text Series,
DOI 10.1007/978-1-4614-9138-5_16, © Springer Science+Business Media New York 2014

Storage conditions that foster preservation processes are subject to Food and Drug Administration (FDA) inspection and enforcement. Preservation from microbial, chemical, and physical contamination, as well as enzymatic activity, is necessary for preserving and extending the shelf life (time a product can be stored without significant change in quality) of food.

Heat Preservation

Heating or cooking foods as a means of preserving them or making them more palatable has been important for centuries. Heating is a vital form of food preservation, and as discussed in this chapter, there are many different methods of heating processes available today.

Foods are heat processed for four main reasons, enumerated below:

- To eliminate pathogens (organisms that cause disease)
- To eliminate or reduce spoilage organisms
- To extend the shelf life of the food
- To improve palatability of the food

Methods of Heat Transfer

Heat may be transferred to a food by conduction, convection, or radiation. Usually, the cooking process involves more than one of these heat transfer methods. Heat may also be generated directly in a food when it is microwaved and directly in the pan when an induction cooktop is used. The following listing offers a brief discussion of conduction (heat transferred through a solid), convection (heat transferred through the air), and radiation (direct heat—from the sun, broiling, grilling, electric range, and so forth). Microwaves and heating by induction are described later in the chapter.

Conduction is the term used for the transfer of heat from molecule to molecule and is the major method of transfer through a solid. Examples of heat transfer by conduction include a saucepan resting on a hot ring. The heat is transferred by direct contact with the heat source. Another example would be transfer of heat from the outside to the center of a large piece of meat. Conduction is a relatively slow method of heat transfer.

Convection occurs when currents are set up in heated liquid or gas. For example, as water is heated in a saucepan, the warmer sections become less dense and therefore rise, whereas the cooler regions flow down toward the bottom of the pan. This sets up a flow or current, which helps to spread the heat throughout the liquid. Heating by convection is therefore faster than heating by conduction.

Radiation is the fastest method of heat transfer. This occurs when heat is transferred directly from a radiant (red-hot) heat source, such as a broiler or a campfire, to the food to be heated. The energy is transferred in the form of electromagnetic waves, which may be transmitted through a gas such as air or through a vacuum. Any surfaces between the heat source and the food being heated reduce the amount of energy transmitted by radiation. The rays fan out as they travel, and so the farther a food is from the heat source, the fewer rays it receives and the longer it takes to get hot. (Heat transfer by radiation involves waves in the infrared region of the electromagnetic spectrum. Microwaves are also electromagnetic waves, yet of a different wavelength, and so they have a different effect on food, as discussed later on in the chapter.)

When food is cooked, more than one of these heat transfer methods is usually involved. For example, when roast chicken is cooked in the oven, heat is transferred to the outside of the chicken by radiation from the red-hot element and by convection due to the air currents in the oven; however, heat is transferred from the outside to the center by conduction.

Heat Treatment Methods: Mild or Severe

Heat treatment methods can be divided into two categories, depending on the amount of heat

Table 16.1 Overview of mild and severe heat treatments

Mild heat treatment	Severe heat treatment[a]
Aims	*Aims*
Kill pathogens	Kill *all* bacteria
Reduce bacterial count (food is *not* sterile)	Food will be commercially sterile
Inactivate enzymes	
Advantages	*Advantages*
Minimal damage to flavor, texture, and nutritional quality	Long shelf life
	No other preservation method is necessary
Disadvantages	*Disadvantages*
Short shelf life	Food is overcooked
Another preservation method must be used, such as refrigeration or freezing	Major changes in texture, flavor, and nutritional quality
Examples	*Examples*
Pasteurization, blanching	Canning

[a]See the section on canning

applied: the heat processing method may be *mild* or *severe*. The aims, advantages, and disadvantages of these two types of heat treatment are different. Depending on the objectives, a food processor may choose to use either a mild or a severe form of heat treatment to preserve a food product. Consumers rely on cooking to uphold conditions of food safety in the home. The two types of heat treatment will be discussed in detail; an overview of the main aims, advantages, and disadvantages of both is shown in Table 16.1.

Mild Heat Treatment

Examples of *mild* heat treatment include pasteurization and blanching.

Pasteurization is a *mild* heat treatment used for milk, liquid egg, fruit juices, and beer. The main purpose of pasteurization is to achieve the following as discussed in subsequent paragraphs below:

- Destroy pathogens
- Reduce bacterial count
- Inactivate enzymes
- Extend shelf life

Pathogens are microorganisms causing foodborne disease, either directly (foodborne infection), by releasing a substance that is toxic (foodborne intoxication), or via a toxin-mediated infection. All *pathogens must be destroyed* so that the food is safe to eat or drink; however, a pasteurized product is *not* sterile, the *bacterial count in a pasteurized product is simply reduced.* Any bacteria that are more heat resistant than those pathogens intended for destruction will not be destroyed, and they are able to grow and multiply in the food. They will cause spoilage of the food after a while, although that is usually obvious, as opposed to the unseen proliferation of pathogens causing contamination.

For a more detailed description of pasteurization of milk, see Chap. 11. Pasteurization of other products may differ in detail, yet the principles are the same. For example, egg white or whole egg is heated to 140–144 °F (60–62 °C) and held for 3.5–4.0 min to prevent growth of *Salmonella.* Fruit juices are also pasteurized, the main aim being to reduce the bacterial count and to inactivate enzymes, as fruit juices do not *normally* carry pathogenic microorganisms.

The mild heat treatment involved in pasteurization is usually sufficient to denature and *inactivate enzymes*. For example, milk contains the enzymes phosphatase and lipase, both of which are denatured during pasteurization (Chap. 11). To ensure that milk has been pasteurized properly, a colorimetric phosphatase test may be performed: if phosphatase is present, it turns a chemical reagent blue, indicating that the heat treatment has been insufficient. Absence of the blue color indicates that the phosphatase has been inactivated and the milk has been adequately pasteurized.

In order to *increase the shelf life* of a pasteurized product, it is necessary to refrigerate it to delay bacterial growth. For example, milk is pasteurized to ensure that it is safe to drink, although harmless bacteria are still present. If the milk is kept out on the kitchen counter on a warm day, the bacteria grow and produce lactic acid, and the milk turns sour within a day or two. However, milk can be stored in a refrigerator for at least a week, and sometimes longer, before it turns sour.

Blanching is another *mild* heat treatment, used mainly for vegetables and some fruits prior to freezing. The main aim of blanching is to *inactivate enzymes* that would cause deterioration of food during frozen storage. This is essential, because freezing does not completely stop enzyme action, and so foods that are stored in the frozen state for many months slowly develop off-flavors and off-colors.

Blanching usually involves dipping the vegetable in boiling or near-boiling water for 1–3 min. Blanching treatments have to be established on an experimental basis, depending on size and shape and enzyme level of the different vegetables. For example, peas, which are very small, require only 1–1.5 min in water at 212 °F (100 °C), whereas cauliflower and broccoli that are broken into small flowerets require 2–3 min. Corn on the cob is blanched for 7–11 min depending on size to destroy the enzymes within the cob itself.

Some, yet not all, destruction of bacteria is achieved during blanching, and the extent depends on the length or the heat treatment. As with pasteurization, blanching does *not* produce a sterile product. Foods that have been blanched require a further preservation treatment such as freezing in order to significantly increase their shelf life.

Severe Heat Treatment

Canning

Canning is a well-known method employed in food preservation. It involves hermetically sealing food in a container and then inhibiting pathogenic and spoilage organisms with the application of heat. Nicolas Appert (1752–1841) is credited with the thermal process of canning (vacuum bottling technique), which was discovered in 1809 as a result of a need to feed Napoleon's troops. Soon afterward, in 1810, Peter Durand received a patent for the tin-plated can. Decades later, Louis Pasteur understood the principle of microbial destruction and was able to provide the explanation for canning as a means of preservation. Samuel Prescott and William Underwood of the United States established further scientific applications for canning, including time and temperature interactions, in the late nineteenth century.

Canning (Table 16.1) is an example of a food processing method that involves *severe* heat treatment. Food is placed inside a cylinder, or body of a can, the lid is sealed in place, and the can is then heated in a large commercial pressure cooker known as a retort. Heating times and temperatures vary, although the heat treatment must be sufficient to sterilize the food (Potter and Hotchkiss 1995). Temperatures in the range 241–250 °F (116–121 °C) are commonly used for canning. Calcium may be added to canned foods as it increases tissue firmness.

The main purpose of canning is to achieve the following:

- Commercial sterility
- Extended shelf life (more than 6 months)

Commercial sterility is defined as "that degree of sterilization at which all pathogenic and toxin-forming organisms have been destroyed, as well as all other types of organisms which, if present, could grow in the product and produce spoilage under normal handling and storage conditions."

Table 16.2 Logarithmic death rate

Time (min)	No. of survivors
1	1,000,000
2	100,000
3	10,000
4	1,000
5	100
6	10
7	1
8	0.1
9	0.01

Commercially sterilized foods may contain a small number of heat-resistant bacterial spores that are unable to grow under normal conditions. However, if they were isolated from the food and given special environmental conditions, they could be shown to be alive (Watanabe et al. 1988).

A good number of commercially sterile foods have a shelf life of 2 years or more. Any product deterioration that occurs over time is due to texture or flavor changes, not due to microbial growth.

In the case of canning fruits and vegetables, the canneries may be located immediately near the field. The raw food is washed and prepared, blanched, placed into containers, perhaps under a vacuum (to mechanically exhaust the air), sealed, sterilized to destroy remaining bacteria, molds, and yeasts (240 °F (116 °C)), then cooled and labeled. Next, the can is sent to the warehouse for storage prior to distribution.

Bottling

The bottling process helps in preserving foods. Using a sterile bottle and subsequent boiling deters or destroys any bacteria. Once opened the bottle contents may begin the spoilage process. If the bottled contents are *high-acid* foods, they are subject to reduced boiling and use of less preservatives.

The Effect of Heat on Microorganisms

Heat denatures proteins, destroys enzyme activity, and, therefore, kills microorganisms. Bacteria are destroyed at a rate proportional to the number present in the food. This is known as the *logarithmic death rate*, which means that at a constant temperature, the same percentage of a bacterial population will be destroyed in a given time interval, irrespective of the size of the surviving population (Table 16.2).

In other words, if 90 % of the bacterial population is destroyed in the first minute of heating, then 90 % of the remaining population will be destroyed in the second minute of heating, and so on. For example, if a food contains one million (10^6) organisms and 90 % are destroyed in the first minute, then 100,000 (10^5) organisms will survive. At the end of the second minute, 90 % of the surviving population will be destroyed, leaving a population of 10,000 (10^4) microorganisms. This is illustrated in more detail in Table 16.2.

If the logarithm number of survivors is plotted against the time at a constant temperature, a graph is obtained like the one shown in Fig. 16.1. This is known as a ***thermal death rate curve*** (Fig. 16.1). Such a graph provides data on the rate of destruction of a specific organism in a specific medium or food at a *specific* temperature.

An important parameter that can be obtained from the thermal death rate curve is the ***D value*** or ***decimal reduction time***. The *D* value is defined as the time in minutes at a specified temperature required to destroy 90 % of the organisms in a given population. It can also be described as the time required to reduce the population by a factor of 10, or by one log cycle.

The *D* value varies for different microbial species. Some microorganisms are more heat resistant than others; therefore, more heat is required to destroy them. The *D* value for such organisms will be higher than the *D* value for heat-sensitive bacteria. A higher *D* value indicates greater heat resistance, because it takes longer to destroy 90 % of the population.

Destruction of microorganisms is temperature dependent. Bacteria are destroyed more rapidly at higher temperatures; therefore, the *D* value for a particular organism decreases with increasing temperature. For a specific microorganism in a specific food, a set of *D* values can be obtained at different temperatures. These can be used to plot a ***thermal death time curve*** (Fig. 16.2) with the

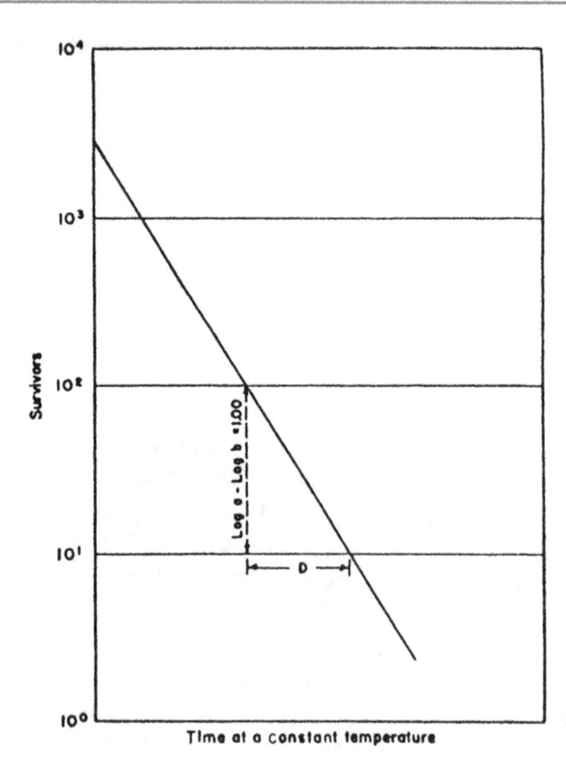

Fig. 16.1 A typical thermal death **rate** curve (*Source*: Stumbo, Thermobacteriology in Food Processing, 2nd ed. Academic Press, NY, 1973)

logarithm of the time plotted on the *Y* axis and the temperature on the *X* axis.

A thermal death time curve provides data on the destruction of a specific organism at different temperatures. The heating time on the graph may be the *D* value or it may be the time to achieve 12*D* values, as will be explained later. The important thing to remember about the thermal death time curve is that every point on the graph represents destruction of the same number of bacteria. In other words, every time–temperature combination on the graph is equivalent in terms of killing bacteria. Such graphs are important to the food processor in determining the best time–temperature combination to be used in canning a particular product and ensuring that commercial sterility is achieved.

Additional parameters shown on the thermal death time curve are beyond the scope of this book, and so will not be explained in detail here. (The *z* value indicates the resistance of a bacterial population to changing temperature and the *F* value is a measure of the capacity of a heat treatment to sterilize.)

Selecting Heat Treatments

All canned food must be commercially sterile and must therefore receive a heat treatment that is sufficient to kill essentially all bacterial vegetative cells and spores. However, such severe heat treatment adversely affects food qualities such as texture, flavor, and nutritional quality. The food processor aims to ensure commercial sterility and to achieve this using the mildest heat treatment possible, so that the food does not taste too "overcooked."

In other words, the optimal heat treatment will do the following:

- Achieve bacterial destruction (commercial sterility)
- Minimize adverse severe heat effects
- Be the mildest heat treatment necessary

In order to select a safe heat preservation treatment, it is important to know the time–temperature

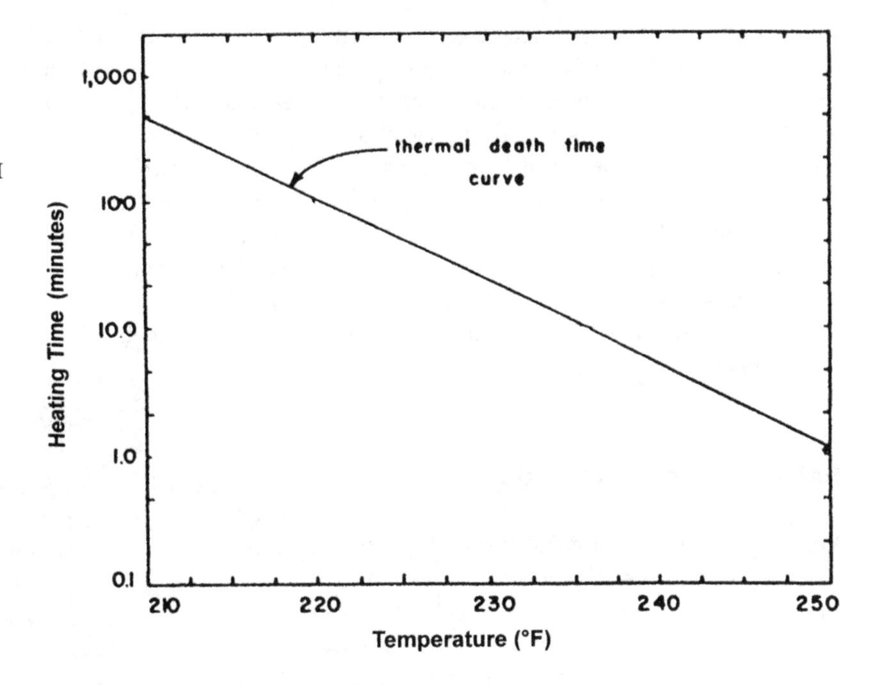

Fig. 16.2 A typical thermal death **time** curve (*Source*: Adapted from Desrosier and Desrosier, Technology of Food Preservation, 4th ed. AVI Publishing Co. Westport, CT, 1977)

combination required to inactivate the most heat-resistant pathogens and spoilage organisms in a particular food. This depends on several factors:

1. *The heat penetration characteristics of the food.* The food in the center of the can must receive sufficient heat treatment to achieve commercial sterility. This may mean that the food toward the outside of the can is overcooked. How fast the heat penetrates to the center of the can depends on the size of the can and also on the consistency of the food. Heat will reach the center of liquid foods, such as soup, much more quickly than solid foods such as meat.
2. *The pH of the food.* Bacteria are more heat resistant at neutral pH than they are in acid. Therefore, high-acid foods, such as tomatoes or fruits, need a less severe heat treatment to achieve sterility.
3. *The composition of the food.* Proteins, fats, and sugar in high concentrations all have a protective effect on bacteria, because they hinder the penetration of wet heat; thus, a more severe heat treatment is needed to sterilize foods that are high in protein, fat, or sugar.
4. The pathogenic and spoilage organisms likely to be present.

In order to ensure commercial sterility, it is important to have thermal death time curve data available for the most heat-resistant microorganisms that may be present in the food. Such data must be obtained for the food to be processed, as the composition of the food affects the heat sensitivity of the bacteria. Thermal death time curves obtained in one food may not apply to the same bacteria in a different medium. Without obtaining thermal death time curves for the specific food, it is impossible to ensure commercial sterility.

As has already been mentioned, every point on the thermal death time curve is equivalent in terms of destruction of bacteria. An increase in temperature greatly reduces the time required to achieve commercial sterility. However, the color, flavor, texture, and nutritional value of foods are not as sensitive to temperature increase. Generally speaking, a 50 °F (10 °C) rise in temperature *doubles* the rate of chemical reactions and causes a *tenfold* increase in the thermal death rate. Therefore, a high-temperature short-time combination is preferred, in order to minimize adverse chemical changes in the food such as loss of flavor, texture, and nutritional quality.

The food processor wants to use the time–temperature combination that causes the least damage to food quality.

Refrigeration Preservation

Refrigeration is another means of food preservation discussed in this chapter. Our ancestors were familiar with placing food in cold cellars, holes in the ground, or natural caves, as these storage sites would assure uniform temperatures in storage and preserve food.

Ice became widespread as a means of cold preservation in the middle 1800s—food was stored in a closed, wooden "ice box" that contained a block of ice in a chamber above the food to keep it cold. *Mechanical* refrigeration was introduced in the later 1800s and has undergone enormous developments since then. Even so, there are persons who may still refer to the refrigerator as the "ice box"!

Refrigerator and freezer temperatures both fail to sterilize food, yet the latter temperatures are more effective in retarding bacterial growth. Refrigerated food is generally held at temperatures below 45 °F (7.2 °C) (or 41 °F (5 °C)) and is subject to *state* or *local* FDA or USDA handling, storage, and transport requirements.

The extended shelf life of refrigerated foods poses microbiological and safety quality issues both at home and at the plant. A food may be better preserved in storage if it is stored under controlled atmospheric (CA) conditions. CA extends shelf life by reducing oxygen and increasing carbon dioxide in the atmosphere surrounding fruits (Chap. 7). Controlling gases in the atmosphere is also useful in providing longer storage of meats (Chap. 9) and eggs (Chap. 10). Meat preservation, for example, involves controlling microbial growth, retarding enzymes, and preventing the development of rancidity through the oxidation of fatty acids.

Packaging materials may be used in conjunction with refrigeration of food in order to preserve foods. Simply covering a food inhibits unwanted dehydration and contamination, yet as a later chapter will address, the correct choice of film material used may also assist in prolonging shelf life.

Problems Associated with Refrigeration

Spoilage, or damage to the edible quality of food, is possible without maintenance of the proper temperatures and humidity, use of FIFO, and regular cleaning.

Cross-contamination, or the transfer of harmful substances from one product to another, is possible without adequate covering or placement of foods. Pathogens from an improperly placed raw product may contaminate other food.

Temperatures. If temperatures are *too cold*, "chill injury" to fresh vegetables or fruits or sugar development in potatoes may result. Low-temperature storage increases the starch content of sweet corn (Chap. 7). *High* refrigerator temperatures or large containers of food that cannot cool quickly can lead to foodborne illness. Potentially hazardous foods must be kept at 41 °F (5 °C) or less and, if refrigerated after preparation, must be cooled to 41 °F (5 °C) or less in 4 h or less. The Centers for Disease Control and Prevention (CDC) report that improper cooling (including improper cooling in the refrigerator) is, by far, the number one cause of bacterial growth leading to foodborne illness (see local jurisdiction).

Odor. Odors may be transferred from some foods, such as onions, to butter, chocolates, and milk. If possible, strong odor foods should be stored separately from other foods. Packaging may be utilized to minimize odor problems.

Table 16.3 Why use liquid nitrogen freezing? (*Source*: Air Products)

Why use liquid nitrogen freezing instead of ammonia or Freon?
• Freezing in seconds instead of hours—LIN is one of the coldest refrigerants on earth
• Enhanced product quality with faster freezing, resulting in smaller ice crystals
• Increased production yields—less dehydration and moisture loss
• Lower capital costs
• Colder LIN temperatures equal smaller equipment footprint

Fig. 16.3 Example of cryogenic freezing of foods—hamburger patties (*Source*: Air Products)

Freezing Preservation

Frozen food is held at colder temperatures than refrigerated food—obviously. As opposed to refrigerator *short-term storage*, freezing is a *long-term storage* form that entails several months or a year. In freezing, water is rendered unavailable for bacteria; thus, bacteria are dormant, and consequently, there is no multiplication of pathogens. Foods freeze as their water component turns to ice, or *crystallizes*.

Freezing Methods

Rapid freezing by commercial freezing methods includes the following procedures including air blast tunnel freezing, plate freezing, and cryogenic freezing as described below:

Air blast procedures utilize convection and cold air. With this method of freezing, foods are placed either on racks that are subsequently wheeled into an insulated tunnel, or on a conveyor belt, where very cold air is blown over the food at a quick speed. When *all* parts of the food reach a temperature of 0 °F (−17.8 °C), the packages are put into freezer storage. The products may be packaged prior to or following freezing.

In *plate freezing*, the packaged food is placed between metal plates, which make full contact with the product and conduct cold, so that *all* parts of the food are brought to 0 °F (−17.8 °C). Automatic, continuous operating plate freezers can freeze the food and immediately deposit it to areas for casing and storage.

Cryogenic freezing may involve either *immersion* or *spraying* the food product with liquid nitrogen (LIN). LIN has a boiling point of −320 °F (−196 °C) and therefore freezes food more rapidly than other mechanical techniques (Table 16.3). Food such as meats, poultry, seafood, fruits, and vegetables, prepared or processed foods, may be preserved by cryogenic freezing (Fig. 16.3).

Cryogenic techniques for freezing include use of tunnel freezers that use LIN sprayed onto food. The LIN vaporizes to nitrogen gas at −320 °F (−196 °C), at the end of the tunnel, and is then recirculated to the tunnel entrance. LIN (Fig. 16.4) is approved by the FDA's Food Safety and Inspection Service (FSIS), for contact and freezing both meat and meat products and poultry and poultry products.

Cooking in a private residential situation, the consumer does not usually have access to these aforementioned options and it is recommended that no more than 2–3 lb of food per cubic foot of food be placed in the freezer at one time.

In order that physical damage to frozen food is minimized, the speed of freezing comes into play. For example, with a *slow* freeze, *extracellular* crystallization occurs prior to *intracellular* crystallization. This slow freezing speed is destructive, and as a result, water is drawn from

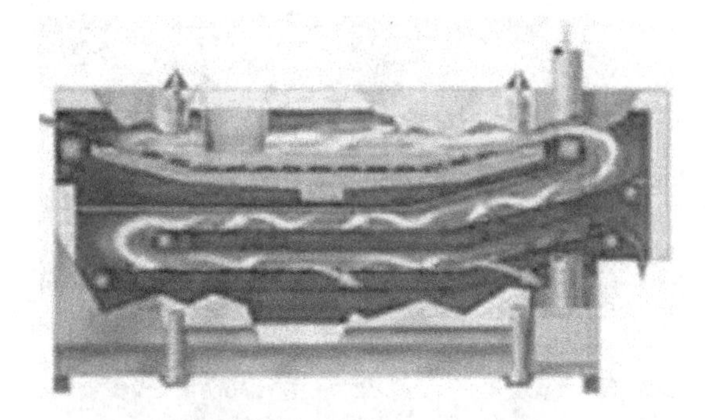

Fig. 16.4 Example of the cryogenic immersion freezing process. LIN immersion technology in a multitiered freezing system (*Source*: Air Products)

the inside of the cell as the external solute concentration increases. Consequently cell walls tear and shrink. At a cellular level, there is physical damage to the food as the water expands and the extracellular ice crystallization separates cells.

In opposition to a slow and damaging freeze, food tissues survive a *rapid* freeze better than a slow freeze. In a rapid freeze situation, water does not have time to migrate to seed crystals or form large, destructive ice crystals.

Problems Associated with Freezing

The preponderance of the problems associated with freezing is due to *physical damage*. This damage may be due to formation of ice crystals. Also, changes in texture and flavor may be caused by the *increased solute concentration* that occurs progressively as liquid water is removed in the form of ice.

These effects of ice crystals, texture and flavor, are minimized by fast freezing methods. Fast freezing minimizes formation of large crystals that would cause the most damage to cell structure and colloidal systems. Ice crystals can actually rupture cell walls, break emulsions, and cause syneresis in gels.

Increases in solute concentration can cause changes in pH, denaturation of proteins, and increased enzyme activity, all of which may lead to deterioration of food quality. Fast freezing shortens the time period during which the concentration effects are important, thereby decreasing their effect on food quality.

Recrystallization may be a problem in maintaining a high-quality product. With refreezing, ice crystals enlarge because they are subject to fluctuating temperatures. Evidence of refreezing is frequently observed with large crystalline formation on the inside of the package.

Freezer burn is the dehydration that may accompany the freezing process. The surface of food may show white patches and becomes tough. This occurs due to sublimation of ice. Solid ice will become a moisture vapor, bypassing the liquid phase, and the vapor pressure differential between the food material and the atmosphere will lead to sublimation and desiccation. The use of moisture-proof freezer wraps is suggested for storage.

Oxidation may lead to the development of off-flavor fats, as the double bonds of unsaturated fats are oxidized. Fruits and vegetables may brown during freezer storage due to enzymatic oxidative browning if enzymes are not denatured before freezing. Vitamin C (ascorbic acid) may be oxidized.

Colloidal substance change in freezing may occur due to the following:

- Starch syneresis—the freezing and thawing cycle may produce "weeping" because in thawing, less water is reabsorbed than what was originally present (Chap. 4).
- Cellulose becomes tougher.
- Emulsions break down and are subject to dehydration and precipitation.

Chemical Changes to Frozen Foods

Chemical changes to foods may occur as they are frozen. Off-odors may develop as acetaldehyde is changed to ethanol. As mentioned earlier, in oxidation, enzymatic oxidative browning is observable as phenols react with available oxygen, and ascorbic acid may become oxidized. It is recommended that blanching occur prior to freezing, as it may prevent oxidation. Pigments such as chlorophyll undergo degradation.

An illustration is apparent, for example, when eggs are frozen. Eggs show an increased concentration of soluble salts in unfrozen portions if freezing is slow. Specifically, the egg *yolks* show granule disruption due to an aggregation of low-density lipoproteins, forming a gummy product.

Moisture Control in Freezing

Drying occurs when a food gives up water to the atmosphere. Such dehydration also draws water from bacterial cells too, which is needed for subsequent growth outside of the freezer.

CULINARY ALERT! In 1930 Clarence Birdseye was awarded a US patent for a "Method of Preparing Food Products," a system that packed fish, meat, and vegetables in waxed cartons that were then flash-frozen. Now the Birdseye name still appears on high-quality frozen food.

An astute naturalist employed by the United States government was the first to take particular notice of how the Eskimos prepared their frozen fish. On duty in the Arctic, Clarence Birdseye watched in fascination as the Arctic ice and the bitter Arctic wind froze the fresh fish almost instantly. More importantly, Birdseye found that when these frozen fish were later thawed, cooked, and eaten, their taste was remarkably similar to the original fresh food. Recognizing that this "flash" or practically instantaneous freezing had commercial potential, Birdseye left his government job and formed Birdseye Seafoods, Inc. in 1924. In 1930 he was awarded a United States patent for a "Method of Preparing Food Products" (#1,773,079), a system that packed fish, meat and vegetables in waxed cartons that were then flash-frozen.

Regarding frozen food research, "For practical purposes, the question was to determine what variance in the ideal temperature a product could withstand without affecting its quality. That is, "what is the tolerance of a frozen food to adverse conditions, measured in terms of time and temperature combinations?" In typical scientific fashion, this title was shortened simply to the T-TT studies (Frozen Foods Research: Time–Temperature Tolerance Studies—American Chemical Society)

Dehydration Preservation

Dehydration is a means of preservation that subjects food to some degree of water removal. The primary intent is to reduce moisture content and preclude the possibility of microbial growth such as bacteria, mold, and yeast. A decrease in relative humidity (RH) leads to a decrease in microorganism growth.

Even as traditional methods of drying are utilized around the world, new drying techniques are being developed.

Methods used for drying foods include the following:

- Natural or sun drying—dries by direct sunlight or dry, hot air.
- Mechanical drying—dries with heated air blown in a tunnel, cabinet, or tray that contains the food (fluidized-bed drying, where hot air passes through the product and picks up moisture, is a special type of hot air drying).
- Drum drying—dries the product on two heated stainless steel drums before it is scraped off. Milk, juices, and purees may be dried in this manner.
- *Freeze drying—freezes* and subsequently *vacuums* to evaporate moisture in the process of sublimation (ice is converted to a vapor without passing through the liquid phase); examples include instant coffee, meats, and vegetables.
- Puff drying—either by *heat* and subsequent *vacuum* (to increase the pressure difference between the internal and external environments) or a combination of vacuuming and steaming. The product may also puff as the temperature of the water in the food is raised above 212 °F (100 °C) and then external pressure is quickly released. Examples are some ready-to-eat puffed cereal products.
- Vacuuming a food environment removes any necessary oxygen for aerobic bacteria and may serve to reduce loss of flavor due to oxidation.
- Smoking—preserves by dehydrating, thus offering microbial control (Chap. 9) and also treats meat to impart flavor by exposure to aromatic smoke.

- Spray drying—dries the product as it is sprayed into a chamber concurrently with hot air. For example, eggs, instant coffee, and milk may be spray-dried.

The result of any dehydration is increased shelf life and a reduction in distribution costs due to less weight.

Deterioration may occur even in dried products. Detrimental color, flavor, or textural changes may result from enzymatic changes, and these may be controlled by deactivating enzymes, by blanching, or adding sulfur compounds prior to dehydrating. Nonenzymatic browning may occur in dried foods either due to caramelization or the Maillard browning. The Maillard reaction products may lead to significant unwelcome browning, development of bitter flavors, less solubility of proteins, as well as diminished nutritive value. Dry milks or eggs and breakfast cereals participate in this reaction. Overall, oxidative spoilage, or chemical changes by oxidation of fats, is the primary cause of deterioration.

Factors needing control in dehydration include atmospheric conditions such as temperature, humidity, pressure, and portion size. The length of storage time is also a factor in the quality of the end product.

Concentration to Preserve Food

Foods are ***concentrated***, primarily in order to reduce weight and bulk. This makes transportation, shipping, and handling easier and less expensive, and so, it is economically advantageous. Many foods are concentrated, including fruit and vegetable juices and purees, milk products, soups, sugar syrups, jams, and jellies, to name a few.

Concentration is not usually considered to be a method of preservation of a food, since the water activity is not reduced sufficiently to prevent bacterial growth (see Chap. 2). The exception to this is jams and jellies, which contain high levels of sugar. Additional preservation methods, such as pasteurization, refrigeration, or canning, are therefore used to prevent spoilage of concentrated foods.

Methods of Concentration

Some of the more common methods of concentration are as follows:

- Open kettles—used to concentrate maple syrup, where the high heat produces the desired color and flavor. They are also used for jellies, jams, and some types of soups. The disadvantage of open kettles is the risk of product burn-on at the kettle wall due to high heat and long processing times.
- Flash evaporators—use heated steam (150 °C), which is injected into the food and later removed, along with water vapor from the food. This reduces heating time, but temperatures are still high, and so foods may lose volatile flavor constituents.
- Thin-film evaporators—enable the food to be continuously spread in a thin layer on the cylinder wall, which is heated by steam. As the food is concentrated (by removal of water vapor), it is wiped from the wall and collected. Heat damage is minimal due to the short time required to concentrate the food.
- Vacuum evaporators—used to concentrate heat-sensitive foods, which would be damaged by high heat. Operation under vacuum allows concentration to be achieved at much lower temperatures.
- Ultrafiltration and reverse osmosis—expensive processes that may be operated at low temperatures and use selectively permeable membranes to concentrate liquids. Different membranes are required for different liquid foods. These processes are used to concentrate dilute protein dispersions

such as whey protein, which cannot be concentrated by traditional methods without being extensively denatured. Ultrafiltration involves pumping the dispersion under pressure against a membrane that retains the protein but allows smaller molecules such as salts and sugars to pass through. Reverse osmosis is similar, but higher pressures are used, and the membrane pores are smaller, and so they are able to hold back various salts and sugars, as well as larger protein molecules.

Changes During Concentration

The product changes that occur during concentration arise primarily due to exposure of food to high heat. A "cooked" flavor may develop, and discoloration may occur. In addition, the product may thicken or gel over time, due to denaturation of proteins. This is a potential problem in evaporated milk. Nutritional quality may also be lost. The extent of the changes depends on the severity of the heat treatment.

Concentration methods that employ low heat or short processing times cause the least damage to food. However, they are also the most costly and may not always be the practical choice for the food processor, who must balance cost against quality.

Added Preservatives

Preservatives may be used along with preservation techniques such as heating, refrigerating, freezing, canning, and so forth. Specific preservatives may be applied to a food to extend shelf life:

- **Acid**—denatures bacterial proteins, preserving food, although not always sufficient to ensure sterility. Acid may be naturally present in foods such as citrus fruits and tomatoes. The combination of acid and heat provides more effective preservation.
- **Sugar and salt**—heavy syrups or brines compete with bacteria for water. By osmosis, the high percentage of water moves out of bacterial cells to equal the lower level of water in the surrounding medium. Other microorganisms, such as the fungi, yeast, and mold, are capable of growing in a high sugar or salt environment. Early US settlers preserved meats using salt and sugar.
  Sugar *syrups* may be used in preserving seasonal fruits, and *crystallized* sugar is found in cooked, candied fruit peels.
- **Smoke**—may contain a preservation chemical such as formaldehyde. Smoke retards bacterial growth due to surface dehydration. Smoking may also be used simply to impart flavor.
- **Vinegar**—used to create an acidic environment. Pickling controls the growth of microorganisms.
- **Chemicals**—subject to FDA approval. The burden of proof for usefulness and harmlessness is on industry. The chemical properties of the foods itself, such as pH and moisture content, affect the growth of microorganisms.

Fermentation—with the addition of nonpathogenic bacteria to a food, acid is produced, the pH is reduced, and the growth of pathogenic bacteria is controlled.

Radiation to Preserve Food

Foods may be heated by radiation, including the use of microwave heat treatment, or the lesser heat of irradiation. As has already been mentioned, these are both different from the radiation that occurs from a red-hot heat source such as a broiler or a fire, because of the frequency of the electromagnetic waves, which determines the effects of the radiation on food

Microwave Heating

Microwave heating is a nonionizing, rapid method of cooking. It may be used for both processing and preservation. It is reported that microwave heating inactivates vitamin B_{12}, which is found in animal products and fortified vegetarian products. The nutrient plays an important role in maintaining the nerve tissue (Watanabe et al. 1988). In combination with newer food packaging technologies, microwaveable foods are plentiful in the marketplace.

Microwaves are high-frequency (2.5 GHz) electromagnetic waves that cause heat to be generated in the food itself, due to the friction generated as polar molecules such as water try to align with the constantly changing electromagnetic field they produce. Microwave heating is fast and efficient, and since heat is generated within the food, the oven does not get hot.

Nevertheless, microwave heating is uneven, and hot and cold spots are generated within the food. The presence of localized cold spots could present a health hazard when cooking poultry or raw meat; it is important to check the internal temperature in several places if poultry is cooked in the microwave, to ensure that the correct temperature has been reached throughout. Microwaves penetrate 1–2 in. into the food; beyond that, heat is transferred by conduction, if the food is solid. Small portions are therefore best suited for microwave cooking. Large portions tend to overcook on the surface before the heat reaches the center.

Microwave-safe containers must be used in a microwave oven. These include containers that transmit microwaves, such as glass, ceramic, and some plastics. Containers that absorb microwaves, and therefore get hot, should not be used. Metal containers should also not be used, since they reflect the microwaves, which can cause arcing, and possibly a fire.

Foods do not generally "brown" when microwaved, since the surface does not get as hot as in a conventional oven. However, special packaging has been developed for some food products that allows for browning; for example, Hot Pockets and some chicken pot pies have metalized coatings that cause reflection of the microwaves back onto the surface, allowing browning of the crust. Ingredients such as commercially available liquid browning sauces and powders may be added for the purpose of fostering browning.

General recommendations to be followed when heating by microwave include the following:

- Turn the container while cooking to avoid "hot spots" of concentrated energy in one spot.
- Include a "rest period" or "standing time" beyond the designated cooking time, in order to continue cooking the food.
- Beware of hot containers from conduction of heat from food to the container.
- Select a low power setting for defrosting. Microwave energy is then sent intermittently into the frozen food.

Several definitions relating to the microwave method of heating include the following:

Hot spots—the nonuniform heating of high-water foods
Molecular friction—the heat generation method of microwave heating
Skin—the surface dehydration and hardening as more microwave energy is absorbed at the surface of the food
Shielding—protection of portions of food such as cylindrical ends of food, which readily overcook
Thermal runaway—differential heating of food without thermal equilibrium

Irradiation

Irradiation is the administration of measured doses of energy that are product-specific. A positive biological effect is that it has a **bactericidal** effect, thus reduces the microbial load of a food, kills insects, and controls ripening. It also inhibits the sprouting of some vegetables.

Irradiation is a cold process of food preservation that does not add heat to the food. In the spectrum of energy waves, radio waves are at one end of the spectrum, microwaves are in the middle, and the gamma rays of irradiation are at the other end of the continuum. Gamma rays are passed through the food to be irradiated, and the food is thus sterilized and preserved as it passes through an irradiation chamber on a conveyor belt. Scientific evidence demonstrates that foods do not become radioactive and that no radiation residue remains in the food.

Irradiation is a process approved by the FDA, for use with specific foods, and only at designated dosages. Gamma rays are the isotope-sourced form of irradiation, previously mentioned. As well, there is a machine-sourced form of irradiation that is electronically generated. It is known as e-beam (Higgins 2000). Foods that may be irradiated include wheat, potatoes, spices, pork, red meat, fruit, poultry, dehydrated enzymes, or vegetable substances, including fresh produce (and bagged salads). So, in looking at dosages needed for irradiating pork, as an example, it is shown that a *low* dose is required to stop *reproduction* of *Trichinella spiralis* (the parasite responsible for causing trichinosis), and a much *higher* dose is needed in order to *eliminate* it from the pork.

Whole food items must be labeled if they are irradiated. A universal symbol of irradiation, namely the radura symbol, is used for recognition of irradiated food. In the United States, the words "Treated with Radiation" or "Treated by Irradiation" may also appear with the symbol. Spices do not require this labeling. Processed foods that contain irradiated ingredients, or restaurant foods prepared using irradiated ingredients, do not require an irradiation label.

Research has been conducted on the sensory aspects of irradiated food. It is reported that "The sensory appeal of foods which are processed with irradiation at levels that are approved for use is quite good. Researchers who have conducted experimental studies using sensory panelists to evaluate such foods found that food freshness, color, flavor, texture, and acceptability are not significantly different from unirradiated foods" (Texas A&M University—Center for Food Safety). Due to its cold process of food preservation, the nutritive value of irradiated food is not significantly different from food subject to alternate methods of preservation, including canning.

Irradiation preserves food by killing insects and pests. It also kills microorganisms. With regard to food safety, food is made safer by the elimination of disease-causing bacteria such as *E. coli*, *Salmonella*, and the parasite *Trichinella spiralis*. Irradiated food lasts longer and there are reduced losses due to spoilage (Texas A&M University—Center for Food Safety).

Low doses of irradiation may be used to slow fruit ripening and control pests, without the use of pesticides. The process of irradiation leaves no residue.

Food irradiation facilities exist for the irradiation of foods, whereby the food product is sent off-site for treatment. As well, in-line irradiation brings the technology to a company's own production line. A large defense contractor that radiates medical supplies is now using electron beams to pasteurize/irradiate meat, including prepared meats, and other foods (Higgins 2000). A patent was awarded to this corporation for development of a miniature version of their chamber that could incorporate the electronic pasteurization into food producer's processing line. Both cost and convenience issues need to be addressed by a company considering irradiation of its products.

Despite the fact that irradiation of meat and poultry has received the approval of every major government and health agency in the United States, consumer health activists have yet to give their stamp of approval. As a result, meat companies are proceeding at a less than full-steam-ahead rate with irradiation (Gregerson 2001).

The General Accounting Office (GAO) reports to the US House Committee on Commerce have stated that the benefits of irradiation outweigh any risks. "Food safety experts believe that irradiation can be an effective tool in helping to control foodborne pathogens and should be incorporated as part of a comprehensive program to enhance food safety."

Irradiation is subject to FDA approval as an *additive*. Only specific foods, dosages, and irradiation sources are approved to kill microorganisms. "The Food and Drug Administration (FDA) announced a final rule ... amending the food additive regulations to provide for the safe use of ionizing radiation for the control of foodborne pathogens and extension of shelf-life in fresh iceberg lettuce and fresh spinach. FDA has determined that this use of ionizing radiation will not adversely affect the safety of the food" (FDA).

Ohmic Heating

Ohmic heat processing of foods is relatively new for food manufacturers. In place of radiant heat, an electrical current is passed through food to heat it rapidly. A continuous heating system reaches the food as it passes between electrodes.

Concerning ohmic heating, the *liquid* portion of the food, such as stew or soup, is heated rapidly and it subsequently conducts heat rapidly to the *inner* portion. In comparison, conventional heating tends to overprocesses the surrounding liquid as it conducts heat to the inner portion; consequently, food quality may be diminished.

> What is ohmic heating?
>
> Ohmic heating is an advanced thermal processing method wherein the food material, which serves as an electrical resistor, is heated by passing electricity through it. Electrical energy is dissipated into heat, which results in rapid and uniform heating. Ohmic heating is also called electrical resistance heating, Joule heating, or electro-heating, and may be used for a variety of applications in the food industry.
>
> How is ohmic heating different from conventional thermal processing?
>
> During conventional thermal processing, either in cans or aseptic processing systems for particulate foods, significant product quality damage may occur due to slow conduction and convection heat transfer. On the other hand, ohmic heating volumetrically heats the entire mass of the food material, thus the resulting product is of far greater quality than its canned counterpart. It is possible to process large particulate foods (up to 1 in.) that would be difficult to process using conventional heat exchangers. Additionally, ohmic heater cleaning requirements are comparatively less than those of traditional heat exchangers due to reduced product fouling on the food contact surface. (Ohio State University Extension Fact Sheet. Food Science and Technology)

Induction Heating

Heat may be transferred, or more correctly, generated, by *induction*. Induction is the transfer of heat energy to a neighboring material without contact. This occurs in some smooth cooktops and is a relatively new and therefore fairly expensive technology. Induction involves use of a powerful, high-frequency electromagnet to generate heat in a ferromagnetic (iron or stainless steel) pan on the surface of the cooktop. The heat is then transferred from the pan to the food it contains by normal heat transfer methods.

Induction cooktops all contain an electromagnetic coil beneath the surface. When switched on, alternating current flows through the coil, generating a fluctuating high-frequency electromagnetic field. When a ferromagnetic pan is placed on the cooktop, the electromagnetic field generates many small electric currents, known as eddy currents, in the pan. Because iron is a poor conductor, or in other words has high resistance, these eddy currents are converted to heat. Since heat is generated directly in the pan, and not in the cooktop itself, heating is even—no "hot spots" are generated—and the process is faster and more efficient than the more conventional methods of heating.

As well, the cooktop does not get hot! In addition, it is possible to instantly and precisely control the amount of heat generated in the pan and therefore transferred to the food within it. The only disadvantage of induction cooking is that iron or steel pots and pans must be used; copper, aluminum, or Pyrex pans will not work. However, this is a minor disadvantage, since these types of pans are readily available.

An induction oven has also been produced, where the heating coil has been replaced by a ferrous plate that is heated by embedded induction coils beneath it. This allows for use of any type of bakeware within the oven (http://theinductionsite.com/how-induction-works.shtml).

As the technology progresses, the application of induction cooking is likely to become more widespread and more common.

At this time, televised commercials advertise this method of home stovetop cooking!

High-Pressure Processing

High-pressure processing, or HPP, is a nonthermal processing method that uses physical pressure to preserve food, instead of heat, chemicals, or irradiation. HPP may be used to destroy harmful foodborne pathogens and extend the shelf life of a wide range of foods, without sacrificing sensory characteristics or nutritional quality. Its effect is instantaneous and uniform throughout the product and does not depend on the size or shape of the package.

The process involves subjecting food to extremely high hydrostatic pressure—up to 87,000 lb per square inch (psi)—for a short period of time. The uniform high pressure destroys vegetative bacteria because it disrupts microbial cellular integrity and metabolism. However, it does not destroy bacterial spores (Ramaswamy et al. 2004). Hence, it is useful to prolong shelf life and to reduce bacterial counts and may be used as a pasteurization technique, although it does not sterilize food; after HPP, foods should be refrigerated in order to effectively prolong shelf life. This is especially true of low-acid foods such as vegetables, milk, or soups. HPP can at least double the refrigerated shelf life of many perishable products.

In a typical HPP process, the product is packaged in a flexible container and then placed in a high-pressure chamber which is filled with water. The chamber is pressurized, and the pressure is transmitted through the package into the food itself, usually for a period of 3–5 min. The processed product is then removed and refrigerated. Because the pressure is uniform on all sides, most foods retain their shape and are not squashed or damaged.

HPP does not break covalent bonds in foods, so no free radicals or chemical by-products are formed, and HPP does not "add" anything to food. Hence, neither the FDA nor the Food Safety Inspection Service of the USDA requires approval for high-pressure processing (Raghubeer 2008). Flavor, texture, color, and nutritional quality of food are unaffected by HPP. The process is very effective on foods with high moisture content, such as ready-to-eat meats and poultry (cold cuts), fresh juice, prepared fruit and vegetable products such as salsa and guacamole, and seafood and molluscan shellfish. It is not effective on dry products, since moisture is needed for microbial destruction. Also, it cannot be used on products with internal air pockets, such as bread, or fruits such as strawberries, because the pressure causes them to implode.

HPP is being used extensively by processors for manufacturing of all-natural products in the ready-to-eat meat category, for shucking and shelling of seafood, and for processing preservative-free fruit and vegetable products, juices, and smoothies. HPP-processed products cost more to produce than thermally processed products, yet consumers benefit from the increased shelf life, quality, and availability of value-added products that are impossible to make using other thermal processing methods.

Other Preservation Techniques

The objective of processing is to slow down or stop spoilage that would otherwise exhibit loss of taste, textural quality, or nutritive value. In order to better achieve this, new techniques are constantly being explored.

> People are always looking for ways to increase the shelf life of their food. We have sought out canning, pickling, freezing, adding preservatives, and many other methods to extend the life of our food supply since the beginning of recorded history. Reduced oxygen packaging (ROP) is one of those methods. This method of preservation has many unique advantages, but comes with significant microbiological concerns. (The Association of Nutrition & Foodservice Professionals (ANFP). http://www.anfponline.org/CE/food_protection/2010_11.shtml)

ROP will be covered in the packaging chapter.

Early Methods of Food Preservation

The preservation of harvested and prepared food for future consumption is one of the oldest practical arts, a necessity that developed from the sheer need to survive in a hostile environment where fresh food was not always available. Techniques for drying foods date back to ancient times, when fruits and vegetables were dried in the sun or on an open stove. Without water present, the dehydrated foodstuffs would not support microorganisms and therefore did not spoil. By 1000 BC, the Chinese were using salt, spices and smoking to create a sterile environment for different food products. Salt also acts as a dehydrating agent and is particularly useful for fish and meat. Salted meat served explorers and military forces well because of its stability and portability, and it was a technique that lasted into the twentieth century.

It was also discovered very early that making cheese could preserve dairy products, grape juice could be fermented into wine that would last for years at normal temperatures, and even cabbage could be preserved by converting it to fermented sauerkraut. North American Indians made pemmican by drying the meat of buffalo or deer and then mixing it with a large amount of fat. This was effective because the fat presumably excluded oxygen. (American Chemistry Society (ACS))

Nutritive Value of Preserved Foods

There is no question regarding the importance of preserving factors such as the appearance, texture, and flavor of food. For example, prolonged or improper storage may have a deleterious effect on food due to the browning caused by the Maillard reaction. Still, in a discussion of food preservation and the extension of a food's shelf life, the preservation of nutritive value also becomes important. For instance, water-soluble vitamins may be lost from a food or high levels of sugar or salt may be added. These, and more, become issues to address with regard to nutritive value of preserved foods.

Irradiated fresh produce, such as bagged salad, may now be a healthful addition to the diet for a multitude of persons, including the young, elderly, pregnant, and immunocompromised individuals. The microbial load can be drastically cut, assuring less likelihood of *Shigella* and *E. coli*.

Safety of Preserved Foods

The safety of foods must be taken into account when seeking to store and extend the shelf life of foods. The processor/manufacturer's Good Manufacturing Practices (GMPs), the FDA's inspection, and the consumer's attentiveness all contribute to ensuring that food is properly preserved, stored, and not held beyond acceptable time parameters (See http://www.science.howstuffworks.com/innovation/edible-innovations/food-preservation.htm).

> Because food is so important to survival, food preservation is one of the oldest technologies used by human beings . . .
>
> A food that is **sterile** contains no bacteria. Unless sterilized and sealed, all food contains bacteria. (science.howstuffworks.com)

See FoodSafety.gov:

Some foods are more frequently associated with food poisoning or foodborne illness. With these foods, it is especially important to:

- CLEAN: Wash hands and food preparation surfaces often. And wash fresh fruits and vegetables carefully.
- SEPARATE: Don't cross-contaminate! When handling raw meat, poultry, seafood and eggs, keep these foods and their juices away from ready-to-eat foods.
- COOK: Cook to proper temperature.
- CHILL: At room temperature, bacteria in food can double every 20 min. The more bacteria there are, the greater the chance you could become sick. So, refrigerate foods quickly because cold temperatures keep most harmful bacteria from multiplying.

Conclusion

One aspect of food processing is preservation of the food. Storage conditions and preservation processes are subject to FDA inspection and enforcement. Consumer vigilance is also necessary in order to preserve food. Environmental control of oxygen and water availability and enzymatic control extend shelf life of food and assist in providing food safety.

Techniques of *food preservation* may occur by heating (e.g., cooking, mild heat treatment methods such as blanching or pasteurization, severe heat treatment such as canning or bottling), cold refrigeration, freezing, freeze drying, dehydration, concentrating, microwave heating, high-pressure processing, or other means. These topics appear in this chapter.

Usage of *additives* (Chap. 17) may be employed in preservation—namely preservatives including those used in fermentation, chemical preservation, irradiation (FDA labels this an additive), salt (e.g., salting), sugar, and vinegar (e.g., pickling). As well, preservation may entail *packaging* (Chap. 18), including modifying or removing oxygen as a preservation technique.

Further advances in safe and effective food preservation are on the horizon.

Notes

CULINARY ALERT!

Glossary

Blanching Mild heat treatment that inactivates enzymes that would cause deterioration of food during frozen storage.

Canning An example of a food processing method that involves *severe* heat treatment. Food is placed inside a can, the lid is sealed in place, and the can is then heated in a large commercial pressure cooker known as a retort.

Commercial sterility Severe heat treatment. A sterilization where all pathogenic and toxin-forming organisms have been destroyed as well as all other types of organisms which, if present, could grow in the product and produce spoilage under normal handling and storage conditions.

Concentration Method of removing some of the water from a food, to decrease its bulk and weight. Concentration does not prevent bacterial growth.

Conduction Transfer of heat from one molecule to another molecule; the major method of heat transfer through a solid.

Convection Flow or currents in a heated liquid or gas.

***D* value** Decimal reduction time; time in minutes at a specific temperature required to destroy 90 % of the organisms in a given population.

Dehydration A means of preservation with the primary intent to decrease moisture content and preclude the possibility of microbial growth such as bacteria, mold, and yeast.

Irradiation The administration of measured doses of energy that are product-specific. It reduces the microbial load of a food, kills insects, controls ripening, and inhibits the sprouting of some vegetables.

Ohmic heat In place of radiant heat, a continuous electrical current is passed through food to heat it rapidly, maintaining quality.

Pasteurization Mild heat treatment that destroys pathogenic bacteria and most nonpathogens. It inactivates enzymes and extends shelf life.

Radiation Fastest method of heat transfer; the direct transfer of heat from a radiant source to the food being heated.

Thermal death rate curve Provides data on the rate of destruction of a specific organism in a specific medium or food at a specific temperature.

Thermal death time curve Provides data on the destruction of a specific organism at different temperatures.

References

What is the difference between food processing and preservation? http://www.wisegeek.com/what-is-the-difference-between-food-processing-and-preservation.htm)

Potter N, Hotchkiss J (1995) Food science, 5th edn. Springer, New York

Watanabe F, Abe K, Fujita T, Goto M, Hiermori M, Nakano Y (1988) Effects of microwave heating on the loss of vitamin B_{12} in foods. J Agric Food Chem 46:206–210

Higgins KT. E-beam comes to the heartland. *Food Engineering*. 2000; October: 89–96.

Gregerson J (2001) Bacteria busters. Food Engineering 101:62–66

Induction cooking: how it works. http://theinductionsite.com/how-induction-works.shtml. Accessed 6/1/2013

Ramaswamy R, Balasubramaniam VM, Kaletun G (2004) High pressure processing. Fact sheet for food processors. Ohio State University Extension Fact Sheet FSE-1-04 http://ohioline.osu.edu/fse-fact/0001.html. Accessed 6/1/2013

Raghubeer EV (2008) The role of technology in food safety. Avure Technologies Inc., Kent, WA

Bibliography

CSPI—Center for Science in the Public Interest. http://www.cspinet.org

http://www.cdc.gov/ncidod/dbmd/diseaseinfo/foodirradiation.htm

http://www.fda.gov—DHHS. FDA. Center for Food Safety and Applied Nutrition. Refrigerator & Freezer Storage Chart

http://www.foodfreshly.com/food-preservation/food-preservation.html

http://www.food-irradiation.com/

http://www.nutrition.gov—Refrigerator Freezer Chart

http://www.usda.gov—USDA. Food Safety and Inspection Service. Freezing and Food Safety

International Food Information Council—IFIC

Food Additives

17

Introduction

This *Food Additives* chapter is in the newly named *Aspects of Food Processing* section of the text. The other chapters covering *food preservation* and *food packaging* components of the food processing section are discussed in Chaps. 16 and 18, respectively.

> There are various methods of food preservation, which include *the addition of chemicals, dehydration, and heat processing*.... (http://www.wisegeek.com/what-is-the-difference-between-food-processing-and-preservation.htm)
> (italics added)

According to the Food and Drug Administration (FDA), a food additive in its broadest sense is any substance added to food. *Legally*, the term refers to "any substance the intended use of which results or may reasonably be expected to result directly or indirectly in its becoming a component or otherwise affecting the characteristics of any food."

Additives are useful in controlling such factors as decomposition and deterioration, nutritional losses, loss of functional properties, and aesthetic value, yet may not be used to disguise poor quality. Their use is subject to regulation in the 1958 Food Additives Amendment to the Food, Drug, and Cosmetic (FD&C) Act with exemptions for prior-sanctioned items and generally recognized as safe (GRAS) substances.

Food processors must petition the federal FDA for approval of a new food additive. FDA approval is then required for use at *specific* levels, only in *specific* products.

Under the condition that a raw ingredient is processed, the processed food represents the change of raw material into food of another form. Food processing may involve the use of specific preservation techniques and also packaging.

The Continuum of Processed Foods (IFIC Foundation).

"Processed foods can be placed on a continuum that ranges from minimally processed items to more complex preparations that combine ingredients such as sweeteners, spices, oils, flavors, colors, and preservatives, with many variations in between."

"... Packaging and use of Additives, namely preservatives including salt, sugar, vinegar (for pickling) and sulfur dioxide are food processing techniques."

Vitamins and minerals are a special category of food ingredients. They are essential for nutrition yet their use *apart from food* is often surrounded with controversy. Their use *in foods* has been *increasing* as they have been associated

V.A. Vaclavik and E.W. Christian, *Essentials of Food Science, 4th Edition*, Food Science Text Series, DOI 10.1007/978-1-4614-9138-5_17, © Springer Science+Business Media New York 2014

with the prevention and/or treatment of at least four of the leading causes of death in the United States. Existing additives, as well as new ones, are utilized in new product development.

Controlling decomposition, nutritional losses, losses of functional properties, aesthetic value, and so forth are issues of utmost importance to our food supply. "Yuck, this stuff is full of ingredients!" says Linus, reading a can label in Charlie Brown. It is true that there was some suspicion of additives in earlier times, yet in this day consumers demand good-looking, nutritious, safe, and tasty foods.

Whether an additive is natural or artificial has no bearing on its safety. (FDA)

Continuously, consumers should be vigilant as to the quality and quantity of food and beverages that they consume. Government agencies have the responsibility for just *part of* the plan.

Definition of Food Additives

The 1958 Food Additives Amendment to the FD&C Act of 1938 legally defined a food additive, and the Committee on Food Protection of the National Research Council (NRC) more simply and practically defined an additive as "A substance or a mixture of substances, other than a basic foodstuff, that is present in a food as a result of an aspect of production, processing, storage, or packaging."

Exempt from food additive regulation are *prior-sanctioned* substances, determined as safe for use prior to the 1958 amendment, such as sodium nitrite and potassium nitrite, and ***generally recognized as safe*** (*GRAS*) substances such as salt, sugar, spices, vitamins, and monosodium glutamate (discussed later).

A food ***additive*** in its *broadest sense* is any substance added to food. *Legally*, additives are classified as *direct* or *indirect*. If they are intentionally or purposely added to foods, these direct additives must be named on food labels. If *indirect*, they are incidentally added to food in very small amounts during some phase of production, processing, storage, packaging, or transportation.

According to the FDA, food additives are substances added to foods for specific physical or technical effects. They may not be used to disguise poor quality yet may aid in preservation and processing, or improve the quality factors of appearance, flavor, nutritional value, and texture (Chap. 1).

The consumer may be skeptical of, or perhaps in opposition to, an uncommon or unfamiliar chemical name of a food additive. Yet, in fact, all additives, including GRAS substances, such as salt, are chemicals. Even water is H and O! Food additives undergo vigorous toxicological analysis before their approval and use in foods.

Food additives have many very important technical functions in foods. Food additives contribute to the overall quality, safety, nutritive value, appeal, convenience, and economy of foods (IFT, 2010; PDF Download). Food and color additives and GRAS substances have been the subjects of research and development, public policy, and regulatory activity as well as public interest for decades.

IFT has addressed food additive topics in numerous ways: through workshops, publications, and expression of scientific viewpoints in numerous forums. IFT is also actively engaged in deliberations on food additives in the scientific and policy arenas (e.g., Codex Committee on Food Additives).—IFT

The Joint FAO/WHO (Food and Agriculture Organization of the United Nations and the World Health Organization) Expert Committee on Food Additives (JECFA) represents an international expert scientific committee that has met since 1956 to evaluate the safety of food additives. Yearly findings are reported by the organization.

One more definition of food additives by the World Health Organization (WHO):

> Food additives are substances that are added to food or animal feed during processing or storage. They include antioxidants, preservatives, colouring and flavouring agents, and anti-infective agents. Most food additives have little or no nutritional value.
>
> The Joint FAO/WHO Expert Committee on Food Additives (JECFA) is an international scientific expert committee that is administered jointly by the Food and Agriculture Organization of the United Nations FAO and the World Health Organization WHO. It has been meeting since 1956, initially to evaluate the safety of food additives. Its work now also includes the evaluation of contaminants, naturally occurring toxicants and residues of veterinary drugs in food.
>
> To date, JECFA has evaluated more than 2600 food additives, approximately 50 contaminants and naturally occurring toxicants, and residues of approximately 95 veterinary drugs. The Committee also develops principles for the safety assessment of chemicals in food that are consistent with current thinking on risk assessment and take account of recent developments in toxicology and other relevant sciences.—JECFA

Everything Added to Food in the United States (EAFUS)

> November 2011
>
> The EAFUS list of substances contains ingredients added directly to food that FDA has either approved as food additives or listed or affirmed as GRAS. Nevertheless, it contains only a partial list of all food ingredients that may in fact be lawfully added to food, because under federal law some ingredients may be added to food under a GRAS determination made independently from the FDA. The list contains many, but not all, of the substances subject to independent GRAS determinations. For information about the GRAS notification program please consult the Inventory of GRAS Notifications. Additional information on the status of Food and Color Additives can be obtained from the Food Additive Status List or the Color Additive Status List (formerly called Appendix A of the Investigations Operations Manual).
>
> [This information is generated from a database maintained by the U.S. Food and Drug Administration]

See http://www.nutrition.gov/whats-food/food-additives

See FDA Food Additive Status List Page Last Updated: 2012

Function of Food Additives

Foods are evaluated based on their appearance, texture, and flavor as previously discussed in Chap. 1. Each attribute is crucial to the food's acceptance and edibility. General categories of food additives include *preservatives*, *nutritional additives*, *sensory agents*, and *processing agents* as noted in this chapter. When new food products are developed, new or existing food additives may be utilized.

> Additives perform a variety of useful functions in foods that are often taken for granted. Since most people no longer live on farms, additives help keep food wholesome and appealing while en route to markets sometimes thousands of miles away from where it is grown or manufactured. Additives also improve the nutritional value of certain foods and can make them more appealing by improving their taste, texture, consistency or color.
>
> Some additives could be eliminated if we were willing to grow our own food, harvest and grind it, spend many hours cooking and canning, or accept increased risks of food spoilage. But most people today have come to rely on the many technological, aesthetic and convenience benefits that additives provide in food. (FDA)

Food manufacturers attempt to increase the shelf life of their products by controlling and preventing deterioration; therefore, additives may be used to *preserve* or combat *microbial* or *enzymatic deterioration*. All living tissue resists microorganism attack to some degree, and additives assist in microbial (pathogens and nonpathogens) protection. The use of additives at the point of manufacture or processing *cannot stop all* foodborne illness though and *cannot guarantee* food safety for the population at large. For example, mishandling of food at restaurants and homes contributes a larger portion of foodborne illness than handling at food processing plants.

A second use of food additives, beside to increase the shelf life, is that they may be included in food *to maintain* or *improve nutritional value*. They *enrich*, *fortify*, or restore what is lost in processing. Additives may add nutrients and correct deficiencies, such as when *iodine* was used to treat goiter, or when the minerals calcium and iron are added to food. Antioxidants such as lemon juice, BHT, BHA, and vitamins A and C are added to control the damaging effect of exposure to oxygen, or vitamin D is added to fortify milk. Many grain products are enriched or fortified with thiamin to prevent the disease beriberi, niacin to control the devastating pellagra, and more recently, with folate to prevent the reoccurrence of neural tube defect. Nutritional fortification is of tremendous benefit to many people.

The first food additive in the United States was iodine. Its function was nutritional—to treat and prevent goiter, common in the Great Lakes and Pacific Northwest regions of the United States. With study it was found that those geographic regions were without *sea*water. Thus the soil, water, and crops in these regions were deficient in iodine, and the problem of goiter in local populations was prevalent. It was added to salt (for the reason that salt was commonly consumed and thus a good vehicle) in 1924 and iodized salt quickly became a common dietary source of iodine.

Food additives play a role in food protection and nutrition fortification. Other roles of additives are as flavor and color *sensory agents*. These agents may be added to food so that it is made *more appealing*. As well, additives may be included in *processing*—for example, as agents to maintain product consistency, to emulsify, stabilize, or thicken a food.

Legislation and Testing for Additives

The *FDA* regulates the inclusion of additives to food products subject to interstate commerce or import, by authority of the Food Additives Amendment (1958) and the Color Additives Amendment (1960) to the Federal FD&C Act of 1938. The *USDA* regulates additives of meat and poultry products.

Approval of Additives

In order to gain approval for use of an additive, manufacturers must petition the FDA and:

- Provide evidence of harmlessness of an additive at the intended level of use
- Provide data from at least 2 years of feeding of at least two animals, male and female (usually dogs and rats)
- Prove the safety, usually by utilizing an outside toxicology laboratory for testing

Manufacturers must show *evidence* that the additive is safe and that it will accomplish its intended effect (show usefulness and harmlessness). Although *absolute safety* of any substance can never be proven, there must be a *reasonable certainty* of no harm from the additive under its proposed use. In the approval evaluation, a "typical" intake level is considered and additives are evaluated on a case-by-case basis.

Beneficial additive approval information includes animal tests and disappearance data. *Animal tests* are conducted to show effects of large doses and lifetime or generational feedings. *Market basket patterns of consumption studies* show disappearance data of food that is both produced and imported. The latter shows, on average, the 7-day intake of an adult male.

If the additive wins approval by the FDA, it is only for use at *specific levels* in *specific products*. For example, certain fat replacements may only be approved for addition to *savory snacks*. Then, subsequent to approval, periodic review of additives based on up-to-date scientific evidence occurs.

The FDA's Adverse Reaction Monitoring System (ARMS) monitors and investigates complaints that are associated with, and related to, food and color additives, specific foods, or vitamin and mineral supplements. The FDA also has an Advisory Committee on Hypersensitivity to Food Constituents. Consumers should read product labels in order to determine specific ingredient information.

Delaney Clause

The ***Delaney Clause*** (named for the congressional sponsor Rep. James Delaney—NY) to the Food Additives Amendment states that *no* additive shown to cause cancer in man or laboratory animals, regardless of the dose, may be used in foods. Proposed additives are not acceptable for use in the food supply if they have been documented to be *carcinogenic* by *any* appropriate test. Such legislation continues to be reviewed, as both finer detection methods to detect minute amounts of a carcinogen, previously *undetected*, and improvements in additive testing over the years have become available.

Such detection and improvement in testing has led to the question of *what is an appropriate test*? For example, is there *any* substance that is *totally* safe at *any* level of ingestion? Or will testing document the presence of carcinogens? The real issue may be in regard to "*risk versus benefit*" of an additive, for an additive may pose a "risk," yet, a risk is *not* a threat to life. On the other hand, a "benefit" of using the additive is that there is an improvement in the condition of a food.

Currently, the FDA must abide by the *mandate* of the Delaney Clause in approving food additives, although the specifics may change in the future. Overall, the goal of *any* ingredient testing is to provide a safe food supply for the public.

Nutrition Labeling Education Act (NLEA)

Further legislation included the Nutrition Labeling and Education Act of 1990 (Chap. 20). It required that all food labels *must list additives*, such as certifiable color additives by the common or usual name. Labels contain valuable information that allows people who may have food or food additive sensitivities to select appropriate food.

Major Additives Used in Processing

Food Ingredients and Packaging Terms.

The FDA Data Standards Council is standardizing vocabulary across the FDA. Therefore, the wording in some terms below may change slightly in the future.

Biotechnology—Refers to techniques used by scientists to modify deoxyribonucleic acid (DNA) or the genetic material of a microorganism, plant, or animal in order to achieve a desired trait. In the case of foods, genetically engineered plant foods are produced from crops whose genetic makeup has been altered through a process called recombinant DNA, or gene splicing, to give the plant desired traits. Genetically engineered foods are also known as biotech, bioengineered, and genetically modified, although "genetically modified" can also refer to foods from plants altered through methods such as conventional breeding. While in a broad sense biotechnology refers to technological applications of biology, common use in the United States has narrowed the definition to foods produced using recombinant DNA. For additional information, see the biotechnology program on the CFSAN Internet.

CEDI/ADI Database—For a large number of food contact substances, CFSAN maintains a database of cumulative estimated daily intakes (CEDIs) and acceptable daily intakes (ADIs). The CEDIs and ADIs are based on currently available information and may be revised when information is submitted or made available. The CEDI/ADI database is updated approximately twice annually. The CEDIs and ADIs are based on currently available information and may be subject to revision on the basis of new information as it is submitted or made available to OFAS.

Color Additive—A color additive is a dye, pigment, or other substance, which is capable of imparting color when added or applied to a food, drug, cosmetic, or the human body. The legal definition can be found in Section 201(t) of the Federal FD&C Act and provides exclusions as well. Color additives for use in food, drugs, and cosmetics require premarket approval. Color additives for use in or on a medical device are subject to premarket approval, if the color additive comes in direct contact with the body for a significant period of time. For additional information, consult the Color Additive Program on the CFSAN Internet.

Colorant—A colorant is a dye, pigment, or other substance that is used to impart color to or to alter the color of a food contact material but that does not migrate to food in amounts that will contribute to that food any color apparent to the naked eye. The term "colorant" includes substances such as optical brighteners and fluorescent whiteners, which may not themselves be colored, but whose use is intended to affect the color of a food contact material (21 CFR 178.3297(a)).

EAFUS—The "Everything Added to Food in the United States" (EAFUS) database is an informational database maintained by CFSAN under an ongoing program known as the Priority-based Assessment of Food Additives (PAFA). PAFA contains administrative, chemical, and toxicological information on over 2,000 substances directly added to food, including substances regulated by the FDA as a direct food additive, secondary direct food additive, color additive, GRAS, and prior-sanctioned substance. In addition, the database contains only administrative and chemical information on approximately 1,000 such substances. Information about the more than 3,000 total substances comprise EAFUS. For a complete listing of EAFUS substances, see the EAFUS list.

Food Additive—A food additive is defined in Section 201(s) of the FD&C Act as any substance the intended use of which results or may reasonably be expected to result, directly or indirectly, in its becoming a component or otherwise affecting the characteristic of any food (including any substance intended for use in producing, manufacturing, packing, processing, preparing, treating, packaging, transporting, or holding food and including any source of radiation intended for any such use), if such substance is not GRAS or sanctioned prior to 1958[1] or otherwise excluded from the definition of food additives.

Food Contact Substance (FCS)—Section 409 of the FD&C Act defines an FCS as any substance that is intended for use as a component of materials used in manufacturing, packing, packaging, transporting, or holding food if such use of the substance is not intended to have any technical effect in such food. Additional information can be found on the Food Contact Substances Notification Program page. There is a hierarchy from **Food Contact Substance** (FCS) through **Food Contact Material** (FCM) to **Food Contact Article** (FCA):

- The **Food Contact Substance** (the subject of an FCN) is a single substance, such as a polymer or an antioxidant in a polymer. As a substance, it is reasonably pure (the chemist's definition of substance). Even though a polymer may be composed of several monomers, it still has a well-defined composition.
- **Food Contact Material** (FCM) is made with the FCS and (usually) other substances. It is often (but not necessarily) a mixture, such as an antioxidant in a polymer. The composition may be variable.
- The **Food Contact Article** is the finished film, bottle, dough hook, tray, or whatever that is formed out of the FCM.

GRAS—"GRAS" is an acronym for the phrase generally recognized as safe. Under sections 201(s) and 409 of the FD&C Act, any substance that is intentionally added to food is a food additive, that is subject to premarket review and approval by FDA, unless the substance is generally recognized, among qualified experts, as having been adequately shown to be safe under the conditions of its intended use, or unless the use of the substance is otherwise excluded from the definition of a food additive. GRAS substances are distinguished from food additives by the type of information that supports the GRAS determination, that it is publicly available and generally accepted by the scientific community, but should be the same quantity and quality of information that would support the safety of a food additive. Additional information on GRAS can be found on the GRAS Notification Program page.

Guidance Document—Guidance documents are documents prepared for FDA staff, applicants/sponsors, and the public that describe the agency's interpretation of or policy on a regulatory issue. Guidance documents include, but are not limited to, documents that relate to the design, production, labeling, promotion, manufacturing, and testing of regulated products; the processing, content, and evaluation or approval of submissions; and inspection and enforcement policies. Guidance documents do not legally bind the public or FDA or establish legally enforceable rights or responsibilities. They represent the agency's current thinking (21 CFR 10.115). A complete listing of CFSAN's guidance documents is available on the Internet.

Indirect Food Additive—In general, these are food additives that come into contact with food as part of packaging, holding, or processing but are not intended to be added directly to, become a component, or have a technical effect in or on the food. Indirect food additives mentioned in Title 21 of the US Code of Federal Regulations (21CFR) used in food contact articles include adhesives and components of coatings (Part 175), paper and paperboard components (Part 176), polymers (Part 177), and adjuvants and production aids (Part 178). Currently, additional indirect food additives are authorized through the food contact notification program. In addition, indirect food additives may be authorized through 21 CFR 170.39.

PAFA—The Priority-based Assessment of Food Additives (PAFA) database is a database that serves as CFSAN's institutional memory for the toxicological effects of food ingredients known to be used in the United States. Currently, PAFA contains oral toxicology information on over 2,100 of approximately 3,300 direct food ingredients used in food in the United States. PAFA also contains minimal information on over 3,200 indirect additives including the names, CAS number, and regulatory information of the indirect additives in the Code of Federal Regulations. The EAFUS list and the Indirect Additive list on the CFSAN Internet consist of selected fields of information generated from PAFA.

Prior-Sanctioned Substance—A substance whose use in or on food is the subject of a letter issued by FDA or USDA offering no objection to a specific use. The prior sanction exists only for a specific use of a substance in food delineating level(s), condition(s), and product(s) set forth by explicit approval by FDA or USDA prior to September 6, 1958. Some prior-sanctioned substances are codified in 21 CFR Part 181.

SCOGS Report—"SCOGS" is the acronym for the Select Committee On GRAS Substances. Beginning in 1972, under a contract with FDA, the Life Sciences Research Office of the Federation of American Societies for Experimental Biology (author added: FASEB) convened the Select Committee, which independently undertook a comprehensive review of the safety and health aspects of GRAS food substances on the FDA's then proposed GRAS list. The Select Committee published its evaluations in a series of reports known as the SCOGS reports. A listing of opinions and conclusions from 115 SCOGS reports published between 1972 and 1980 is available on the CFSAN Internet.

Secondary Direct Food Additive—This term is in the title of 21 CFR 173, which was created during recodification of the food additive regulations in 1977. A secondary direct food additive has a technical effect in food during processing but not in the finished food (e.g., processing aid). Some secondary direct food additives also meet the definition of a food contact substance. For more on food contact substances, consult the Food Contact Substance Notification Program.

Threshold of Regulation (TOR) Exemption—A substance used in a food contact article may be exempted from the requirement of a food additive listing regulation if the use in question has been shown to meet the requirements in 21 CFR 170.39. For details, see 21 CFR 170.39. For a complete listing of the TOR exemptions, consult the Threshold of Regulation inventory on the CFSAN Internet.

[1]This is a shortened definition. The full definition is found in the FD&C Act

See Title 21, Code of Federal Regulations (21 CFR), and the Federal Register containing the regulations and rules appropriate for Food and Color Additive Petitions and Food Ingredient and Food Packaging Notifications.

Additives may be *naturally occurring* or *synthetic*. The additives most commonly used in the United States are ***generally recognized as safe*** (GRAS) flavor agents, along with baking soda, citric acid, mustard, pepper, and vegetable colors, that total greater than 98 % by weight, of all additives in the United States.

Foods may contain ingredients that are added prior to retail sale. Vegetables and fruits, for example, may be treated with pesticides before harvesting, and dyes, fungicides, and waxes may be applied to retard ripening or promote sales. *Sodium nitrites* may be deliberately added to prevent *Clostridium botulinum* growth and preserve color, and *sodium phosphates* may be added for purposes of retaining texture and preventing rancidity.

An additive may serve multiple purposes and therefore be listed in several classes of additives. Some of the major additives used in processing are identified in the following text.

Anticaking Agents and Free-Flow Agents

Anticaking agents and free-flow agents inhibit or prevent lumping and caking in crystalline or fine powders, including salt and baking powder. Various silicates such as *aluminum calcium silicate*, *calcium silicate*, and *silicon dioxide*, as well as *tricalcium phosphate*, are examples of anticaking agents that may be added to powdered food.

Antimicrobials

Antimicrobials inhibit the growth of pathogenic or spoilage organisms. Most recognizable is *sodium chloride*, common table salt. Organic acids such as *acetic acid*, *benzoic acid*, *propionic acid*, and *sorbic acid* are utilized with low pH level food items. *Nitrites* and *nitrates* are used as antimicrobials in meat products to combat *C. botulinum*, while *sulfites* and *sulfur dioxide* are added to fruit juices and wine.

Antioxidants

> "Antioxidants—minerals, vitamins and phytochemicals, such as flavonoids—may protect against the free radicals that are implicated in cell damage and aging, heart disease, cancer and other diseases. . . . They are found in fruits and vegetables, nuts, grains, milk products, teas, legumes, spices, herbs and some meats, poultry and fish.
>
> They are defined as substances that may protect human cells against the effects of highly unstable free radicals and, therefore, potentially disease-producing cell damage (U.S. National Library of Medicine). (Spano 2012)"

The presence of oxygen may result in rancidity, including the formation of toxic products, or deterioration of color, flavor, and nutritional values. Thus, antioxidants are added to combine with available oxygen to halt oxidation reactions. For example, antioxidants prevent or limit rancidity in fats and foods containing *fats*, stabilizing foods by preventing or inhibiting *oxidation of unsaturated fats and oils*, colors, and flavorings. Also, antioxidants prevent browning of cut *fruit*, by enzyme-catalyzed oxidation in *enzymatic oxidative browning* (Chap. 7).

Many antioxidants occur *naturally*, such as *ascorbic acid* (vitamin C), the *tocopherols* (vitamin E), *citric acid*, and some *phenolic compounds*. The most widely used *synthetic* antioxidants are *BHA* (butylated hydroxyanisole), *BHT* (butylated hydroxytoluene), *TBHQ* (tertiary butylhydroquinone), and *propyl gallate*. These four synthetic antioxidants may be used alone or in combination with one another, or another additive to control oxidation. They may be used to prevent oxidation in fat-containing *food* (up to 0.02 % of the fat level) or in *food packaging*, such as in whole grain cereal boxes.

Many meat processors add *sodium ascorbate* or *sodium erythorbate* (*erythorbic acid*) to cured meat to maintain processed meat color and to inhibit the production of nitrosamines from nitrites. (Nitrites are antioxidants that are also used as curing agents.) Ethylenediaminetetraacetic acid (*EDTA*) may be used as an antioxidant or for other purposes in food (see Sequestering Agents).

Bleaching and Maturing Agents

As freshly milled flour ages, it naturally whitens and improves in baking quality. Bleaching and/or maturing agents are *added* to flour either during or after the milling process to whiten and/or speed up the aging process. *Benzoyl peroxide* is added to bleach the yellowish carotenoid pigment to white, and *chlorine dioxide* is added to mature flour for better baking performance. *Bromates* may also be common as is *hydrogen peroxide* that is used to whiten milk for certain types of cheese manufacture.

Bulking Agents

Bulking agents such as *sorbitol*, *glycerol*, or *hydrogenated starch hydrolysates* are used in small amounts to provide body, smoothness, and creaminess that supplement the viscosity and thickening properties of hydrocolloids (colloidal material that binds water). They provide an oily or fatty mouthfeel and are frequently used in foods where sugar is reduced or absent. *Polydextrose* (Chap. 12) is an example of a bulking agent used when calories are limited in foods. It contains 1 cal g and is made from a mixture of glucose, sorbitol, and citric acid (89:10:1).

Coloring Agents

A color additive is any dye, pigment, or substance that imparts color when added or applied to a food, drug, cosmetic, or to the human body. The term "FD&C" is applied to *food color additives* approved by the FDA for food, drug, and cosmetic usage, "D&C" is used for approved *drug and cosmetic coloring agents*, and "External D&C" is granted to approved *color additives applied externally*. The *synthetic* coloring agents are assigned FD&C classifications by initials, the shade, and a number; for example, FD&C Red #40 and FD&C Yellow #5.

Table 17.1 Colors exempt from certification (uncertified)/natural

Annatto extract (yellow-reddish orange)	Grape color extract[a]
B-Apo-8′-carotenal[a] (orange)	Grape skin extract[a]
Beta-carotene	Paprika
Beet powder	Paprika oleoresin
Canthaxanthin	Riboflavin
Caramel color (brown)	Saffron
Carrot oil	Titanium dioxide[a]
Cochineal extract (red)	Turmeric (yellow)
Cottonseed flour, toasted, partially defatted, cooked	Turmeric oleoresin
Ferrous gluconate[a]	Vegetable juice
Fruit juice	

[a]Restricted to specific uses

(See http://www.cfsan.fda.gov/~lrd/cfr70-3.html.) It may be synthesized, isolated, or extracted.

Coloring agents are added to foods because of the sensory appeal they provide, for the purpose of making processed foods look more appetizing. Some argue that ADHD, food allergies, and brain cell damage may be a result of added colorants.

Natural colors may be lost in processing and therefore others may be added. For example, colors are used in baked products; candies; dairy products such as butter, margarine, and ice cream; gelatin desserts; jams; and jellies in order to improve their appearance. It has been said that people "eat with their eyes" as well as their palates!

There are thousands of foods that use colors to make them look appetizing and attractive.

The primary reasons for adding coloring agents include the following:

- Offsetting color loss due to exposure to air, light, moisture, and storage
- To correct natural variations in color or enhance color
- To provide visual appeal to wholesome and nutritious foods
- To provide color to foods that would otherwise be colorless, including "fun foods" and special foods for various holidays

Pigments may be derived from *natural* sources such as plant, mineral, or animal sources primarily the former (Table 17.1) and, if so, are *exempt* from FDA *certification* (see below), although they are still subject to *safety* testing prior to their approval for use in food. Some of the same ingredients added to foods for their health benefits also offer "natural" (uncertified) coloring. These include anthocyanins, carotenoids, chlorophylls, foods such as beets (betalains), cabbage, tomatoes (lycopene), and a number of other flowers, fruits, and vegetables.

Browning attributable to the *Maillard reaction* may successfully occur with *acids* and *proteins*. Browning may also occur due to *sugars* in *caramelization* (Brantley 2012).

Synthetic coloring agents are generally less expensive than natural colorants, are more intense, and have better coloring power, uniformity, and stability when exposed to environmental conditions such as heat and light. They may be water-soluble, or made water *in*soluble by the addition of aluminum hydroxide. Only small quantities of granules, a paste, powder, or solution are needed in foods to achieve the desired coloring effect. Each batch of synthetic food color must be tested by both the manufacturer and the FDA prior to gaining approval. Such testing assures safety, quality, consistency, and strength of the color additive.

The FDA permitted nine "*certifiable*" colors in the 1906 Food and Drug Law under a *voluntary* certification program; seven of those were

Table 17.2 Certifiable colors—synthetic

FD&C Blue No. 1 (bright blue)	FD&C Yellow No. 5 (lemon yellow)—tartrazine; second most widely used food dye
FD&C Blue No. 2 (royal blue)—Indigotine	
FD&C Green No. 3 (sea green)—minimally used	FD&C Yellow No. 6 (orange) Orange B[a] Citrus Red No. 2[a]—used on some orange skins
FD&C Red No. 3 (orange-red)—erythrosine	
FD&C Red No. 40 (cherry-red)—most widely used food dye	

[a]Restricted to specified uses other than foods

approved for addition to *food* (Table 17.2). Certification became *mandatory* in 1938, with authority for testing passing to the USDA. Today, the term "certifiable" food color refers to color additives that are synthetic, or *man-made*, not natural.

In 1982, The National Institutes of Health concluded that there was no scientific evidence to support the claim that coloring agents or other food additives caused hyperactivity. Later, in 2007 a report in the *Lancet* journal by researchers at England's University of Southampton again implicated some synthetic coloring agents, and it led to a European warning label (currently not adopted by the United States) on foods and beverages containing the suspect colors (Decker 2012).

Varieties of fruit and vegetables may be "…processed with only water, resulting in a full spectrum of colors, from yellow and red to blue and green" (Food Product Design 2012).

A reminder from above:

Colorant—A colorant is a dye, pigment, or other substance that is used to impart color to or to alter the color of a food contact material but that does not migrate to food in amounts that will contribute to that food any color apparent to the naked eye. The term "colorant" includes substances such as optical brighteners and fluorescent whiteners, which may not themselves be colored, but whose use is intended to affect the color of a food contact material (21 CFR 178.3297(a)).

Artificial colors have been the cause of controversy since the 1970s when a pediatrician first identified a correlation of intake to children's behavior." Later, studies "…showed a correlation between artificial food colors and additives and exacerbated hyperactive behavior in children. Even though many medical experts questioned the study's protocol, it still stirred up consumer controversy. (Berry 2013)

Curing Agents

Curing agents impart color and flavor to foods. *Sodium nitrate* and *nitrite* contribute to the color, color stability, and flavor of cured meats such as bacon, frankfurters, ham, and salami. They also have antimicrobial properties to control *Clostridium botulinum* bacteria, and in fact, since nitrite use in cured meats became common, there have been no reported cases of botulism linked with cured meats. Nitrites also inhibit the growth of *C. perfringens and Staphylococcus aureus*, as well as other nonpathogens during storage of cured meats.

(A nitrite concern is that nitrites react with certain amines to produce carcinogenic nitrosamines. Repeated testing of cured meat products showed that nitrosamines were absent or present at very low levels. In fact, endogenous human saliva usually contains more nitrite than has been detected in cured meat. Therefore, with these test results, the FDA permits use of nitrates, and nitrites at low levels, yet has encouraged research on alternative ways of preserving meats.)

Dough Conditioners/Dough Improvers

Dough conditioners modify the starch and protein (thus, gluten) components of flour. They promote the aging process and improve both dough handling (such as in bread-making machinery) and baking qualities (more uniform grain and increased volume). *Ammonium chloride*, *potassium bromate* (*bromination*), *diammonium phosphate*, and *calcium* or *sodium stearoyl lactylate* are processing aids that are examples of dough conditioners.

Some enzymes that provide food for yeast and emulsifiers may be included in this class of additives. Conditioners may include iodates, a dietary source of iodine.

Edible Films

Edible films become part of food and are therefore subject to FDA approval. Examples include some *casings*, such as in sausage, and *edible waxes*, such as those applied to fruits and vegetables. The waxes function to improve or maintain appearance, prevent mold, and contain moisture while allowing respiration. Food may be coated with a thin layer of polysaccharides such as *cellulose*, *pectin*, *starch*, and *vegetable gums* or proteins such as *casein* and *gelatin* in order to extend shelf life.

Emulsifiers

Emulsifiers are a class of surface-active agents that improve and maintain texture and consistency in a variety of foods containing fat/oil and water. They contain both a hydrophobic molecular end, which is usually a long-chain fatty acid, and a hydrophilic end and so act as a bridge between two immiscible liquids, forming an emulsion. For example, emulsifiers maintain a uniform dispersion by keeping water and oil fractions of a mixture together and they prevent large crystalline formations in products such as ice cream.

Lecithin, *monoglycerides*, *diglycerides*, and *polysorbates* (polysorbates 60, 65, and 80) are examples of emulsifiers.

Enzymes

Enzymes are nontoxic protein substances that occur *naturally* in foods or may be produced by *microorganisms* or *biotechnology* to catalyze various reactions. They are easily inactivated by a specific pH and temperature. Although the presence of some enzymes may produce negative quality changes, other enzymes are often intentionally added to foods for their beneficial effect. Microorganisms are responsible for producing some of the enzymes desired in foods; thus, the microbes may be intentionally added to food.

Some examples of enzymes that are additions to other foods include *bromelain* (from pineapples), *ficin* (from figs), and *papain* (from papayas). These enzymes act as meat tenderizers of muscle tissue or connective tissue. *Amylases* hydrolyze starch in flour and are used along with acids in the production of corn syrup. *Invertase* is used to hydrolyze sucrose and prevent its crystallization. *Pectinases* clarify pectin-containing jellies or juices; *proteases* may be used as meat tenderizers, to create cheeses from milk (rennin), and to produce soy sauce. *Glucose oxidase* is added to foods such as egg whites in order to prevent the Maillard browning.

Fat Replacers

Fat replacers include carbohydrate, fat, or protein-based replacers. They include the following:

- Hydrocolloids, the long-chain polymers that thicken or gel
- Hemicelluloses, a plant polysaccharide
- β-Glucans, a subgroup of hemicelluloses (used in milk)
- Soluble bulking agents
- Microparticulated cellulose, egg, milk, or whey proteins (such as Simplesse®)
- Composite materials such as gums with milk or whey solids, or cellulose in combination with carboxymethyl cellulose
- Various functional blends of corn syrup solids, emulsifiers, modified food starch, nonfat milk solids, and vegetable protein
- Sucrose polyester (Olean®)

These ingredients are fat replacers that can be used in combination to provide a fatty or oily mouthfeel, or used alone.

The manipulation of omega-6-rich oils from corn, cottonseed, safflower, sunflower, and soy oils may create unhealthy convenience foods which may lead to brain inflammation and blood clots.

Firming Agents

Firming agents, such as *calcium chloride*, act on plant pectins to control the softening that may accompany the canning process of fruits or vegetables.

Flavoring Agents

Flavoring agents are the *largest single group* of food additives. Food and beverage applications of flavors include dairy, fruit, nut, seafood, spice blends, vegetables, and wine flavoring agents. They may complement, magnify, or modify the taste and aroma of foods. There are over 1,200 different flavoring agents used in foods to create flavor or replenish flavors lost or diminished in processing, and hundreds of chemicals may be used to simulate natural flavors. *Alcohols*, *esters*, *aldehydes*, *ketones*, *protein hydrolysates*, and *MSG* are examples of flavoring agents.

Natural flavoring substances are extracted from plants, herbs and spices, animals, or microbial fermentations. They also include essential oils and oleoresins (created by solvent extract with solvent removed), herbs, spices, and sweeteners.

Synthetic flavoring agents are chemically similar to natural flavorings and offer increased consistency in use and availability. They may be less expensive and more readily available than the natural counterpart, although they may not adequately simulate the natural flavor. Some examples of synthetic flavoring agents include *amyl acetate*, used as banana flavoring, *benzaldehyde*, used to create cherry or almond flavor, *ethyl butyrate* for pineapple, *methyl anthranilate* for grape, *methyl salicylate* for wintergreen flavor, and *fumaric acid*, which is an ideal source of tartness and acidity in dry foods.

Flavor enhancers such as *monosodium glutamate* (MSG) intensify or "bring out," enhance, or supplement the flavor of other compounds in food; they have a taste outside of the basic sweet, sour, salty, or bitter. Monosodium glutamate was chemically derived from seaweed in the early 1900s and may be manufactured commercially by the fermentation of starch, molasses, or sugar.

MSG is a GRAS substance to which a small percentage of the public *is* severely allergic. It is important for some individuals to omit all MSG. Therefore, MSG must be identified on food labels when it is included in food. However, it may be a portion of another compound, such as "hydrolyzed vegetable protein" (which may contain up to 20 % MSG); in that case, a wise consumer should know that MSG is not always stated on the label. In a summary of several decades of research and one century of use, arguments *against* the GRAS status of MSG are overwhelmed by evidence *in support of* MSG.

Other examples of flavoring agents include the common salt (sodium chloride) and sugar (sucrose), corn syrup, aspartame (also a nutritive sweetener), autolyzed yeast (also a flavor enhancer), essential plant oils (such as citrus), ethyl vanillin and vanillin (a synthetic flavor compound), extracts (such as vanilla), glycine, mannitol (nutritive sweetener), saccharin (non-nutritive sweetener), and sorbitol (nutritive sweetener).

Fumigants

Fumigants are used to control pests and molds.

Humectants

Humectants are added to foods such as candy or shredded coconut in order to retain moisture content. The four polyhydric alcohols that are used to improve texture and retain moisture due to their affinity for water are *glycerol*, *mannitol*, and *sorbitol* which also taste moderately sweet and *propylene glycol*. These may be functional in diet beverages, candy, stick chewing gum, and other foods to provide texture and sweetness. In order to prevent moisture absorption, *mannitol* may be "dusted" on chewing gum. *Glycerol monostearate* is another commonly used humectant.

Irradiation

Irradiation is subject to FDA approval as an *additive*. Only specific foods, dosages, and irradiation sources are approved to kill microorganisms such as *E. coli*, *Salmonella*, *Trichinella*, and insects.

Leavening Agents

Leavening agents leaven batters and doughs by producing or stimulating CO_2. Inorganic salts such as ammonium or phosphate salts such as *sodium acid pyrophosphate* may be combined with yeast, a biological leaven. *Sodium bicarbonate* (baking soda) or *ammonium bicarbonate* reacts with an acid to chemically produce CO_2.

Lubricants

Lubricants such as mineral hydrocarbons (nonstick cooking spray) may be added to food contact surfaces to prevent foods from sticking and may become a part of the food.

Nutrient Supplements

Nutrient supplements are added to provide essential nutrients that are lost in processing or lacking in the diet. The addition of nutrients maintains or improves the nutritional quality of food. Common foods such as milk, cereal, flour, margarine, and salt contain added vitamins and minerals. For example, iodine may be added to salt, and B vitamins (including thiamin, riboflavin, niacin, folate) and iron are added to grain products. This improves the nutritional status of individuals who might not otherwise obtain essential nutrients. Oftentimes, the nutrients may be encapsulated for addition to food products and thus more intact.

Some examples of nutrient supplementation include the following: vitamins A, B, C, D, and E; beta-carotene; minerals such as *iodine* and *zinc*, calcium pantothenate, and other forms of calcium; and ferrous gluconate as a nutrient and coloring agent—*ferric orthophosphate*, *ferric sodium pyrophosphate*, and *ferrous fumarate* to supply iron. *Ferrous lactate* is also a GRAS color fixative for ripe olives. Phosphates such as *phosphoric acid* acidifies and flavors food; *sodium aluminum phosphate* is a leavening agent; *calcium* and *iron phosphate* act as mineral supplements to food.

Nutrient Supplementation.

December 2009 (Volume 109, Issue 12, Pages 2073–2085)

ABSTRACT

It is the position of the American Dietetic Association that the best nutrition-based strategy for promoting optimal health and reducing the risk of chronic disease is to wisely choose a wide variety of nutrient-rich foods. Additional nutrients from supplements can help some people meet their nutrition needs as specified by science-based nutrition standards such as the Dietary Reference Intakes. The use of dietary supplements in general, and nutrient supplements in particular, is prevalent and growing in the United States, with about one third of adults using a multivitamin and mineral supplement regularly. Consumers may not be well informed about the safety and efficacy of supplements and some may have difficulty interpreting product labels. The expertise of dietetics practitioners is needed to help educate consumers on the safe and appropriate selection and use of nutrient supplements to optimize health. Dietetics practitioners should position themselves as the first source of information on nutrient supplementation. To accomplish this, they must keep up to date on the efficacy and safety of nutrient supplements and the regulatory issues that affect the use of these products. This position paper aims to increase awareness of the current issues relevant to nutrient supplements and the resources available to assist dietetics practitioners in evaluating the potential benefits and adverse outcomes regarding their use.—The Academy of Nutrition and Dietetics

Dietary supplements may be foods, drugs, or natural health products according to a country's specific definition. They are defined under the Dietary Supplement Health and Education Act of 1994 (DSHEA) as a product that is intended to *supplement the diet* and contains dietary ingredients such as amino acid, mineral, vitamin, an herb, or other botanicals. These supplements may be concentrates or extracts.

The form of the dietary supplement could be a capsule, pill, tablet, powder, or liquid form, and the product must be labeled as a "dietary supplement." It is a different item than food and is not for use as a conventional food. As a *supplement*, it is not supposed to be the sole item of a meal or diet.

See CSPI, *Center for Science in the Public Interest*—http://www.cspinet.org. USDA Food and Nutrition Information Center—http://www.nal.usda.gov/fnic.

pH Control Substances

Natural or synthetic *acid* or *alkali* ingredients change or maintain the initial pH of a product. For example, acidulents flavor, preserve, and regulate pH. The acid ingredients regulate by lowering the pH and preserve foods by inhibiting microbial growth. Regardless of the acid level of food ingredients, food acids are incorporated into foods in order to maintain a constant acid level. *Natural acids* include the following: *acetic acid* or vinegar; *citric acid*, from citrus, that controls unwanted trace metals otherwise catalyzing oxidation reactions; *malic acid* (an organic acid from apples and figs); and *tartaric acid* (a weak acid). These acids may be added to foods to impart flavor and control tartness.

Lactic acid, present in almost all living organisms, is an acidity regulator and is used in balancing the acidity in cheesemaking, as well as adding tartness to many other foods. The acid salt *calcium propionate* is added to control pH of breads. *Sodium lactate* (the salt of lactic acid) may be used in processed meat and poultry products.

Examples of *alkaline* ingredients include *sodium bicarbonate* (baking soda), an ingredient that balances the acid component of baking powder, *sodium hydroxide*, used in modified starches, and *potassium hydroxide*. Alkaline compounds are used to neutralize excess acid that could otherwise produce unwelcome flavors. In food they leaven and soften hard water.

Preservatives

Preservatives are classified as either *antimicrobial* agents that prevent microbial growth (bacteria, mold, yeast), or *antioxidants* that halt undesirable oxidative changes in food. They are used to delay natural deterioration, thus extending the shelf life of foods.

A preservative may be used alone or in combination with other additives or preservation techniques such as cold temperature storage, heat preservation, or dehydration. Preservation is offered by the use of *salt* or *sugar*, which competes with bacteria for water and therefore lowers the water activity (A_w) of the food. *Calcium or sodium propionates* and *potassium sorbate* are additives used to control mold and bacterial growth, such as the bacilli that causes "rope" in breads. *Sodium benzoate* inhibits mold and yeast growth in condiments, fruit juices, and preserves.

Sulfur dioxide inhibits unwanted enzymatic oxidative browning in fruits and vegetables; it also prevents wild yeast growth in the wines used to produce vinegar and is legally used on grapes. However, a small number of individuals are severely allergic to sulfites, thus, they were banned in 1986 from addition to products not subject to further cooking (e.g., many salad bar items). They are still used in dehydrated fruits such as dry apples to prevent browning, but they are added prior to dehydration.

Nitrites, such as *sodium nitrite*, are effective preservatives, preventing the growth of *C. botulinum*, adding flavor and retaining meat color in cured meats. *Antibiotics* are incorporated into animal feed and function as preservatives; however, they may *not* be added to human food. Some weak acids or *acid salts* and *chelating agents* (which tie up unwanted traces of metal) are utilized as preservatives.

Pre- and Probiotics

Prebiotics are nondigestible carbohydrates that act as food for probiotics. When probiotics and prebiotics are combined, they form a synbiotic. Fermented dairy products, such as yogurt and kefir, are considered synbiotic because they contain live bacteria and the fuel they need to thrive. [the term "synbiotic" should be used only if the net health benefit is synergistic—the United Nations Food & Agriculture Organization (FAO)]

Probiotics are found in foods such as yogurt, while prebiotics are found in whole grains, bananas, onions, garlic, honey and artichokes. In addition, probiotics and prebiotics are added to some foods and available as dietary supplements. (http://www.mayoclinic.com)

Propellants

Propellants and aerating agents provide *pressure* needed for aerosol can products to be expelled and to *add air* to a product. Carbon dioxide, nitrogen, nitrous oxide, and other gases may be used in aerosol containers, such as containers of "whipped topping."

Sequestrants

Sequestrants are known as chelating agents or metal scavengers. They are substances that *bind* or combine with trace amounts of unwanted metals such as copper and iron, making them unavailable for participating in negative reactions such as deterioration in food. They form *inactive complexes* with metallic ions during processing and storage and therefore prevent metals from catalyzing reactions of fat oxidation, pigment discoloration, and flavor or odor loss and from causing cloudy precipitates in beverages such as tea. They protect vitamins from oxidation.

Examples of sequestrants include citric and *malic acid*, *ethylenediaminetetraacetic acid* (*EDTA*), and *polyphosphates*.

Solvents

Solvents are used to separate substances by dissolving a substance in the solvent and removing the solvent.

Stabilizers and Thickeners

Stabilizers and thickeners function variously in food. For example, they provide consistency and texture to many foods. They are water-holding substances added to stabilize, gel, or thicken foods by absorbing some of the water present in foods. They increase viscosity of the final product, prevent ice crystal formation, or form gels. Stabilizers or thickeners are added for appearance and mouthfeel, to protect emulsions, and to contain volatile oils that would otherwise evaporate. Most are polysaccharides, and pectin, gums, and mucilages are a source of soluble fiber to the diet.

Soluble fiber in the daily diet is of benefit to health in several manners:

1. Reduces "bad" LDL cholesterol—as soluble fiber travels through the GI tract, it binds to bile acids which are made of cholesterol and therefore limits the **amount of cholesterol absorbed, hence lowering blood cholesterol.**
2. Aids digestive system—manages "regularity" and protects against colon and colorectal cancer, diverticular disease, irritable bowel syndrome, and gallstone development.
3. Manages diabetes mellitus—slows the absorption of glucose from the small intestine.

This category of food additives includes the following:

- *Alginates* (from kelp).
- *Carrageenan* (used in foods since 400 A.D., a seaweed derivative) (see Chap. 5).
- To provide visual appeal to wholesome and nutritious foods.
- *Cellulose* may be reacted with derivatives of acetic acid to form sodium carboxymethyl cellulose (CMC), which is used to prevent sugar crystallization.
- *Dextrins*.
- The hydrocolloids (colloidal material that holds water); *gelatin* (the protein from animal bones, hoofs, etc.); *gums* such as arabic, guar, and locust beans; and *pectin*. Some beneficial effects of these include an inclusion of *soluble fiber* in the diet. Manufacturers use *pectin* as a stabilizer and thickener in candies, frozen desserts, jams, jellies, and preserves. *Gums* and *mucilages* offer soluble fiber too. They are found in oatmeal, oat bran, and dried beans.
- Propylene glycol.
- *Protein derivatives* such as casein or sodium caseinate, and hydrolyzed vegetable protein.
- *Starches* (including amylose and modified starches) allow oils, water, acids, and solids to remain well mixed by the addition of natural or chemically modified starches.

Various stabilizers and thickeners are commercially available for use by the healthcare industry in preparation of food items such as pureed foods. The stabilizers xanthan and/or carob bean and/or guar gum may be utilized in commercial food production.

Regarding guar gum, it has been reported that food manufacturers are facing the dilemma of using it in their formulations as the demand for it has rapidly risen as much as 1,600 %. This is attributed to its nonfood use in the fluids used in hydraulic fracturing (fracking) for energy production (Adkins and Nicoll 2013). Water-soluble gelling agents such as guar gum increase viscosity of the fracturing fluid to render it more efficiently in hydraulic fracturing—the process of fracturing various layers of rock by a pressurized liquid.

Hydraulic fracturing (also known as "frac'ing") is the process of creating fissures in underground formations to allow natural gas to flow. During frac'ing, water, sand and other additives are pumped under high pressure into the shale formation to create fractures. Frac fluid is approximately 99.5 % water and sand, with a small amount of special-purpose additives. The sand is used to "prop" open the newly created fractures, which allows the natural gas to flow into the wellbore and up to the surface.—***Barnett Shale Energy Education Council [BSEEC]***

The FSIS allows specific additions to meat. For example, *carrageenan* and locust bean gum may be used.

Carrageenan is a hydrocolloid from red seaweed—Rhodophyceae. Several common species are used either to provide gel strength or viscosity. Carrageenan works best in a neutral pH environment such as milk. It holds insoluble ingredients in suspension and works well in freeze/thaw applications. It also has "... applications, including baked goods, processed cheeses, pasteurized whole egg products, dairy-based salad dressings, gelatin-replacement applications, batters and confections" (Akins and Nicoll 2012).

Surface-Active Agents

Surface-active agents are organic compounds that are used in food systems to reduce the surface tension or forces at the surface of a liquid. Dispersion into food mixtures is resisted if the forces attracting surface molecules of a liquid to one another is not reduced. Classifications of surface-active agents include wetting agents, lubricants, dispersing agents, and emulsifiers. For example, *wetting agents* reduce surface tension in chocolate milk mixes, so that particles to be mixed may absorb water more easily and mix into the milk. *Emulsifiers* enable two ordinarily immiscible substances to combine—for example, oil and water—and they also improve texture.

Sweeteners

Sweeteners are added to many foods and beverages. The disaccharide *sucrose* (table sugar) is a common food additive. *Fructose* is one of the components of sucrose. It is two times as sweet as sucrose and will not crystallize out of solution as does sucrose. It is the most water-soluble sugar and is used to create a very sweet solution. It is hygroscopic and, therefore, may function as a humectant.

Lactose (milk sugar) and *maltose* (malt sugar) are often food additives. *Corn syrup*, *high-fructose corn syrup*, *honey*, *maple syrup sugar*, and *molasses* are other examples of sweeteners used as food additives. *Invert sugar* is the 50:50 mixture of glucose and fructose produced by enzymatic or acid treatment of sucrose. Its use prevents sugar crystallization, for example, in the liquid center of chocolate-covered cherries.

Sweeteners, Alternative (Chap. 14)

An assortment of low-calorie, artificial sweeteners and sugar alcohols that are added to foods includes those listed in the following. For more detailed information on sweeteners, see Chap. 14.

Acesulfame K—used alone or in combination with other sweeteners. It is a synthetic derivative of acetoacetic acid that is not metabolized by the body and is therefore excreted unchanged. Acesulfame K is stable in heat and is many times sweeter than sucrose, with no bitter aftertaste.

Aspartame—a synthetic dipeptide (aspartic acid, phenylalanine). It is in products used by diabetics; however, it cannot be used by phenylketonuric (PKU) individuals. It is not stable in heat unless encapsulated and cannot substitute for sucrose when the texture of the food depends on solids content.

Cyclamates—of sodium, calcium, magnesium, and potassium, approximately 30 times sweeter than sucrose (not used in the United States).

Saccharin—used as the calcium or sodium salt. It cannot be used as a substitute for sucrose when the texture of the food depends on solids content.

Sugar alcohols (polyols)—include mannitol, sorbitol, xylitol, and erythritol: Nectresse®.

Stevia—the leaves of the stevia plant.

Sucralose—trichloro derivative of sucrose; the only *noncaloric* sweetener made from sugar.

Nutrient Supplements in Food

As previously acknowledged, the nutritive value of foods may be improved either by replacing nutrients lost in processing or lacking in the diet. The addition of *vitamins*, such as vitamin C, and *minerals*, such as calcium, is often made to common foods. Further detail appears in this section.

Food processors may choose to add any number of nutrient additives, at varying levels, to their food products. Products are enriched when nutrients that were *lost during processing* are replaced to levels comparable to the original levels. ***Enrichment*** is designed to prevent inadequacies in certain segments of the population, and it is the addition of nutrients to achieve established concentrations specified by the standards of identity. ***Fortification*** is the addition of nutrients (the same or different ones) at levels *higher* than those found in the original or comparable food. It can correct existing deficiencies in segments of the population, such as with the addition of calcium. Breakfast cereals, breakfast bars, and fruit drinks are prominent examples of fortification that provide needed nutrients to many individuals.

In addition to the enjoyment of eating, consumption of a varied diet offers many health benefits to the consumer, including nutrients such as vitamins and minerals as well as nonnutrient compounds (such as phytochemicals) that play an important role in reducing risks of certain diseases. For additional information, the reader is referred to the FDA's regulation of dietary supplements as seen in part below (Food and Drug Administration):

The Regulation of Dietary Supplements.

How are dietary supplements regulated?

The U.S. Food and Drug Administration (FDA) is responsible for regulating dietary supplements through its Center for Food Safety and Applied Nutrition. The Food Drug and Cosmetic Act (FDCA) as amended by the Dietary Supplement Health and Education Act of 1994 (DSHEA) is the law that regulates dietary supplements. Under FDCA, the United States Pharmacopeia and National Formulary (USP–NF) are specifically recognized as providing specifications for dietary supplements. Adherence to these standards, however, is voluntary. Dietary supplement manufacturers are not legally required to meet these specifications.

The FDCA regulates dietary supplements as foods. Under the law, supplement regulations are the same as those that cover conventional foods.

- Dietary supplement manufacturers do not need to register with FDA, or obtain FDA approval before producing or selling their products.
- Prior to marketing a product, manufacturers are responsible for ensuring that a dietary supplement (or a new ingredient) is safe before it is marketed. FDA has the authority to take action against unsafe dietary supplement products.
- Manufacturers must ensure that their product label information is truthful and not misleading.

More information
http://www.cfsan.fda.gov/~dms/supplmnt.html
FDA is responsible for regulating dietary supplements through its Center for Food Safety and Applied Nutrition.

Endorsement of Nutrient Supplementation in Foods

The American Medical Association Council on Foods and Nutrition Board has set the following recommendations for endorsement of nutrient supplementation in foods:

1. The intake of the nutrient is below the desirable level in the diets of a significant number of people.
2. The food used to supply the nutrient is likely to be consumed in quantities that will make a significant contribution to the diet of the population in need.
3. The addition of the nutrient is not likely to create an imbalance of essential nutrients.
4. The nutrient added is stable under proper conditions of storage.
5. The nutrient is physiologically available from the food.
6. There is a reasonable assurance against excessive intake to a level of toxicity.

See http://www.uth.tmc.edu/nutrition/history.htm for additional information. "The word diet is derived from the Greek *daita*, which translates to healthful living according to proper selection of food."

Vitamins and Minerals Manufactured for Addition to Foods

Numerous vitamins and minerals are prepared for addition to foods. Among the category of *water-soluble vitamins*, ascorbic acid (vitamin C) may be added to foods in order to prevent oxidation, to prevent the formation of nitrosamines in cured meats, and to improve nutritive value, especially in beverages. The B-complex vitamins may be used as nutrient additives. *Thiamin hydrochloride* and *thiamin nitrate* (B_1), *riboflavin* and *riboflavin-5′-phosphate* (B_2), and several sources of *niacin* (B_3) are FDA approved and commercially available to processors for addition to foods.

Vitamin B_6 (pyridoxine, pyridoxal, and pyridoxamine), B_{12} (cyanocobalamin), pantothenic acid, folacin, and biotin are also manufactured for use in food. Folic acid (folacin, folate) addition has been required by the FDA, for addition to flour and flour-based products. Folate has been shown to prevent neural tube defect.

The *fat-soluble vitamins* that may be added to foods are the carotenoids, vitamin A precursors, tocopherols (vitamin E), and vitamin D.

Additionally, among *minerals*, there are three major minerals and six trace minerals used in foods. Of the major minerals, calcium, magnesium, and phosphorus may be added to a number of foods. Calcium is now more commonly added to orange juice. Copper, fluoride, iodine, iron, manganese, and zinc are trace minerals used as food additives. Chromium, potassium, molybdenum, selenium, and sodium may not have Reference Daily Intakes (RDIs), although rather "safe and adequate intakes." Labels do not state these values, yet, rather, labels may state "reduced sodium" and so forth if the statement is in compliance with NLEA regulations.

Functional Foods (See More in Appendices)

Eating modified foods may supply health benefits beyond the traditional nutrients a food contains. Therefore, food products may be modified by the addition of nutrients not inherent to the original counterpart. A newly evolving area of food and food technology is ***functional foods***, which are defined as:

A food ingredient or modified food that may provide a health benefit beyond any traditional nutrients it contains.

See http://www.functionalfoodscenter.net/index.html http://www.eatright.org/.

The Journal of *Functional Foods in Health and Disease* (*FFHD*) is a peer-reviewed open access journal that covers various aspects of functional foods, bioactive compounds, and chronic diseases and scientific policies related to functional foods. The journal will strive to develop concepts that help to understand the mechanisms of disease and creation of specific functional and medicinal foods in the prevention and management of various diseases. This is an existing serial which has been published under the title Functional Foods in Health and Disease since December 3, 2010 (http://www.functionalfoodscenter.net/).

Phytochemicals (See More in Appendices)

Phytochemicals—Non-nutrient substances from plants (plant chemicals) may become useful as food additives because they may play an important role in reducing risk of cancers. They are naturally available in the diet and are currently in supplement form. Phytochemicals are defined as:

Functional food components, along with phytochemicals, have been associated with the treatment and/or prevention of at least four of the leading causes of death in the United States—cancer, diabetes, cardiovascular disease, and hypertension. They have been associated with the treatment and/or prevention of other medical maladies including neural tube defect and osteoporosis, as well as abnormal bowel function and arthritis. (ADA Position of the American Dietetic Association 1995)

It is the position of the American Dietetic Association (ADA) that specific substances in foods (e.g., phytochemicals and naturally occurring components and functional food components) may have a beneficial role in health as part of a varied diet. The Association supports research regarding the health benefits and risks of these substances. Dietetics professionals will continue to work with the food industry and government to ensure that the public has accurate scientific information in this emerging field (Academy of Nutrition and Dietetics).

Nutraceuticals (more in Appendices)

The term nutraceutical is not recognized by the FDA and is outside FDA regulations because of the following:

Foods are defined as "products primarily consumed for their taste, aroma, or nutritive value."

Drugs are defined as "intended for use in the diagnosis, cure, mitigation, treatment or prevention of disease or to affect the structure or a function of the body." (Hunt 1994)

Nutraceutical Information—American Nutraceutical Association.
"What is a nutraceutical?"

Q: What is the definition of the word "nutraceutical?" A: The term "nutraceutical" was coined from "nutrition" and "pharmaceutical" in 1989 by Stephen DeFelice, MD, founder and chairman of the Foundation for Innovation in Medicine (FIM), Cranford, NJ. According to DeFelice, "A nutraceutical is any substance that is a food or a part of a food and provides medical or health benefits, including the prevention and treatment of disease." Such products may range from isolated nutrients, dietary supplements and specific diets to genetically engineered designer foods, herbal products, and processed foods such as cereals, soups and beverages. In the United States the term nutraceutical is commonly used in marketing has no regulatory definition.

Examples: beta-carotene, lycopene

Formulating a New Product with Vitamin or Mineral Addition

Food processors may choose to use any additive, including nutrients or non-nutrient supplements, in the manufacture of food products. Regardless of what is used, they must comply with all Nutrition Labeling and Education Act (NLEA) regulations regarding the contents and stated ***health claims*** of their products. They must use vitamin and mineral additives judiciously (not just to enhance the values on their food label) and then only make label claims regarding nutritional benefits that are allowed.

An older yet timely publication states food technologists formulate a new product with a number of considerations in mind regarding vitamin and mineral addition.

Some of the added nutrient considerations include the following:

- Overall product composition—such as pH, water activity, fat, fiber, protein, because the flavor and color of foods may change
- Ingredient interactions—of a vitamin or mineral combination
- Processing considerations—blanching, washing, stability to heat
- Shelf life and packaging—protection from oxidation or light
- Cost factors—price of the nutrient, overages due to loss, and costs of (the above) processing and packaging needs (Giese 1995)

So, what is "natural"? "That's a question current FDA regulations largely lob right back at producers. Far from burdening them with overreaching rules as to what constitutes a "natural" fiber or what level of vitamin fortification aligns with a "natural" designation, 'FDA policy on "natural" is not to restrict the use of the term except for added color, synthetic substancs and flavors'" reports Solae's Director of QA (Decker 2013).

Safety

Issues of safety were addressed throughout this chapter. Government testing and approval is ongoing. However, continued vigilance is most definitely welcome in order to remain in good health. Two comments related to safety:

> Shopping was easy when most food came from farms. Now, factory-made foods have made chemical additives a significant part of our diet. (CSPI)

> Consumers are demanding cleaner ingredient statements without unfamiliar and "chemical sounding" names while expecting increased shelf life and product quality. (Tessier 2001)

Conclusion

Additives are "a substance or a mixture of substances, other than a basic foodstuff, that is present in a food /as a result of an aspect of production, processing, storage or packaging." Additives function in foods to combat microbial and enzymatic deterioration, to maintain or improve nutritional value and product consistency, and to make food more aesthetically appealing. *Less* rancidity, spoilage, contamination, and overall waste and *more* nutritional value and ease of preparation are possible with the use of additives. Many additives are natural food ingredients, used strictly for imparting flavor and color.

The Food Additives Amendment of the Federal FD&C Act of 1938 contains legislation regarding the safety of additives. The Delaney Clause to the Food Additives Amendment requires testing of proposed additives in the United States for carcinogens. Salt, sugar, and corn syrup are the three most commonly used food additives in the US food supply.

Key additives used in food processing include alternative sweeteners, anticaking agents, antioxidants, bleaching and maturing agents, bulking agents, coloring agents, curing agents, dough conditioners, emulsifiers, enzymes, fat replacers, firming agents, flavoring agents, fumigants, humectants, irradiation, leavening agents, lubricants, nutrient supplements, pH control substances, preservatives, propellants, sequestrants, solvents, stabilizers and thickeners, surface-acting agents, and sweeteners.

The nutritional value of foods may be increased to exceed nutrient levels inherent in the traditional product. Food nutrients may be fortified or enriched. Specific vitamins and minerals are manufactured with the purpose of addition to foods. Functional foods are foods that are modified to provide a health benefit beyond the traditional product and may be used to treat/

prevent disease. They along with phytochemicals and nutraceuticals are a newly evolving area of food and food technology.

Further advances in new and existing safe and effective food additives are on the horizon.

Notes

CULINARY ALERT!

Glossary

Additive Substance added to foods for specific physical or technical effects.

Delaney Clause Clause added to Food Additives Amendment stating that no additive shown to cause cancer in man or laboratory animals could be used in foods.

Drugs Intended for use in the diagnosis, cure, mitigation, treatment, or prevention of disease or to affect the structure or function of the body.

Enrichment The addition of nutrients to achieve established concentrations specified by the standards of identity.

Foods Products primarily consumed for their taste, aroma, or nutritive value.

Fortification The addition of nutrients at levels higher than those found in the original or comparable food.

Functional foods Any modified food or food ingredient that may provide a health benefit beyond that obtained by the original food; the term has no legal or general acceptance in the United States but is accepted by some as food for specified health use.

Generally Recognized As Safe (GRAS) Substances in use, not shown to be unsafe.

Health claims Describe an association between a nutrient or food substance and disease or health-related condition.

Nutraceuticals The name given to a proposed new regulatory category of food components that may be considered a food or part of a food and may supply medical or health benefits including the treatment or prevention of disease; a term not recognized by the FDA.

Phytochemicals Plant chemicals; natural compounds other than nutrients in fresh plant material that function in disease prevention; they protect against oxidative cell damage or facilitate carcinogen excretion from the body and exhibit a potential for reducing the risk of cancer.

References

(2012) Plant-based colorants. Food Product Des 76
ADA Position of the American Dietetic Association (1995) Phytochemicals and functional foods. J Am Diet Assoc 95(4):493–496
Adkins M, Nicoll H (2013) Guar gum. Food Product Des 18–20
Akins M, Nicoll H (2012) About carrageenan. Food Product Des 16–18
Berry D (2013) Exempt yet? Making the switch to natural colors. Food Product Des 46–61
Brantley B (2012) Browning and the maillard reaction in product development. Food Product Des 45–48
Decker KJ (2012) Natural colors in the spotlight. Food Product Des 37–44
Decker KJ (2013) A natural approach to fortification. Food Product Des 66–73
Giese J (1995) Vitamin and mineral fortification of foods. Food Technol 49(5):110–122
Hunt J (1994) Nutritional products for specific health benefits—foods, pharmaceuticals, or something in between? J Am Diet Assoc 94:151–154
Spano M (2012) All about antioxidants. Food Product Des 118–122
Tessier J (2001) Increasing shelf-life without preservatives. Bakers' J. http://www.gftc.ca/articles/2001/baker07.cfm
What is the difference between food processing and preservation? http://www.wisegeek.com/what-is-the-difference-between-food-processing-and-preservation.htm

Bibliography

CSPI—Center for Science in the Public Interest. http://www.cspinet.org
Food and beverage flavor companies—view specifics as needed
Food and Drug Administration
Food and Nutrition Board. National Academy of Science
Food Science Publisher (previously known as D&A Inc./FF Publishing since 2004)
Gale Encyclopedia of Food & Culture. http://www.answers.com/topic/additive
http://www.mayoclinic.com/health/probiotics/AN00389—Is it important to include probiotics and prebiotics in a healthy diet?
IFT—Institute of Food Technologists
International Food Information Council Foundation (IFIC), Washington, D.C.
JECFA—The Joint FAO/WHO Expert Committee on Food Additives
Nutrition.gov http://www.nutrition.gov/whats-food/food-additives

18 Food Packaging

Introduction

This chapter is in the newly named *Aspects of Food Processing* section of the text. The chapters covering *food preservation* and *food additives* components of the food processing section are discussed in Chaps. 16 and 17, respectively.

Adequate packaging is an industry technique that may be used along with preservation and has the intent of slowing down or stopping spoilage that would otherwise exhibit loss of taste, textural quality, or nutritive value of food. *Crops* as well as *animal products* produce market-ready and adequately long shelf life food products if adequately packaged.

As presented in the chapter on food preservation, processed foods represent the change of raw material into food of another form. Food processing involves preservation and also packaging.

For added clarification, and a succinct explanation, the following is utilized.

> Food processing and preservation are two techniques that are used to maintain the quality and freshness of foods. In terms of how they are performed, food processing and preservation are different; food preservation is just part of the entire procedure of processing foods. Food processing mostly involves both *packaging* and preservation, while food preservation is concerned with the control and elimination of the agents of food spoilage... (http://www.wisegeek.com/what-is-the-difference-between-food-processing-and-preservation.htm)
> (italics added)

Packaging as part of food processing assists in preserving food against spoilage and contamination as well as extending the shelf life. It provides containment (holding the product), protection (quality, safety, freshness), information (graphics, labels), and utility of use or convenience (The Society of the Plastics Industry (SPI), Washington, DC). Yet, packaging offers much more than these benefits to the manufacturer and consumer. Packaging protects food, may modify atmosphere, thus extending shelf life, gets a message across, aids in marketing, provides added content security, and so forth. It may provide portion control, combating "portion distortion," convenience of use, and convenient transport. This assists the child as well as the adult consumer.

Packaging may simply involve having clean hands, a sanitary pair of foodservice tongs, and a piece of tissue in a bag served across a bakery counter or, it may involve adherence to a specific time and temperature heat application in a foodservice can that is transported along continents. There is great variance in the idea of packaging food to be discussed in this chapter.

V.A. Vaclavik and E.W. Christian, *Essentials of Food Science, 4th Edition*, Food Science Text Series, DOI 10.1007/978-1-4614-9138-5_18, © Springer Science+Business Media New York 2014

There exist various packaging materials, including films, and package oxygen levels that protect foods from air. Packaging may also maintain time-sensitive foods and use dating or doneness indicators. It may be used as a promotion tool on store sales shelves.

Packaging materials for food include metal, glass, paper, plastics, foil, wood crates, cotton, or burlap (jute). Food may be vacuum packaged, subject to controlled or modified atmospheric packaging, or be aseptically packaged. Manufacturers must adhere to FDA regulations regarding both the method and materials of packaging.

Routinely, we see that consumer-convenient packaging such as microwaveable packages, single-serve products, tubs, and zippered pouches tampers evidence, and package atmosphere has become increasingly important as a packaging selection (Sloan 1996). Packaging functionality is a demand of both consumers and food companies alike, who want packaging/materials that meet their needs. Packaging can certainly be very creative.

Types of Packaging Containers

Packaging containers are classified as primary, secondary, and tertiary. Specific packaging materials are discussed later. One packaging material though, namely, plastic wrap, may function as all three types. A ***primary*** container is the bottle, can, drink box, and so forth that contains food. It is a *direct* food contact surface and is, therefore, subject to approval by the Food and Drug Administration (FDA), which tests for the possible migration of packaging materials into food.

Several primary containers are held together in ***secondary*** containers, such as corrugated fiberboard boxes (commonly, yet not correctly, referred to as cardboard), and do *not* have *direct* food contact. In turn, several secondary containers are bundled into ***tertiary*** containers such as corrugated boxes or overwraps that prepare the food product for distribution or palletizing. This offers additional food protection during storage and distribution where errors, such as dropping and denting or crushing cartons, may occur. Tertiary containers prevent the brunt of the impact from falling on the individual food container.

See foodandbeveragepackaging.com.

Packaging Functions

The functions of packaging are numerous and include such purposes as protecting *raw* or *processed* foods against *spoilage* and *contamination* by an array of external hazards. Packaging serves as a barrier in controlling potentially damaging levels of light, oxygen, and water. It facilitates ease of use, offers adequate storage, conveys information, and provides evidence of possible product tampering.

Packaging achieves its functions/goals by assisting in the following manners:

- Preserving against spoilage of color, flavor, odor, texture, and other food qualities.
- Preventing contamination by biological, chemical, or physical hazards.
- Controlling absorption and losses of O_2 and water vapor.
- Facilitating ease of using product contents—such as packaging that incorporates the components of a meal together in meal "kits" (e.g., tacos).
- Offering adequate storage before use—such as stockable, resealable, pourable.
- Preventing/indicating tampering with contents by tamper-evident labels.
- Communicating information regarding ingredients, nutrition facts, manufacturer name and address, weight, bar code information, and so forth via package labeling.
- Marketing—standards of packaging, including worldwide acceptability of certain colors and picture symbols, vary and should be known by the processor. Packages themselves may promote sales. They may be rigid, flexible,

metallized, and so forth and may also carry such information as merchandising messages, health messages, recipes, and coupons.

Packaging Materials

Packaging materials for food may differ in commercial and retail operations. Either type of operation though may include some of the same materials for food packaging. Included as packaging materials may be paper, glass, plastic, metal, cloth, including burlap, paper, poly or mesh bags for 5# or 10# of fresh potatoes. As well, there exist packaging/shipping materials such as bottles, jugs, jars, or cans; plastic 4- and 6-pack rings for securing cans or bottles; boxes for items such as pizza, cakes, pies, cupcake inserts, confections, or take-out foods, disposable deli tissue, insulated "cold packs," tubes, pails, drums, paper or plastic wraps or bags, and shipping foam "peanuts," just to name a few items!

In choosing the appropriate packaging for their product, packers must consider many variables. For example, *canners* must make packaging choices based on cost, product compatibility, shelf life, flexibility of size, handling systems, production line filling and closing speeds, processing reaction, impermeability, dent and tamper resistance, and consumer convenience and preference (Sloan 1996).

Processors who use *films* for their product must select film material based on its "barrier" properties that prevent oxygen, water vapor, or light from negatively affecting the food. As an example, the use of packaging material that prevents light-induced reactions will control degradation of the chlorophyll pigment, bleaching or discoloration of vegetable and red meats, destruction of riboflavin in milk, and oxidation of vitamin C.

The most common food packaging materials include metals, glass, paper, and plastic. Some examples of these leading materials appear in the following text.

Metal

Metals such as steel and aluminum are used in cans and trays. A metal can forms a hermetic seal, which is a complete seal against gases and vapor entry or escape, and it offers protection to the contents. The trays may be reusable, or disposable recyclable trays, and either steam table or No. 10 can size. Metal is also used for bottle closures and wraps.

Steel

Steel has a noncorrosive coating of tin inside, thus the name "tin can," whereas *tin-free steel* (TFS) relies on the inclusion of chromium or aluminum in place of tin. Steel is manufactured into the traditional *three-piece* construction can, which includes a base, cylinder, and lid, and also a *two-piece* can, consisting of a base and cylinder in one piece without a seam and a lid. The latter are lightweight and stackable. The vast majority of the many billion cans used annually in the United States are made of steel.

In addition to steel cans and trays, tens of billions of beverage bottle crowns (closures with crimped edges) made of steel are used annually in the United States. The five primary types of steel vacuum closures include side seal caps, lug caps, press–twist caps, snap-on caps, and composite caps.

Aluminum

Aluminum is easily formed into cans with hermetic seals. It is also used in trays and for wraps such as aluminum foil, which provide an oxygen and light barrier. Aluminum is lighter in weight than steel and resists corrosion.

Glass

Glass is derived from metal oxides such as silicon dioxide (sand). Glass is used in forming bottles or jars (which subsequently receive hermetic seals) and thus protects against water vapor or oxygen loss. The thickness of glass must be sufficient to prevent breakage from internal pressure, external impact, or thermal stress. Glass that is *too* thick increases weight and thus freight costs and is subject to an increased likelihood of thermal stress or external impact breakage.

Technological advances in glass packaging have led to improvements in strength and weight, as well as color and shape. A resurgence of glass may be noted on supermarket shelves. The product is commercially sterile, yet the see-through glass tends to denote "fresh" to the consumer.

Glass coatings, similar to eyeglass coatings of silicons and waxes, may be applied to glass containers in order to minimize damage-causing nicks and scratches.

Paper

Paper is derived from the pulp of wood and may contain additives such as aluminum particle laminates, plastic coating, resins, or waxes. These additives provide burst strength (strength against bursting), wet strength (leak protection), and grease and tear resistance, as well as barrier properties that assure freshness, protect the packaged food against vapor loss and environmental contaminants, and increase shelf life.

Varying thicknesses of paper may be used to achieve thicker and more rigid packaging.

- *Paper* is thin (one layer) and flexible, typically used in bags and wrappers. Kraft (or "strong" in German) paper is the strongest paper. It may be bleached and used as butcher wrap or may remain unbleached and used in grocery bags.
- *Paperboard* is thicker (although still one layer) and more rigid. Ovenable paperboard is made for use in either conventional or microwave ovens by coating paperboard with PET polyester (see Plastic).
- Multilayers of paper form *fiberboard*, which is recognized as "*cardboard*."

When packaging serves as a *primary* container for food, it is a food contact surface and must be coated or treated accordingly. For example, *paper* bags or wraps for bakery products (thus a food contact surface) may be laminated to improve burst or wet strength, grease, and tear resistance or prevent loss of product moisture. *Paperboard* may be lined and formed to hold fluid milk. It may be formed into canisters with foil linings and resealable plastic overwraps, to provide convenience, protection, and extended shelf life. Another example is *corrugated paperboard* which may be waxed in order to package foods such as raw poultry.

Dual-ovenable trays are designed to be *microwaveable* and also able to be placed in a *conventional* oven. As with all new processing and packaging technology, the use of these trays is a new concept for many people and may require consumer education, including written instructions provided by the food manufacturers.

Recycled papers may contain small metal fragments that could be unacceptable in packaging used for microwave cooking. The sparks, generated as the microwaves are reflected by metal, may "arc" and start a fire in the microwave oven. Yet, paper may be purposely manufactured to designated specifications and deliberately contain areas with small particles of aluminum, which form a "*susceptor*."

Susceptors are desirable for browning and crisping microwaveable foods such as baked goods, french fries (often placed in individual compartments of a susceptor), and pizzas. They are also used in packages of microwaveable popcorn. Due to the fact that the metal reflects microwaves, which subsequently heat the surface of the food, the browning and crisping can occur.

Paper may be used in combination with metal, such as aluminum, to produce fiber-wound tubing. An example of fiber-wound tube containers used for refrigerated biscuits is shown in Fig. 18.1.

Paper and metallized films are increasingly chosen for food applications. The appearance, and its barrier properties to grease, and moisture are desirable for packaging specific foods. These materials may also contain plastic, which is discussed in the following chapter section.

Fig. 18.1 An example of fiber-wound tubes (*Source*: Sara Lee Corporation)

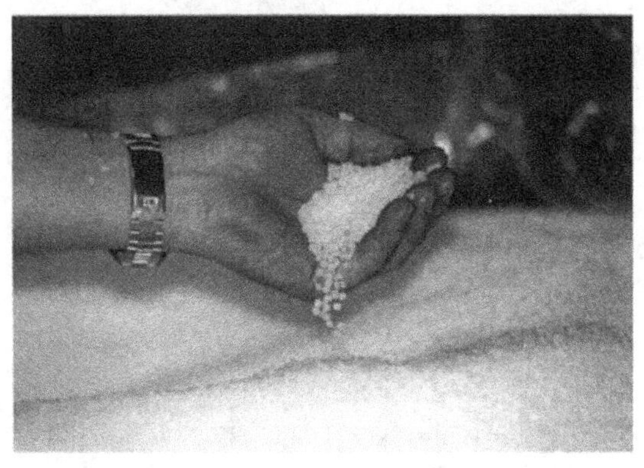

Fig. 18.2 Polyethylene pellets (*Courtesy*: Rodeo Plastic Bag and Film, Inc., Mesquite, TX)

Fig. 18.3 Blown film bubble coming off the die (*Courtesy*: Rodeo Plastic Bag and Film, Inc., Mesquite, TX)

Plastic

Plastic has shrink, nonshrink, flexible, semirigid, and rigid applications and varies in its degree of thickness. In making plastics, pellets and blown bubbles may be seen in Figures18.2 and 18.3.

Important *properties* of the many types of plastics that make them good choices for packaging material include the following:

- Flexible and stretchable
- Lightweight
- Low-temperature formability
- Resistant to breakage, with high burst strength
- Strong heat sealability
- Versatile in its barrier properties to O_2, moisture, and light

Basic hydrocarbon building blocks such as ethane and methane, which are derived from natural gas and petroleum, form organic chemical compounds called *monomers*. These are then chemically linked to form plastic molecular chains, or *polymers*. Their manufacture represents a small percentage of total US energy consumed. Plastic has multiple *functions* as a packaging material, including use in bottles and jars, closures, coatings, films, pouches, tubs, and trays (The Society of the Plastics Industry (SPI), Washington, DC). It may also be used *in combination* with other packaging materials such as metal (for lining cans), paper, and so forth.

Choices of Plastics for Packaging. The food industry must provide packaging with barrier protection (against moisture, light, air, grease, and so forth) and must be familiar with what level of barrier performance is warranted for the foods they are packaging. *Insufficient* packaging, attributable in part to a high cost of materials, is not satisfactory, and *too much* performance (excesses in packaging contribution), with excessive barrier protection, is unnecessary.

Among the thousand types of plastics that are created, less than two dozen are polymers utilized in food packaging (see Table 18.1). Some of the more commonly used plastics for food products are briefly discussed in the following.

Table 18.1 Example of the repeating units of common packaging polymers

Polymer	Repeating unit
Polyester (PET)	
Polyethylene (PE)	
Polypropylene (PP)	
Polystyrene (PS)	

Source: The Society of the Plastics Industry

Polyethylene (PE): Polyethylene is the most common and the least expensive plastic, comprising the largest percentage of total plastic packaging. It is a water vapor (moisture) barrier and prevents dehydration and freezer burn. Polyethylene pellets are used in producing plastic, such as plastic bags, "zipper" seals, and plastic storage containers. Use of this PE may be recommended for less plasticizer migration into food.

Polyethylene with ethyl vinyl acetate (EVA): Polyethylene with ethyl vinyl acetate creates "freezer wrap," which offers moisture-loss protection without getting brittle in low temperatures. Polyethylene terephthalate (PET) has been used widely, including as a tube which dispenses food. Some advantages of PET is that it withstands high-temperature foods and is lighter in weight than the glass that it replaces. Polyethylene naphthalate (PEN) received FDA approval in 1996 for use in food packaging. It provides a barrier against gas, moisture, and ultraviolet light. As bottled beverages, including waters, teas, and juices, continue to appear in the marketplace, the use of plastic bottle containers made of PET and PEN may be increasing.

Polypropylene (PP): Polypropylene has a higher melting point and greater tensile strength than polyethylene. It is often used as the inside layer of food packages that are subject to higher temperatures of sterilization (e.g., retort pouches or tubs).

Polystyrene (PS): Polystyrene is a versatile, inexpensive packaging material and represents less than 10 % of total plastic packaging. When foamed, its generic name is expandable polystyrene (EPS). This styrofoam has applications in disposable packaging and drinking cups. It offers thermal insulation and protective packaging. EPS is used in "clam shell" fast-food packaging, egg cartons, bowls, cups, and meat trays and is the "peanuts" in packages. Substantially less energy is required to form polystyrene cups than paperboard cups.

Polyvinyl chloride (PVC or vinyl): Polyvinyl chloride comprises less than 10 % of total plastic packaging. It blocks out air and moisture, preventing freezer burn, and offers low permeability to gases, liquid, flavors, and odors. PVC prevents the transfer of odor and keeps food fresh by controlling dehydration and is capable of withstanding high temperatures without melting. PVC has good puncture resistance and "cling" properties. It is used to prevent splattering in microwave food preparation.

Polyvinylidene chloride (PVDC, Saran®): Polyvinylidene chloride is a thermoplastic resin used for household wraps and has excellent barrier properties. *Cryovac* is a Saran film used in vacuum-sealing (*Kryos* = cold, *Vacus* = empty in Latin).

Polyvinylidene chloride (PVDC) and ethylene vinyl alcohol (EVOH) are also utilized as barrier plastics. They have properties related to oxygen and water vapor permeability.

Purpose of Packaging: Food and Agriculture Organization (FAO) of the United Nations.

The basic purpose of packaging is to protect meat and meat products from undesirable impacts on quality including microbiological and physio-chemical alterations. Packaging **protects** foodstuffs during processing, storage and distribution from:

- **Contamination by dirt** (by *contact with surfaces and hands*)
- **Contamination by micro-organisms** (*bacteria, moulds, yeasts*)
- **Contamination by parasites** (*mainly insects*)
- **Contamination by toxic substances** (*chemicals*)

- **Influences affecting colour, smell and taste** (*off-odour, light, oxygen*)
- **Loss or uptake of moisture** (*evaporation* or *water absorption*)

Adequate packaging can prevent the above listed secondary contamination of meat and meat products. But the further growth of microorganisms, which are already present in meat and meat products, cannot be interrupted through packaging only. To halt or reduce microbial growth, packaging has to be **combined** with other treatments, such as **refrigeration**, which will slow down or stop the further growth of microorganisms, or with **heating/sterilization**, which will reduce or completely eliminate contaminating microorganisms.—FAO

(Other foods too)

Many manufacturers specify proprietary molded and shaped bottles to hold their specific food contents. The appropriate plastic may be chosen to satisfy this highly specialized demand.

Manufacturers may also use *food-based* materials to produce thermal plastic resins. They are made from natural sugars found in corn and other plants. For example, wheat starch and corn sugar are being developed for packaging purposes as biodegradable materials that will compost down fully in around 30–60 days (Higgins 2000).

Metal, glass, paper, and plastic, the most common food packaging materials, have been discussed briefly. At this point, let us move on to view some other packaging materials.

Other Packaging Materials

Cotton or **burlap** (jute) may be used for grains, flour, legumes, and some vegetables, primarily in transport.

Edible films are subject to FDA approval because they become part of the food. Natural edible films extend shelf life, although for shorter time than synthetic nonedible packaging materials. Edible films are a unique type of packaging material.

These films are "... defined as a thin layer of edible material formed on a food as a coating or placed (pre-formed) on or between food components. Its purpose is to inhibit migration of moisture, oxygen, carbon dioxide, aromas, and lipids, etc.; carry food ingredients (e.g. antioxidants, antimicrobials, flavor); and/or improve mechanical integrity or handling characteristics of the food" (Krochta & DuMulder-Johnston 1997).

Antimicrobials may be included in films or containers. Antimicrobial activity may be due to the addition of specific substances, radiation, or gas flushing. Irradiation sterilization of packaging materials may be forthcoming with FDA approval.

Examples of edible films include those used as the sugar shell on individual chocolate-covered candies (M&Ms®), *casings*, such as in sausage, and *edible waxes*, such as those applied to fruits and vegetables. Serving in the role of edible films, the casings "contain," and the waxes function to improve or maintain appearance, prevent mold, and contain moisture while still allowing respiration. As well, food may be coated with a thin layer of polysaccharides such as cellulose, pectin, starch and vegetable gums, or proteins, such as casein and gelatin. Cut, dried, fruit pieces are often sprayed with an edible film prior to their inclusion into items such as breakfast cereal (see "Active Packaging Technologies" section).

Bindings may be applied to a food's surface to be an adhesive for seasoning. Other coatings may significantly improve appearance (and reduce microbial contamination) by replacing egg washes and acting as a glaze.

Foil is a packaging material that may be used in snack bags (chips) or as a laminate in aseptic packaging (see Aseptic Packaging). It is used as a wrapping for dry, refrigerator, or freezer storage. It provides a moisture-proof and vapor-proof barrier.

Laminates are multilayers of foil, paper, or plastics that may be utilized selectively according to the specific food packaging need.

In combination, the various laminates may provide more strength and barrier protection than the individual laminate material. Laminates provide barriers useful in controlling O_2, water vapor, and light transmission, and they provide good burst strength. The laminates may resist pinholes and flex cracking. Retort pouches are examples of laminates used in packaging and contain polyester film, aluminum foil, and polypropylene.

Resins are used for sealing food packages. They must withstand the stress of processing and offer seal integrity that prevents product contamination.

Wood may be used in the manufacture of crates that contain fresh fruits and vegetables.

Bag in a box is now offered in five-gallon bags with snap-on caps over a 1″ polyethylene spout. There is a high barrier film, with heat resistance up to 190 °F.

Regardless of the materials that are selected for use, *source reduction*, *reuse*, and *recycling* should be important considerations of packaging manufacturers. The food industry challenge is to provide the appropriate materials to accomplish packaging functions at reasonable cost.

Controlling Packaging Atmosphere

Reduced *temperature* remains as the primary means of food protection. However, controlling the *other* known elements in the package environment, such as O_2 (controlled or modified atmosphere in packaging), CO_2, water vapor, and ethylene concentration, may also reduce spoilage and contamination (e.g., enzymatic, biological), thus extending shelf life. The material that follows in this chapter will address controlling the internal package environment and modification of gases.

The following are the significant manners of controlling packaging atmosphere. (ROP is defined as any packaging procedure that results in a reduced oxygen level in a sealed package.) The term is often used because it is an inclusive term and can include other packaging options such as (1) *cook-chill*, (2) *controlled atmosphere packaging (CAP)*, (3) *Modified Atmosphere Packaging (MAP)*, (4) *sous vide, and* (5) *vacuum packaging*.

A definition and examples of reduced oxygen packaging follow.

What is ROP?

(http://www.anfponline.org/CE/food_protection/2010_11.shtml)

According to AFNP

Packaging using an ROP method can be used to describe any packaging process in which a sealed product has an environment that is reduced in oxygen. ROP is an all-inclusive term used to describe methods such as Controlled Oxygen Packaging (CAP), Modified Atmosphere Packaging (MAP), Cook-Chill, Vacuum Packing (VP), and Sous Vide. Each form of ROP has its unique methods and outcomes, but all have one thing in common: the final product will be in a sealed package in which there is little or no oxygen present. . .

The method of Sous Vide is a specific process of ROP utilized for food that requires refrigeration/freezing after packaging—usually potentially hazardous foods (PHF). The process of Sous Vide does reduce the initial bacterial load of a product to lower levels, but not low enough to make the food shelf stable. The process generally has several steps: preparation of the raw materials (which may include partial grilling or a similar step); packaging the product by use of vacuum sealing; cooking/pasteurizing the product to the desired cooking temperature while in the package; rapid cooling/freezing; reheating to 165 °F for hot holding or any temperature for immediate service. This method is said to retain the color, texture, moisture, and flavor of the final product.—AFNP

See the following for ***Definitions:***

http://www.cfsan.fda.gov/~dms/fcannex6.html

FDA (B) Definitions.

The term ROP is defined as any packaging procedure that results in a reduced oxygen level in a sealed package. The term is often used because it is an inclusive term and can include other packaging options such as:

1. *Cook-chill* is a process that uses a plastic bag filled with hot cooked food from which air has been expelled and which is closed with a plastic or metal crimp.
2. *Controlled atmosphere packaging (CAP)* is an active system which continuously maintains the desired atmosphere within a package throughout the shelf life of a product by the use of agents to bind or scavenge oxygen or a sachet containing compounds to emit a gas. Controlled atmosphere packaging (CAP) is defined as packaging of a product in a modified atmosphere followed by maintaining subsequent control of that atmosphere.
3. *Modified Atmosphere Packaging (MAP)* is a process that employs a gas flushing and sealing process or reduction of oxygen through respiration of vegetables or microbial action. Modified Atmosphere Packaging (MAP) is defined as packaging of product in an atmosphere which has had a one-time modification of gaseous composition so that it is different from that of air, which normally contains 78.08 % nitrogen, 20.96 % oxygen, and 0.03 % carbon dioxide.
4. *Sous vide* is a specialized process of ROP for partially cooked ingredients alone or combined with raw foods that require refrigeration or frozen storage until the package is thoroughly heated immediately before service. The sous-vide process is a pasteurization step that reduces bacterial load but is not sufficient to make the food shelf stable. The process involves the following steps:
   (a) Preparation of the raw materials (this step may include partial cooking of some or all ingredients).
   (b) Packaging of the product, application of vacuum, and sealing of the package.
   (c) Pasteurization of the product for a specified and monitored time/temperature.
   (d) Rapid and monitored cooling of the product at or below 3 °C (38 °F) or frozen.
   (e) Reheating of the packages to a specified temperature before opening and service.
5. *Vacuum packaging* reduces the amount of air from a package and hermetically seals the package so that a near-perfect vacuum remains inside. A common variation of the process is vacuum skin packaging (VSP). A highly flexible plastic barrier is used by this technology that allows the package to mold itself to the contours of the food being packaged.

The creation of a packaging environment with little or no oxygen has beneficial applications for the food industry. However, microbiological concerns arise simultaneously. As will be discussed, proper controls need to be in place for reduced oxygen packages.

FDA: Benefits of ROP

ROP can create a significantly anaerobic environment that prevents the growth of aerobic spoilage organisms, which generally are Gram-negative bacteria such as Pseudomonads or aerobic yeast and molds. These organisms are responsible for off-odors, slime, and texture changes, which are signs of spoilage.

ROP can be used to prevent degradation or oxidative processes in food products. Reducing the oxygen in and around a food retards the amount of oxidative rancidity in fats and oils. ROP also prevents color deterioration in raw meats caused by oxygen. An additional effect of sealing food in ROP is the reduction of product shrinkage by preventing water loss.

These benefits of ROP allow an extended shelf life for foods in the distribution chain, providing additional time to reach new geographic markets or longer display at retail. Providing an extended shelf life for ready-to-eat convenience foods and advertising foods as

"Fresh-Never Frozen" are examples of economical and quality advantages.

Providing oxygen control in packaging is needed by fruits and vegetables. They continue to breathe and require oxygen after harvesting and processing; thus, the package must contain oxygen. Yet it needs to be controlled, as *too high* a level causes oxidation and spoilage and *too low* a level leads to anaerobic spoilage. In extending shelf life of fruit, oxygen levels should approximate 5 % and carbon dioxide at 1–3 % (with refrigeration maintained at temperature-specific levels). Packaging environments must match the respiration rate as closely as possible.

The function of CO_2 addition in packaging is to inhibit growth of many bacteria and molds. The O_2 maintains respiration and color and inhibits growth of anaerobic microorganisms. Nitrogen (N_2) is used to flush the package and rid it of air (O_2 specifically). Nitrogen also prevents a collapse of the loose-fitting packaging material.

Cook-Chill

Cook-chill is defined as a packaging procedure that also results in reduced oxygen levels. By the FDA definition, it "is a process that uses a plastic bag filled with hot cooked food from which air has been expelled and which is closed with a plastic or metal crimp." Such a system is one that may be frequently employed in hospital foodservice operations as an alternative to a more conventional foodservice operation.

Modified Atmosphere Packaging

Modified Atmosphere Packaging (MAP) modifies the internal package atmosphere of food. It replaces the air in the package with nitrogen or carbon dioxide, and the shelf life of the product can increase by as much as 200 %. Gas flushing and sealing reduces oxygen coming through respiration of vegetables. MAP is a *one-time modification* of gases so that it is different from air, which normally contains 78.08 % nitrogen, 20.96 % oxygen, and 0.03 % carbon dioxide.

MAP is *primarily* applied to fresh or minimally processed foods that are still undergoing respiration, and it is used for the packaging of a variety of foods. Such foods include baked goods, coffees and teas, dairy products, dry and dehydrated foods, lunch kits, and processed meats (to keep the meat pigment looking desirable). It is also used for nuts, snack food applications, and pasta packaging. This type of packaging with high CO_2 levels inhibits many aerobic bacteria, molds, and yeasts.

MAP is one of the most widely used packaging technologies, as it functions to enhance appearance, minimize destructive waste, extend shelf life, and reduce the need for artificial preservatives. Its use thus expands a product's ability to reach new markets. Nitrogen is used in bread products, while carbon dioxide is best suited to high-fat products.

Following the packaging of foods, a machine vacuums out *all* of the package air and then, through the same package perforations, *evenly inserts* the new, desired gas combination. Since MAP contains the food under a gaseous environment that differs from air (some other percentage), it controls normal product respiration (consuming O_2 and generating CO_2, water vapor, and perhaps ethylene) and growth of aerobic microorganisms. For example, the change in CO_2 level shows an inhibitory effect on aerobic microorganisms. This effect is dependent upon conditions such as the level of CO_2 (a high level in proportion to air is more effective), moisture, pH, and temperature.

The initial mix of packaging atmosphere changes over time as a result of factors such as product respiration, the aerobic and anaerobic bacterial load, respiration of bacteria, permeation of gases through the packaging materials/seals, temperature, light, and time (Labuza 1996).

The addition of *nitrogen* gas, which is odorless, tasteless, colorless, nontoxic, and nonflammable, is introduced into the food package *after* all atmosphere has been removed from the pouch and vacuum chamber, and just *prior* to hermetic sealing of the package. It increases the package's internal pressure. This modification, by a predetermined dose of liquid nitrogen (LIN), offers protection from spoilage, oxidation, dehydration, weight loss, and freezer burn and extends shelf life, as nitrogen consumes oxygen.

Unlike vacuum packaging, the high barrier film (used to keep air out and to prevent the modified atmosphere from escaping) used for MAP remains loose-fitting. This avoids the crushing effects of skintight vacuum packaging. When used in combination with aseptic packaging, which reduces the microbial load, MAP becomes a more effective technology. Most new and minimally processed foods use MAP in combination with aseptic technology and reduced temperature.

Controlled Atmosphere Storage and Packaging

Both controlled atmosphere (CA) in *storage* environments and ***controlled atmosphere packaging*** (CAP) are utilized in order to permit controlled oxygen and carbon dioxide exchange, thus preserving foods. As well, CAP is a prime alternative to pesticides and preservatives. When storage temperatures and conditions of distribution vary in fresh and processed foods, CAP and MAP assist in standardizing these variables and maintaining product quality.

The FDA defines CAP as "an active system which *continuously* maintains the desired atmosphere within a package throughout the shelf-life of a product by the use of agents to bind or scavenge oxygen or a sachet containing compounds to emit a gas." Controlled atmosphere packaging (CAP) is defined as a packaging of a product in a modified atmosphere followed by maintaining subsequent control of that atmosphere.

However, at any given time, and under variable environments, there is no continual "control" that the food technologist would describe as "ideal." The question then becomes: how much control *is* there in the package environment? Is it then more likely that the atmosphere is *modified*? This form of packaging also utilizes a high barrier film (or pouch), which may be EVOH high barrier polymers, or polyamide, a form of nylon.

Many packaged food products undergo respiration and microbial growth, requiring oxygen, while producing CO_2 and water. The carbohydrate molecule, in the presence of oxygen, $C_6H_{12}O_6 + O_2$, for example, yields $CO_2 + H_2O$ + heat. Therefore, CA or CAP containers offer control by reducing the available O_2, elevating CO_2, and controlling water vapor and ethylene concentration. The worldwide distribution and marketing of produce depends on CAP for high-quality food. A benefit is that less senescence and maintenance of nutritional value is observed.

C. botulinum is an anaerobic bacterium that grows in the absence of available oxygen. Therefore, it may grow in anaerobic packaging environments. To retard its growth in CAP food products, foods must have short-storage times and be held at cold temperatures. Control of water activity (A_w) and salt is also necessary to prevent growth as sodium competes with the bacteria for water absorption.

Food production has shown a rising use, thus the demand for various industrial gases such as CO_2 and N_2. Perhaps this increase in demand may be attributed to more convenient foods and packagings that provide a longer shelf life, CAP, and MAP.

Certainly, if the packaging material is a *poor* barrier, then the nitrogen or carbon dioxide will be replaced with the surrounding oxygen due to diffusion. Considering the opposite effect, if the packaging offers a *good* barrier, then the gases will remain in the package for a longer period of time, protecting the product.

Sous Vide

Sous vide ("under vacuum") packaging involves mild, partial precooking of food prior to vacuum packaging. Once again, according to the FDA definition in the 1999 Guidelines for ROP, "Sous Vide is a specialized process of ROP for partially cooked ingredients alone or combined with raw foods that require refrigeration or frozen storage until the package is thoroughly heated immediately before service. The sous

Fig. 18.4 An example of small vacuum-packaging machinery. Countertop (C200 Courtesy of *Multivac, Inc.*)

Fig. 18.5 An example of very large floor standing vacuum-packaging machinery (C800 Courtesy of *Multivac, Inc.*)

vide process is a pasteurization step that reduces bacterial load but is not sufficient to make the food shelf-stable." Since *some* of the ingredients may be partially cooked, and other ingredients may be raw, the product requires refrigeration or freezing, and then heating through prior to service.

The product package will have its levels of oxygen *reduced* and CO_2 *raised* in the packaging environment in order to reduce the microbial (aerobic pathogens) load and extend the shelf life. Sous-vide products are pasteurized, yet are *not* sterile, and may contain heat-resistant microorganisms and spores. Therefore, strict temperature regulation in production, as well as in the distribution process, is necessary to assure product safety. Food products must be kept cold to prevent the growth of bacteria.

According to FDA guidelines, guidelines related to the sous-vide process include the following: some cooking, packaging, pasteurization, proper cooling and reheating.

Vacuum Packaging

Vacuum packaging modifies the atmosphere surrounding the food by removing oxygen, and it extends shelf life. Further explained by the FDA, in its guidelines for reduced oxygen packaging (ROP), "Vacuum packaging reduces the amount of air from a package and hermetically seals the package so that a near-perfect vacuum remains inside... A highly flexible plastic barrier is used by this technology that allows the package to mold itself to the contours of the food being packaged."

With the removal of oxygen, vacuum packaging controls rancidity that occurs with the oxidation of fatty acids. Vacuum-packaging machines are available for small-, medium-, or large-scale production capacity (Figs. 18.4 and 18.5) and may be used to successfully package a variety of food sizes and forms such as small cheese blocks, large primal cuts of meat, or liquids.

In order to get an idea of the sizes:

Countertop Model C200 or smaller—countertop use

Floor Model C800 machine dimensions: width 1,650 mm, depth 1,050 mm, height 1,070 mm, weight approx. 720 kg

The procedure used for vacuum packaging is to place the food in a flexible film and barrier pouch and put it inside a vacuum-packaging chamber, where oxygen is removed. This creates a skintight package wall and protects against the entry or escape of gases such as air and CO_2 or water vapor. It assures inhibition of microbial growth that would alter microbial and organoleptic properties such as appearance and odor. Water weight loss and freezer burn are also inhibited with this packaging method. The transparent, vacuum-packaging film allows product visibility from all angles.

Controls Needed for Vacuum Packaging. The FDA recommends that local regulatory agencies prohibit vacuum packaging in retail stores unless the following six controls are all in effect:

- Foods must be limited to those that do not support growth of *Cl. botulinum* (as it is an anaerobe).
- Temperatures of 45 °F (7 °C) and below are maintained at all times. Anaerobic pathogens increase their growth rate exponentially with an increase in temperature.
- Consumer packages are prominently labeled with storage–temperature requirements and shelf life.
- Shelf life must neither exceed 10 days nor extend that labeled by the initial processor.
- Detailed, written in-store procedures must be developed, observed, and carefully monitored. These should be HACCP based and include records subject to review by regulatory authorities.
- Operators must certify that individuals responsible are qualified in the equipment, procedures, and concepts of safe vacuum packaging.

Good manufacturing practice to prevent contamination with pathogens is still needed.

FDA: Safety Concerns

The use of ROP with some foods can markedly increase safety concerns. Unless potentially hazardous foods are protected inherently, simply placing them in ROP without regard to microbial growth will increase the risk of foodborne illnesses. ROP processors and regulators must assume that during distribution of foods or while they are held by retailers or consumers, refrigerated temperatures may not be consistently maintained. In fact, a serious concern is that the increased use of vacuum packaging at retail supermarket deli-type operations may be followed by temperature abuse in the establishment or by the consumer. Consequently, at least one barrier or multiple hurdles resulting in a barrier need to be incorporated into the production process for products packaged using ROP. The incorporation of several subinhibitory barriers, none of which could individually inhibit microbial growth but which in combination provide a full barrier to growth, is necessary to ensure food safety.

(1) *Refrigerated Holding Requirements for Foods in ROP*

Safe use of ROP technology demands that adequate refrigeration be maintained during the entire shelf life of potentially hazardous foods to ensure product safety.

Active Packaging Technologies

Active packaging began as "*smart*" *packaging* in the 1980s and was referred to as "*interactive*" *packaging* almost from the start. All three terms describe the same thing, which is packaging that could "sense" changes in the internal environment and respond by adapting as necessary. Included in a food package are small packets in order to control elements such as ethanol, oxygen, or microbes (Brody 2000).

By means of its inherent design, packaging typically serves in a *passive* role by protecting food products from the external environment. It provides a physical barrier to external spoilage, contamination, and physical abuse in storage and distribution. Today, packaging more *actively* contributes to the product's development, controls product maturation and ripening, helps in achieving the proper color development in meats, and extends shelf life. Thus, it is considered to play an *active* (not passive) role in protecting foods. Yet, despite the many attributes and benefits of smart/interactive/active packaging, it generally does not actually "sense" the environment conditions and change accordingly.

Examples of ***active packaging*** technologies are listed in the following text.

Active packaging for *fresh and minimally processed* foods provides the following:

- Edible moisture or oxygen barrier (to control loss of moisture and enzymatic oxidative browning in fresh-cut fruits and vegetables and to provide controlled permeability rates matched to the respiration rate of the fruit)
- Edible antimicrobial (biocidal) polymer films and coatings (which may release controlled amounts of chlorine dioxide into the food, depending on temperature and humidity, or destroy *E. coli* 0157:H7 in meats and prevent mold growth in fruits)
- Films that are scavengers of off-odors
- Oxygen scavengers for low-oxygen packaging

Active packaging for *processed* foods provides the following:

- Edible moisture barrier
- O_2, CO_2, and odor scavenger

Other active packaging technologies include the following:

- Microwave doneness integrators (indicators)
- Microwave susceptor films to allow browning and crispness (french fries, baked products, popcorn)
- Steam release films
- Time–temperature indicators (TTI) which are unable to reverse their color when the product has been subject to time–temperature abuse

Specifically, the predictability of the behavior of a living, breathable fruit or vegetable, or even meat, is quite different from a nonfood item that is packaged. There are numerous interactions between the food, any internal gas in the package atmosphere, and the material used for packaging. Sachets or films may release their intended effect at a controlled rate.

The FDA gave the "go ahead" for a type of active packaging that releases chlorine dioxide gas to kill harmful bacteria and spoilage organisms (Higgins 2001a).

Aseptic Packaging

In order to destroy any *C. botulinum* spores and extend the shelf life of low-acid foods, ***aseptic packaging*** may be utilized. Independent sterilization of both the *foods* and *packaging material*, with assembly under *sterile environmental* conditions, is the rule for aseptic packaging.

In an aseptic system of packaging, the packaging material consists of *layers* of polyethylene, paperboard, and foil (Fig. 18.6). It is sterilized by heat (superheated steam or dry hot air) or a combination of heat and hydrogen peroxide and then roll-fed through the packer to create the typical brick/block shape (Fig. 18.7).

The container is filled with a sterile (no pathogens or spores) or commercially sterile (no pathogens, although *some* spores) liquid food product and sealed in a closed, sterile chamber. Once packed, the product requires no refrigeration. Liquids such as creamers, milk, or juices may be packed in this manner. Triple or multiple packs of flavored milk and juice, with attached straws, are available on grocery shelves. The market leaders of aseptic packages have introduced easy-open, easy-pour features into their cartons. The plastic devices are injection molded and adhere to the package tops.

The sterility of packaging material has formerly relied on *chemical* technologies of sterilization (principally heat with hydrogen peroxide). *Nonchemical* techniques have been explored in order to avoid chemical sterilant residues. Ionizing and nonionizing radiation have been tested for use in aseptic packaging.

Flexible Packaging

Flexible packaging is available for packaging use in the foodservice industry and is finding more applications at the *retail* level, including packaging for bagged cereal candies, poultry, red meat, and sliced deli meat. Nonrigid packaging

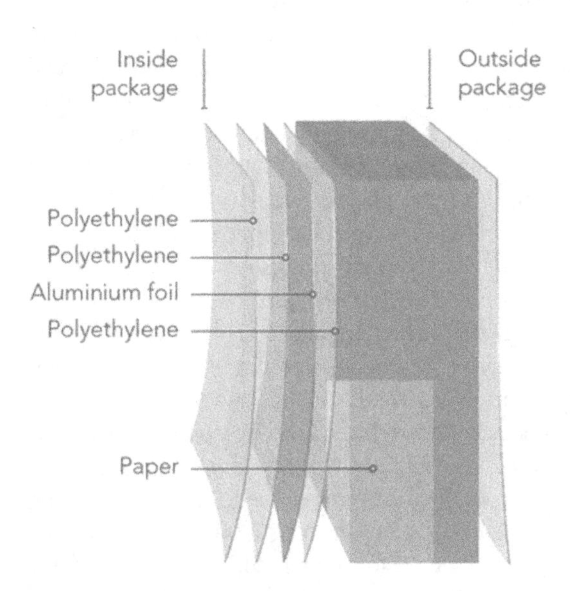

Fig. 18.6 Tetra Brik® aseptic packaging material layers (*Source*: Tetra Pak Inc.)

containers such as stand-up pouches or tubes and zippered bags are examples of flexible packaging used for peanuts, peanut butter, or produce such as fresh-cut lettuce and peeled baby carrots. The same packaging might also need to be *resealable* to meet consumer demands and may require zipper handles or spouts with *easy-open* screw-off tops.

"Flexible packaging uses less material. It keeps products fresh by providing flavor and aroma barriers, which keep outside odors out and flavors in. It is used for fresh fruits and vegetables and matches respiration rate as closely as possible. Overall, the Flexible Packaging Industry encourages innovation and excellence in part by giving packaging achievement awards based on factors including Packaging Excellence, Technical Innovation, Sustainability and Environmental Achievement, and Printing Achievement."

Manufacturers are offering more food products in flexible packaging and find that cost savings and environmental concerns are some of the driving forces behind the switch to flexible packaging.

Fig. 18.7 Different Tetra Brik® aseptic package sizes (*Source*: Tetra Pak Inc.)

Freezer Packaging Protection

Freezing foods is a means of preservation, although foods may spoil due to desiccation or cavity ice if they are not adequately protected. Therefore, a moisture-barrier film, such as a freezer wrap, is needed in packaging material. Tear strength and wet strength are also needed in packaging material for freezer storage.

Freezer Burn

Pronounced desiccation occurs as water diffuses from the product to the atmosphere. This results in ***freezer burn*** with its resultant change in appearance, flavor, texture, and weight.

Cavity Ice

Cavity ice is the ice formation within the food package due to water condensation. Therefore, it is important to use moisture-proof and vapor-proof packaging.

Tamper-Evident Banding and Sleeve Labeling

Tamper-evident banding and sleeve labeling may assist manufacturers and consumers by providing protection and offering the security that the package contents are unviolated. Today, tamper-evident neckbands and shrink-film sleeves are made in a number of colors and may be custom printed. Technology for full-body shrink-labels over glass and plastic bottles is more apparent, as it has become more affordable and attractive. Pull-tabs and perforations provide ease of use.

While the majority of rigid packaging includes tamper-evident attributes, not all food is packaged in this manner. Considering security issues, especially susceptible may be bakery and dairy products (Higgins 2001b).

Manufacturing Concerns in Packaging

Selection of Packaging Materials

Environmental conditions in package transport and government regulations may dictate the materials a company uses in shipping food containers domestically or overseas. Many components of the food industry demand that packaging material is biodegradable, recyclable, strong, and waste-to-energy efficient. The food processor must choose materials that effectively preserve shelf life and are environmentally friendly and affordable. The packaging material needs to meet all criteria of shipment, labeling, marketing, and other purposes of packaging.

A list of food and beverage packaging associations appear in the 2012 Food & Beverage Packaging Buyer's Guide. Food&Beverage Packaging. 2012;(12):82–83. foodandbeverage-packaging.com. see Associations and References at the end of this chapter.

Migration from Packaging Materials

The packaging industry recognizes that the migration of substances from the packaging into the food *could* be harmful to the consumer or have an adverse effect on the acceptability of the food. Therefore, compliance with limits set on migration of packaging materials and control of additives at the point of manufacture is ensured.

Plastics have a greater likelihood of imparting their "plastic taste" and odor to a food than paper. They may contain many additives, including antioxidants, antistatics, plasticizers (to improve the flexibility of some "cling" films), and stabilizers to improve the functional properties of the plastics. Although there has been no report of danger to human health from plasticizer additives, the plastics industry has reformulated some grades of films that contain plasticizers and continues to offer polyethylene plastic wraps with low levels of plasticizers

The FDA has responded to the stories about the dangers of chemicals leaching from plastics into microwaved food. Any plastic used as a "food contact surface" needs FDA approval that it is safe for its intended use (usefulness and harmlessness) prior to being approved for use. According to the FDA, "It's true that substances used to make plastics can leach into food. But as part of the approval process the FDA considers the amount of a substance expected to migrate into food and the toxicological concerns about the particular chemical." While the FDA finds that levels of migrating material are well within the established margin of safety, the issue will continually be revaluated in light of new materials or new data.

Diethylhexyl adipate (DEHA) is commonly used in polyvinyl chloride (PVC) food wrap as a plasticizer. PVC does not use the plasticizer with phthalates. A close analysis of DEHA indicated *no* toxic affect in animal studies. It is approved as a food contact surface.

Migration from packaging materials is more likely to occur at high temperatures with fatty foods; therefore, *industry* packaging of microwave foods is designed to be safe for microwave use at high temperatures. *Consumers* who use packaging films for cooking or reheating in the microwave should be aware that "microwave safe" criteria may not be established for packaging films that can have direct contact with food during reheating in a microwave. Therefore, using *glass* containers may be preferable choices for microwave reheating.

In addition to plastics, the printing *ink* on the package must be controlled, as it too imparts undesirable flavor to packaged food, and may stain the surface of material it contacts while hot (i.e., microwave oven).

The use of recycled plastics and paper reduces control over contaminants that may be in the second-hand materials. Further research on the use of recycled materials must be conducted and brought forth, before it is recommended that recycled materials be used in food contact applications (due to the possible migration of contaminants).

There is another concern regarding dioxins in plastic food wrap. According to the FDA, a concern for *dioxins* in plastic is not warranted. "The FDA has seen no evidence that plastic containers or films contain dioxins and knows of no reason why the would" (FDA—Machuga).

Packaging Lines at Processing Plants and Foodservice Operations

Packaging lines at processing plants may operate efficiently when correctly managed, or be down and hold up production. As well, *product shelf life and safety*; *consumer and environmental concerns*, including ease of use and recycling; as well as *economics* of packaging are important food packaging ideas. Future packaging ideas continue to be explored and utilized.

Although labels are discussed in another chapter of this text, they will be mentioned here, in packaging. Paper and perhaps full sleeve, heat shrink PVC labels may be applied to food containers. They offer graphics, assistance as a marketing tool, provide tamper evidence and information, and more. The latest in development and marketing of packaging technologies is available online and at various trade shows.

Packaging with Radio-Frequency Identification Tags

Radio-frequency identification (RFID) tags may be contained in packaging. RFID is more than an inventory or packaging/labeling technology as it assists manufactures and users track packaged food throughout the supply chain. Although it may be required (mandated) of vendors delivering to suppliers, such as major food clubs, trials are still ongoing at this point. Currently, improvement is desired in readability rates, tag costs and availability, tag application, accurate customer recording, and other aspects of this newer technology (Food Engineering).

The RFID Journal is also available to packers.

Selecting the Right RFID Tag becomes important. For example, does the packaging company need high-frequency or ultrahigh-frequency tags, passive or active technology? There is a rapid evolution of RFID technology.

"RFID tags contain a microchip and antenna, and come in a wide variety of sizes and form factors. Some are as small as a grain of rice and encased in glass, while others are enclosed in plastic and the size of a key fob or credit card. Still others, known as smart labels, are embedded in paper. Some are disposable, while others can be reused. Costs vary widely, too, depending on the form factor, the amount of data the tag can store and the volume of tags purchased.

With the increasing use of RFID technology and the adoption of standards for some categories of tags by the International Orga nization for Standardization (ISO) and EPCglobal, tag prices have been falling. Electronic Product Code (EPC) passive UHF tags, widely used in supply-chain applications, cost about 7 cents each. Active (battery-powered) tags, commonly deployed for tracking assets over longer distances, generally cost upwards of $50."

(RFID Journal—How to Select the Right RFID Tag)

Packaging as a Communication and Marketing Tool

Packages contain and protect food during storage, shipment, and sale and serve other functions, as discussed, such as to provide convenience and utility of use. They also *communicate* important consumer information on their labels. For example, information regarding ingredients, nutrition facts, manufacturers' name and address, contents weight, bar coding, and so forth appear on package labels. The food processor must be aware of worldwide differences in acceptability of packaging format, including use of colors and picture symbols, before a product is marketed in another culture.

Packaging serves as a *marketing tool*. The package and label design are significant in attracting potential customers, and many labels may carry recipes, coupons, mail-in offers, or announcements of special upcoming events. It may be that a change in packaging greatly increases sales. Yet changes must not confuse consumers who have built product loyalty by familiarity over the years. For example, milk cartons may not be readily accepted if changed, yet a difference in packaging material, such as cereal without a cardboard box, might be well accepted and profitable.

It is reported that consumers use more of a product at a time when it comes in a *larger* package. This may be attributed to (1) the buy-more, use-more phenomenon, as consumers perceive food products as less expensive when purchased in larger quantity (although this is not always true), (2) less concern with running out, and (3) desire to finish the food product as the large size occupies excessive shelf space (Tufts University Diet & Nutrition Letter 1997). At the other end of the spectrum, *single* servings of products are also popular in the marketplace.

Environmental Considerations and Packaging

Sustainable package development shows environmental consideration and responsibility; *reduce, reuse, and recycle* must be kept in mind from initial packaging development to discards of those packages. Responsibility dictates that packaging materials should be environmentally friendly.

Safety Considerations and Packaging

Keeping the food supply a safe one is an important consideration of food packaging. Packaging is part of food processing, and it assists in preserving food against spoilage and contamination as well as extending the shelf life. Adequate packaging is an industry technique that may be used along with preservation and has the intent of slowing down or stopping spoilage that would otherwise exhibit loss of taste, textural quality, or nutritive value of food. *Crops* and *animal products* produce market-ready and adequately long shelf life food products if adequately packaged.

Of interest to some readers might be irradiation and packaging, addressed by the FDA as follows:

Effects of Irradiation

Irradiation can cause changes to a packaging material that might affect its integrity and functionality as a barrier to chemical or microbial contamination. Radiation does not generally affect all properties of a polymer or adjuvant to the same degree. Two concepts are important here. **First**, most food packaging materials are composed of polymers that may be susceptible to chemical changes induced by ionizing radiation that are the result of two competing reactions, cross-linking (polymerization) and chain scission (degradation). Radiation-induced cross-linking of polymers dominates under vacuum or an inert atmosphere. Chain scission dominates during irradiation of polymers in the presence of oxygen or air. Both reactions are random, generally proportional to dose, and depend on dose rate and the oxygen content of the atmosphere in which the polymer is irradiated. The idea of cross-linking predominating under vacuum or in an inert atmosphere is important because it served as the basis for granting recent exemption requests under 21 *CFR* 170.39 for packaging materials irradiated in contact with food either in a verifiably oxygen-free environment or while frozen and contained under vacuum.

Second, in the presence of an oxygen atmosphere, radiation-induced degradation of both the base polymer and adjuvants, such as antioxidants or stabilizers, is likely to occur and result in the formation of radiolytic products. The radiolytic products formed upon irradiation may be present at significant levels such that they could migrate into food and affect the odor, taste, or safety of the irradiated food. For example, it is well known that certain adjuvants are prone to degradation during polymer processing. During irradiation they would be expected to degrade preferentially over the

> polymer and result in the formation of radiolytic products in the polymer that could potentially migrate into food. Therefore, the migration of both base polymers and adjuvants, as well as migration of their radiolytic products, must be evaluated in the premarket safety assessment of new packaging materials prior to their use, especially at high dose levels and in the presence of oxygen.—FDA

Packaging serves as a safety barrier in controlling potentially damaging levels of light, oxygen, and water. It facilitates ease of use, offers adequate storage, conveys information, and provides evidence of possible safety issues of unsafe product tampering.

Food Traceability

Beginning in 2005, FDA required certain food facilities to maintain records identifying the sources, recipients, and transporters of food products. The purpose of these records is to allow FDA to trace an article of food through each stage of the food supply chain—from a retail shelf back to a farm—if FDA has a reasonable belief that a food product is adulterated and presents a serious health threat. Traceability is the ability to follow the movement of a food product through the stages of production, processing, and distribution. Traceability includes both traceback and trace forward. Traceback is the ability to trace a food product from the retail shelf back to the farm.

Conversely, trace forward is the ability to trace a food product from the farm forward to the retail shelf. Traceability is often needed to identify the sources of food contamination and the recipients of contaminated food in product recalls and seizures. this study refers to such a situation as a 'food emergency.' (Department Of Health And Human Services office of Inspector General 2009) Traceability In The Food Supply Chain

Packaged Food and Irradiation

> CFR 179 specifies that the irradiation of both food and packaging materials in contact with food is subject to premarket approval before introduction of the food into interstate commerce (Reports REGULATORY REPORT | December 2007/January 2008 Food Safety magazine in an article entitled *Irradiation of Food Packaging Materials*).

"The U.S. Food and Drug Administration (FDA) allows the use of irradiation as a means for improving food safety and extending the shelf life of certain foods. Although not yet widely used, irradiation can kill the bacteria responsible for foodborne illness and food spoilage, as well as insects and parasites that may be present on food. Additionally, in certain fruits and vegetables irradiation can inhibit sprouting and delay ripening. For example, irradiated strawberries stay unspoiled up to 3 weeks, versus 3–5 days for untreated berries. Foods are typically packaged in final form prior to being irradiated, thus reducing the likelihood that new pathogens will be introduced after the irradiation step. This means that the packaging materials are being exposed to the same irradiation source as the food itself. In an effort to ensure the safety of the food, one must be certain that an otherwise safe packaging material is not being altered in a fashion that causes a chemical in the packaging to be added indirectly to the food.

Over the years many food packaging materials have been approved for irradiation. Likewise, new food packaging products, such as oxygen barrier materials, have also been introduced. The safety assessment of such complex materials has presented new challenges to FDA and the food industry. This article describes FDA regulations pertaining to packaging materials that are in contact with food during irradiation, the effects of irradiation on new food packaging materials, and the premarket safety assessments of these materials....

Improved microbiological safety of food may be attained by using irradiation in the production of several types of raw or minimally processed foods such as poultry, meat and meat products, fish, seafood, fruits, and vegetables. In fact, following the recent outbreaks of foodborne pathogens in fresh produce, there has been increased interest in using irradiation for improving the safety of fresh produce. However, food manufacturers must ensure that both the irradiated food and packaging materials used during the irradiation process are authorized for the proposed use.

Most significantly, 21 *CFR* 179 specifies that the irradiation of both food and packaging materials in contact with food is subject to premarket approval before introduction of the food into interstate commerce. It further specifies that the current good manufacturing practice for irradiated foods includes three things. First, manufacturers must comply with the general requirements of the current good manufacturing practice for manufacturing, packaging or holding human food, found in 21 *CFR* 110. Second, manufacturers must ensure that the radiation dose used is the minimum dose required to achieve the intended technical effect and does not exceed the level specified in the regulations. Third, the packaging materials used during the irradiation treatment must comply with the requirements found in 21 *CFR* 179.45, the specifications for an effective food contact notification, or a threshold of regulation (TOR) exemption. Foods currently permitted to be irradiated are listed in 21 CFR 179.26(b) [Table 1]. Finally, the irradiated food must be properly labeled.

Although many packaging materials and their components are available for non-irradiation uses, their utilization during the irradiation of prepackaged food is considered a new use and they must pass a premarket safety evaluation before they can be legally used in direct contact with food during irradiation. . . .

Irradiating New Materials

FDA's safety assessment relies on evaluating probable consumer exposure to a food contact substance, including all constituents or impurities as a result of the proposed use and on the available toxicological information. It is important to understand that the safety assessment focuses on those substances that would be expected to become components of food as a result of the proposed use of the food contact substance. The safety information required in a new submission to the Center for Food Safety and Applied Nutrition's Office of Food Additive Safety generally includes chemical, toxicological and environmental components. The chemistry data, include the identity and amounts of migrants, as well as other data to allow the calculation of dietary exposures for the migrants under the intended conditions of use. FDA's toxicological assessment is based on a tiered approach and, therefore, the recommended toxicological data depend on the exposure estimates for the radiolytic products and other migrants from the proposed use. FDA recommends that all data and information be generated in accordance with the available guidance documents. The identities, residue and migration levels, and consumer exposures to the radiolytic products that are generated in the packaging materials may be of concern depending on the regulatory status of the substances, as well as the presence or absence of oxygen. . .

Currently the packaging materials that comply with the provisions of 21 *CFR* 179.45, the specifications of an effective food FCN [food contact notification] or a TOR [threshold of regulation] exemption may be used in direct contact with food during irradiation. Materials that do not fall under one of these umbrellas require a premarket safety assessment before being used in foods. In instances where food packages contain oxygen, such as in modified atmosphere packaging to maintain the quality of fresh produce, the safety assessments can be quite complex... FDA encourages individuals to discuss proposed studies with FDA prior to the submitting a petition or notification to ensure that the studies will address FDA's safety concerns, and provide adequate data. ...

Conclusion
Improved microbiological safety of food may be attained by using irradiation in the production of several types of raw or minimally processed foods such as poultry, meat and meat products, fish, seafood, fruits, and vegetables. In fact, following the recent outbreaks of foodborne pathogens in fresh produce, there has been increased interest in using irradiation for improving the safety of fresh produce. However, food manufacturers must ensure that both the irradiated food and packaging materials used during the irradiation process are authorized for the proposed use."

http://www.foodsafetymagazine.com/magazine-archive1/December-2007january-2008/irradiation-of-food-packaging- materials

Government Concerns in Packaging

Government concerns must be addressed as they arise, and prevention must be practiced as necessary to avoid errors that lead to illness.

FDA Packaging & Food Contact Substances (FCS)

http://www.fda.gov/Food/IngredientsPackagingLabeling/PackagingFCS/default.htm

FDA Irradiated Food & Packaging

http://www.fda.gov/Food/IngredientsPackagingLabeling/IrradiatedFoodPackaging/default.htm

FDA is responsible for regulating the use of irradiation in the treatment of food and food packaging. This authority derives from the 1958 Food Additives Amendment to the Federal Food, Drug, and Cosmetic Act (FD&C Act) where ***Congress explicitly defined a source of radiation as a food additive*** (Section 201(s) of the FD&C Act). FDA

FDA Environmental Decisions

http://www.fda.gov/Food/IngredientsPackagingLabeling/EnvironmentalDecisions/default.htm

The National Environmental Policy Act of 1969 (NEPA) requires all agencies of the Federal Government to take, to the fullest extent possible, environmental considerations into account in the planning and making of their major and final agency decisions. In a manner that is consistent with the FDA's authority under the Federal Food, Drug, and Cosmetic Act, the Public Health Service Act, and with its NEPA-implementing regulations in 21 CFR 25, the agency assesses, as an integral part of its decisionmaking process, the environmental impacts of its actions and ensures that the interested and affected public is informed of environmental analyses. Such actions include approval of a food additive, color additive, or (Generally Recognized as Safe) GRAS affirmation petitions, the granting of a request for exemption from regulation as a food additive under 21 CFR 170.39, or allowing a notification submitted under 21 USC 348(h) to become effective. FDA

"Packaging or equipment that contacts food may be subject to FDA regulation if their chemical components are deemed by FDA to be "indirect food additives," also known as "Food Contact Substances." (FCS) The determination of how a particular food contact substance is regulated by FDA depends on its chemical composition." http://www.registrarcorp.com/fda-food/contact-substances/index.jsp?lang=en

Packaging of the Future

http://www.futureofpackaging.com/ represents Packaging Technology Integrated Solutions (PTIS).

"The Future of Packaging: 2013–2023 is an invitation-only event and is limited to industry leaders from across the packaging value chain. We keep the number of sponsors to a small group to enable an environment rich with dialogue, content and value. Proprietary research and insights, strategies, and tools for driving growth will be shared exclusively with sponsors and will not be made available at any other time.

> The Power of Packaging is unleashed at The Future of Packaging: 2013–2023. If you are responsible for ensuring that packaging drives bottom-line impact and growth for your business, request your invitation today to gain a long-term strategic, yet practical understanding of packaging that will pay dividends today and for years to come. You'll come away with:
>
> - An understanding of why Holistic Packaging Design is central to packaging in the future
> - Cutting-edge, proprietary research on packaging innovation, global consumer trends, mass customization, emerging markets, and the role of mobile and social technology
> - A global view of scenarios affecting packaging over the next decade
> - A 10-year strategic packaging roadmap—PTIS

In its year-end e-mail message, consulting company Packaging Technology Integrated Solutions (PTIS) believe the Top 9 trends for packaging in 2013 will be:

10. **Clean machine design** helps drive OEM growth and meet new regulations.
9. **Digital printing** continues to make advances and becomes a reality.
8. **Sustainable packaging and waste to value** propositions get more clarity and harmonization.
7. **Interactive and intelligent packaging** becomes more real and gets commercialized.
6. **Flexible packaging** growing in new markets and categories globally.
5. **Companies ramp up new processes** to look at risk and anticipatory issues.
4. **Digital media and package design:** Social media and zero moment of truth will offer big opportunities for packaging going forward.
3. **Packaging enabled innovation and growth:** More companies targeting a percentage of their growth driven by packaging.
2. **Consumer insight for packaging:** Companies recognize the importance of consumer insights for packaging and how it can vary, and expect more data from suppliers."

PTIS asks readers to submit ideas for the **number-one** trend in packaging for 2013!

See more at

http://www.packworld.com/trends-and-issues/oee-amp-lean-manufacturing/top-9-packaging-trends-2013

Conclusion

Primary, secondary, and tertiary packaging protects raw and processed food against spoilage and contamination while offering convenience and product information to the consumer. A variety of packaging materials, such as metal, glass, paper, and plastics or combinations of these, are used for packaging if they meet with FDA requirements. A number of professional trade shows are dedicated to packaging technologies.

Packaging films and atmospheres may be selected according to a food product's storage and distribution needs. They may eliminate damaging levels of oxygen, light, and temperature and prevent water vapor loss while at the same time protecting the food from spoilage and contamination. Compliance with limits set on migration of packaging materials and control of additives in the materials must be ensured. A variety of packaging technologies, including vacuum packaging, flushing the package with gas, or *active* packaging, may be used to contribute to effective packaging of food.

Packaging protects food; may modify atmosphere, thus extending shelf life; gets a message across; aids in marketing; provides added content security; and so forth. It may provide portion control, combating "portion distortion," convenience of use, and convenient transport. This assists the child as well as the adult consumer.

Tamper-evident banding is a protection against external hazards and may be viewed as essential by the manufacturer and consumer. The package label may communicate important information to the consumer serving as a marketing tool. Packaging today provides protection for food products that was unavailable in the past.

Reduce, reuse, and recycle are important to the environment. More advances in such considerations as safe, effective, convenient, and creative packaging are on the horizon.

Packaging may simply involve having clean hands, a sanitary pair of foodservice tongs, and a piece of tissue in a bag served across a bakery counter. It may involve adherence to a specific time and temperature heat application in a foodservice can that is transported along continents. There is great variance in the idea of packaging food as discussed in this chapter.

(some articles are old, yet are retained as they are classic in their content)

Notes

CULINARY ALERT!

Glossary

Active packaging Packaging that makes an active, not passive, contribution to product development or shelf life by such techniques as providing an oxygen barrier, or odor and oxygen scavenger.

Aseptic packaging Independent sterilization of foods and packaging with assembly under sterile conditions.

Cavity ice Ice formation with the frozen food package due to water condensation and freezing.

Controlled atmosphere packaging (CAP) Controls O_2, CO_2, water vapor, and ethylene concentration.

Flexible packaging Nonrigid packaging such as stand-up pouches, tubes, or zippered bags.

Freezer burn Desiccation of frozen food product as the water diffuses from the frozen food to the atmosphere.

Modified Atmosphere Packaging (MAP) Or gas flush packaging—modification of O_2, CO_2, water vapor, and ethylene concentration by flushing with nitrogen gas.

Polyethylene Most common, least expensive plastic film used in packaging material.

Polystyrene Plastic type that is typically foamed to create expandable polystyrene or styrofoam.

Polyvinyl chloride (PVC or vinyl) Plastic packaging film.

Polyvinylidene chloride (PVDC or Saran®) Plastic packaging film.

Primary container A direct food contact surface as in bottle, can, or drink box that contains food or beverage.

Secondary container Does not have food contact but holds several primary containers in materials such as corrugated fiberboard, boxes, or wraps.

Sous vide Mild, partial precooking to reduce the microbial load, followed by vacuum packaging to extend the shelf life.

Tamper-evident banding Sleeves or neckbands providing protection and offering security by indicating evidence of tampering with the product.

Tertiary container Holds several secondary containers in corrugated fiberboard boxes, overwraps, and so forth.

Vacuum packaging Removes all atmosphere from the pouch and creates a skintight package wall.

References

(1997) The bigger the package, the more you eat. Tufts Univ Diet & Nutr Lett 14(11):1–2

Brody AL (2000) Smart packaging becomes Intellipack®. Food Technol 54(6):104–106

Higgins KT (2000) Not just a pretty face. Food Eng 72 (5):74–77

Higgins KT (2001a) Active packaging gets a boost. Food Eng 73(10):20

Higgins KT (2001b) Security takes center stage. Food Eng 73(12):20

Krochta JM, DuMulder-Johnston C (1997) Edible and biodegradable polymer films: challenges and opportunities. A publication of the Institute of Food Technologists' Expert Panel on Food Safety and Nutrition. Food Technol 51(2):61–74

Labuza TP (1996) An introduction to active packaging for foods. Food Technol 50(4):68–71

Sloan AE (1996) The silent salesman. Food Technol 50 (12):25

What is the difference between food processing and preservation? http://www.wisegeek.com/what-is-the-difference-between-food-processing-and-preservation.htm)

Bibliography

21 CFR 179.45 Table 2 Packaging materials listed for use during irradiation of prepackaged foods

21 CFR 179.26 (b) Table 1 Foods permitted to be irradiated (as of Oct. 2007)

Association of Nutrition and Foodservice Professionals (ANFP)

Food and Agriculture Organization of the United Nations (FAO)

Global Supplier Quality Assurance (GSQA)

Sonoco Products Company. Hartsville, South Carolina (Sonoco was named the top global packaging company for sustainability and corporate responsibility in the 2011 and 2012 the Dow Jones Sustainability World Index)

Associations and Organizations

A list of food and beverage packaging associations appears in the 2012 Food & Beverage Packaging Buyer's Guide. Food & Beverage Packaging. 2012; (12):82–83. foodandbeveragepackaging.com

Part VIII

Food Safety

Food Safety

19

Introduction

"Food is our common ground, a universal experience," says James Beard. However, we must keep it safe! Food safety is an important issue today as there are many demands on the food production system and a variety of food handlers serving numerous individuals who are immunocompromised.

Providing safe food is the responsibility of many individuals/groups. For example, federal agencies such as the U.S. Food and Drug Administration (FDA) and the U.S. Department of Agriculture (USDA), Centers for Disease Control and Prevention (CDC), as well as the state and local counterparts, numerous professional organizations, food processors, and consumers are all interested in preventing the occurrence of foodborne illness. Further discussion of government regulation and labeling appears in a later chapter (see Chap. 20). The FDA ranking of food safety concerns, according to risk ranks foodborne illness as the primary concern, followed by nutritional adequacy of foods, environmental contaminants, naturally occurring toxicants, pesticide residues, and food additives (FDA).

While efforts are made to educate the consumer regarding food safety, hazards in the food supply may be controlled/prevented *before* foods reach the consumer. It is said that the achievements of scientists are not readily explained to the public (Stier 2006), yet it is known that the United States has one of the most diverse and safe food supplies in the world.

The effective use of the Hazard Analysis and Critical Control Point (HACCP) method of food safety, practiced in the food processing industry, has been shown to yield safer foods. Irradiation is also employed to reduce the incidence of disease.

As well, there are numerous aspects of preservation and processing, additives, packaging, and government regulation that contribute to food safety. These are discussed in Chaps. 17–20.

Many tables and much data are included in this chapter! Food safety is addressed in a plethora of reference articles, some of which are included in this chapter. Relevant internet sites may appear in order to emphasize specific points. The food safety of individual food commodities is also discussed in appropriate chapters throughout this text.

> Food & Drug Administration (FDA) has jurisdiction over 80 percent of the food supply, including seafood, dairy and produce. The US Dept of Agriculture (USDA) regulates meat, poultry and processed egg products, while FDA regulates all other food products.
>
> - FDA, USDA, National Oceanic and Atmospheric Administration (NOAA) Statements on Food Safety

Examples that appear in this chapter on foodborne illness and food safety are varied and the reader information may be applied in the

V.A. Vaclavik and E.W. Christian, *Essentials of Food Science, 4th Edition*, Food Science Text Series,
DOI 10.1007/978-1-4614-9138-5_19,

retail foodservice operation and commercial warehouse or at home.

Foodborne Illness

Since the FDA ranks foodborne illness as the *primary* food safety concern (Department of Health and Human Services, Public Health Service, Food and Drug Administration), this chapter will focus on its *causes* and *prevention*. *Foodborne illness* represents disease carried to people by food and is the result of various biological, chemical, or physical hazards to the food supply. These hazards will be addressed in this chapter.

Foodborne illness is typically due to ingestion of contaminated *animal* products, yet *plant* foods may be implicated as a result of airborne, water, soil, insect, or even human contamination when they are grown or raised. Recently, national foodborne illness cases, including one death, were the result of eating contaminated bagged spinach.

Natural and synthetic foods that support the growth of microorganisms are classified by the FDA as *potentially hazardous foods* (phf), defined as follows:

(a) "Potentially hazardous food" means a FOOD that is natural or synthetic and that requires temperature control because it is in a form capable of supporting:
  - (i) The rapid and progressive growth of infectious or toxigenic microorganisms
  - (ii) The growth and toxin production of *Clostridium botulinum*
  - (iii) In shell eggs, the growth of *Salmonella enteritidis*

(b) "Potentially hazardous food" includes an animal FOOD (a FOOD of animal origin) that is raw or heat-treated, a FOOD of plant origin that is heat-treated or consists of raw seed sprouts, cut melons, and garlic and oil mixtures that are not acidified or otherwise modified.

(c) "Potentially hazardous food" does not include:
  - (i) An air-cooled hard-boiled egg with shell intact
  - (ii) A FOOD with a WATER ACTIVITY (A_w) value of 0.85 or less
  - (iii) A FOOD with a pH level of 4.6 or below when measured at 75 °F (24 °C)
  - (iv) A FOOD, in an unopened HERMETICALLY SEALED CONTAINER, that is commercially processed to achieve and maintain commercial sterility under conditions of nonrefrigerated storage and distribution
  - (v) A FOOD for which laboratory evidence demonstrates that rapid and progressive growth of infectious and toxigenic microorganisms or the growth of *S. enteritidis* in eggs or *C. botulinum* cannot occur, such as a food that has an A_w and a pH that are above the levels specified above that may contain a preservative or other barriers to growth
  - (vi) A FOOD that may contain an infectious or toxigenic microorganism or chemical, physical contaminant at a level sufficient to cause illness but that does not support the growth of microorganisms as specified in the definition of potentially hazardous food

While *prevention* policies are the first line of defense against hazards (avoidance in the first place), control of biological or chemical agents, or physical objects as well as *rapid detection* of contaminants, is imperative to food safety. Any risk of disease must be controlled throughout the steps of manufacturing, processing, storage, and distribution as well as final cleanup of foods, equipment and utensils, and the food prep areas.

Some *examples* of potentially hazardous foods are products that contain:

• Meat	• Shellfish
• Poultry	• Some synthetic ingredients
• Eggs	• Tofu
• Milk	• Baked potatoes
• Fish	• Cut melon

Wise actions are needed in combating foodborne illness. The government must regulate the food supply, and both the manufacturer and consumer play vital roles in food safety.

Food manufacturers are significantly involved in food safety. "Quality control and anti-tampering measures developed by the food industry in cooperation with government agencies over the past 2 decades have made the U.S. food supply the safest in the world. Since the September 11 [2001] attacks, our industry has recognized that we must take additional proactive measures to ensure safety of consumers. The safeguards that we developed to address long-standing food safety issues and past tampering incidents are being reexamined, strengthened and enforced with vigilance in light of these recent events." (International Foodservice Distributors Association, IFDA, Mclean, VA)

Biological (Microbiological) Hazards to the Food Supply

Biological hazards that cause foodborne illness include *microorganisms* such as bacteria, viruses, fungi, and parasites. These may be small in size, yet they can cause serious foodborne illness or death. Biological hazards to food are controlled by the following:

- Temperature—adequate cooking, cooling, refrigeration, freezing, and handling
- The avoidance of cross-contamination
- Enforcement of personal hygiene among food handlers

Bacteria: The Major Biological Foodborne Illness

Bacteria are the *primary* microbiological hazard organism implicated in foodborne disease and are therefore the *primary* microbial concern of many consumers, food processors, microbiologists, and other personnel responsible for producing and serving safe food.

Bacteria cause foodborne illnesses by one of three manners: infection, intoxication, or toxin-mediated infection as noted in the following:

Foodborne *infection* results from ingesting *living*, pathogenic bacteria such as *Salmonella, Listeria monocytogenes*, or *Shigella* (see Fig. 19.1).

Foodborne *intoxication* results if a preformed *toxin* (poison) is ingested, such as that produced by *Staphylococcus aureus*, *Clostridium botulinum*, and *Bacillus cereus* present in the food (Fig. 19.1).

A *toxin-mediated infection* is caused by ingestion of *living*, infection-causing bacteria such as *C. perfringens* and *E. coli* O157:H7 that also produce a *toxin* in the intestine (Figs. 19.1 and 19.2).

The Educational Foundation of the National Restaurant Association has compiled data on the most common *pathogenic,* or disease-causing bacteria in foods (The Educational Foundation of the National Restaurant Association, Chicago, IL). In Fig. 19.1, the bacteria name, incubation period, duration of illness, symptoms, reservoir, foods implicated, and means of prevention in foods are presented. An astute manager of a food manufacturing operation (as well as the consumer at home!) understands the benefit of having this knowledge and applying this food safety information to their own food products. Such understanding and practice promotes customer goodwill and prevents foodborne illness.

Major foodborne diseases of bacterial origin							
	Salmonellosis Infection	Shigellosis Infection	Listeriosis Infection	Staphyloccal Intoxication	Clostridium Perfringens Toxin Mediated Infection	Bacillus Cereus Intoxication	Botulism Intoxication
Bacteria	*Salmonella* (facultative)	*Shigella* (facultative)	*Listeria monocytogenes* (reduced oxygen)	*Staphylococcus aureus* (facultative)	*Clostridium perfringens* (anaerobic)	*Bacillus cereus* (facultative)	*Clostridium botulinum* (anaerobic)
Incubation Period	6–72 hours	1–7 days	1 day to 3 weeks	1–6 hours	8–22 hours	1/2–5 hours; 8–16 hours	12–36 hours + 72
Duration of Illness	2–3 days	Indefinite, depends on treatment	Indefinite, depends on treatment, but has high fatality in the immuno-compromised	24–48 hours	24 hours	6–24 hours; 12 hours	Several days to a year
Symptoms	Abdominal pain, headache, nausea, vomiting, fever, diarrhea	Diarrhea, fever, chills, lassitude, dehydration	Nausea, vomiting, headache, fever, chills, backache, meningitis	Nausea, vomiting, diarrhea, dehydration	Abdominal pain, diarrhea	Nausea and vomiting; diarrhea, abdominal cramps	Vertigo, visual disturbances, inability to swallow, respiratory paralysis
Reservoir	Domestic and wild animals; also humans, especially as carriers	Human feces, flies	Humans, domestic and wild animals, fowl, soil, water, mud	Humans (skin, nose, throat, infected sores); also, animals	Humans (intestinal tract), animals, and soil	Soil and dust	Soil, water
Foods Implicated	Poultry and poultry salads, meat and meat products, milk, shell eggs, egg custards and sauces, and other protein foods	Potato, tuna, shrimp, turkey and macaroni salads, lettuce, moist and mixed foods	Unpasteurized milk and cheese, vegetables, poultry and meats, seafood, and prepared, chilled, ready-to-eat foods	Warmed-over foods, ham and other meats, dairy products, custards, potato salad, cream-filled pastries, and other protein foods	Meat that has been boiled, steamed, braised, stewed or roasted at low temperature for a long period of time, or cooled slowly before serving	Rice and rice dishes, custards, seasonings, dry food mixes, spices, puddings, cereal products, sauces, vegetable dishes, meat loaf	Improperly processed canned goods of low-acid foods, garlic-in-oil products, grilled onions, stews, meat/poultry loaves
Spore Former	No	No	No	No	Yes	Yes	Yes
Prevention	Avoid cross-contamination, refrigerate food, cool cooked meats and meat products properly, avoid fecal contamination from foodhandlers by practicing good personal hygiene	Avoid cross-contamination, avoid fecal contamination from foodhandlers by practicing good personal hygiene, use sanitary food and water sources, control flies	Use only pasteurized milk and dairy products, cook foods to proper temperatures, avoid cross-contamination	Avoid contamination from bare hands, exclude sick foodhandlers from food preparation and serving, practice good personal hygiene, practice sanitary habits, proper heating and refrigeration of food	Use careful time and temperature control in cooling and reheating cooked meat dishes and products	Use careful time and temperature control and quick chilling methods to cool foods, hold hot foods above 140°F (60°C), reheat leftovers to 165°F (74°C)	Do not use homecanned products, use careful time and temperature control for sous-vide items and all large, bulky foods keep sous-vide packages refrigerated, purchase garlic-in-oil in small quantities for immediate use, cook onions only on request

Fig. 19.1 Major foodborne diseases of bacterial origin (*Source*: Reprinted with permission from *Applied Foodservice Sanitation: A Certification Coursebook, Fourth Edition*, © 1992, The Educational Foundation of the National Restaurant Association)

As well, the CDC has compiled extensive data. Some of that data is available at the end of this chapter.

Beyond the immediacy of illness, there is also USDA increasing evidence that foodborne gastrointestinal (GI) pathogens may give rise to other illness such as chronic joint disease, i.e., arthritis (Agricultural Research Service, USDA, Washington, DC).

As is the case of *bacteria in general*, the *bacteria causing foodborne illness may require* the following elements for growth:

Campylobacteriosis Infection	*E. coli* 0157: H7 Infection/Intoxication	Norwalk Virus Illness
Campylobacter jejuni	*Escherichia coli*	Norwalk and Norwalk-like viral agent
3–5 days	12–72 hours	24–48 hours
1–4 days	1–3 days	24–48 hours
Diarrhea, fever, nausea, abdominal pain, headache	Bloody diarrhea; severe abdominal pain, nausea, vomiting, diarrhea, and occasionally fever	Nausea, vomiting, diarrhea, abdominal pain, headache, and low-grade fever
Domestic and wild animals	Humans (intestinal tract); animals, particularly cattle	Humans (intestinal tract)
Raw vegetables, unpasteurized milk and dairy products, poultry, pork, beef, and lamb	Raw and undercooked beef and other red meats, imported cheeses, unpasteurized milk, raw finfish, cream pies, mashed potatoes, and other prepared foods potatoes, and other prepared foods	Raw vegetables, prepared salads, raw shellfish, and water contaminated from human feces
No	No	No
Avoid cross-contamination, cook foods thoroughly	Cook beef and red meats thoroughly, avoid cross-contamination, use safe food and water supplies, avoid fecal contamination from foodhandlers by practicing good personal hygiene	Use safe food and water supplies, avoid fecal contamination from foodhandlers by practicing good personal hygiene, thoroughly cook foods

Fig. 19.2 Emerging pathogens that cause foodborne illness (*Source:* Reprinted with permission from *Applied Foodservice Sanitation: A Certification Coursebook*, 4th ed., © 1992, The Educational Foundation of the National Restaurant Association)

- **Protein** (or sufficient nutrients)
- **Moisture** [water activity (A_w) above 0.85]
- **pH** (above pH of 4.5, generally neutral—pH 7)
- **Oxygen** (if aerobic)
- A general **temperature** 40–140 °F (4–60 °C), the *temperature danger zone* (TDZ) (consult your local jurisdiction for specific temperature requirements that may differ)

Bacterial growth is portrayed in Fig. 19.3. Bacteria vary in their *temperature* requirements—e.g., they may be *thermophiles* (high temperatures needed for survival), *mesophiles,* or *psychrotrophs* (cooler temperatures of 50–70 °F [10–20 °C] requirements). Bacteria also vary in their *nutrient needs.*

Once bacteria are in the TDZ, they remain in the LAG phase of bacterial growth for approximately 4 h (cumulative); there is generally no increase in number. Yet, due to unsafe temperatures of holding food, or especially

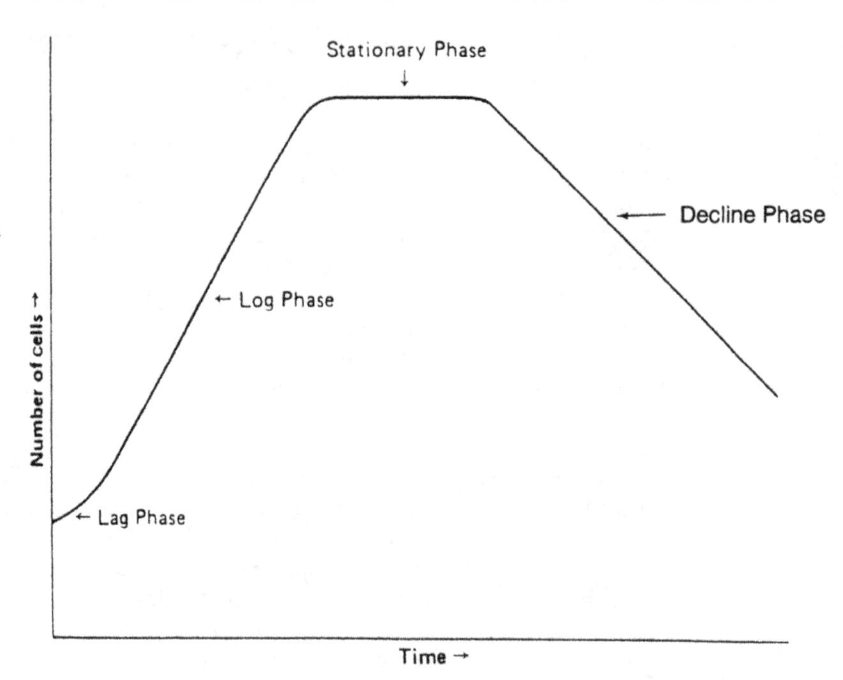

Fig. 19.3 Bacterial growth curve (*Source*: Reprinted with permission from *Applied Foodservice Sanitation: A Certification Coursebook*, 4th ed., © 1992, The Educational Foundation of the National Restaurant Association)

improper cooling, bacterial growth becomes exponential. Then, after the LAG phase, the unicellular structure undergoes binary fission, and rapid growth occurs in foods which are phfs. This rapid growth or multiplication of bacteria is termed the LOG (logarithmic) phase of bacterial growth. It is followed by the STATIONARY phase, where growth rates approximate death rates and there is no net change in the number of pathogens in the food, and, subsequently, by the DECLINE phase of bacterial growth, where the level of bacteria is reduced.

It is important to recognize the fact that although the end of the DECLINE phase may show *less* than the original amount of bacteria, this phase may contain *more* harmful waste products or toxins that *cannot* be destroyed by cooking. In addition to toxins, the two *Clostridium* bacteria and *Bacillus cereus* may contain *spores* (unlike the spores of mold), which are highly resistant formations in bacterium that remain in food, even after vegetative cells are destroyed.

As mentioned, this makes *prevention* of food contamination the *primary* defense against foodborne illness—harmful substances or possibly their waste products *cannot* be destroyed later through cooking. A careful time–temperature control of potentially hazardous food is required. Refrigeration, for example, *slows* growth, and freezing *halts* growth; however, neither *destroys* bacteria.

The CDC reports that *improper cooling* of large quantities of food is the number one cause of foodborne illness. (Despite the fact that large quantities of food may be in a pot placed in the refrigerator at cold temperatures, bacteria do not "know" that they are in the refrigerator. Rather, they are in a large, warm stockpot or steam table pan and LOVE it!)

Adherence to specific heat application temperatures aids in promoting food safety and is required of food preparation staff. The topic of food temperature requires important consideration. There is the traditional food thermometer that may be used as part of ensuring food safety. With this instrument, food temperature is read when the thermometer is inserted up to the "dimple" mark appearing on the thermometer stem. The thermometer reading presented represents an average temperature of all food contacting the length of the thermometer stem up to the "dimple."

In another mention of temperature, use of a device known as a TTI (**time–temperature indicator**) or a smart label shows the accumulated time–temperature history of a product. The indicator is commonly used to indicate exposure to excessive temperature (and length of time at temperature). The approximate dimensions of the TTI labels are 47 × 78 mm or 1.8″ × 3″. Time–temperature indicator labels are designed to be used during processing, storage, and shipping. The labels change color irreversibly if there is an unacceptable temperature exposure.

As well as using a TTI or a smart label, there is a wide array of products, including the plastic *pop-up* type of temperature indicators that were once only used for cooking the traditional holiday turkey—with their use, temperature control in cooking is better assured. The *Food Temperature Indicator Association* works with the USDA and food manufacturers to conduct studies regarding temperature and food safety. As mentioned, such specialized pop-up timers may be on items such as turkeys, yet also on a variety of meats and fish to indicate doneness. Paper thermometers to save in files or dispose of after each use are available. Paper may be used to measure temperatures with food or dishware.

Reading thermometers is important. A pop-up plastic thermometer may indicate doneness by a simple *binary* indicator. (Believe it or not, it may be remelted, pushed closed, cooled, and then used again!) *Thermocouple thermometers* measure in a matter of seconds by utilizing two wires located at the tip of the probe inserted into the food. A thermocouple is designed to be used near the end of the cooking cycle. *Thermistors* are another temperature tool. They are designed for use outside of the oven and take approximately 10 s to register the food temperature. Due to the fact that the semiconductor used to measure temperature is located in the tip, either thin or thick foods may be "temped."

A *thermometer fork* is another tool used to monitor a safe food supply—usually it had outdoor cooking applications. The fork uses a thermocouple or thermistor in the fork tines to read the food temperature which is then displayed on the fork handle. These forks or perhaps spatulas are frequently used in outdoor grilling applications and are able to accurately measure even thin foods.

Today, the USDA's Food Safety and Inspection Service (FSIS) requires establishments that slaughter cattle, chicken, swine, and turkeys to test specifically for bacteria *E. coli*. They must verify the adequacy of process controls for the prevention and removal of fecal contamination and associated bacteria. Additionally, the FSIS has extended such testing to establishments that slaughter species including ducks, equines, geese, goats, guineas, and sheep.

According to the manufacturer of one TTI, "From supplier through warehousing and distribution all the way to the consumer, the ThermoTrace TTI system gives customers the exact data they need to assess product quality and shelf life." Its size is approximately 47 × 78 mm or 1.8″ × 3″ (Fig. 19.4).

Viruses

In addition to bacteria, although with lesser incidence, viruses may also be responsible for an unsafe food supply and foodborne illness. A virus does *not* multiply in food, as do bacteria, yet it can *remain* in food if it is insufficiently cooked. Subsequently, viruses infect individuals who ingest it. It is possible for *spot contamination* of food to occur, so that only those individuals consuming the contaminated portion of the food become ill.

A virus of concern to the consumer, or a food processing and handling operation, is the hepatitis A virus. A person will become infected with the virus 15–50 days following ingestion of a contaminated product and will shed the virus unknowingly, contaminating *other* people or food *prior* to displaying symptoms of illness. Although the actual infection may last several

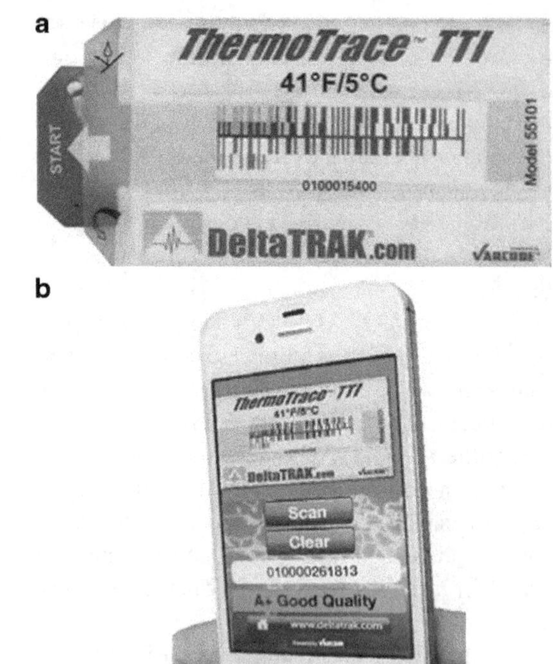

Fig. 19.4 (**a**) DeltaTRAK's ThermoTrace time temperature indicator (TTI) (Source: "courtesy of DeltaTRAK Inc. http://www.deltatrak.com".). (**b**) ThermoTrace time temperature indicator (TTI) used on a smart phone ® (Source: "courtesy of DeltaTRAK Inc. http://www.deltatrak.com".)

weeks or months and exhibit symptoms such as abdominal pain, jaundice, and nausea, there are possible widespread and long-lasting financial implications of this illness to the business that is responsible for its spread.

Two sources of the hepatitis A virus are (a) raw shellfish from polluted water where sewage is discarded and (b) feces (and urine) of infected persons. To control the first listed source, the growth and harvesting of raw shellfish (clams, oysters, mussels) is subject to inspection and regulation by the FDA. The FDA inspects the water beds from which shellfish is harvested. Also, a *tag* must appear on commercial fresh shellfish to show its source. The tag needs to be retained by the receiver for 90 days. Unfortunately, some unreputable suppliers may obtain their shellfish supply from "off-limits" contaminated water, thus harvesting a contaminated product.

Control for the second source of the hepatitis virus is that consumers at home and food handlers in food processing or assembly operations must practice *good personal hygiene*. Just minute amounts of feces may spot contaminate food, causing foodborne illness when the food is ingested. A number of state or local health department jurisdictions require the use of disposable gloves by food handlers responsible for handling food that is not subject to further cooking.

Hepatitis A is of major concern. Another virus of significance to the consumer and food processing operation is HIV. The CDC states that there is no evidence that this can be transmitted by food (CDC).

Fungi

Mold and yeast are *fungi* that may be responsible for *spoilage* in the food supply. Details on each appear below.

Mold

The (accidental) ingestion of mold is not known to cause gastrointestinal distress. Rather, it has been implicated in other long-term illness, such as liver cancer, in animals that have been fed moldy crops. Mold obviously causes food spoilage. It also causes loss of food, dissatisfied consumers, and waste of money.

Mold is a multicellular fungus that reproduces by spore formation. After spores form, they are then dispersed through the air and may replicate when in contact with food (mold spores are unlike bacterial spores). Mold is the unwanted blue, green, white, and black fuzzy growth on food. It may be considered acceptable in medicine, such as penicillin, or some cheeses such as blue cheese. A small percent of persons may fatally suffer from allergies to molds.

Yeast. Yeast is a unicellular structure that grows by the budding process. It causes food *spoilage*, as is evidenced by the formation of pink patches on moist cheeses or cloudy liquid in condiment (such as olives) jars. Foodborne yeast has *not* been shown to cause illness, yet, nonetheless, undesirable growth of yeast must be controlled, or food is damaged and wasted. Yeast is *generally* shown to have beneficial uses in the food industry such as when it leavens baked products or is used in fermentation to produce alcoholic beverages.

Parasites

Parasites may be a source of foodborne illness. Parasites are tiny organisms that depend on living hosts for their nourishment and life. Undercooked pork products, for example, may carry the parasite *Trichinella spiralis*, which causes the disease trichinosis. Two days to 28 days following ingestion of the Trichinella parasite, an individual may exhibit nausea, vomiting, abdominal pain, and swelling of tissue surrounding the eye. Fever and muscular stiffness then develop. Since pork may be contaminated with *Trichinella spiralis*, all pork products must be cooked to 155 °F (68 °C) [or 170 °F (77 °C) if cooked in a microwave oven], and all equipment used in its preparation should be sanitized.

Fish from *unapproved* sources may carry the parasite Anisakis and result in the parasitic disease *anisakiasis*. Reputable suppliers to processing plants are the best assurance that the product has been handled safely. Freezing for the correct time and temperature actually can kill Anisakis. On the other hand, when fresh fish is served *raw*, food safety takes on new significance!

Contamination and Spoilage

Contamination and spoilage are not the same. The latter *may not* ever cause illness due to the fact that the consumer *sees* signs of spoilage and that spoiled food never gets eaten! Illness is more likely to be the result of ingesting unseen microbial (or chemical) contamination. Thus, it is *contaminated* food that is truly "bad" food—is *not* apparent to the eye. Impure or harmful substances may be too small or unnoticeable. *Spoiled* food has visible damage to the eating quality of food and is *not* the primary cause of foodborne illness.

In order to maintain food safety, any chance of initial contamination should be prevented and then subsequent growth controlled. *Cross-contamination*, or the transfer of germs from one contaminated food or place to another by hands, equipment, or other foods, should be avoided.

Beside those pathogens mentioned, there is the possibility of contamination by other *emerging* pathogens. Their incidence has increased within the last few years or they threaten to increase in the near future.

Chemical Hazards to the Food Supply

All food is made of chemicals and is expected to be safe for consumption. However, a chemical hazard to the food supply may occur when dosages or levels of specific chemicals reach toxic levels. Hazards may be accidental, caused by additives, by toxic metals, or naturally occurring.

As mentioned previously, chemical contamination includes accidental chemical contamination, such as when contents of a container, perhaps unlabeled, are mistakenly used in food. Excessive quantities of additives become problematic especially when an individual has a specific allergy.

As well, included in the list of chemical hazards are toxic metals such as galvanized iron. Steel may permanently bond to the metal zinc through galvanizing. Such zinc-coated material may be beneficially used for building fabrication or for shelving, however, should be avoided as a food contact surface since it is highly reactive with acids. In the past, containers used for beverages, temporary working surfaces, and shelving made of toxic galvanized steel had been part of many restaurant operations.

One additional type of chemical hazard is the animal/plant foodstuff itself. Naturally occurring toxins in various foods such as the puffer fish or different mushrooms may cause severe illness.

Control of chemical hazards prior to receipt or use, and control in inventory, storage, and

Table 19.1 Control of chemical hazards (Watson)

I.	*Control before receipt*. Raw material specifications; vendor certification/guarantees; spot checks—verification
II.	*Control before use*. Review purpose for use of chemical; ensure proper purity, formulation, and labeling; control quantities used
III.	*Control storage and handling conditions*. Prevent conditions conducive to production of naturally occurring toxicants
IV.	*Inventory all chemicals in facility*. Review uses and records of use

Source: Ref. Watson DH. *Safety of Chemicals in Food: Chemical Contaminants*

Table 19.2 Main materials of concern as physical hazards and common sources

Material	Injury potential	Sources
Glass	Cuts, bleeding; may require surgery to find or remove	Bottles, jars, light fixtures, utensils, gauge covers
Wood	Cuts, infection, choking; may require surgery to remove	Fields, pallets, boxes, buildings
Stones	Choking, broken teeth	Fields, buildings
Metal	Cuts, infection; may require surgery to remove	Machinery, fields, wire, employees
Insects, other filth	Illness, trauma, choking	Fields, plant postprocess entry
Insulation	Choking, long term if asbestos	Building materials
Bone	Choking, trauma	Fields, improper plant processing
Plastic	Choking, cuts, infection; may require surgery to remove	Fields, plant packaging materials, pallets, employees
Personal effects	Choking, cuts, broken teeth; may require surgery to remove	Employees

Source: Adaptation from Corlett (Pierson & Corlett 1992)

handling are identified in Table 19.1. Care must be taken to avoid chemical hazards.

Cleaning and sanitizing solutions must be safely stored and utilized. Of course they need to be appropriately measured for strength to be effective. Whether the solutions are in a cleaning and sanitizing bucket; spray bottle; sink, such as a three-compartment sink; or dish machine, safety is important to the facility, the health inspector, and the public.

Physical Hazards to the Food Supply

Physical hazards to the food supply are any *foreign objects* found in food that may contaminate it. Certainly, they are unwanted by the consumer! Certainly they should not be deliberate! They may be present due to harvesting, or some phase of manufacturing, or they may be intrinsic to the food, such as bones in fish, pits in fruits, eggshells, and insects or insect parts.

Animals or crops grown in open fields are subject to physical contamination, although hazards may enter the food supply due to a variety of incidences that range from faulty machinery, to packaging wraps, to human error. An astute manager prevents the chance of physical contamination by following good manufacturing practices (GMPs) and using his/her observational skills.

The foremost materials of concern as physical hazards include foreign bits and pieces such as glass, wood, metal, plastic, stones, insects and other filth, insulation, bones, and personal effects (Katz 2000) (Table 19.2). Modern optical scanning technologies are capable of sorting difficult, potential problem products and are designed to minimize such contamination at the processing plant. Devices such as screen, filters, magnets, and metal detectors may be used online or throughout the manufacturing plant to search for foreign objects and avert health disasters or product recalls. X-ray units are reliable in detecting a variety of objects (Fig. 19.5).

Metal detectors are designed to detect metals in liquid, solid, granular, or viscous food products, in various packaging trays and wraps. The use of the common X-ray, a 40-year-old technology, is now a quality assurance tool in the inspection of finished food products. Its use may be a requirement of vendors supplying their foods to a warehouse club. Continual developments make it more affordable, compact, and faster to use in the manufacturing plant (Higgins 2006).

Personal effects, such as jewelry, may not be worn in the production areas. Personnel rules such as "no gum chewing" and "cover hair" need to be enforced in the workplace.

Physical hazards may harm a consumer's health and cause psychological trauma or dissatisfaction. Ill, upset, or dissatisfied consumers may call or write to the responsible manufacturer or processor, contact the foodservice establishment, or involve the local health department in investigating their complaint. Any chance of physical objects getting into the food supply should be prevented.

Foreign substance laboratories in food manufacturing companies as well as personnel in foodservice establishments need to look out for and be informed of any reported food safety problems—chemical or physical, so that they may investigate and prevent possible problems. Consumers benefit from this prevention and incidences of contamination are reduced.

See current and recent discussions on *food safety*: http://www.foodproductdesign.com/reports.aspx.

See later in this chapter

http://www.choosemyplate.gov/food-groups/downloads/TenTips/DGTipsheet23BeFoodSafe.pdf.

Food Protection Systems

Numerous agency names appear in this section of text, with each agency addressing food protection systems. While the listing may be lengthy, the names actually represent only a portion of the many groups responsible for the United States' food safety—considered the safest in the world. The CDC, the FDA, the USDA's FSIS, and state and county health departments have regulatory authority for food protection, and they provide education to the public. There are also numerous trade associations and professional organizations involved in providing education and protecting the public from foodborne illness.

A coordination of inspection, enforcement, and research may all contribute to food safety. Many food companies also maintain extensive food protection systems. Eliminating or reducing the biological, chemical, and physical hazards to the food supply is the goal of food safety.

In the quest to destroy pathogens and better protect foods, the FDA approved irradiation of meat in 1997, as it was shown to yield safer meat than meat that is not irradiated (Crawford and DVM). Irradiation, both off-site and online, is also being utilized as a means of food safety for a wide array of food items (Higgins 2003).

> Food safety and defense is mission critical. Through its focus on issues of concern and hot topics, IFT provides viewpoints and technical resources that will enhance your understanding of additive and ingredient safety, allergens, novel technologies, and microbial and chemical contaminants. (IFT)

With regard to the newer term *traceability*, safety guidelines are crucial and the following is said:

> Supply chain management in the food industry, and more specifically supply chain quality management, encompasses all types of raw materials, products and items. Suppliers and manufacturers must be able to track many shipments, and to do so effectively, they need complete histories of where each material comes from and where it is delivered to. Food traceability is vital when dealing with farm-to-fork food shipments. . .
>
> Proper safety guidelines need to be followed at each point of the farm-to-fork supply chain, beginning with the growing stage. . . . *Food traceability* allows you to further track your shipments of

Fig. 19.5 Metal detectors (*Source*: Advanced Detection Systems)

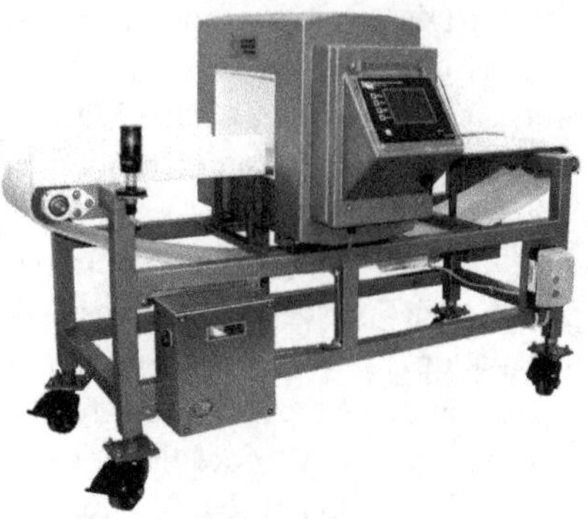

Conveyor metal detection system for packaged products.

Pipeline metal detection system with reject valve for pumped liquid and slurry applications

Gravity drop metal detection system with divert valve for bulk powder application.

processed and packaged food to ensure they are delivered on time.

(Global Supplier Quality Assurance) [GSQA]

An alternative safety mechanism processing plants may use in their food protection programs, although perhaps not providing *real-time* information is *statistical analysis*. It may be used to define acceptable upper and lower control limits and then to improve product quality. Statistical Process Control (SPC) provides advances in microbiological analyses that in turn will allow the manufacturing process to integrate testing with quality improvement and productivity (Hussain 2000). SPC may also be integrated with the HACCP method of food safety (discussed in a later section of this chapter, below).

In addition to a corporate level *manufacturing concern*, many retail *foodservice operations* have sent at least one manager to local training in Food Protection Management. Numerous news shows, newspapers, and magazines have addressed the issue of food safety for consumers, and receiving training can supply added assurance to the wary customer that the *restaurant* food supply is served safely.

FDA (See Chap. 20)

The FDA, the CDC of the U.S. Department of Health and Human Services (HHS), and the FSIS of the USDA release new edition of the Food Code every 4 years.

See Food Safety Research Information Office (FSRIO). Also see FDA Food Code:

http://fsrio.nal.usda.gov/faq-page/regulations-standards-and-guidelines

http://www.fda.gov/Food/FoodSafety/RetailFoodProtection/FoodCode/default.htm

The reader should remain current on foodborne illness prevention and produce safety.

The New FDA Food Safety Modernization Act (FSMA)

The House of Representatives and the Senate passed the FDA Food Safety Modernization Act. It was signed into law on January 4, 2011.

Some of the provisions in the new law:

- **Issuing recalls:** For the first time, FDA will have the authority to order a recall of food products. Up to now, with the exception of infant formula, the FDA has had to rely on food manufacturers and distributors to recall food voluntarily.
- **Conducting inspections:** The law calls for more frequent inspections and for those inspections to be based on risk. Foods and facilities that pose a greater risk to food safety will get the most attention
- **Importing food:** The law provides significant enhancements to FDA's ability to oversee food produced in foreign countries and imported into the United States. Also, FDA has the authority to prevent a food from entering this country if the facility has refused US. inspection.
- **Preventing problems:** Food facilities must have a written plan that spells out the possible problems that could affect the safety of their products. The plan would outline steps that the facility would take to help prevent those problems from occurring.
- **Focusing on science and risk:** The law establishes science-based standards for the safe production and harvesting of fruits and vegetables. This is an important step forward. These standards will consider both natural and man-made risks to the safety of fresh produce.
- **Respecting the role of small businesses and farms:** The law also provides some flexibility, such as exemptions from the produce safety standards for small farms that sell directly to consumers at a roadside stand or farmer's market as well as through a *community supported agriculture* program (CSA).

Later came the **Current Good Manufacturing Practice and Hazard Analysis and Risk-Based Preventive Controls for Human Food Proposed Rule**.

http://www.gpo.gov/fdsys/pkg/FR-2013-01-16/html/2013-00125.html

"The rule has two major features. First, it contains new provisions requiring hazard analysis and risk-based preventive controls. Second, it would revise the existing Current Good Manufacturing Practice (CGMP) requirements found in 21 CFR part 110" rule.

The proposed rule:

Summary

The Food and Drug Administration (FDA or we) is extending the comment period for the proposed rule, and for the information collection related to the proposed rule, "Standards for the Growing, Harvesting, Packing, and Holding of Produce for Human Consumption" that appeared in the Federal Register of January 16, 2013. We are taking this action in response to requests for an extension to allow interested persons additional time to submit comments on the proposed rule. We also are taking this action to keep the comment period for the information collection provisions associated with the rule consistent with the comment period for the proposed rule.

Dates

The comment period for the proposed rule published January 16, 2013, at 78 FR 3504, is extended. In addition, the comment period for the information collection issues in the proposed rule, extended February 19, 2013, at 78 FR 11611, is further extended. Submit either electronic or written comments on the proposed rule by September 16, 2013. (FDA)

USDA Food Protection (See Chap. 20)

The USDA FSRIO has created a website for the general public and food safety researchers. The site contains educational, professional, and foreign government links for food safety (http://www.nal.usda.gov/fsrio).

See more specifics in the chapters on milk, eggs, and poultry.

The HACCP System of Food Protection: USDA

In order to assess and reduce foodborne illness risks from biological, chemical, or physical hazards, the *Hazard Analysis and Critical Control Point—HACCP* (pronounced hassip)—system of food safety may be required for use by food processors and foodservice operations (Table 19.3). The system depends upon *prevention*, rather than strictly *inspection*.

The "Mega-Reg" or Pathogen Reduction: HACCP System regulation was signed by the US President in 1996. It codified principles for the prevention and reduction of pathogens and required both the development of Sanitation Standard Operating Procedures (SSOPs) and a written HACCP plan that is monitored and verified by inspectors of various food processing plants. Compliance deadlines were phased in, depending upon the size of the company.

FSIS tests raw livestock and poultry in the slaughtering processes for *Salmonella* and *E. coli*. Seafood HACCP became effective in 1997. Juice processors must have HACCP. Later, HACCP plans for protection against L. monocytogenes in ready-to-eat meat products were required by the FSIS. Currently, egg processors, as well as dairy plants, and additional industries have implemented HACCP.

HACCP traces the flow of food from entry into an operation through exit. It does more

Table 19.3 Steps of a hazard analysis and critical control point (HACCP) program

I. Assessing the hazards
Hazards are assessed at each step in the flow of food throughout an operation
II. Identifying critical control points (CCPs)
Identify CCPs regarding hygiene, avoiding cross-contamination, and temperatures and procedures for cooking and cooling. A flowchart of preparation steps is developed, showing where monitoring is necessary to prevent, reduce, or eliminate hazards
III. Setting up control procedures and standards for critical control points
Establish standards (criteria) for each CCP and measurable procedures such as specific times and temperatures, moisture and pH levels, and observable procedures such as hand washing
IV. Monitoring critical control points
Checking to see if criteria are met is one of the most crucial steps in the process. Assigning an employee to monitoring temperatures of storage, cooking, holding, and cooling is necessary to see if controls against hazards are in place
V. Taking corrective action
Observe if there is a deviation between actual and expected results. Correct the procedures by using an alternate plan if a deficiency or high-risk situation is identified in using the original procedure. This may be accomplished by a trained employee empowered to initiate corrective action without a supervisor being present
VI. Develop a record-keeping system to document HACCP
Time–temperature logs, flowcharts, and observations are used for record keeping
VII. Verify that the system is working
Make use of time and temperature logs completed during preparation, holding, or cooling. Observe

See Ref on HACCP

than *detect and correct* errors *after* they have occurred; as mentioned, it is a program that *prevents* errors regarding food safety before they occur. By definition of a Quality Assurance [QA] Program, HACCP may not fit QA. Sanitation is typically a separate program at a food manufacturing facility and is not put into the same category as other aspects of food quality. Sanitation is a 24/7 duty and may compose the entire third shift of a 24-h food manufacturing plant.

HACCP was originally designed by the Pillsbury Company, in cooperation with Natick Laboratories of the US Army and the U.S. Air Force Space Laboratory Project Group. The system was designed for use by the National Aeronautics and Space Administration (NASA) Program. HACCP has been used as a food safety system in the food industry since 1971, and it offers practical food protection techniques that are needed anywhere food is prepared or served.

The National Advisory Committee for Microbiological Criteria for Foods (NACMCF) (National Advisory Committee for Microbiological Criteria for Foods) has identified seven major steps involved in the HACCP system (Table 19.3). HACCP can be used in multiple spots within the food chain, for example, in growing, harvesting, processing, preparing, or serving of foods.

With the establishment of an HACCP Program, new terminology may be used; some selected HACCP definitions are given in Table 19.4. Many food companies must have an HACCP system in place.

Using the HACCP system, the HACCP team must first identify phf's that are prepared in their operation. Then, they must observe the flow of those phf's from the acquisition of raw ingredients to completion of the finished product, especially studying the flow of sensitive ingredients known to have been associated with a hazard and for which there is a reason for concern. This observation leads to the development of a flowchart.

After identifying those foods that are potentially hazardous and creating a flowchart, management needs to identify specific, measurable Critical Control Points (CCPs). In the absence of

Table 19.4 Selected HACCP definitions (health and human service)

Control point: any point, step, or procedure at which biological, physical, or chemical factors can be controlled
Corrective action: procedures to be followed when a deviation occurs
Critical control point (CCP): a point, step, or procedure at which control can be applied and a food safety hazard can be prevented, eliminated, or reduced to acceptable levels
Critical limit: a criterion that must be met for each preventive measure associated with a critical control point
Deviation: failure to meet a critical limit
HACCP Plan: the written document which is based on the principles of HACCP and which delineates the procedures to be followed to assure the control of a specific process or procedure
HACCP system: the result of the implementation of the HACCP plan
HACCP team: the group of people who are responsible for developing an HACCP plan
Hazard: a biological, chemical, or physical property that may cause a food to be unsafe for consumption
Monitor: to conduct a planned sequence of observations or measurements to assess whether a CCP is under control and to produce an accurate record for future use in verification
Risk: an estimate of the likely occurrence of a hazard
Sensitive ingredient: an ingredient known to have been associated with a hazard and for which there is reason for concern
Verification: the use of methods, procedures, or tests in addition to those used in monitoring to determine if the HACCP system is in compliance with the HACCP plan and/or whether the HACCP plan is working

CCPs, food is subject to *unacceptable* risks or likelihood of a hazard.

Subsequently, control procedures and criteria for critical limits must be established and then *monitored* by the individual assigned responsibility for tracking CCP procedures. The CCPs may include temperature of the food product and processing equipment, time of processing, package integrity, and more. Measurements and observational skills are employed in order to reveal any unacceptable deviations between actual and expected results. Deviation may require corrective action in order to prevent foodborne illness.

A view of the headings on the foodservice HACCP of chicken salad and ribs, HACCP (Figs. 19.6 and 19.7) indicates several major concepts. *First*, the flow process of foods is drawn from the point of receiving food until it is discarded. *Secondly*, the CCPs are identified, and *next*, criteria for control are established and briefly stated for ease of understanding. Criteria for control specify such factors as minimum and maximum temperatures that must be reached, correct storage procedures, instructions for personal hygiene and equipment sanitation, and discard rules.

Note that monitoring and verifying the HACCP program includes instructions to follow for assurance of compliance with criteria. It may be taking temperatures, measuring time to complete preparation, measuring depth of storage pans, or observing procedures that are used in preparation or storage. The entire HACCP process also states the action to be taken if the criteria are not met. HACCP systems require that a designated individual using reliable tools/instruments must monitor the CCPs. The reliability of instruments such as thermometers or thermocouples must be validated.

Handling a phf requires knowledge of the flow process and how to keep food safe. For example, in *receiving* chicken, the corrective action to take if established criteria are not met would be to reject the products upon delivery. In *storage*, the product may require a lower air temperature if the established criteria for maximum temperature were not met.

A further *cooking* criterion requires that chicken reach a minimum temperature of 165 °F (74 °C). If the temperature is not met at an initial check for doneness, the chicken must continue to be cooked until that temperature is reached. Thus, the HACCP *continues* with corresponding action to take for *each* criterion if the criterion is not met and includes corrective practices for handling, personal hygiene, equipment sanitation, food storage, and discarding food.

Identifying such items as the flow process of food, stating CCPs and criteria for control, monitoring and verifying, and specifying action to

HAZARD ANALYSIS CRITICAL CONTROL POINT FLOW PROCESS FOOD: CHICKEN SALAD

FLOW PROCESS	*CCP*	*CRITERIA FOR CONTROL*	*MONITOR & VERIFY*	*ACTION TO TAKE IF CRITERIA NOT MET*
Receive refrigerated whole chickens		-Maximum 45°F internal temperature	-Take internal meat temperature with metal stem thermometer	-Reject product
Store in Walk-in-Cooler (WIC)		-Maximum 45°F internal temperature	-Observe proper storage practices -Monitor air temperature each shift-record on log	-Store chicken in approved manner -Lower air temperature
		-Store chickens off floor -Prevent cross-contamination -Air temperature 40°F or less		
Boil chickens	CCP	-Minimum 165°F internal temperature of meat	-Take internal meat temperature -Observe cooking time	-Cook chicken until temperature is reached
Cool to debone (30 minutes in WIC)		-Do not cover chickens	-Observe storage in WIC	-Store chicken to allow rapid cooling to debone
Debone/dice chicken meat	CCP	-Clean hands or gloves used to handle meat	-Observe handling procedures -Inspect employee hands daily	-Instruct workers to wash hands or use gloves
		-No infected wounds or bandages on hands	-Observe proper cleaning of equipment	-Remove worker or require gloves -Have equipment rewashed
		-Wash and sanitize equipment used after completion		
Mix ingredients (mayo, sour cream, relish, spices, meat)		-Use utensils for mixing	-Observe use of utensils	-Correct practice
		-Limit time for preparation of meat salad – refrigerate when completed	-Measure time to complete preparation process	-Modify procedures to limit time at room temperature
		-Use refrigerated ingredients	-Observe use	-Change practice
Store 1/3 of salad in prep cooler	CCP	-Cool to 45°F within 4 hours after preparation	-Measure salad temperature periodically to determine cooling rate	-Remove excess salad from pan
Store 2/3 of salad in WIC (Use 1/2 each day at prep cooler)			-Lower air temperature	
		-Maximum 45°F internal temperature in storage	-Take internal temperature	-Store in coolest part of cooler
		Store salad 3 inches or less in pans -Air temperature 40°F or less	-Measure depth of salad stored in pans -Monitor air temperature each shift – record on log	
Sell Discard after 3 days from preparation		-Old salad not mixed with fresh salad	-Observe storage process	-Correct practice
		-Discard remaining salad	-Observe salad discarded	-Discard salad

Fig. 19.6 HACCP flow process. Food: chicken salad (*Source:* Alvin Black, R.S. City of Farmers Branch, Environmental Health Division. Farmers Branch, TX)

take if the criteria are not met, all function to assist management in controlling the spread of disease. Applying an HACCP system to food manufacturing or the foodservice operation is an effective means of reducing the likelihood of foodborne illness. HACCP is so much more than inspection, and to be effective, it requires dedication and perseverance on the part of employees and management.

In *food manufacturing plants*, manufacturers must take steps to ensure that food is safe! They must do so, by law. As a result of taking those critical steps, only a *small* percentage of all foodborne illness cases is linked to poor processing practices. A *greater* number of cases are the result of faulty practices in foodservice operations and the home. Many state and local health departments have adopted rules for *foodservice establishments* also, requiring knowledge of foodborne illness and HACCP principles. These foodservice establishments include hospitals, restaurants, retail grocery stores, and schools.

HAZARD ANALYSIS CRITICAL CONTROL POINT FLOW PROCESS FOOD: BBQ RIBS

FLOW PROCESS	*CCP*	*CRITERIA FOR CONTROL*	*MONITOR & VERIFY*	*ACTION TO TAKE IF CRITERIA NOT MET*
Frozen beef ribs		-Received frozen	-Feel if frozen upon delivery	-Reject if thawed
Thaw in walk-in-cooler (WIC)		-Meat thawed under refrigeration	-Observe ribs stored in WIC	-Store ribs properly to prevent contamination or cross-contamination
		-Store meat off floor	-Observe proper storage practices	
		-Prevent cross-contamination		
Cook in oven (add BBQ sauce)	CCP	-Minimum 140°F internal temperature of ribs	-Take internal meat temperature with metal stem thermometer -Observe cooking time and oven temperature	-Cook ribs until temperature is reached
Hold at steam table with overhead heat lamp	CCP	-Minimum 140°F internal temperature	-Take temperature of meat every 2 hours – record on log	-Reheat ribs
				-Discard ribs if held below 130°F over 2 hours -Check equipment
Sell				
Leftover ribs cooled in WIC overnight	CCP	-Cool from 140° to 45°F within 4 hours	-Measure meat temperature periodically to determine cooling rate in WIC	-Remove excess ribs from pan
				-Lower air temperature
		-Store meat 3 inches or less in pans	-Measure depth of meat stored in pan	Remove covers
			-Monitor air temperature of WIC each shift-record on log	-Eliminate stacking
		-No tight cover during cooling process		-Move ribs to coolest part of WIC
			-Observe meat uncovered during cooling process	-Discard inadequately cooled ribs
		-Do not stack pans -Store meat close to fans in WIC -Air temperature 40°F or less		
Reheat in convention oven next morning	CCP	-Minimum 165°F internal temperature within 2 hours	-Take internal meat temperature with metal stem thermometer	-Reheat meat until temperature is reached
		-Leftovers not mixed with fresh ribs	-Observe reheating time and oven temperature	-Discard meat
		-Discard remaining ribs	-Observe meat discarded	
Steam table		-Same instructions as above		
Sell				
Discard by 6:00 p.m.		-Leftovers not mixed with fresh ribs	-Observe storage process	-Correct practice
		-Discard remaining ribs	-Observe meat discarded	-Discard meat

Fig. 19.7 HACCP flow process. Food: BBQ Ribs (*Source:* Alvin Black, R.S. City of Farmers Branch, Environmental Health Division. Farmers Branch, TX)

An example of HACCP plans for foodservice operations has been shown for chicken salad and BBQ beef ribs.

As well as the HACCP Flow Process charts mentioned, samples of *two written recipes* that incorporate the HACCP principles appear in Figs. 19.8 and 19.9. These recipes demonstrate ways in which a foodservice operation may include CCPs in preparation steps and flowcharts. For example, an acceptable method of defrosting, cooking, and holding is stated after labeling the preparation step as a CCP, and CCPs are highlighted in the flowcharts.

Today, *numerous* foods are processed in manufacturing plants (Table 19.5) and are distributed to operations such as retail grocery stores, hotels, restaurants, or institutional operations. These foods must provide assurance of food quality, including *microbiological (M), chemical (C), and physical (P)* safety, and have critical limits, including meeting all safety specifications prior to shipping, measuring temperatures of incoming and chilled ingredients with calibrated instruments, using microbiological tests for food contact surfaces and the environmental area, sanitizing equipment, storing, refrigerating palletizing products, distribution, and labeling.

A revised HACCP model designed by the FSIS has shown improvements in food safety. Young chicken plants have demonstrated a

Fig. 19.8 HACCP. Basic beef chili (*Source*: La Vella Food Specialists St. Louis, MO)

Basic Beef Chili

Ingredients	Amount	25	50	100
Lean Ground Beef	Lbs	7	14	28
Canned Tomatoes	Qts	1 ½	3	6
Canned Kidney Beans	Qts	1 ¾	3 ½	7
Tomato Paste	Cups	1 ¾	3 ½	7
Water	Gals	½	1	2
Dehydrated Onions	Ozs	1	2	4
Chili Powder	Tbsp	3	6	12
Sugar	Tbsp	1 ¼	2 ½	5
Cumin	Tbsp	2	4	8
Garlic Powder	Tbsp	1	2	4
Onion Powder	Tbsp	1	2	4
Paprika	Tbsp	1	2	4
Black Pepper	Tbsp	½	1	2

Preparation

1. **CCP** Thaw ground beef under refrigeration (41°F, maximum 1 day).

2. Place ground beef in steam kettle or in large skillet on stove top. Cook meat using medium high heat until lightly browned (15 minutes). While cooking, break meat into crumbs of about ½" to ¼" pieces.

3. Drain meat well, stirring while draining to remove as much fat as possible. If desired, pour hot water over beef and drain to remove additional fat.

4. Mash or grind canned tomatoes with juice. Add to kettle or stock pot with cooked ground beef. Add remaining ingredients to mixture and stir well.

5. **CCP** Simmer chili mixture for 1 hour, stirring occasionally. Temperature of cooked mixture must register 155°F or higher.

6. Remove from heat and portion into service pans.

7. **CCP** Cover and hold for service (140°F, maximum 1 hour).

8. Portion: 1 cup (8 ounces) per serving.

Service:

1. **CCP** Maintain temperature of finished product above 140°F during entire service period. Keep covered whenever possible. Take and record temperature of unserved product every 30 minutes. Maximum holding time, 4 hours.

Storage:

1. **CCP** Transfer unserved product into clean, 2-inch deep pans. Quick-chill. Cooling temperature of product must be as follows: from 140° to 70°F within 2 hours and then from 70° to 41°F or below, within an additional 4 hour period. Take and record temperature every hour during chill-down.

2. **CCP** Cover, label, and date. Refrigerate at 41°F or lower for up to 10 days (based on quality maintained) or freeze at 0°F for up to 3 months.

Reheating:

1. **CCP** Thaw product under refrigeration, if frozen (41°F).

2. **CCP** Remove from refrigeration, transfer into shallow, 2-inch deep pans and immediately place in preheated 350°F oven, covered. Heat for 30 minutes or until internal temperature reaches 165°F or above.

Discard unused product.

Fig. 19.9 HACCP. Chicken stew (*Source*: La Vella Food Specialists St. Louis, MO)

Chicken Stew

Ingredients	Amount	25	50	100
Chicken Pieces, 8 cut, frozen	Lbs	10	20	40
Carrots, fresh, peeled, cut in ¾ inch pieces	Lbs	2 ½	5	10
Onions, chopped	Qts	½	1	2
Potatoes, peeled, cut into ¾ inch pieces	Lbs	3 ¾	7 ½	15
Green Peas, frozen	Lbs	2	4	8
Margarine	Cups	½	1	2
Flour	Cups	1 ½	3	6
Chicken Stock	Qts	1	2	4
Salt	tsp	1	2	4
Pepper	tsp	1	2	4

Preparation

1. **CCP** Thaw raw chicken pieces under refrigeration (41°F, 1 day).

2. **CCP** Wash carrots, onions, and potatoes under cool running water. Cut as directed. Use immediately in recipe or cover and refrigerate until needed (41°F, maximum 1 day).

3. Place chicken pieces on sheet pans. Cover and bake in preheated 350°F conventional (325°F convection) oven for 30 minutes.

4. Cook potatoes, carrots, and peas separately in steamer or on stovetop, until tender (4–15 minutes).

5. Remove chicken from oven; dr ain off juices and fat. Place in 4 -inch deep steamtable pans, cover, and return to heated oven (while preparing gravy).

6. In stockpot over medium heat, melt margarine and sauté onions until tender. Add flour and stir until smooth. Add chicken drippings, s tirring well. Add chicken broth as needed for gravy-like consistency. Season with salt and pepper.

7. Add cooked vegetables and gravy to chicken pieces. Cover, place back in 350°F conventional (325°F convection) oven and bake for 30 minutes or until chicke n pieces are tender and sauce is flavorful.

8. **CCP** Internal temperature of cooked stew must register 165°F for 15 seconds at end of cooking process.

9. **CCP** Cover and hold for service (140°F, maximum 1 hour).

10. Portion: 1 –2 pieces of chicken, ½ cup vegeta bles with gravy (10 ounces) per serving

Service:

1. **CCP** Maintain temperature of finished product above 140°F during entire service period. Keep covered whenever possible. Take and record temperature of unserved product every 30 minutes. Maximum holding time, 4 hours.

Storage:

1. **CCP** Transfer unserved product into clean, 2 -inch deep pans. Quick -chill. Cooling temperatures of product must be as follows: from 140° to 70°F within 2 hours and then from 70° to 41°F or below, within an additional 4 hours. Take and record temperature every hour during chill-down.

2. **CCP** Cover, label, and date. Refrigerate at 41°F or lower for up to 10 days (based on quality maintained) or freeze at 0°F for up to 3 months.

Reheating:

1. **CCP** Thaw product under refrigeration, if frozen (41°F).

2. **CCP** Remove from refrigeration, transfer into shallow, 2 -inch deep pans and immediately place in preheated 350°F oven, covered. Heat for 30 minutes or until internal temperature reaches 165°F or above.

Discard unused product.

Table 19.5 Ingredients of refrigerated chicken salad (Pierson and Corlett) (Pierson & Corlett 1992)

CCP number	CCP description	Critical limit(s) description
1-MPC	Hazard controlled	1.1 Sanitary condition
	Microbiological, physical, and chemical	1.2 Refrig. material ≤45 °F
		1.3 Frozen material ≤32 °F
	Point or procedure: incoming inspection	1.4 Vendor met all safety specifications before shipping
2-T	Hazard controlled: microbiological	2.1 Material internal temperature not to exceed 45 °F
	Point or procedure: refrigerated	
	Ingredient storage	2.2 Calibrate temperature-measuring devices before shift
3-M	Hazard controlled: microbiological	3.1 Comply with USDA sanitation requirements
	Point or procedure: sanitation requirements in	
		3.2 Sanitation crew trained
	• Preparation area	3.3 Each area must pass inspection before shift start-up
	• Staging area	
	• Filling/packaging area	
	Hazard controlled	3.4 Food contact surface: microbiological test
	Point or procedure: *Listeria*	
		3.5 Environmental area: microbiological tests (USDA methodology for 3.4 and 3.5)
4-M	Hazard controlled: microbiological	Application of alternative approved treatments
	Point or procedure: controlled treatment to reduce microbiological contamination on raw celery and onions	
		4.1 Wash product with water containing
		• Chlorine, or
		• Iodine, or
		• Surfactants, or
		• No process additives
		4.2 Hot water or steam blanch followed by chilling
		4.3 Substitute processed celery or onions
		• Blanched frozen
		• Blanched dehydrated
		• Blanched canned
5-M	Hazard controlled: microbiological	5.1 Not to exceed 45 °F
	Point or procedure: chilled storage temperature of prepared celery, onions, and chicken	5.2 Refrigerator not to exceed 45 °F
		5.3 Daily calibration of temperature-measuring devices
6-MPC	Hazard controlled: microbiological, physical, and chemical	6.1 Physical barrier in place
	Point or procedure: physical barrier to prevent cross-contamination from raw material preparation area	6.2 Doors kept closed when not in use
		6.3 Color-coded uniforms
		6.4 Supervision in place
7-M	Hazard controlled: microbiological	7.1 Comply with USDA sanitation requirements

(continued)

Table 19.5 (continued)

CCP number	CCP description	Critical limit(s) description
	Point or procedure: cross-contamination prevention from transfer equipment from raw material area	7.2 Prevent entry of soiled pallets cart wheels, totes, and other equipment
8-M	Hazard controlled: microbiological	8.1 Time limit not to exceed 4 h for any materials in staging area
	Point or procedure: time limit for in-process food materials	
9-M	Hazard controlled: microbiological	9.1 Product pH must not exceed a pH of 5.5
	Point or procedure: maximum pH limit on finished salad before packaging	9.2 pH meter must be calibrated with approved standards before each shift
10-M	Hazard controlled: microbiological	10.1 Internal temperature not to exceed 45 °F
	Point or procedure: chilled product storage temperature and time before packaging	10.2 Product must not be held more than one shift before filling/packaging
11-P	Hazard controlled: physical	11.1 Ferrous metal detection device for individual packages
	Point or procedure: metal detector for packages	11.2 Calibration or inspection not to exceed every 4 h
12-M	Hazard controlled: microbiological	12.1 Physical barrier in place
	Point or procedure: physical barrier to prevent cross-contamination from warehouse area	12.2 Doors kept closed when not in use
		12.3 Color-coded uniforms
		12.4 Supervision in place
13-M	Hazard controlled: microbiological	13.1 Product internal temperature not to exceed 45 °F in 4 h
	Point or procedure: refrigerated storage of cased/palleted finished product	13.2 Temperature-measuring devices calibrated before shift
14-M	Hazard controlled: microbiological	14.1 Shipping compartments must be precooled to 45 °F or less before loading product
	Point or procedure: truck and shipping containers for distribution of finished product	
15-M	Hazard controlled: microbiological	15.1 Each package or bulk case shall have label instructions
	Point or procedure: label instructions	
		15.2 Each label shall include
		• Keep refrigerated
		• Code
		• Storage instructions

Source: Pierson MD, Corlett DA
HACCP principles and applications
M microbiological hazard, *P* physical hazard, *C* chemical hazard

greater achievement of performance standards in FSIS verification checks over traditional slaughter inspection (FSIS).

Using research data from testing in large plants over a 2-year period, the FSIS reports that there are substantial reductions in the prevalence of *Salmonella* compared to pre-HACCP baseline figures in raw meat and poultry.

Traditionally, lower moisture ingredients and grain-based products are not typically considered potentially hazardous foods. However, unless products are pasteurized, companies need to set up a plan that includes hazard identification and risk analysis from farm to fork to deal with potential food safety issues that may arise. (Kuntz 2012)

For more, see HACCP Principles & Application Guidelines. Last Updated: 03/15/2013 (Application Guidelines 1997).

Surveillance for Foodborne Disease Outbreaks

The FDA estimates of foodborne diseases have been reported to be in the tens of millions, while the actual report of cases to the CDC is in the thousands. Since all illnesses are *not* reported, the true number is unknown.

For more than a quarter century, since 1973, the CDC has maintained surveillance data regarding the occurrence and causes of foodborne disease outbreaks (FBDOs). Now, the CDC actively surveys emerging foodborne diseases. The Foodborne Diseases Active Surveillance Network (FoodNet) is the primary foodborne disease component of the CDC's Emerging Infections Program (EIP). It began in the mid-1990s with five states and now includes many more states, representing over 25.4 million persons (more than 10 % of the United States population).

"The Foodborne Diseases Active Surveillance Network, or FoodNet, has been tracking trends for infections commonly transmitted through food since 1996. FoodNet provides a foundation for food safety policy and prevention efforts. It estimates the number of foodborne illnesses, monitors trends in incidence of specific foodborne illnesses over time, attributes illnesses to specific foods and settings, and disseminates this information." (CDC.gov)

The tracking is an "active" reporting system where public health officials frequently contact laboratory directors for data that is then electronically transmitted to the CDC. It has five components as shown below:

- Active laboratory-based surveillance
- Survey of clinical laboratories
- Survey of physicians
- Survey of the population
- Epidemiologic studies

The reporting data is tabulated and appears several years after the occurrence of illness. Recent summary statistics are found at the CDC:

See http://www.cdc.gov/foodborneoutbreaks/outbreak_data.htm (CDC's Outbreak Response Team).

See http://www.cdc.gov/foodborneburden/PDFs/FACTSHEET_B_TRENDS.PDF (1996–2010).

The number of FBDOs is reported by state and territorial health departments to the CDC on a standard reporting form.

Various food safety CDC data in Morbidity and Mortality Weekly Report (MMWR) is in the following and is cited in relevant text material:

http://www.cdc.gov/mmwr/preview/mmwrhtml/mm6203a1.htm?s_cid=mm6203a1_w%22

During 2009–2010, a total of 1,527 foodborne disease outbreaks (675 in 2009 and 852 in 2010) were reported, resulting in 29,444 cases of illness, 1,184 hospitalizations, and 23 deaths. Among the 790 outbreaks with a single laboratory-confirmed etiologic agent, norovirus was the most commonly reported, accounting for 42 % of outbreaks. *Salmonella* was second, accounting for 30 % of outbreaks.

CDC.gov (Fig. 19.10)

Many individuals who become ill do *not* relate it to food consumption or report this incident to appropriate authorities. Therefore, perhaps only a small percentage of actual FBDOs are reported. Nonetheless, surveillance data provide "an indication of the etiologic agents, vehicles of transmission, and contributing factors associated with FBDO and help direct public health actions."

Persons most "at risk" for, or likely to become ill with, a foodborne illness include the elderly (the largest risk segment of the US population), pregnant and nursing women, school-age children, and infants. These are represented as "highly susceptible populations." As well, increasing numbers of persons testing positive

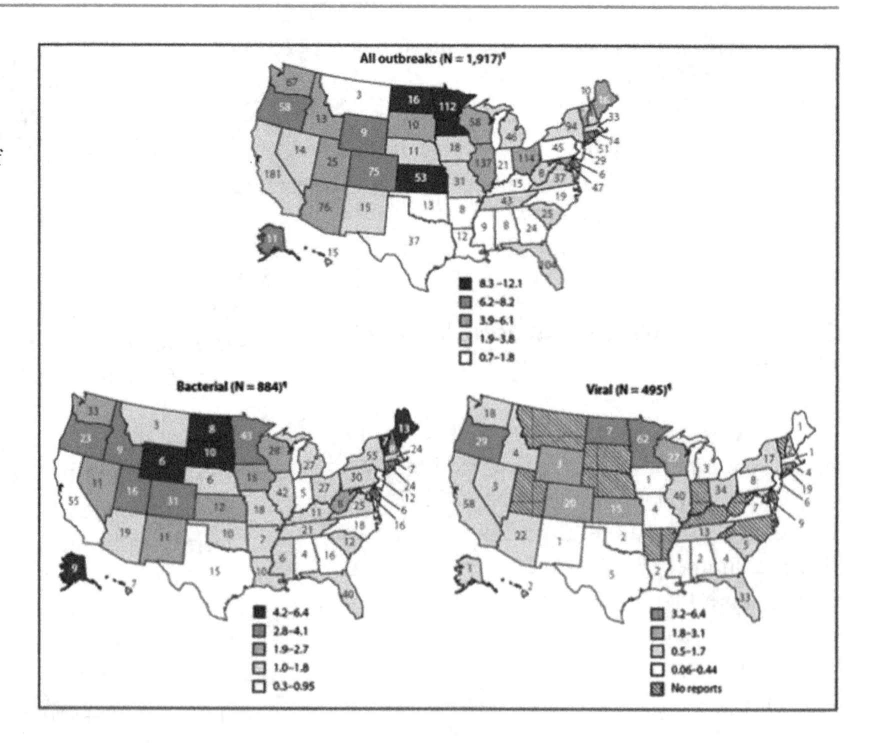

Fig. 19.10 Average annual rate of reported foodborne disease outbreaks per one million population and number of outbreaks, by state and major etiology group—Foodborne Disease Outbreak Surveillance System, United States, 2009–2010

for the human immunodeficiency virus (HIV), and persons with acquired immunodeficiency syndrome (AIDS), or persons with a weakened immune system due to pharmaceutical or radiological treatment are very susceptible to illness.

The growth of the *number* of persons at risk, coupled with a greater *number* of meals eaten away from home, provides *increasing* opportunities for the occurrence of foodborne illness. Controlling hazards and ensuring a safe food supply is possible through such methodologies as the use of an HACCP system and employee training. It is recommended that FSIS should seek authority to impose monetary penalties for violations and do a better job of monitoring test procedures (Food Eng 2000).

The CDC Guide to Confirming a Diagnosis in Foodborne Disease is located in:

http://www.cdc.gov/foodborneoutbreaks/guide_fd.htm

http://www.cdc.gov/outbreaknet/

The USDA FSIS is *science-based* and responsible for "ensuring the quality of the nation's meat, poultry, and egg products is a process that is increasingly dependent upon science." They state, "learn more about how the agency is protecting public health through science."

Other Causes of Spoilage and Contamination

In addition to the biological, chemical, and physical hazards that may contaminate the food supply, enzymatic activity and exposure to excessive moisture spoil food. Pests also contaminate food, perhaps making it noticeably nonedible. Cockroaches and insects carry germs, and some insects regurgitate on food with their acidic saliva in order to break it down prior to ingestion. Rodents, for example, do not have bladders and may contaminate all surfaces with which they make contact.

Throughout the world, in developed and underdeveloped countries, food spoilage may be responsible for a lot of food waste. Diligence in care is a must. Depending upon resources and beliefs, preservation, the use of additives, and packaging (see appropriate chapters for each topic) are all methods that may be used for controlling undesirable spoilage or contamination.

For further study and exploration of disease related to foods, see *food toxicology*. One example is seen here: "**Food toxicology** is the study of the nature, properties, effects, and detection of toxic substances in food, and their disease manifestation in humans. . . .including dose–response relationships, absorption of toxicants, distribution and storage of toxicants, biotransformation and elimination of toxicants, target organ toxicity, teratogenesis, mutagenesis, carcinogenesis, food allergy, and risk assessment. . .examine chemicals of food interest such as food additives, mycotoxins, and pesticides, and how they are tested and regulated . . . the etiology of foodborne disease related to naturally-occurring toxins and . . . the ecology of food." (The University of Idaho)

Responsibility for Food Safety

Governments, food companies, foodservice establishments, and consumers are each responsible for safe food. There are increasing numbers of "at-risk" populations that complicate the prevention of foodborne and waterborne illnesses. The food supply should be safe.

Governments around the world regulate their own food supply (Chap. 20) to assure its safety and wholesomeness, by fostering science-based regulation, inspection and enforcement services, education, and research. In the United States, the FDA and the USDA's FSIS have a history of providing their numerous, well-researched documents (also available on their websites) on foodborne illness and risk assessment to both the food manufacturer and consumer. Below is a .gov reference that becomes the resolve of consumers.

The *USDA* continues to collaborate with states and private companies to protect food. "Food safety remains top priority at USDA" (USDA Homeland Security). Food defense exercises are taking place throughout the United States to coordinate government, nongovernment, and the private sector alike.

The Food and Drug Center for Food Safety and Applied Nutrition cites seven critical hindrances to maintaining a sanitary operation. These areas include the aforementioned microorganisms: bacteria and mold; chemical contamination; pests including birds, insects, and rodents; and ignorance/carelessness. Thus, targeted employee training in these critical areas is crucial to food safety.

In the food *industry*, a loss of human life, loss of brand loyalty, or loss of the company itself may propel personnel to maintain the right attitude and do things right. Crisis management teams and crisis management plans must emphasize prevention.

Food manufacturing and processing industries, as well as foodservice operations, including hospitals, nursing homes, and restaurants, must comply with government regulations. For example, as a means of food protection, foods companies have reevaluated strategies to provide required plant sanitation and prevent product recalls. Increases in both time and financial resources allocated to food safety are apparent, as is the hiring of plant design engineers trained in sanitation (Van Milligen 2001).

Food processors including slaughterhouses are subject to close scrutiny from the FSIS so that there is no risk of mad cow disease in the nation's meat supply. Inspectors continually receive audits and training to ensure safety. Food safety websites provide the latest in food safety news and recall information consumer advice, instructions on reporting possible foodborne illnesses, and more. (http://www.foodsafety.gov/presidentscouncil)

The Center for Science in the Public Interest (CSPI) has the following to say regarding food safety (http://www.cspinet.org/foodsafety):

CSPI Mission Statement

Foodborne illness, commonly called "food poisoning," causes an estimated 48 million illnesses, 128,000 hospitalizations, and 3,000 deaths in the United States annually. These illnesses can range from troubling cases of nausea, vomiting, and diarrhea to life-threatening illnesses that require hospitalization. Most of these illnesses are entirely preventable with care throughout the food chain. From the farmer to the chef and from the food processor to the guy at the grill, everyone has a role to play in making safe food.

Food safety is a key area of focus for the Center for Science in the Public Interest. **The mission of CSPI's Food Safety Program is to reduce the burden of foodborne illnesses.** We provide solid food safety advice to consumers on our website and in Nutrition Action Healthletter, and our staff of experts encourages policymakers, government regulators, and the food industry to work harder to protect American consumers from contaminated food.

The Food Safety Program's lawyers and public health researchers work with Congress and State legislatures to strengthen food safety laws and to provide funding for the federal agencies that protect our food supply and public health. We encourage the Food and Drug Administration and the U.S. Department of Agriculture to improve federal food safety programs and increase oversight of industry practices. We provide those agencies with our best thinking on food safety policy through petitions, comments, and participation in public hearings. We also communicate directly with industry leaders on ways they can ensure the safety of the food products they market.

The Food Safety Program provides useful, up-to-date research to the public, policymakers, and regulators on current and emerging food safety issues. CSPI's publication Outbreak Alert! is an ongoing compilation of foodborne illnesses and outbreaks, organized by food categories. It is used by scientists and policymakers around the world. CSPI's Food Safety Program has been in the forefront of advocating for a unified food safety system and for tougher laws governing meat, poultry, and seafood production.

Our work doesn't stop at the border. In an increasingly global marketplace, the Food Safety Program represents consumers at many international food safety meetings, and we manage the work of the International Association of Consumer Food Organizations (IACFO), which represents consumers in every world region.

CSPI

After initial passage of the "sweeping," new law implementation was slow to move according to Gannett News: (http://www.usatoday.com/story/news/nation/2012/12/15/promise-of-food-safety-law-largely-unfulfilled/1772261/)

As the President of a food safety and sanitation consulting firm noted, "Sanitation is an attitude, not a process." "The bulk of sanitation cannot be done during production, that's the way the rule is and that's not going to change" (Van Milligen 2001).

Also, food companies may literally move "sanitation" tasks from the third shift to the first or second shift, reflecting different priorities and the adoption of a greater emphasis on sanitation and the safety of their food products. Perhaps it is insufficient to just say that a food is safe. Data must support the claim.

In an article entitled "Why can't scientists communicate science?—poor media coverage and a lack of consumer education feed fear about our nation's food supply," an important question was raised. It asks the question "...how can an industry that can produce such bounty have a problem when it comes to communicating safety and efficiency?" Perhaps safety is not sensational enough for some media reporters. Yet, there might not be anything wrong with "touting your own success" or "blowing your own horn" to tell the public how good things are in science and technology and food safety. "Our food supply is potentially the healthiest in the world. We should tell people how we do it" (Stier 2006).

Consumers must ultimately be responsible for the consumption of safe foods that they themselves prepare or that are processed/prepared by others in the food supply. The consumer must be vigilant and become educated on matters concerning food safety because it may be literally in the hands of the food handler!

The large body of food and nutrition professionals, represented by The Academy of Nutrition and Dietetics, has stated the following:

> ... the public has the right to a safe food and water supply. The Association supports collaboration among dietetics professionals, academics, and representatives of the food industry and appropriate government agencies to ensure the safety of the food and water supply by providing education to the public and industry, promoting technological innovation and applications, and supporting further research. (AND)

More specifically, "Clean, Separate, Cook, Chill" advice from previous chapters now follows with detail that is appropriate in this chapter.

Food Safety Advice

Clean: Wash Hands and Surfaces Often

Bacteria can be spread throughout the kitchen and get onto hands, cutting boards, utensils, counter tops, and food.

- Wash your hands with warm water and soap for at least 20 s before and after handling food and after using the bathroom or changing diapers.
- Wash your hands after playing with pets or visiting petting zoos.
- Wash your cutting boards, dishes, utensils, and counter tops with hot soapy water after preparing each food item and before you go on to the next food.
- Consider using paper towels to clean up kitchen surfaces. If you use cloth towels, wash them often in the hot cycle of your washing machine.
- Rinse fresh fruits and vegetables under running tap water, including those with skins and rinds that are not eaten.
- Rub firm-skinned fruits and vegetables under running tap water or scrub with a clean vegetable brush while rinsing with running tap water.
- Keep books, backpacks, or shopping bags off the kitchen table or counters where food is prepared or served.

Separate: Don't Cross Contaminate

Cross-contamination is how bacteria can be spread. When handling raw meat, poultry, seafood, and eggs, keep these foods and their juices away from ready-to-eat foods. Always start with a clean scene—wash hands with warm water and soap. Wash cutting boards, dishes, countertops, and utensils with hot soapy water.

- Separate raw meat, poultry, seafood, and eggs from other foods in your grocery shopping cart, grocery bags, and in your refrigerator.
- Use one cutting board for fresh produce and a separate one for raw meat, poultry, and seafood.
- Use a food thermometer, which measures the internal temperature of cooked meat, poultry, and egg dishes, to make sure that the food is cooked to a safe internal temperature.
- Never place cooked food on a plate that previously held raw meat, poultry, seafood, or eggs.

Cook: Cook to Proper Temperatures

Food is safely cooked when it reaches a high enough internal temperature to kill the harmful bacteria that cause foodborne illness. Use a food thermometer to measure the internal temperature of cooked foods.

- Use a food thermometer, which measures the internal temperature of cooked meat, poultry, and egg dishes, to make sure that the food is cooked to a safe internal temperature.
- Cook beef roasts and steaks to a safe minimum internal temperature of 145 °F. Cook pork to a minimum of 160 °F. All poultry should reach a safe minimum internal temperature of 165 °F throughout the bird, as measured with a food thermometer.

- Cook ground meat to 160 °F. Information from the Centers for Disease Control and Prevention (CDC) links eating undercooked ground beef with a higher risk of illness. Remember, color is not a reliable indicator of doneness. Use a food thermometer to check the internal temperature of your burgers.
- Cook eggs until the yolk and white are firm, not runny. Don't use recipes in which eggs remain raw or only partially cooked. Casseroles and other dishes containing eggs should be cooked to 160 °F.
- Cook fish to 145 °F or until the flesh is opaque and separates easily with a fork.
- Make sure there are no cold spots in food (where bacteria can survive) when cooking in a microwave oven. For best results, cover food and stir and rotate for even cooking. If there is no turntable, rotate the dish by hand once or twice during cooking.
- Bring sauces, soups, and gravy to a boil when reheating. Heat other leftovers thoroughly to 165 °F.
- Use microwave-safe cookware and plastic wrap when cooking foods in a microwave oven.

Chill: Refrigerate Promptly!

Refrigerate foods quickly because cold temperatures slow the growth of harmful bacteria. Do not overstuff the refrigerator. Cold air must circulate to help keep food safe. Keeping a constant refrigerator temperature of 40 °F or below is one of the most effective ways to reduce the risk of foodborne illness. Use an appliance thermometer to be sure the temperature is consistently 40 °F or below. The freezer temperature should be 0 °F or below.

- Refrigerate or freeze meat, poultry, eggs, and other perishables as soon as you get them home from the store.
- Never let raw meat, poultry, eggs, cooked food, or cut fresh fruits or vegetables sit at room temperature more than 2 h before putting them in the refrigerator or freezer (1 h when the temperature is above 90 °F).
- Never defrost food at room temperature. Food must be kept at a safe temperature during thawing. There are three safe ways to defrost food: in the refrigerator, in cold water, and in the microwave using the defrost setting. Food thawed in cold water or in the microwave should be cooked immediately.
- Always marinate food in the refrigerator.
- Divide large amounts of leftovers into shallow containers for quicker cooling in the refrigerator.
- Use or discard refrigerated food on a regular basis.

FDA, USDA, NOAA Statements on Food Safety
See http://www.fda.gov/newsevents/publichealthfocus/ucm248257.htm.

- The United States enjoys one of the world's safest food supplies. The U.S. Food and Drug Administration (FDA), the U.S. Department of Agriculture (USDA), and the National Oceanic and Atmospheric Administration (NOAA), working with the U.S. Customs and Border Protection, have systems in place to assure that our food supply, both domestic and imported, is safe to eat.

- If the government has any reason to believe that food coming into or produced in the United States has been tainted, we will keep it from entering into the stream of commerce.
- FDA has jurisdiction over 80 % of the food supply, including seafood, dairy, and produce. USDA regulates meat, poultry, and processed egg products, while FDA regulates all other food products.

FDA's Core Messages

- FDA has a team of more than 900 investigators and 450 analysts in the Foods program who conduct inspections and collect and analyze product samples.
- Altogether, FDA screens all import entries and performs multiple analyses on about 31,000 import product samples annually. During Fiscal Year (FY) 2010, the Agency performed more than 175,000 food and feed field exams and conducted more than 350 foreign food and feed inspections.
- FDA works to inspect imports that may pose a significant public health threat by carrying out targeted risk-based analyses of imports at the points of entry.
- If unsafe products reach our ports, FDA's imports entry reviews, inspections, and sampling at the border help prevent these products from entering our food supply.
- Although FDA doesn't physically inspect every product, the Agency screens shipments of imported foods products before they reach our borders. Based on Agency risk criteria, an automated system alerts FDA to any concerns. Then inspectors investigate further and, if warranted, do a physical examination of the product.
- FDA also works cooperatively with U.S. Customs and Border Protection and other agencies to help identify shipments that may pose a threat.

NOAA's Core Messages

- Less than 2 % of the seafood consumed in the United States is imported from Japan.
- Federal seafood safety experts, including FDA and NOAA, are working together to closely monitor the situation in Japan. These experts will continue to ensure that imported seafood remains safe.
- In the unlikely scenario that airborne pollutants could affect the US fishermen or fish landed in the United States, NOAA will work with the FDA to ensure frequent testing of seafood caught in those areas, and inspection of facilities that process and sell seafood from those areas.

USDA's Core Messages

- USDA ensures the safety of meat, poultry, and processed egg products both domestically and from countries approved to export product to the United States.
- Since April 21, 2010, Japan has not been eligible to export raw beef products, which have been the only USDA-regulated products they had exported to the United States prior to April 2010.
  - USDA issued an import alert that banned importation of commodities from Japan that could harbor Foot and Mouth Disease virus.

10 tips
Nutrition Education Series

be food safe

10 tips to reduce the risk of foodborne illness

A critical part of healthy eating is keeping foods safe. Individuals in their own homes can reduce contaminants and keep food safe to eat by following safe food handling practices. Four basic food safety principles work together to reduce the risk of foodborne illness—**Clean, Separate, Cook, and Chill**. These four principles are the cornerstones of Fight BAC!®, a national public education campaign to promote food safety to consumers and educate them on how to handle and prepare food safely.

CLEAN

1 wash hands with soap and water

Wet hands with clean running water and apply soap. Use warm water if it is available. Rub hands together to make a lather and scrub all parts of the hand for 20 seconds. Rinse hands thoroughly and dry using a clean paper towel. If possible, use a paper towel to turn off the faucet.

2 sanitize surfaces

Surfaces should be washed with hot, soapy water. A solution of 1 tablespoon of unscented, liquid chlorine bleach per gallon of water can be used to sanitize surfaces.

3 clean sweep refrigerated foods once a week

At least once a week, throw out refrigerated foods that should no longer be eaten. Cooked leftovers should be discarded after 4 days; raw poultry and ground meats, 1 to 2 days.

4 keep appliances clean

Clean the inside and the outside of applicances. Pay particular attention to buttons and handles where cross-contamination to hands can occur.

5 rinse produce

Rinse fresh vegetables and fruits under running water just before eating, cutting, or cooking. Even if you plan to peel or cut the produce before eating, it is important to thoroughly rinse it first to prevent microbes from transferring from the outside to the inside of the produce.

SEPARATE

6 separate foods when shopping

Place raw seafood, meat, and poultry in plastic bags. Store them below ready-to-eat foods in your refrigerator.

7 separate foods when preparing and serving

Always use a clean cutting board for fresh produce and a separate one for raw seafood, meat, and poultry. Never place cooked food back on the same plate or cutting board that previously held raw food.

COOK AND CHILL

8 use a food thermometer when cooking

A food thermometer should be used to ensure that food is safely cooked and that cooked food is held at safe temperatures until eaten.

9 cook food to safe internal temperatures

One effective way to prevent illness is to check the internal temperature of seafood, meat, poultry, and egg dishes. Cook all raw beef, pork, lamb, and veal steaks, chops, and roasts to a safe minimum internal temperature of 145 °F. For safety and quality, allow meat to rest for at least 3 minutes before carving or eating. Cook all raw ground beef, pork, lamb, and veal to an internal temperature of 160 °F. Cook all poultry, including ground turkey and chicken, to an internal temperature of 165 °F (www.isitdoneyet.gov).

10 keep foods at safe temperatures

Hold cold foods at 40 °F or below. Keep hot foods at 140 °F or above. Foods are no longer safe to eat when they have been in the danger zone between 40-140 °F for more than 2 hours (1 hour if the temperature was above 90 °F).

Go to www.ChooseMyPlate.gov for more information.
Go to www.fsis.usda.gov for food safety information.

DG TipSheet No. 23
October 2012
USDA is an equal opportunity provider and employer.

- Japan has not exported any beef products to the United States for nearly a year.
- Japan is not eligible to export any poultry products or processed egg products to the United States since USDA has not determined Japan to be equivalent in these two commodities.

Also see:

- ***FDA—Keep It Safe to Eat:***
  - Separate raw, cooked and ready-to-eat foods.
  - Do not wash or rinse meat or poultry. Wash cutting boards, knives, utensils and counter tops in hot soapy water after preparing each food item and before going on to the next one.
  - Store raw meat, poultry and seafood on the bottom shelf of the refrigerator so juices don't drip onto other foods.
  - Cook foods to a safe temperature to kill microorganisms. Use a meat thermometer, which measures the internal temperature of cooked meat and poultry, to make sure that the meat is cooked all the way through.
  - Chill (refrigerate) perishable food promptly and defrost foods properly. Refrigerate or freeze perishables, prepared food and leftovers within 2 h.
  - Plan ahead to defrost foods. Never defrost food on the kitchen counter at room temperature. Thaw food by placing it in the refrigerator, submerging air-tight packaged food in cold tap water (change water every 30 min), or defrosting on a plate in the microwave.
  - Avoid raw or partially cooked eggs or foods containing raw eggs and raw or undercooked meat and poultry.
  - Women who may become pregnant, pregnant women, nursing mothers, and young children should avoid some types of fish and eat types lower in mercury.—FDA

Keeping food safety in the mix is what product developers can and should do (Kuntz 2012).

Sanitizing in the Workplace

Having already discussed many aspects of food safety, in the following there will be an emphasis on proper sanitizing and its documentation in food production. It can be seen that proper temperature is crucial to food safety.

In addition to the importance of temperature control for food handling, it is also of great significance in warewashing. There are *manual* sinks for washing, rinsing, and sanitizing dishes and utensils with or without right or left drain boards off to one side. The sink is two- and three-compartment with hot water or chemical means of sanitizing (one-compartment sinks are more of a *prep* sink). As well, there are many sizes and styles of *automatic* dish machines—including both high-temperature and low-temperature dishwashers. There are also pot washing sinks, spray bottles, and buckets of sanitizing agents—all capable of cleaning and sanitizing jobs in the food preparation workplace. Each has its own strength, time, and temperature requirement.

The following are some examples of useful tools that may be utilized as a means of controlling disease in the important tasks of temperature regulation in cooking and cleaning. See Figs. 19.11 and 19.12.

It is crucial to move the food, especially phf's quickly through the food facility, and to follow the principle of proper stock rotation—first-in-first-out, or FIFO. Product "use-by" date stickers available in multiple languages and styles may adhere to a food product and easily dissolve to come off of the container when a food container is washed.

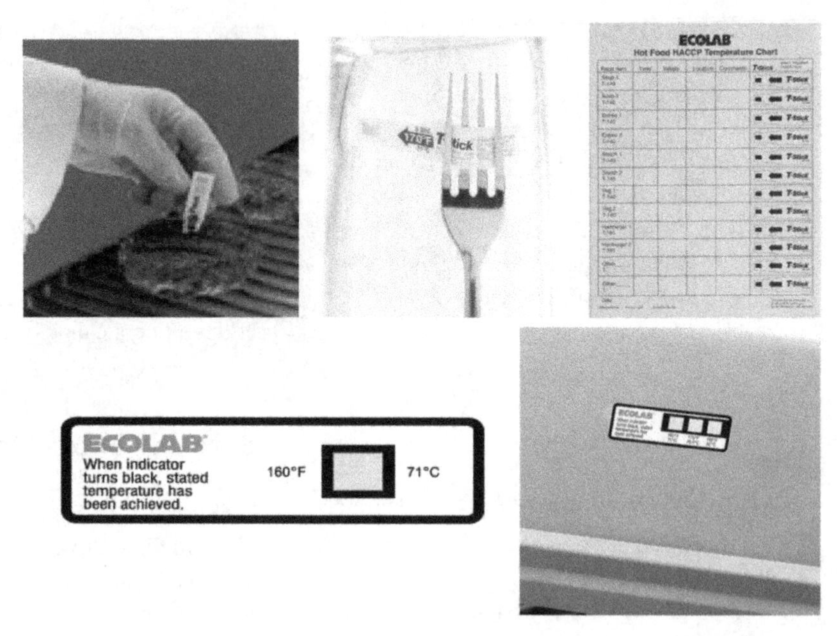

Fig. 19.11 Various means of temperature checks for controlling food safety. ECOLAB Eagan, MN. http://www.FoodSafetySolutions.com

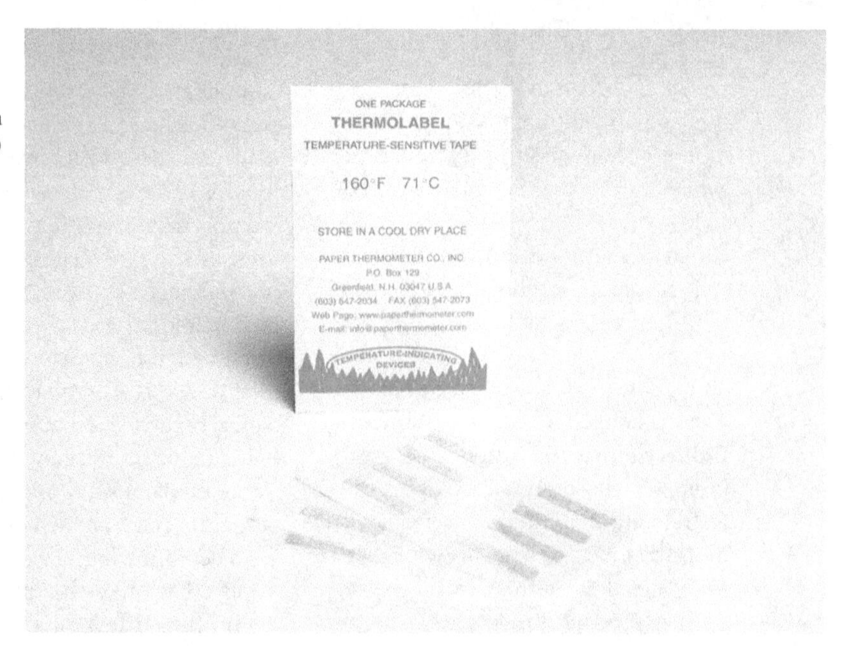

Fig. 19.12 Example of paper thermometer (dishwasher temperature labels) to adhere to items in dish machine.

Labeling as a Means of Assuring Food Safety

Dating

Other than product dissolvable labels, placed by food-handling personnel, open dates may be placed by food processors on the food package. This provides the end user with information regarding optimum time periods to keep foods on hand. However, dating is not a guarantee against spoilage or harmful contamination. Personnel in processing and foodservice operations, as well as consumers, must observe foods for any possible deterioration and not use damaged products.

Open date labeling is *mandatory* for dairy products, but other foods may *voluntarily* have open date labeling. Foods may also display a *code date*, which can be read only by the manufacturer. Some examples of the types of dating that may appear on foods include the following:

- **"Best if used by" date**—informs consumers of the food's optimal period for retention of high quality.
- **Expiration date**—indicates a deadline for recommended use.
- **Pack date**—indicates when the food was packaged.
- **Pull date**—signifies the last day the food may be sold as fresh. All food products should be made available for use only within established time frames.

USDA Meat & Poultry Hotline 1-888-MPHotline (1-888-674-6854)

"The Food Safety and Inspection Service (FSIS) is the public health agency in the U.S. Department of Agriculture responsible for ensuring that the nation's commercial supply of meat, poultry, and egg products is safe, wholesome, and correctly labeled and packaged."

Food Safety Information USDA: FSIS

Food Product Dating

"Sell by Feb 14" is a type of information you might find on a meat or poultry product. Are dates required on food products? Does it mean the product will be unsafe to use after that date? Here is some background information which answers these and other questions about product dating.

What is dating?

"Open Dating" (use of a calendar date as opposed to a code) on a food product is a date stamped on a product's package to help the store determine how long to display the product for sale. It can also help the purchaser to know the time limit to purchase or use the product at its best quality. It is not a safety date. After the date passes, while not of best quality, the product should still be safe if handled properly and kept at 40 °F or below for the recommended storage times listed on the chart (see below).

If product has a "use-by" date, follow that date.

If product has a "sell-by" date or no date, cook or freeze the product by the times on the chart below.

Is dating required by federal law?

Except for infant formula (see below), product dating is not generally required by Federal regulations. However, if a calendar date is used, it must express both the month and day of the month (and the year, in the case of shelf-stable and frozen products). If a calendar date is shown, immediately adjacent to the date must be a phrase explaining the meaning of that date such as "sell by" or "use before." There is no uniform or universally accepted system used for food dating in the United States. Although dating of some foods is required by more than 20 states, there are areas of the country where much of the food supply has some type of open date and other areas where almost no food is dated.

What types of food are dated?

Open dating is found primarily on perishable foods such as meat, poultry, eggs and dairy products. "Closed" or "coded" dating might appear on shelf-stable products such as cans and boxes of food.

Types of Dates

- A "Sell-By" date tells the store how long to display the product for sale. You should buy the product before the date expires.
- A "Best if Used By (or Before)" date is recommended for best flavor or quality. It is not a purchase or safety date.

- A "Use-By" date is the last date recommended for the use of the product while at peak quality. The date has been determined by the manufacturer of the product.
- "Closed or coded dates" are packing numbers for use by the manufacturer.

Safety After Date Expires

Except for "use-by" dates, product dates don't always refer to home storage and use after purchase. "Use-by" dates usually refer to best quality and are not safety dates. But even if the date expires during home storage, a product should be safe, wholesome and of good quality if handled properly and kept at 40 °F or below.

See the accompanying refrigerator charts for storage times of dated products. If product has a "use-by" date, follow that date. If product has a "sell-by" date or no date, cook or freeze the product by the times on the chart below.

Foods can develop an off odor, flavor or appearance due to spoilage bacteria. If a food has developed such characteristics, you should not use it for quality reasons. If foods are mishandled, however, foodborne bacteria can grow and cause foodborne illness—before or after the date on the package. For example, if hot dogs are taken to a picnic and left out several hours, they might not be safe if used thereafter, even if the date hasn't expired.

Other examples of potential mishandling are products that have been: defrosted at room temperature more than 2 h; cross contaminated; or handled by people who don't use practice sanitation. Make sure to follow the handling and preparation instructions on the label to ensure top quality and safety.

Dating Infant Formula

Federal regulations require a "use-by" date on the product label of infant formula under FDA inspection.

What do can codes mean?

Cans must exhibit a packing code to enable tracking of the product in interstate commerce. This enables manufacturers to rotate their stock as well as to locate their products in the event of a recall. These codes, which appear as a series of letters and/or numbers, might refer to the date or time of manufacture. They aren't meant for the consumer to interpret as "use-by" dates. There is no book or website that tells how to translate the codes into dates. Cans may also display "open" or calendar dates. Usually these are "best if used by" dates for peak quality. In general, high-acid canned foods such as tomatoes, grapefruit and pineapple will retain best quality on the shelf for 12–18 months; low-acid canned foods such as meat, poultry, fish and most vegetables will retain best quality on the shelf for 2–5 years—if the can remains in good condition and has been stored in a cool, clean, dry place.

Dates on Egg Cartons

Use of either a "Sell-by" or "Expiration" (EXP) date is not federally required, but may be State required, as defined by the egg laws in the State where the eggs are marketed. Some State egg laws do not allow the use of a "sell-by" date. Many eggs reach stores only a few days after the hen lays them. Egg cartons with the USDA grade shield on them must display the "pack date" (the day that the eggs were washed, graded, and placed in the carton). The number is a three-digit code that represents the consecutive day of the year starting with January 1 as 001 and ending with December 31 as 365. When a "sell-by" date appears on a carton bearing the USDA grade shield, the code date may not exceed 45 days from the date of pack.

Always purchase eggs before the "Sell-By" or "EXP" date on the carton. After the eggs reach home, refrigerate the eggs in their original carton and place them in the coldest part of the refrigerator, not in the door. For best quality, use eggs within 3–5 weeks of the date you purchase them. The "sell-by" date will usually expire during that length of time, but the eggs are perfectly safe to use.

UPC or Bar Codes

Universal Product Codes appear on packages as black lines of varying widths above a series of numbers. They are not required by regulation but manufacturers print them on most product labels because scanners at supermarkets can "read" them quickly to record the price at checkout.

Bar codes are used by stores and manufacturers for inventory purposes and marketing information. When read by a computer, they can reveal such specific information as the manufacturer's name, product name, size of product and price. The numbers are not used to identify recalled products...." **(fsis.usda.gov)**

Regarding labeling, a "clean" label, on a product line is: "... its ability to offer a clean label and broad-spectrum solution without impacting the sensory characteristics of the target food or beverage." For example, in combating mold, yeast, and bacteria, applications of a natural antimicrobial "... proved a more effective shelf-life enhancer than sodium benzoate and potassium sorbate, while also improving the overall flavor profile of the tested formulas" (Food Product Design 2012).

Allergen-Free Labeling

Allergens are an issue that relates to food safety (Chap. 20). The FDA is responsible for ingredient labeling and has given notice to food processors that exemptions from ingredient labeling would not be tolerated. A food product must contain what it states on the label, and it should *not* contain an ingredient it does *not* disclose. Life-threatening allergens must be reported on the food label, and in uncertain cases, statements such as "may contain" are displayed as a safeguard.

The eight major foods to which people have allergies include milk, eggs, peanuts, tree nuts (e.g., almonds, cashews), soy, wheat, fish, and shellfish. These are responsible for 90 % of food allergic reactions and therefore represent ingredients that should thus be isolated in the production process. Severe allergic reactions can cause anaphylaxis or death (Food Allergy Network) (J Am Diet Assoc 2000). Food safety is being redefined to include allergen-free and pathogen-free (Higgins 2000).

If an allergen is detected following product distribution, product recalls may be necessary. Doing things right the first time is a more sensible alternative! Either an independent lab or allergen test kits may authenticate that products are allergen-free. Testing is part of industry's GMPs.

According to the Director of the Office of Scientific Analysis and Support at FDA's Center for Food Safety and Applied Nutrition, "... both FDA and food companies are looking harder for allergens...allergic consumers are becoming more aware of the allergens in foods, and ... [there are] improved allergen-detection methods (FDA)."

Some of the best practices for allergen control relate to the following:

- R&D/product development
- Engineering and system design with dedicated production lines
- Vendor certification of raw materials and ingredients
- Production scheduling to include longer production runs
- Rework segregated
- Labeling and packaging with the right product going into the right package and ingredients listing to match the actual food product!
- Sanitation. An HACCP-like approach
- Training (Morris 2002)

Conclusion

"I just want to eat" says the consumer! The consumer expects safe food and protection from microbial, chemical, and physical hazards to the food supply. They do not expect to have food safety be an issue for them to experience. Yet, foodborne illness *may be* their experience. Unfortunately, illness could originate from bacteria, viruses, molds, parasites, and naturally occurring chemicals in food (such as toxins), accidental chemical contamination, toxic levels of additives or preservatives, and foreign objects. That illness may severely or fatally impact the health and welfare of a food company, hospital, restaurant, or them as the consumer—even at home.

The HACCP is a food safety system that focuses on foodborne disease prevention and ensures a greater likelihood of food safety. Various segments of the food industry apply the HACCP system of food protection to their handling of hazardous ingredients. HACCP team members promote food safety by assessing potential hazards in the flow of foods through their operation and by establishing limits or controls for the identified hazards. HACCP is applied to all steps of handling, including processing, packaging, and distribution. This same process may be followed by foodservice operations.

The CDC monitors and reports FBDOs. See Fig. 19.4. Open and code dating are utilized. Food allergens are monitored by the FDA. It is the consumer (the one who consumes!) who must ultimately be responsible for the consumption of safe foods.

Simulated food defense including training exercises with all levels of the government, non-government agencies, and the private sector allow better preparation for, and protection against, possible contamination of the food supply from terrorist threats. Bioterrorism preparedness training is crucial to food science and foodservice professionals (http://www.usda.gov/homelandsecurity). See more in chapter addendum.

The USDA FSRIO has created a website for the general public and food safety researchers. The site contains educational, professional, and foreign government links for food safety (http://www.nal.usda.gov/fsrio).

Addendum

Bioterrorism Threat to Food Safety

Responsibility for food safety has been discussed. Specifically, agencies related to food safety and emergency preparedness are American Red Cross, CDC, FEMA, FDA, OSHA, USDA, and state and local agencies. Several agency reports are cited below. Needless to say the agencies are no replacement for personal vigilance.

The Bioterrorism Preparedness and Response Act of 2002 ("the Bioterrorism Act") was issued and contained the following:

The events of Sept. 11, 2001, reinforced the need to enhance the security of the United States. Congress responded by passing the Public Health Security and Bioterrorism Preparedness and Response Act of 2002 (the Bioterrorism Act), which President Bush signed into law June 12, 2002.

http://www.fda.gov/oc/bioterrorism/bioact.html

The Bioterrorism Act is divided into five titles:

- Introduction
- Title I—National Preparedness for Bioterrorism and Other Public Health Emergencies
- Title II—Enhancing Controls on Dangerous Biological Agents and Toxins
- Title III—Protecting Safety and Security of Food and Drug Supply
- Title IV—Drinking Water Security and Safety
- Title V—Additional Provisions

The FDA is responsible for carrying out certain provisions of the Bioterrorism Act, particularly Title III, Subtitle A (Protection of Food Supply) and Subtitle B (Protection of Drug Supply).

In the interim final rule reported on September 28, 2005, registration of food facilities is required. The rule requires domestic and foreign facilities that manufacture, process, pack, or hold food for consumption in the United States to register with FDA.

The FDA answers some consumer questions on issues of food safety and terrorism, such as on September 11, 2001. The discussion appears in the following website:

- http://www.cfsan.fda.gov/~dms/fsterrqa.html

The USDA is also charged with the duty of protecting the nation's food supply from terrorist threats. Dr. Richard Raymond, USDA Undersecretary for food safety states "We remain steadfast in our commitment to work with our federal, state and private sector partners so that we can keep or agricultural commodities safe."

The USDAs FSIS celebrated 100 years of protecting consumers, "As we stand on the threshold of the second century of ensuring the safety of America's meat, poultry and egg products, we take pride in our achievements in public health protection and look forward to strengthening our commitment to safeguarding future generations." (Agriculture Deputy Secretary Chuck Conner. USDA) The USDA has said:

- "Forming the USDA's Homeland Security Council was the first step in a series of organizational changes aimed at improving the Department's ability to perform homeland security-related activities."—former Secretary Ann M. Veneman, Statement—September 9, 2003

An interesting USDA article entitled "Keeping Food Safe During an Emergency" may be seen at the following address:

- http://www.fsis.usda.gov/fact_sheets/keeping_food_Safe_during_an_emergency/index.asp (Keeping Food Safe During an Emergency)

The CDC includes information on Biological and Chemical Terrorism:

http://www.bt.cdc.gov/agent/agentlist-category.asp

For various biologic agents, causes, systems affected, routes of transmission, and stages of clinical presentation, also in Spanish, locate agents alphabetically A–Z or by category A, B, and C:

- http://www.cdc.gov/mmwr/preview/mmwrhtml/rr4904a1.htm

The International Foodservice Distributors Association (IFDA, Mclean, VA) has the following to say regarding vigilance and food safety since September 11 attacks:

"The intent of those who store and deliver food for foodservice operations is to prevent or minimize the potential for attacks on the food supply. Their aim is to have a safe and secure system. Each individual program should be equipped to minimize threats "to the greatest extent possible."

For foodservice operations, deliveries and Memorandum of Understandings (MOUs) are significant. They greatly assist in emergencies.

MOU's are ... written agreements are used to specify what is transported, when, to where, and so forth. Foodservice Distributors serving the healthcare industry and/or others may provide arrangements for the provision of food and food related products (including bottled drinking water)."

Another reference for Bioterrorism threats and food safety is two bioterrorism training computer modules (UTHSCSA) related to food safety. The modules contain the following information:

"If/when disasters and emergencies happen, the foodservice operation must be prepared. As best as possible, all contamination should be prevented; the foods should be comforting, aligned with culture, ethnicity, religion, nutrition and so forth. Everyday practice of precautions, or "the right way" to do something in a non-emergency situation, lessens any panic in an emergency!" (University of Texas Health Science Center San Antonio (UTHSCSA))

The professionals agree that even the best-laid plans and training will not address all situations. The key factor in addressing unexpected situations is to maintain a calm demeanor; make decisions, whether they are right or wrong; and be flexible in changing the decisions that do not work (Cody 2002)

Some relevant food science facts appear below:

Terrorism and Food Science

Relevant Food Science Facts

Government Agencies Protecting the Food Supply

- Emergency preparedness has been increased following 9/11.
- Students in Nutrition, Dietetics (RD), Food Science, and Culinary Arts are expected to "expect the unexpected" and follow all rules and regulations and current information of governmental agencies such as the FDA, USDA, FEMA, CDC, and health departments.

Food Processing

- Processors, both consumers at home and corporations, must protect foodstuffs that they are processing.
- Foods are protected against hazardous external conditions by canning, etc..

Food Safety and Spoilage

- Foods may become contaminated by hazards including biological, chemical, or physical. Prevention of hazards is a key to food safety.
- Microbiological hazards include bacteria, viruses, fungi, and parasites.
- Specific bacteria are listed in the CDC website.
- Control measures for bacterial hazards include proper temperatures for storage, cooking, and holding with avoidance of the TDZ, eliminating all cross-contamination; practice meticulous personal hygiene.
- Chemical hazards include intentional and unintentional substances and may be accidental, additives, toxic metals, or toxic substances.
- Physical hazards are foreign objects in food. They may harm health, cause psychological trauma or dissatisfaction, and must not be viewed merely as a manufacturing inconvenience. (University of Texas Health Science Center San Antonio (UTHSCSA))

Shelf Life of Foods for Storage (FEMA)

General guidelines for rotating common emergency foods:

Use within 6 months:

- Powdered milk (boxed)
- Dried fruit (in metal container)
- Dry, crisp crackers (in metal container)
- Potatoes

Use within 1 year:

- Canned condensed meat and vegetable soups
- Canned fruits, fruit juices, and vegetables
- Ready-to-eat cereals and uncooked instant cereals (in metal containers)
- Peanut butter
- Jelly
- Hard candy, chocolate bars, and canned nuts

May be stored indefinitely (in proper containers and conditions):

- Wheat
- Vegetable oils
- Corn
- Baking powder
- Soybeans
- Instant coffee, tea
- Cocoa
- Salt

- Noncarbonated soft drinks
- White rice
- Bouillon products
- Dry pasta
- Powdered milk (in nitrogen-packed cans)

Safety Considerations

- In a workplace that is equipped for emergency preparedness, there must be a constant check and consideration for safety.
- Safety may be threatened by accidental and intentional activities.
- In-house regulation by Safety Committee or an outside agency contributes to the regulation of safety in the foodservice
- Intentional bioterrorist attack may introduce foreign objects into the workplace.
- Questionable chemicals should not be used or handled, and suspect persons or packages should be reported." (University of Texas Health Science Center San Antonio (UTHSCSA))

Mildred M. Cody, PhD, RD, author of the American Dietetic Association's Food Safety for Professionals, agrees that adhering to basic food safety behaviors will help reduce the risk of foodborne illnesses. However, considering the fact that a bioterrorism attack would be silent and the effects might not be visible for several days, Cody also emphasizes following these additional standard food safety guidelines on a day-by-day basis (Cody 2002):

- Accept only food from reputable vendors to take advantage of public controls from regulatory agencies
- Check for intact packaging

Wash cans before opening to keep debris from can lids from falling into foods."

"In the case of water for drinking, cooking or cleaning, it is essential that it is safe. Some FEMA suggestions for purified water follow, however there may still be some dangerous chemical or physical residue.

- Boiling . . .
- Chlorination uses liquid chlorine bleach (5.25 % sodium hypochlorite as the only ingredient) to kill microorganisms. . .
- Purification tablets release chlorine or iodine . . . (FEMA)

"In addition to provisions for the work environment, Cody advises individuals to keep 3 days or more worth of food, water (One gallon of water [drinking] per person per day) and prescription medicine available, because while safe food and water will be made available after a bioterrorism emergency, the distribution of these items may take some time to coordinate and organize.

As reported by Puckett and Norton, ideas for training within the disaster plan of a foodservice operation include:

Review of plans and employee duties before, during and after an emergency
Security procedures
Location of water, food, emergency supplies, first aid supplies, fire fighting equipment, water purification system, key employees
Safe food handling
Sanitizing procedures
Record keeping
Physical layout of the foodservice area and entire facility
Physical security
ID badges
Important names and numbers, including vendor data" (Puckett and Norton 2003)

For More Information

- Congressional Inquiries: (202) 720-3897
- Constituent Inquiries: (202) 720-8594
- USDA Meat and Poultry Hotline: (800) 535-4555 or (202) 720-3333
- Consumer Inquiries: Call USDA's Meat and Poultry Hotline 1-800-535-4555.

In the Washington, DC, area, call (202) 720-3333. The TTY number is 1-800-256-7072.

Emergency Point of Contact:
U.S. Food and Drug Administration
5600 Fishers Lane
Rockville, MD 20857

If a food establishment operator suspects that any of his/her products that are regulated by the FDA have been subject to tampering or criminal or terrorist action, he/she should notify the FDA 24-h emergency number at 301-443-1240 or call their local FDA District Office. FDA District Office telephone numbers are listed at:

http://www.fda.gov/ora/inspect_ref/iom/iomoradir. html. The operator should also notify local law enforcement.

Contact Us:

- Centers for Disease Control and Prevention
  Atlanta, GA 303331600 Clifton Rd
  800-CDC-INFO or (800-232-4636)

. . . .

Partnership for Food Safety Education: Fight BAC!® (http://www.fightbac.org)

- Academy of Nutrition and Dietetics (formerly American Dietetic Association)
- American Egg Board and Egg Safety Center
- Association of Food and Drug Officials
- Consumer Federation of America
- Food Marketing Institute
- Food Temperature Indicator Association
- Institute of Food Technologists
- International Food Information Council
- International Fresh-Cut Produce Association
- National Association of State Departments of Agriculture
- National Chicken Council
- National Pork Board
- National Turkey Federation
- NSF International
- Produce Marketing Association
- School Nutrition Association
- The Soap and Detergent Association
- United Fresh Fruit and Vegetable Association Federal Government Liaison
- U.S. Department of Agriculture
- U.S. Food and Drug Administration
- U.S. Department of Health and Human Services, CDC
- U.S. Environmental Protection Agency

Notes

CULINARY ALERT!

Glossary

Biological hazard Microbiological hazard from bacteria, viruses, fungi, and parasites.

Chemical hazard Toxic levels of a specific chemical that may occur by accident, use of toxic level additives, or toxic metals.

Contaminated Presence of harmful substances.

Cross-contamination Transfer of harmful microorganisms from one food to another by way of another food, hands, equipment, or utensils.

Emerging pathogens Pathogens whose incidence has increased within the last few years or which threaten to increase in the near future.

Foodborne illness Disease carried to people by food.

Fungi Microorganisms that include mold and yeast.

HACCP Hazard Analysis and Critical Control Point system of food safety.

Infection Illness that results from ingesting living, pathogenic bacteria such as *Salmonella*, *Listeria*, or *Shigella*.

Intoxication Illness that results from ingesting a preformed toxin such as that produced by Staphylococcus aureus, Clostridium botulinum, or Bacillus cereus.

Pathogenic Disease-causing agent.

Physical hazard Foreign object found in food; may be due to harvesting or manufacturing; may be intrinsic to the food (bone, shell, pit).

Potentially hazardous food Natural or synthetic food in a form capable of supporting the rapid and progressive growth of infectious or toxigenic microorganisms; the growth and toxin production of *C. botulinum* or, in shell eggs, the growth of *S. enteritidis*.

Spoiled Damage to the eating quality.

Spore Thick-walled formation in a bacterium that is resistant to heat, cold, and chemicals; it remains in food after the vegetative cells are destroyed and is capable of becoming vegetative cell.

Temperature danger zone (TDZ) Temperature range within which most bacteria grow and reproduce 40–140 °F (4–60 °C).

Toxin Poison produced by a microorganism while it is alive; may remain in food and cause illness after the bacteria is killed.

Toxin-mediated infection Infection/intoxication illness that results from ingestion of living, infection-causing bacteria that also produce a toxin in the intestine, such as *C. perfringens* or *E. coli* 0157:H7.

References

(2000) Dietitians face the challenge of food allergies. J Am Diet Assoc 100:13

(2000) Eye on Washington. Food Eng 72:16

(2012) All-natural sanitizers and preservatives. Food Prod Des 2012:54

Crawford LM, DVM (2000) Food irradiation's advantages will not escape public attention. Food Technol 52(1):55

HACCP Principles & Application Guidelines (1997) Adopted August 14, 1997. Page Last Updated: 03/15/2013

Food and Drug Administration. Department of Health and Human Services. Public Health Service

Higgins KT (2000) Food safety is being redefined to include allergen-free as well as pathogen-free. Food Eng 72(6):75–82

Higgins KT (2003) E-beam comes to the heartland. Food Eng 75(10):89–96

Higgins KT (2006) Beam me through, Scotty. Food Eng 78(1):107–112.0

Hussain SA (2000) ConAgra refrigerated prepared foods, technical services. Surak JG, Clemson University, Cawley JL, Northwest Analytical. Butterball integrates SPC with HACCP. Food Eng 72(10):82

Katz F (2000) Research priorities move toward healthy and safe. Food Technol 54(12):42–46

Kuntz LA (2012) Keeping food safety in the mix. Food Prod Des: 2–13

Morris CE (2002) Best practices for allergen control. Food Eng 74(3):33–35

National Advisory Committee for Microbiological Criteria for Foods (NACMCF). HACCP principles for food production. USDA—FSIS Information Office, Washington, DC

Pierson MD, Corlett DA (eds) (1992) HACCP principles and applications. Chapman & Hall, New York (reprint 2012)

Stier RF (2006) Why can't scientists communicate science? Food Eng 78(3):25

Van Milligen D (2001) Sanitation 101. Food Eng 73 (1):55–60

Addendum

CDC Table B 1-4 Guidelines for confirmation of foodborne-disease outbreaks. http://www.cdc.gov/mmwr/preview/mmwrhtml/ss4901a3.htm
Cody MM (2002) Food safety for professionals. The American Dietetic Association, Chicago, IL
Partnership for Food Safety Education—Fight BAC!® (http://www.fightbac.org)
Puckett RP, Norton LC (2003) Disaster and emergency preparedness in foodservice operations. The American Dietetic Association, Chicago, IL
University of Texas Health Science Center San Antonio (UTHSCSA)

Bibliography

CDC Surveillance for Foodborne Disease Outbreaks—United States (2009–2010)
Center for Science in the Public Interest (CSPI)
Centers for Disease Control and Prevention (CDC)
Current Good Manufacturing Practice and Hazard Analy sis and Risk-Based Preventive Controls for Human Food Proposed Rule. (http://www.gpo.gov/fdsys/pkg/FR-2013-01-16/html/2013-00125.htm)
FDA, USDA, National Oceanic and Atmospheric Administration (NOAA) Statements on Food Safety
FDA Center for Food Safety and Applied Nutrition
Food Safety and Inspection Service (FSIS)
Food Seminars International
Grocery Manufacturers Assoc. http://www.gmaonline.org
http://producesafetyalliance.cornell.edu
http://www.cdc.gov/foodborneoutbreaks/outbreak_data.htm
http://www.cdc.gov/mmwr/preview/mmwrhtml/mm6203a1.htm?s_cid=mm62031_w\
Institute of Food Technologists' Expert Panel on Food Safety and Nutrition. Scientific Status Summary, Foodborne illness: Role of home food handling practices
LaVella B, Bostik JL (1994) HACCP for food service. LaVella Food Specialists, St. Louis, MO
Medeiros LC, Kendall P, Hillers V, Chen G, DiMascola S (2001) Identification and classification of consumer food-handling behaviors for food safety education. J Am Diet Assoc 101(1326–1332):1337–1339
Model Food Code
National Restaurant Association (1992) The Educational Foundation. Applied Foodservice Sanitation, 4th edn. Wiley, New York
Texas A&M University—Center for Food Safety. College Station, TX
University of Idaho—http://www.webpages.uidaho.edu/foodtox/
USDA ChooseMyPlate.gov
USDA Food Safety Research Information Office (FSRIO)
Watson DH (1993) Safety of chemicals in food: chemical contaminants. Ellis Horwood, New York

Associations and Organizations

American Public Health Association (APHA). Washington, DC
Association of Food and Drug Officials (AFDO). York, PA
Cooperative Extension Service (CES), throughout the United States
Council for Agriculture Science and Technology (CAST). Ames, IA
Food Marketing Institute (FMI). Washington, DC
Gannett News
Institute of Food Technologists (IFT). Chicago, IL
International Association of Milk, Food, and Environmental Sanitarians (IAMFES). Des Moines, IA
International Council of Hotel, and Restaurant Industry Educators (CHRIE). Washington, DC
International Foodservice Distributors Association (IFDA). Mclean, VA
International Food Manufacturers Association (IFMA). Chicago, IL
National Center for Nutrition and Dietetics (NCND)
National Environmental Health Association (NEHA). Denver, CO
National Sanitation Foundation (NSF) International. Ann Arbor, MI
The Academy of Nutrition and Dietetics National Center for Nutrition and Dietetics (NCND). Chicago, IL
The Educational Foundation of the National Restaurant Association. Chicago, IL
United States Environmental Protection Agency, (EPA), Washington, DC
University of Texas Health Science Center San Antonio (UTHSCSA). San Antonio, TX

Part IX

Government Regulation of the Food Supply

Government Regulation of the Food Supply and Labeling

20

Introduction

Consumers want the assurance that they have a sure, safe, and sanitary food supply. They want deceptive claims and fraudulence to be nonissues for them to face in everyday life. Therefore, for centuries, governments throughout the world have regulated the food supply. Federal, state, and local government, their regulation, enforcement, as well as the educational materials they offer, assist in providing a safe food supply. The intent of this chapter is to view government regulation of the food supply and labeling. However, a safe and sound food supply is still dependent, not alone on a government agency or program yet also upon the individual!

All the way through this food science textbook, the role of government has been addressed. One of the major regulatory agencies protecting the food supply is the *Food and Drug Administration* (*FDA*). Their basic purpose is to protect the public from foodborne illness. The FDA regulations known as "Good Manufacturing Practices" or GMPs are in operation at food plants. Of course, maintaining plant sanitation and food safety (see Chap. 19) are ongoing duties of the food processing plant's own personnel—hopefully well trained and motivated!

The FDA's Federal Food, Drug, and Cosmetic Act of 1938 (FD&C Act) is the main law that regulates the food supply in the United States. They are responsible for public health encompassing safety, specific safe drugs, and cosmetics as well as biological products and medical devices. It ensures the safety of all food except for meat, poultry, and some egg products.

The FDA Code of Federal Regulations (CFR) is cited several times in this chapter, with the hope of better portraying and understanding the government ruling on an issue.

Interstate transport of food, food packaging, and labeling are regulated, and grading standards and ordinances that specify sanitation for the food environment are enforced. *Intrastate* transport is regulated by each *state's* Department of Agriculture that may adopt their own, more strict regulations than the federal.

Another federal regulatory agency with influence and enforcement over the food supply is the *United States Department of Agriculture* (*USDA*). This agency has responsibility for inspecting animal products, including meat, poultry, and eggs; processing plants for meat and poultry; as well as voluntary grading.

The two federal agencies, FDA and USDA, despite friction at times, work together to maintain food safety and consumer health. It may be

V.A. Vaclavik and E.W. Christian, *Essentials of Food Science, 4th Edition*, Food Science Text Series,
DOI 10.1007/978-1-4614-9138-5_20, © Springer Science+Business Media New York 2014

proactive or reactive responses that are necessary for the well-being of the US citizens.

Of course, in addition to the government's regulation of the food supply, *industry* and *consumers* must be vigilant and play their part in assuring a safe food supply! Food safety is still dependent upon the individual!

Additionally, general labeling, nutrition labeling, health claims, food allergen labeling, and labeling for foodservice are discussed in this chapter.

The Food and Drug Administration

The FDA is a public health agency. The agency regulates approximately 25 % of every dollar spent annually by American consumers—over $1 trillion worth of products (FDA)—and does so at a taxpayer cost of just dollars per individual. The FDA inspects food—to assure that it is safe and wholesome, it also inspects cosmetics, medicines and medical devices, radiation-emitting devices (such as microwave ovens), animal feed, and drugs.

> **Biological products** often represent the cutting edge of medical science and research. Also known as biologics, these products replicate natural substances such as enzymes, antibodies, or hormones in our bodies.
>
> Biological products can be composed of sugars, proteins, or nucleic acids, or a combination of these substances. They may also be living entities, such as cells and tissues. Biologics are made from a variety of natural resources—human, animal, and microorganism—and may be produced by biotechnology methods.—FDA

> Today, the FDA regulates $1 trillion worth of products a year. It ensures the safety of all food *except for meat, poultry and some egg products*; ensures the safety and effectiveness of all drugs, biological products (including blood, vaccines and tissues for transplantation), medical devices, and animal drugs and feed; and makes sure that cosmetics and medical and consumer products that emit radiation do no harm. (FDA)

FDA Federal Food, Drug, and Cosmetic Act: 1938

"The Food and Drugs Act of 1906 was the first of more than 200 laws that constitute one of the world's most comprehensive and effective networks of public health and consumer protections. The Federal *Food, Drug, and Cosmetic Act of 1938* officially passed after a legally marketed toxic elixir killed 107 people, including many children. The FD&C Act completely overhauled the public health system. Among other provisions, the law authorized the FDA to demand evidence of safety for new drugs, issue standards for food, and conduct factory inspections." (FDA)

Since the origin of this law, there have been numerous amendments. This law replaced the 1906 Federal Food and Drug Act, or "Pure Food Law," and is assigned to regulate many packaged or processed food products. The regulation includes the necessity for adequate and truthful labels if the food is subject to import or interstate commerce. Additionally, a federal Code of Regulations was written to cover specific rules for the food industry.

The FDA has several thousand researchers, inspectors, and legal staff in approximately 150 cities throughout federal, regional, and local offices in the United States, including scientists (over 2,000), chemists (approximately 900), and microbiologists (approximately 300). Agents of the FDA may work with public affairs or small business as well as any laboratory personnel. They interpret law and monitor the manufacture, import, transport, and storage of products both prior to and following sale on the market. Products are examined for construction integrity, and labels must be truthful.

Among the varied activities of federal FDA agents includes advising state and local agencies in general duties and prevention of disasters. The FDA has both a regulatory arm of enforcement and cooperative programs of partnership with industry. The latter, for example, helps train employees in preventing foodborne illness. Despite budgetary constraints and a transition of the FDA to a Hazard Analysis Critical Control Point (HACCP) focus, the role of this government agency remains to protect the public.

Voluntary correction of public health problems is necessary, although when warranted, *legal sanctions* may be brought to bear against manufacturers or distributors. Recalls of faulty products are generally the fastest and most effective way to protect the public from unsafe products on the market.

Amendments to the Food, Drug, and Cosmetic Act

Several major amendments to the Food, Drug, and Cosmetic Act that were introduced and became the US law include the following:

- *1954 Pesticide Chemical Amendment*: The use of pesticides is subject to FDA approval. Raw agriculture products are prohibited from containing pesticide residues above a certain level.
- *1958 Food Additives Amendment*: With this amendment, the burden of proof for usefulness and harmlessness of an additive was shifted to industry. Exempt from this proof were Generally Recognized as Safe (GRAS) substances already in common use with no proof of cancer (see section "GRAS Substances" below).
- The *Delaney Clause* (1966) of the Food Additives Amendment states that an additive *cannot* be used if it leads to cancer in man or animals or if the carcinogen is detectable by any appropriate test.

In recent years, a much-debated question on the necessity of the Delaney Clause has arisen. For example, what *is* an appropriate test to determine the level of a food additive that induces cancer? Finer detection of minute amounts of agents responsible for cancer has become available. Thus, the question is: At what level is the presence of a carcinogen indicative of the need to remove that item from the food supply? There is *no* food item that is, or can be, totally safe at any level of ingestion. (Simply consuming too much *water* has landed people in the hospital!) The future will offer more debate and regulation of this matter.

- *1960 Color Additives Amendment*: The use of food colors is subject to FDA approval.
- *1966 Fair Packaging and Labeling Act* requires all consumer products in interstate commerce to contain accurate information on the package, facilitating better control of misinformation. Consumers benefit in that they can use the label information on packages, in making purchasing and value comparisons.
- *1990 Nutrition Labeling and Education Act* (NLEA) was passed by Congress, and the FDA then wrote regulations for compliance covering extensive labeling changes, including mandatory nutrition labels, uniform use of product health claims, and uniform serving sizes.

This was an attempt to protect the consumer against misinformation and fraud. New "nutrition facts" labels appeared on food products in May of 1994.

GRAS Substances

GRAS substances, according to the General Provisions of the CFR, Title 21 (21CFR582), Sec. 582.1, are discussed as follows:

"It is impractical to list all substances that are generally recognized as safe for their intended use. However, by way of illustration, the Commissioner regards such common food ingredients as salt, pepper, sugar, vinegar, baking powder, and monosodium glutamate as safe for their intended use."

Standards for Interstate Transport of Food

The FDA has mandatory standards, identified in the following:

Standard of Identity. The FDA describes food and lists both required and optional ingredients that are included in manufacture. Examples of products that follow a Standard of Identity included foods such as mayonnaise, white bread, and jelly.

When initially introduced as law, a food product followed a Standard of Identity in its manufacture, and many required and optional ingredients were *not* listed on labels, as it was understood that the consumer was familiar with ingredients that composed basic foods. In time, however, it became apparent that this familiarity with foods was not widespread! As a result, after 1967, *optional* ingredients of foods were *required* to be included on labels, even if the product followed the Standard of Identity. A standard was continually reviewed and revised as new additives are approved for food use.

Currently, manufacturers are *required* to state *all* ingredients on the product label, including required and optional ingredients. This change to the complete identification of food ingredients benefits consumers who are unfamiliar with food ingredients that make up a food, as well as those with food allergies or intolerances.

Standards of Minimum Quality. The FDA states the minimum quality standards for specific characteristics in a food, such as color, defects, and tenderness. (Color, tenderness, blemishes, clarity of liquid, and product size are some of the criteria used at the wholesale and retail level for evaluation.) A food must state "below standard in quality" if the minimum level of a particular quality descriptor is not obtained.

For example, we see processors of canned vegetables and fruits follow this standard. Substandard does not signify safety hazards.

Standard of Fill of Container. This FDA standard ensures that the headspace/void volume of packaged food offered for sale does not interfere with the *weight* of the product as stated on the label. It assures that the product offers the correct weight even if the package is only partially full! For example, packages of cereal, crackers, and potato chips may not appear full due to extra air space in the package that is needed to prevent food breakage, yet, this fact is taken into account, and the food is sold by the *weight*, *not* by the *volume*. Food products packed in a liquid medium, such as canned fruits or vegetables, must contain the stated weight of the product.

Adulterated and Misbranded Food

Adulterated and misbranded foods are defined as follows:

Adulterated food may *not* be offered for sale. According to the FDA, a food is adulterated if it:

- Is poisonous or harmful to health at detrimental concentrations
- Contains filth or is decomposed
- Contains a food or coloring agent that is not approved or certified
- Was prepared or packed under unsanitary conditions, making it contaminated
- Is derived from a diseased animal
- Contains any excessive levels of residue
- Was subject to radiation, other than where permitted
- Has any valuable constituent omitted
- Substitutes a specified ingredient with an unspecified ingredient
- Is damaged or conceals defects
- Is increased in bulk weight or reduced in its strength, making it appear better than it is

According to the FDA, a food is *misbranded* if it:

- Is labeled falsely or misleadingly
- Is offered for sale under the name of another food
- Is an imitation of another food, without stating "imitation" on the label
- Is packaged (formed or filled) so as to be misleading
- Fails to list the name and address of the manufacturer, packer, and distributor and a statement of net contents on the label

- Fails to declare the common name of the product and the names of each ingredient or has label information that is not legible and easily understood
- Is represented as a food for which there is a Standard of Identity but the food does not conform with an accurate statement of quantity or ingredients
- Is represented to conform to a quality standard or to a fill of container and does not conform
- Is represented with a nutritional claim or for special dietary use but the label fails to provide information concerning dietary properties of the food, as required by law
- Lacks proper nutrition labeling

Food Safety Modernization Act (FSMA) Proposed 1/2013

" . . .The rules follow extensive outreach by the FDA to the produce industry, the consumer community, other government agencies and the international community. Since January 2011, FDA staff have toured farms and facilities nationwide and participated in hundreds of meetings and presentations with global regulatory partners, industry stakeholders, consumer groups, farmers, state and local officials, and the research community.

"The FDA Food Safety Modernization Act is a common sense law that shifts the food safety focus from reactive to preventive," said Health and Human Services Secretary Kathleen Sebelius. "With the support of industry, consumer groups, and the bipartisan leadership in Congress, we are establishing a science-based, flexible system to better prevent foodborne illness and protect American families." . . .

See *Food Safety Modernization Act*. The focus of legislation changed from *responding* to a problem to its *prevention*.

In the FSMA through the FDA, two new rules were proposed. One requires HACCP, risk-based preventive controls and plans for correcting any domestic food problem that arise. "The second rule proposes science- and risk-based standards for the safe production and harvesting of produce on farms" (Kuntz 2013) (see more on this law in Chap. 19).

The FDA also enforces *the Public Health Service Act* to maintain sanitary standards at retail foodservice establishments and in milk processing and shellfish operations. The FDA monitors food for safety and wholesomeness on interstate carriers such as planes and trains. As well, the FDA has a Seafood HACCP (program, which is aimed at controlling pathogens and foodborne illness from seafood).

If the FDA determines that a product poses a serious risk to public health, the FDA inspectors will submit Form 482c *Notice of Inspection-Request for Records* in order to conduct an emergency food contamination inspection. The FDA is allowed to obtain needed records, and the form must be submitted in writing to the owner, operator, or agent in charge of the company.

Maintaining business, protecting profits, as well as learning how to recover from disasters are duties of the food plant. Each of these goals must be protected. As mentioned in a recent article *Building your plant's ark*, "Noah may have been among the first to plan for impending natural disaster. Don't let him be the last" (Stier 2006).

Developing an emergency plan is "more than putting words on paper. Map out how your plant will react to a variety of disasters: hurricanes, earthquakes, tornados, fires, chemical spills/leaks, terrorism or other potential problems. At the very least, you need an evacuation plan to get workers to a safe location" (Stier 2006).

The United States Department of Agriculture (USDA)

The USDA is another major government agency regulating and with enforcement powers, the food supply in the United States. At the helm is the Secretary of Agriculture, and it is a full federal government department. It is responsible for inspection of meat, poultry, agricultural products, including milk, eggs, fruit, and vegetables, and also meat and poultry processing plants. The USDA also has involvement in the protection of the United State's natural resources and environment.

While the ***inspection service***, including bacterial counts, is *mandatory*, the ***grading service*** is *voluntary* and is paid for by the manufacturer, marketer, or packer. Accommodations such as a desk, telephone, and parking space should be made available for the USDA inspector who is routinely or regularly present at a plant to assure safe food handling and plant sanitation. Of course, it needs to be stressed once more—food safety is still dependent upon the individual!

The USDA, or the individual State Departments of Agriculture (states may exceed, however, at least *meet* federal standards), inspects meat and stamps it with an abbreviation of "Inspected and Passed," containing a number that identifies the plant from which it came. Every *carcass*, although not every *cut* of meat, requires this stamp (made using nontoxic vegetable dye) as proof of sanitary quality and wholesomeness. The stamp is required for shipment in interstate commerce. The label stating ***wholesome*** indicates that no signs of illness were found, *not* that the meat is free from pathogenic microorganisms.

The Federal Meat Inspection Act of 1906, Federal Poultry Products Inspection Act of 1957, and the Wholesome Poultry Products Act of 1968 are enforced by the *Food Safety and Inspection Service* (*FSIS*) of the USDA. The inspection, labeling, and handling of poultry and poultry products are similar to the meat inspection process. Processed poultry products do not undergo a mandatory inspection.

The FSIS conducts activities such as the following to ensure the *safety* of meat and poultry products consumed in the United States:

- The USDA inspectors and veterinarians conduct slaughter inspection of all carcasses at meat and poultry slaughtering plants for disease and other abnormalities and sample for the presence of chemical residues.
- The USDA conducts processing inspection for sanitation and cleanliness, labeling, and packing at facilities where meat and poultry is cut up, boned, cured, and canned.
- Scientific testing in support of inspection operations is performed by USDA/FSIS laboratory services to identify the presence of pathogens, residues, additives, diseases, and foreign matters in meat and poultry.
- Inspection systems in countries exporting meat and poultry products to the United States are reviewed by the USDA as part of the import–export inspection system.
- The USDA is placing increased emphasis on pathogen reduction and HACCP in the entire meat and poultry production chain. This involves developing new methods for rapid detection of pathogenic microorganisms, new production, and inspection practices to reduce bacterial contamination and educating consumers on safe food-handling practices.
- The USDA's Meat and Poultry Hotline is a toll-free service where consumers, educators, researchers, and the media can speak with experts in the field of food safety.

The USDA also has a Food and Nutrition Service (FNS).

The USDA FNS administers the food and nutrition assistance programs in the U.S. Department of Agriculture. FNS provides children and needy families with better access to food and a more healthful diet through its programs and nutrition education efforts.

Program and Service Highlights includes:

- Women, Infant, and Children (WIC) Program—The FNS administers several programs that provide healthy food to children including the National School Lunch Program, the School Breakfast Program, the Child and Adult Care Food Program, the Summer Food Service Program, the Fresh Fruit and Vegetable Program, and the Special Milk Program. Administered by state agencies, each of these programs helps fight hunger and obesity by reimbursing organizations such as schools, child care centers, and after-school programs for providing healthy meals to children.
- Supplemental Nutrition Assistance Program—SNAP offers nutrition assistance to millions of eligible, low-income individuals and families and provides economic benefits to communities. SNAP is the largest program in the domestic hunger safety net. The FNS works with state agencies, nutrition educators, and neighborhood and faith-based organizations to ensure that those eligible for nutrition assistance can make informed decisions about applying for the program and can access benefits. FNS also works with state partners and the retail community to improve program administration and ensure program the integrity.
- School Meals—The FNS administers several programs that provide healthy food to children including the National School Lunch Program, the School Breakfast Program, the Child and Adult Care Food Program, the Summer Food Service Program, the Fresh Fruit and Vegetable Program, and the Special Milk Program. Administered by state agencies, each of these programs helps fight hunger and obesity by reimbursing organizations such as schools, child care centers, and after-school programs for providing healthy meals to children.
- Food Distribution Programs—The FNS Food Distribution Programs' mission is to strengthen the nation's nutrition safety net by providing food and nutrition assistance to school children and families and support American agriculture by distributing high-quality, 100 % American-grown USDA Foods.
- Disaster Assistance—Nothing is more important than providing food when people find themselves suddenly, and often critically, in need following a storm, earthquake, flood, or other disaster emergency. The USDA makes sure that people have enough to eat.
- Child and Adult Care Food Program—CACFP plays a vital role in improving the quality of day care for children and elderly adults by making care more affordable for many low-income families.

Through CACFP, more than 3.3 million children and 120,000 adults receive nutritious meals and snacks each day as part of the day care they receive.

- Summer Food Service Program—During the school year, many children receive free and reduced-price breakfast and lunch through the School Breakfast and National School Lunch Programs. What happens when school lets out? Hunger is one of the most severe roadblocks to the learning process. Lack of nutrition during the summer months may set up a cycle for poor performance once school begins again. Hunger also may make children more prone to illness and other health issues. The Summer Food Service Program is designed to fill that nutrition gap and make sure children can get the nutritious meals they need.
- Farmers' Market Nutrition Programs—The WIC Farmers' Market Nutrition Program (FMNP) is associated with

the Special Supplemental Nutrition Program for Women, Infants and Children, popularly known as WIC. The WIC Program provides supplemental foods, health care referrals, and nutrition education at no cost to low-income pregnant, breastfeeding, and non-breastfeeding postpartum women and to infants and children up to 5 years of age, who are found to be at nutritional risk.

The WIC FMNP was established by Congress in 1992, to provide fresh, unprepared, locally grown fruits and vegetables to WIC participants and to expand the awareness, use of, and sales at farmers' markets. Women, infants (over 4 months old), and children that have been certified to receive WIC program benefits or who are on a waiting list for WIC certification are eligible to participate in the WIC FMNP. State agencies may serve some or all of these categories. A variety of fresh, nutritious, unprepared, locally grown fruits, vegetables, and herbs may be purchased with FMNP coupons. State agencies can limit sales to specific foods grown within state borders to encourage FMNP recipients to support the farmers in their own states.

- Nutrition Education—NS provides children and adults of all ages with nutrition education materials on how to improve their diets and their lives.—USDA

There are many USDA programs. In order to face the complex nutrition issue in the twenty-first century, there may be a need for researchers, policymakers, and both private and public sector organizations to define and implement a strategy for action agenda.

Presently, there are many nutrition programs, as mentioned, such as the Food Stamp Program and Special Supplemental Nutrition Program for Women, Infants, and Children. There are also Dietary Guidelines, NLEA, etc., the "building blocks" over the past few decades, yet they do not represent a national nutrition policy (Crockett et al. 2002).

The most recent USDA Food Code is discussed in the chapter on Food Safety (Chap. 19).

The USDA's FSIS has a Food Biosecurity Action Team (F-BAT). Its intent is to protect agriculture and the food supply, ensure employee safety, have adequate capacity and security at agency laboratories, ensure that essential USDA functions can continue, and be able to pass on necessary information (to employees, consumers, industry, the media, Congress, and other agencies) in a single, consistent message (USDA).

The USDA Undersecretary for Food Safety formed the F-BAT to coordinate and facilitate all activities pertaining to biosecurity, countering terrorism, and emergency preparedness with FSIS. F-BAT also serves as FSIS' voice with other governmental agencies and internal and external constituents on biosecurity issues (USDA).

Unfortunately the FSIS has managed many recalls of food products. However, they have decided that during food recalls, distribution lists, which are usually confidential, may be made available to state and federal agencies. Such lists would not be subject to public disclosure.

See

Current Good Manufacturing Practice and Hazard Analysis and Risk-Based Preventive Controls for Human Food Proposed Rule.

[Federal Register Volume 78, Number 11 (2013)] [www.gpo.gov]

State and Local Health Departments

As previously mentioned, the *federal* agencies (FDA, USDA) regulate interstate food supplies, and it is the task of *state* agencies, such as state FDAs and state Agriculture Departments, to regulate *intra*state food supplies. In some states, the State Health Department has complete authority over all food operations, whereas in other states, county or city health departments adopt their own specific foodservice regulations.

Additional Agencies Regulating the Food Supply

The Federal Trade Commission (FTC) protects against unfair and deceptive advertising practices of products, including food.

The National Marine Fisheries Service (NMFS) of the Commerce Department is responsible for voluntary grading of seafood.

The Occupational Health and Safety Administration (OSHA) regulates health hazards in the workplace (such as food manufacture, processing, or retail foodservice) and determines compliance with regulations.

The Environmental Protection Agency (EPA) sets environmental standards. This agency regulates air and water pollution by plants, toxic substances, pesticides, and the use of radiation.

Education and Training

Education and training on the part of the government and industry is significant in regulating the food supply. Each segment/person must be properly trained and motivated to do their part in maintaining a safe food supply and seeing that they adhere to proper labeling. The public should do their part in adhering to proven government safety and labeling strategies (see Chap. 19).

General Labeling

General labeling requires that complete information about food must be supplied on food packages. It must include the following:

- Name of product; name and place of business
- Net weight—ounces (oz.), or pounds and ounces
- Ingredients—listed by weight in descending order on ingredients list of label (not Nutrition Facts portion)
- Company name and address
- Product date if applicable to product
- Open date labeling—voluntary types able to be read by the consumer
- Expiration date—deadline for recommended eating (i.e., yeast)
- "Best if used by" date—date for optimum quality, QA, or freshness
- Pack date—date food was packaged
- Pull date—last day sold as fresh (i.e., milk, ice cream, deli)
- Code date—read only by manufacturer
- Nutrition information—"Nutrition Facts" on nearly all labels
- Nutrient content claims substantiated
- Health claims used only as allowed
- Other information
  - Religious symbols—such as Kosher (if applicable)
  - Safe handling instructions—such as on meats
  - Special warning labels—alcohol, aspartame that may affect select consumers
  - Product code (UPC)—bar code

Labeling Basics (Reported Concisely Herein by a Labeling Company)
"Labeling regulations exist to ensure that consumers know what they are buying. Improperly labeled products may be deemed misbranded or adulterated and subject to regulatory agency action. Avoid receiving an FDA warning letter or cyber letter and other costly labeling errors by ensuring that your labels are compliant with FDA & FTC regulations. FDALabels.com helps you with all your labeling needs.

Every product sold in the United States must include the following information on the label:

- Statement of identity or standard product name (What is it?)
- Net quantity of contents statement (How much is in the package?)
- Component/Ingredient statement (What is it made of?)
- Signature line or name and place of business of the US manufacturer, packer or distributor (Who made it and who should be contacted if something goes wrong?)

In addition, each type of FDA-regulated product has other distinct labeling requirements:

Functional Foods—Food and functional food products sold in the United States are subject to FDA labeling regulations found in the CFR. In addition to the requirements that labels include a statement of identity; net quantity of contents statement, ingredient list and signature line, food labels must also include a Nutrition Facts Box and allergen statement.

"The contents of the Nutrition Facts Box must comply with thresholds of declaration and rounding rules outlined in 21 CFR 101.9. FDA regulations allow for the use of calculated values to declare the levels of calories, fat, trans fat, protein, carbohydrates, sodium, cholesterol and other nutrients. Food products that meet specific guidelines may be labeled with nutrient content claims and certain health claims."

"Food products containing meat or poultry are regulated by the USDA. While the labeling regulations for these products mirror the requirements for foods regulated by FDA, USDA labels must be submitted to the USDA Food Safety and Information Service for review prior to use. Organic labeling is also administered by the USDA and subject to prior certification."—*FDALabels.com—a site providing product development and regulatory affairs consultation for FDA and USDA regulated products*

"The U.S. Federal Food, Drug and Cosmetic Act (FFDCA) defines food 'labeling' as all labels and other written, printed, or graphic matter upon any article or any of its containers or wrappers, or accompanying such article. The term 'accompanying' is interpreted liberally to mean more than physical association with the food product. It extends to posters, tags, pamphlets, circulars, booklets, brochures, instructions, websites, etc.

The Nutrition Labeling and Education Act (NLEA), which amended the FFDCA requires most foods to bear specific nutrition and ingredient labeling and requires food, beverage, and dietary supplement labels that bear nutrient content claims and certain health messages to comply with specific requirements. Furthermore, the Dietary Supplement Health and Education Act (DSHEA) amended the FFDCA, in part, by defining "dietary supplements," adding specific labeling requirements for dietary supplements, and providing for optional labeling statements."

http://www.registrarcorp.com/fda-food/labeling/regulations.jsp?lang = en

Radio Frequency Identification Tags

Radio Frequency Identification (RFID) tags appear on many food product labels. A number of consumer packaged goods, retail operations, transportation, defense, and pharmaceuticals use RFID. A number of retailers require it of their suppliers. It is more than an inventory or packaging/labeling technology (Higgons 2006). It assists manufactures and users track packaged food throughout the supply chain. For example, benefits of RFID may include better consumer safety and security and improved operating efficiencies for packaging, manufacturing, distribution, and sales.

Since it may be required of various vendors delivering to suppliers, training in its benefits and uses may assist users/potential users of the technology. Training can help hardware and software providers, and both the public and private sectors, as well as educators and researchers.

Nutrition Labeling

Food products intended for human consumption are subject to mandatory ***nutrition labeling***, regulated by the FDA. As a result of the Nutrition Labeling and Education Act of 1990 (NLEA), there are regulations that specify information food processors must include on their labels, including "Nutrition Facts." The purpose of the NLEA is the following:

- Assist consumers in selecting foods that can lead to a healthier diet
- Eliminate consumer confusion
- Encourage production innovation by the food industry

NLEA regulations became effective in 1994, and approximately 595,000 food products had to meet these regulations, according to the FDA and USDA.

Consumers benefit from the educational component of the labeling law, as the information on labels is easy to read and may be useful in planning healthful diets. The label provides consumers with consistency under mandatory "Nutrition Facts" which appears on most products offered for sale in the United States. Voluntary information for cuts of meat, raw fish, and the 20 most commonly eaten fruits and vegetables may appear on package bags, brochures, or posters at the point of sale. Labeling values for produce and fish have been revised since initially required and further revisions will be proposed every 4 years.

The FDA has set 139 reference serving sizes for use on "Nutrition Facts" labels that more closely approximate amounts consumers actually eat than previous labeling. The serving size indicates values, such as the number of ounces in a beverage or the ounces and number of cookies or crackers per serving; the nutrient content of a food is based on this reference-serving size and stated on the label. In packaged food, a food is still labeled as a *single serving* if the amount of food is greater than 50 % and less than 200 % of the designated single-serving size. Also see the following website: http://vm.cfsan.fda.gov/~lrd/cf101-12.html.

Portion sizes are thus designated by the FDA. (See choosemyplate.gov) View Food Gallery) to see photos of actual foods and portion sizes. Depending on personal intake, the individual nutrient consumption may be more or less than that FDA "one serving." Of course that is acceptable as long as the person who desires to either limit or attain certain nutrients realistically knows what constitutes that "one serving"! In example, a serving of ice cream is one scoop, not one bowlful! Thus calories, fat, cholesterol, and so forth are calculated accordingly. ("Portion distortion" is what is sometimes referred to as a person's mistaken idea of what equals an actual portion!)

With the passage of the NLEA, the FDA set regulations stating that a food label must express nutrient information in terms of recommended daily intake, in grams (or milligrams) or as a percentage, thus the "% Daily Values" or "DV." It shows how a serving of the food fits into a total day's diet.

Two sets of values were included in the establishment of Daily Values. One is the Reference Daily Intakes (RDI), which is based on former "U.S. RDA" (derived from 1968 RDA) labeling values. The second is Daily Reference Values (DRV) for nutrients, such as fat, sodium, cholesterol, and total carbohydrates including dietary fiber and sugars, which do not have an RDA yet have a significant health impact. The DV reference values are based on a 2,000- or 2,500-calorie diet, and consumers ingesting more or less calories should adjust numbers accordingly.

Numerous values are provided on nutrition labels. For example, the total calories and calories from fat, the total fat, and the saturated fat (perhaps monounsaturated and polyunsaturated fat if the processor wants to include these) and trans fat are stated. Cholesterol and sodium are stated in milligrams. The total carbohydrate, sugar, and dietary fiber are also reported on Nutrition Facts. Protein is expressed as a quantity that takes into account the completeness of amino acids (complete = having all essential amino acids in the needed amount). Food processors have the option of reporting protein as a %DV on a label, and, if they do, they must determine the quality of the protein to ascertain which Daily Value of protein to use as a comparison.

As mentioned earlier, consumers may be attempting to limit or attain specific quantities of certain nutrients in their diet. For example, a consumer may desire to limit fat or cholesterol, or they may want to increase their intake of vitamins and minerals commonly needed in the United States, such as vitamins A and C, calcium, and iron. A nutrition label can help the consumer know what nutrients are in food.

Examples of terms allowed on food labels appear in Table 20.1. Terms are consistent among products, and manufacturers and food processors must abide by these definitions on their product labels. Yet, when merchandising a product through the various forms of advertisement, there exists no FDA regulation of terms.

The label information intended to assist consumers in making informed food choices, and it did not come cheaply to food processors. The product analyses, as well as label redesign and printing costs, were incurred. In a survey conducted by the National Food Processors Association, it was estimated that well over $1 billion would be spent by the food industry as it implemented NLEA in an 18-month period.

Methods of analyses for nutrition labeling are available to food processors from the AOAC International and the Food Chemicals Codex (FCC). "Whole food" and "ingredient" databases assisted in providing the necessary nutrient information for labels.

In a land of plenty, with an increasing concern of managing personal weight, the USDA has released the 2005 Dietary Guidelines for Americans. They include the following: the USDA and Health and Human Services (HHS) publication emphasizes both lifestyle and dietary measures for health. Food Technology reports: "So, the news for the food sector is to continue to improve processes and formulations where appropriate, and help consumers avoid foodborne illness and excess, while keeping excitement at the table" (Katz 2000). Other nations have adopted similar dietary guidelines for their population.

Dietary Guidelines for Americans

"Dietary Guidelines for Americans is published jointly every 5 years by the Department of Health and Human Services (HHS) and the Department of Agriculture (USDA). The Guidelines provide authoritative advice for people two years and older about how good dietary habits can promote health and reduce risk for major chronic diseases." (USDA)

Current recommendations are found in www.healthierus.gov/dietaryguidelines:

- Integrate better eating habits into your life.
- Integrate better activity habits into your life.
- Set realistic goals.
- Take small steps to meet them.

Table 20.1 Some examples of terms allowed on food labels

General descriptive terms
• *Free—negligible amount of the nutrient*
• Good source of—between 10 and 19 % of the Daily Value of the nutrient
• Healthy—low-fat, saturated fat, cholesterol, and sodium food with at least 10 % of the Daily Value for vitamins A and C, protein, iron, calcium, or fiber
• Low—not meeting Daily Values with frequent consumption
• High—20 % or more of the Daily Values for a nutrient per serving
• Light or lite—one-third fewer calories, or one-half the fat of the comparison food
• More—at least 10 % more of the Daily Value than a comparison food
• Less—at least 25 % or less of a nutrient than the comparison food
Energy/calories
• *Free—fewer than 5 cal per serving*
• Low calorie—40 cal or less per serving
• Reduced calorie—at least 25 % fewer calories per serving than a comparison food
• Light—one-third less calories than the comparison food
Fat and cholesterol
• *Fat*
• *Fat-free—less than 0.5-g fat per serving*
• Low fat—3 g or less fat per serving
• Percent (%) fat-free—only if low fat or fat-free, calories based on 100-g portions
• Less fat—25 % or less fat than a comparison food
• Light—50 % less fat than a comparison food
Saturated fat
• *Saturated fat-free—less than 0.5 g of saturated fat and trans-fatty acid per serving*
• Low saturated fat—1 g or less saturated fat per serving
• Less saturated fat—25 % or less saturated fat than a comparison food
Cholesterol
• *Cholesterol-free—less than 2-mg cholesterol and 2 g or less saturated fat per serving*
• Low cholesterol—20 mg or less cholesterol and 2 g or less saturated fat per serving
• Less cholesterol—25 % or less cholesterol than a comparison food, and 2 g or less saturated fat per serving
• Extra lean—less than 5 g of fat, 2-g saturated fat, and 95-mg cholesterol per serving and per 100 g of meat, poultry, and seafood
• Lean—less than 10-g fat, 4.5-g saturated fat, and 95 mg of cholesterol per serving and per 100 g of meat, poultry, and seafood
Carbohydrates: fiber and sugar
• High fiber—5 g or more fiber per serving, with 3 g or less of fat per serving (low fat) unless a higher level of fat is specified
• Sugar-free—less than 0.5-g sugar per serving
Sodium
• *Sodium-free—less than 5-mg sodium per serving*
• Low sodium—140 mg or less per serving
• Light—50 % less sodium, in a low-calorie or low-fat food
• Very low sodium—35 mg or less per serving

Health Claims (More in Appendices)

In order to make the approved health claims (Table 20.2), a food must contain no more than 20 % of the Daily Value for total fat, saturated fat, cholesterol, or sodium, and the food must naturally contain at least 10 % of the Daily Value for either vitamins A and C, protein, fiber, calcium, or iron.

Table 20.2 Examples of approved model health claims used on food labels

• Calcium and lower risk of osteoporosis
• Sodium and a greater risk of hypertension (high blood pressure)
• Saturated fat and cholesterol and a greater risk of coronary heart disease (CHD)
• Dietary fat and a greater risk of cancer
• Fiber-containing grain products, fruits, and vegetables and a reduced risk of cancer
• Fruits, vegetables, and grain products that contain fiber (particularly soluble fiber) and a reduced risk of CHD
• Fruits and vegetables and a reduced risk of cancer
• Folate and reduced risk of neural tube defect
• Sugar alcohols and reduced risk of tooth decay
• Soluble fiber from whole oats and psyllium seed husk and reduced risk of CHD
• Soy protein and reduced risk of CHD
• Whole grains and reduced risk of CHD and certain cancers
• Plant sterol and plant stanol esters and reduced risk of CHD
• Potassium and reduced risk of high blood pressure and stroke

Examples of approved health claims appear in Table 20.2. Currently, the FDA is considering greater flexibility in the use of health claims on foods; yet, other claims outside of these may not be used on food products. Health claims for dietary supplements are being constructed (http://www.cfsan.fda.gov).

Labeling for Food Allergens

Food product legislation for more simple wording and common sense labeling is supported by the Food Allergy Initiative (FAI), the Food Allergy and Anaphylaxis Network, and the Center for Science in the Public Interest (CSPI). It has been suggested that perhaps labels should just say "wheat" or say "milk products." This is in part due to food allergies. Additional information on food allergens is found in the chapter on food safety.

Allergen food labeling is required after or adjacent to the ingredients list if a food may/does contain allergens. **The Food Allergen Labeling and Consumer Protection Act of 2004** (**FALCPA**) requires that food manufacturer identify foods that contain the presence of protein derived from crustacean shellfish, eggs, fish, milk, peanuts, soybeans, tree nuts, or wheat. Use of any ingredients that may contain protein from these eight major allergens must be clearly stated for the consumer.

Immediately after or adjacent to the list of ingredients, put the word "Contains" followed by the name of the food for each of the major food allergens present in the food's ingredients.

For example:

Contains Wheat, Milk, Egg, and Soy—USDA

See FARE—Food Allergy Research & Education, Inc. (http://www.foodallergy.org\).

See Chap. 19 on Food Safety; also see Food Safety Research Information Office (FSRIO) National Agricultural Library (NAL)—Frequently Asked Questions—Regulations, Standards, and Guidelines:

http://fsrio.nal.usda.gov/faq-page/regulations-standards-and-guidelines

Labeling for Foodservice

The inclusion of material in this labeling for foodservice chapter section is intended to clarify labeling requirements of food served for immediate consumption. While this section addresses the *menu*, and *not labels* on packaged foods, it may be of less concern to the food scientist. Yet foods eaten at a foodservice operation represent a significant portion of the buying public's consumption and therefore deserve attention.

The FDA encourages foodservice operations to provide nutrition and health claims to consumers, and further regulations may be forthcoming. Yet, nutrition analysis testing and Nutrition Facts labeling are *not* required of food service.

Any nutrient content or health claims appearing on menus must be substantiated by the foodservice operation, either verbally or in written form, to consumers who request such information. Claims must meet established FDA criteria—specified in the CFR, a reliable cookbook or computer software program may be used as a reference, and preparation methods must support the claim, or the menu item must be removed from the menu.

The CFR (21CFR101) specifies the following with regard to labeling for foodservice: "A nutrient claim used on food that is served in restaurants or other establishments in which food is served for immediate human consumption or which is sold for sale or use in such establishments shall comply with the (same) requirements of this section..."

Menu-related obstacles to nutrition labeling include menu variations, use of daily specials, limited page space, and loss of flexibility. There are also *personnel-related* obstacles, for instance, difficulty in training employees and a shortage of time.

Today there are information options. Nutrition expertise and labeling assistance could be provided to companies by dietitians as many restaurants already provide.

The National Center for Nutrition and Dietetics of the Academy of Nutrition and Dietetics has a *hotline number* (800-366-1655) that offers messages and personally answers consumer questions about food labeling (The National Center for Nutrition and Dietetics of the Academy of Nutrition and Dietetics, Chicago, IL).

Supermarket Savvy Information and Resource Service® (order@supermarketsavvy.com) is an example of a *service* that provides new product information. A newsletter is included as one part of its service. It is written for the health professional and designed to provide information about new products (especially the healthier ones) so that the health professional can answer his/her clients' questions about new foods and guide his/her clients to better food choices in the supermarket and health/natural foods store (McDonald. Information and resource service).

Conclusion

Concluding this issue is difficult to do! Perhaps this chapter cannot close! However, suffice it to say that government regulation, industrial compliance, and consumer education are all means of ensuring a safe food supply to consumers. Food safety is still dependent upon the individual! We need to then act on what we know.

The FDA is a public health agency that regulates food, cosmetics, medicines, medical devices, and radiation-emitting products, such as microwave ovens. The Food, Drug, and Cosmetic Act of 1938 and its amendments were introduced to regulate the processing of many products subject to interstate commerce or import. Food inspections are the responsibility of the FDA, with meat product inspection regulated by the USDA. Food packaging and labeling is regulated by the FDA and USDA for their respective products. The USDA administers the FSIS and numerous food programs.

The NLEA is an attempt to protect the consumer against fraud and misinformation. Labeling terms, "Nutrition Facts," and health claims are regulated by the FDA. The purpose of the NLEA is to assist consumers in selecting foods that can lead to a healthier diet, eliminate confusion, and encourage production innovation by the food industry. With greater knowledge of nutrients, nutrient interactions, and promotion of health, greater health benefits may be provided with the formulation of new food products.

Additionally, general labeling, nutrition labeling, health claims, food allergen labeling, and labeling for foodservice were discussed in this chapter.

Of course, in addition to the government's regulation of the food supply, *industry plants* and *consumers* must be vigilant and play their part in assuring a safe food supply!

Extra: Food Security and an Emergency Plan

In the aftermath of the September 11, 2001 terrorist attacks on the United States, the FDA has urged industry to take necessary steps to ensure better food security. For example, farms, processors, grocery stores and restaurants can better protect the nation's food supply by requiring criminal background checks of all workers, and closely checking all food and water sources. New guidelines were issued by the FDA, and addressed by the ADA. (FDA, ADA)

Bioterrorism Preparedness and Response Act of 2002 ("The Bioterrorism Act")

The events of Sept. 11, 2001, reinforced the need to enhance the security of the United States. Congress responded by passing the Public Health Security and Bioterrorism Preparedness and Response Act of 2002 (the Bioterrorism Act), which President Bush signed into law June 12, 2002. (http://www.fda.gov/oc/bioterrorism/bioact.html)

Notes

CULINARY ALERT!

Glossary

Daily Value (%DV) Two sets of values used on nutrition labels, including Reference Daily Intakes (RDI), based on former US RDAs and Daily Reference Values (DRV) of nutrients that do not have an RDA but have a significant health impact.

Generally Recognized As Safe (GRAS) Substances (food ingredients) generally recognized as safe for their intended use.

Grading Service Conducted as a voluntary service of the USDA, paid for by packers.

Health Claims Describe an association between a nutrient or food substance and disease or health-related condition.

Inspection Service Of the USDA or state Department of Agriculture inspects and stamps inspected meat with a circle containing the abbreviations for "inspected and passed."

Nutrition Labeling For the purpose of assisting consumers in selecting foods that can lead to a healthier diet, to eliminate consumer confusion, and to encourage production innovation by the food industry. Labeling expresses nutrients in terms of Reference Daily Intakes (RDI) and Daily Reference Values (DRV), both comprising the Daily Values.

Standard of Fill of Container FDA standard that the volume of packaged food offered for sale does not interfere with the weight of the product as stated on the label.

Standard of Identity FDA list of required and optional ingredients that are included in manufacture.

Standards of Minimum Quality FDA minimum quality standards for specific food characteristics–color, etc.

Wholesome The carcass and viscera of the animal were examined, and no signs of illness were indicated, and conditions met sanitary standards.

References

Crockett SJ, Kennedy E, Elam K (2002) Food industry's role in national nutrition policy: working for the common good. J Am Diet Assoc 102:478–479

Higgons K (2006) RFID making the right moves. Food Eng 78(2):44–48

Katz F (2000) 2000 IFT annual meeting and food expo. How food technologists react to the new dietary guidelines for Americans. Food Technol 54(8):64–68

Kuntz LA (2013) FSMA strikes back. Food Prod Des (Jan/Feb):10

McDonald L. Information and resource service. Publisher of the SUPERMARKET SAVVY®, Houston, TX

Stier R (2006) Building your plant's ark. Food Eng 78(1):29

Bibliography

Center for Disease Control (CDC)

Center for Food Safety and Applied Nutrition. http://www.cfsan.fda.gov (search for Health Claims)

Food and Drug Administration (1995) Focus on food labeling. FDA consumer. Food labeling, questions and answers, vol 2. U.S. Dept. of Health and Human Services, Washington, DC

Model FDA Food Code

Packaging Technology Integrated Solutions (PTIS)

The Food Marketing Institute. Consumer Affairs Department. Washington, DC

USDA ChooseMyPlate.gov

Appendices

Introduction

There are several parts to these Appendices: I–III.

I. In this portion of the Appendices: Considering the frequency with which terms are used (and used interchangeably!) on labels and in the press, a brief discussion and explanation of terms is provided in these Appendices that follow:

Appendix A—Biotechnology: Genetically Modified Organisms (GMOs)
Appendix B—Functional Foods
Appendix C—Nutraceuticals
Appendix D—Phytochemicals
Appendix E—Medical Foods

II. In food companies the food scientist who develops new products, the technical staff, and the marketer of these foods must stay abreast of the health concerns of consumers who are making dietary changes in managing their personal healthcare. An expertise in such areas as foods, culinary ideas, consumer food acceptability, food engineering, food laws, ingredient technology, nutrition, and more positions the food company for product success. (So says the Research Chefs!)

Appendix F—USDA ChooseMyPlate.gov
Appendix G—Food Label Heath Claims
Appendix H—Research Chefs Association Certification as a Culinary Scientist, etc.

III. New topics continuing to expand on the horizon:

Appendix I—Human Nutrigenomics
Appendix J—Product Development: Innovation

All of these topics hold great significance for the food scientist today and in the future. Before we forget though, despite what foods *should* and *could* do—taste still rules! A food must please the palate to remain on the shopping list!

V.A. Vaclavik and E.W. Christian, *Essentials of Food Science, 4th Edition*, Food Science Text Series,
DOI 10.1007/978-1-4614-9138-5, © Springer Science+Business Media New York 2014

Appendix A

Biotechnology: Including GMOs

Modern methods of ***biotechnology*** or *genetic engineering* have led to the production of specific desired traits in plant material. There exists breeding of new types of produce, disease-resistant strains, and longer shelf life. Much of biotechnology addresses *crops*, such as corn, soy, cotton, canola, pepper, and squash; however, biotechnology is not limited to fruits and vegetables. Beyond these plant applications, biotechnology produces other specific desired traits in *animals* and *microorganisms* too. For example, according to The International Food Information Council (IFIC), rennet, an enzyme for making cheeses, and yeast for breads are commonly produced by biotechnology.

The Food and Drug Administration (FDA) definition of biotechnology is as follows:

The FDA Data Standards Council is standardizing vocabulary across the FDA. Therefore, the wording in some terms below may change slightly in the future.

> Biotechnology—refers to techniques used by scientists to modify deoxyribonucleic acid (DNA) or the genetic material of a microorganism, plant, or animal in order to achieve a desired trait. In the case of foods, genetically engineered plant foods are produced from crops whose genetic makeup has been altered through a process called recombinant DNA, or gene splicing, to give the plant desired traits. Genetically engineered foods are also known as biotech, bioengineered, and genetically modified, although "genetically modified" can also refer to foods from plants altered through methods such as conventional breeding. While in a broad sense biotechnology refers to technological applications of biology, common use in the United States has narrowed the definition to foods produced using recombinant DNA. For additional information, see the Biotechnology Program on the CFSAN Internet.
> http://www.fda.gov/Food/FoodIngredients-Packaging/ucm064228.htm

Also see: Chap. 7—a statement by the FDA Biotechnology Coordinator regarding food Biotechnology.

Modern genetic engineering, in practice since the 1970s, is a biotechnology development that inserts a desired gene into another crop's chromosomes. The resultant cells may be grown into plants, and then conventional breeding techniques follow to yield crops with specific desirable traits. The entire crop is *not* representative of clones of one original plant; rather, the individual plants of a crop are *unique*.

Prior to use of modern biotechnology, rennet was obtained from the intestinal tract of calves' stomach. Now, the specific gene is available once it is removed and subsequently reproduced in bacteria.

USDA regulation includes:

1. **USDA's Animal and Plant Health Inspection Service (APHIS)**—According to the **USDA**, companies or organizations, who wish to field test a genetically engineered crop, must obtain permission from the USDA.

   Low-risk traits for familiar crops may have streamlined approval through a *notification procedure*, while *high-risk* traits used for producing pharmaceuticals or industrial compounds require a *permit*. Regardless, field sites are inspected and records audited by APHIS officers.
2. **Biotechnology Regulatory Services (BRS)**, part of APHIS, protects and promotes the US agricultural health by ensuring the safe development and use of agricultural biotechnology products. In June 2002, APHIS created BRS "to place increased emphasis on our regulatory responsibilities for biotechnology. However,

while BRS was established fairly recently, APHIS has a long history of regulating agricultural biotechnology products, overseeing the safe conduct of more than 10,000 field tests of genetically engineered crops and the deregulation, or removal from government oversight, of more than 60 products.

"While biotechnology holds enormous potential for reducing herbicide use, increasing crop health and production, and manufacturing medicines and industrial products, the challenges posed by biotechnology highlight the importance of the regulation of this technology. APHIS BRS is committed to ensuring a dynamic, robust regulatory system based on science and risk which ensures safe field testing and product development in the US, and is mindful of the global implications of our work." (APHIS) www.aphis.usda.gov/

Field sites are inspected and records audited by APHIS officers. Some infractions, investigation, and deliberation result in civil fines and compensation for damage or remediation. Biotechnology should *improve*, *not harm*, the environment or its people. *Currently* there is strict legislation for use and safety; however, according to some environmentalists, this may not have appeared to be the case with *initial* genetic engineering a decade ago.

Genetically Modified Organisms (GMOs)

To date, many GMOs have been approved, and seeds have received approval for planting. For example, genetically modified seeds for crops including soy, maize, and cotton are *routinely* planted in the United States. Papaya, potatoes, squash, tomatoes, and more are also produced.

In the United States, GMOs have USDA, FDA, EPA, and other *independent agency oversight* for environmental and food safety protection allowing consumers to be confident and accept usage of GMOs. However, this acceptance is not universal (Huffer 2012). Many European consumers, for example, do *not* have independent regulatory agencies which function independently from the industry that they regulate. This poses a dilemma for acceptance of GMOs by the consumer. Ultimately, it is the consumer's decision as to what they will consume; however, the food industry is responsible for promoting safe and environmentally sound practices in utilizing GMOs.

Proposition 37, the California ballot measure that would have required (by law) food companies to label genetically engineered foods was defeated in November 2012. Arguments were made, and continue to be made, by both sides. *Proponents* saw passage of Prop 37 as key to the health of Americans. *Opponents* of the Act feared that consumers would "interpret a GE label as a warning label device despite no conclusive data on hazardous effects of GM foods on the market" (Huffer 2012).

The *initial* wave of GMO concentrated on insect resistance and herbicide tolerance. The *next* wave, already seen in the United States, includes developing select attributes—fat type and so forth. The yield is functional or designer foods. In the *future*, entire manufacturing facilities may be dedicated to the production of genetic materials for medical, pharmaceutical, and foods use. (IdentiGEN Genetic Testing Services)

Today, GMOs are banned in *organic* agriculture. The "Non-GMO Project Verified" seal also indicates that the food is not contaminated with GMOs. Also, non-GMO foods by law are grown without pesticides.

Appendix B

Functional Foods

Hopefully all foods are "functional" in that they provide aroma, taste, nutritive value, and perhaps "comfort." Yet, the term "functional food" indicates a different connotation—it is that those named foods provide benefit *beyond* that of basic nutrition. Functional foods may be modified by the addition of nutrients not inherent to the original counterpart (Peter Pan Peanut Butter, Fullerton, CA).

Functional foods are a newly evolving area of food and food technology (Chap. 20), which are defined as:

> Any modified food or food ingredient that may provide a health benefit beyond the traditional nutrients it contains (Jenkins 1993; Goldberg 1994a).

The term functional foods has *no* legal or general acceptance in the United States. However, it is defined by the Institute of Medicine's Food and Nutrition Board (IOM/FNB), and is accepted by some, as a modified food or food ingredient for specified health use (Goldberg 1994b; Hasler 1998; Sloan 2000).

The IFIC defines such foods as "...foods or dietary components ..."; "foods that provide health benefits beyond basic nutrition" (International Food Information Council (IFIC)). (So, it may be seen that "beyond" is the operative word!)

According to the IFIC report in "Functional Foods" the *simplest* functional foods are *unmodified* foods such as fruits and vegetables (that Americans do not eat enough of by the way!) (Chap. 7). Examples include broccoli, garlic, oats, purple grapes, soy food, tea, and tomatoes. For example, tomatoes are rich in the food ingredient lycopene, and carrots are rich in beta-carotene. Other functional foods may be *modified* foods including *fortified* foods, and foods *enriched* with components such as phytochemicals. Thus these foods are supportive of health beyond basic nutrition.

The idea of functional foods originated in Japan, in the mid 1980s. Foods processed to contain specific ingredients significant to health and disease prevention were studied. The aim was solving medical problems such as high blood pressure. Today *in Japan*, products must meet eligibility requirements of the Japanese Ministry of Health and Welfare to bear the approval stamp *FOSHU—Foods for Specified Health Use*. Today, *in the United States*, the functional foods category is *not* recognized legally however; many foods are created to target diseases such as cancer, diabetes, heart disease, hypertension, and more.

Such foods have been associated with the treatment and/or prevention of other medical maladies including neural tube defect and osteoporosis, as well as abnormal bowel function and arthritis (International Food Information Council (IFIC)).

Use of Functional Foods

While research shows beneficial properties of specific substances, such as iron and vitamins, their *survival* in the food manufacturing process and their *contribution* to appearance, texture, and flavor are *also* important considerations. Usage of functional food components by an individual and/or company must consider the risk:benefit ratio, and follow acceptable scientific guidelines with regard to toxicity (ADA Position of the American Dietetic Association 1995). With availability of these types of foods, greater health benefits may be provided by the formulation of food products with added nutrients/nutrient combinations (Pszczola 1998).

Functional Foods

Volume 109, Issue 4, Pages 735–746 (April 2009)

This position paper has expired and it has been reaffirmed to be updated. The updated position paper is under development.

Functional Food Abstract

All foods are functional at some physiological level, but it is the position of the American Dietetic Association (ADA) that functional foods that include whole foods and fortified, enriched, or enhanced foods have a potentially beneficial effect on health when consumed as part of a varied diet on a regular basis, at effective levels. ADA supports research to further define the health benefits and risks of individual functional foods and their physiologically active components. Health claims on food products, including functional foods, should be based on the Significant Scientific Agreement (SSA) standard of evidence and ADA supports label claims based on such strong scientific substantiation. Food and nutrition professionals will continue to work with the food industry, allied health professionals, the government, the scientific community, and the media to ensure that the public has accurate information regarding functional foods and thus should continue to educate themselves on this emerging area of food and nutrition science. Knowledge of the role of physiologically active food components, from plant, animal, and microbial food sources, has changed the role of diet in health. Functional foods have evolved as food and nutrition science has advanced beyond the treatment of deficiency syndromes to reduction of disease risk and health promotion. This position paper reviews the definition of functional foods, their regulation, and the scientific evidence supporting this evolving area of food and nutrition. Foods can no longer be evaluated only in terms of macronutrient and micronutrient content alone. Analyzing the content of other physiologically active components and evaluating their role in health promotion will be necessary. The availability of health-promoting functional foods in the US diet has the potential to help ensure a healthier population. However, each functional food should be evaluated on the basis of scientific evidence to ensure appropriate integration into a varied diet.

Academy of Nutrition and Dietetics. Formerly The ADA.

Thus, functional foods include those foods whose nutritional value is enhanced by natural ingredient addition, and they may offer health benefits when consumed as part of a varied diet (ADA).

Functional foods may be derived from *plant and animal sources*. The Scientific Status Summary of the Institute of Food Technologists reviewed the literature for the primary plant and animal foods linked with healthful benefits. The review focused on foods, rather than specific compounds isolated from foods. (Scientific Status Summary of the Institute of Food

Technologists Functional Foods: Their Role in Disease Prevention and Health Promotion. *Food Technology* 1998. 52 (2): 57–62.)

"Although the term 'functional foods' may not be the ideal descriptor for this emerging food category, focus-group research conducted by the IFIC showed that this term was recognized more readily and was also preferred by consumers over other commonly used terms such as 'nutraceutical' or 'designer foods.' Widespread use and general acceptance of the term 'functional foods' by the media, scientists, and consumers have led the ADA to work within this framework rather than introduce a new, more descriptive term."

Several parties identified in the following have contributed to input in public hearings.

- The Institute of Food Technologists (IFT)
  - Advocate of functional food category for several years. A *new* category is needed so

that food marketers could describe items as functional foods, as long as labels were reflective of scientific evidence.
  - "Under existing regulatory policies, some food label claims cannot be factual and still accurately represent the science. This limits the scope and accuracy of consumer information and hinders the development and marketing of functional foods." (IFT)
- The Center for Science in the Public Interest (CSPI).
  - Wary of the creation of a new category. In *theory* it may help consumers. In *practice*, industry and government regulators may not make it workable.
  - "The food industry is pressuring the ... Administration to extend already weak standards for ... ingredients and label claims ... about as dependable as nineteenth-century snake oil."

  Functional Food Center (Dallas, TX) has adopted a new definition of functional foods: **Functional Food** is a "natural or processed food that contains known or unknown biologically-active compounds; which in defined quantitative and qualitative amounts, provide a clinically proven and documented health benefit, and thus are important sources in the prevention, management and treatment of chronic diseases in the modern age." (2013 http://www.functionalfoodscenter. "Providing research expertise for further development of functional food innovations.")

See also: http://www.mdheal.org/articles/word2/functionalfoods2.htm Functional Foods. Leo Galland—Director, Foundation for Integrated Medicine

Functional Foods Fact Sheet: Probiotics and Prebiotics (IFIC)

Wise food choices may increase control of personal health. Terms such as "functional foods" or "nutraceuticals" (Further discussed in the next Appendices) are widely used in the marketplace. Such foods are regulated by FDA under the authority of the Federal Food, Drug, and Cosmetic Act, even though they are not specifically defined by law.

Appendix C

Nutraceuticals

Nutraceuticals is the name given to a proposed new regulatory category of food components that may be considered a food or part of a food. Although they may supply medical or health benefits including the treatment or prevention of disease, the FDA does *not* recognize the term. Such foods, as "functional foods" are regulated by FDA under the authority of the Federal Food, Drug, and Cosmetic Act, even though they are not specifically defined by law.

The term nutraceutical was originally defined by Dr. Stephen L. DeFelice, founder and chairman of the Foundation of Innovation Medicine (FIM), New Jersey. Since the term was initially coined by Dr. DeFelice, its meaning has been modified. The word is created from using the words *nutrition* and *pharmaceutical*.

A ***nutraceutical*** is *not* a food or drug; therefore, it is not recognized by the FDA. It falls outside FDA regulations because of the following: ***Foods*** are defined as "products primarily consumed for their taste, aroma, or nutritive value." The category of food is further divided into *conventional food* and *dietary supplements*. ***Drugs*** are defined as "intended for use in the diagnosis, cure, mitigation, treatment or prevention of disease or to affect the structure or a function of the body." Unlike prescription drugs or over-the-counter medicines, dosages and composition of some nutraceuticals do *not* need to meet a quality control standard. Of course this leads to skepticism and may be harmful to users who believe and follow claims made by the manufacturer of nutraceuticals.

Nutraceuticals are defined by the Foundation for Innovation in Medicine as:

> Any substance that may be considered a food or part of a food and provides medical or health benefits, including the prevention or treatment of disease. Nutraceuticals may range from isolated nutrients, dietary supplements, and diets to genetically engineered 'designer' foods, herbal products, and processed products, such as cereals, soups, and beverages (Report 2001).

So, they may range from isolated nutrients to processed food products with a lot in between! Foods may also be known as *designer foods* or may even be referred to as *functional foods*.

According to the *American Nutraceutical Association*, nutraceuticals are functional foods with properties which are potentially disease-preventing and health-promoting. They also include naturally occurring dietary substances in forms similar to pharmaceutical dosages—capsules, etc., and "dietary supplements" as defined by the Dietary Supplement Health and Education Act of 1994 (DSHEA).

The Nutraceuticals Institute is a joint partnership of Rutgers (State University of New Jersey) and St. Joseph's Philadelphia Jesuit University. Nutraceuticals are defined as "natural, bioactive chemical compounds that have health promoting, disease preventing or medicinal properties." Their mission is to involve universities, government and industry in research, development of safe products and link with the health care industry, and develop markets.

Appendix D

Phytochemicals

Phytochemicals (phyto = plant) are important non-nutrients in food that may be responsible for disease prevention such as reduction of cancer. While many frequently consumed foods including grains, legumes, seeds, fruits and vegetables, as well as green tea are *naturally* a source of phytochemicals, a product may contain *added* phytochemicals (Chaps. 7 and 17). If added, the label must state on the food package that the product contains phytochemicals; however, no nutritional claim may be made other than stating the already approved (Table 20.2) nutritional or medical benefits that are based on sound scientific data.

Therefore, whether naturally available in the diet, added, or in supplement form, phytochemicals are defined as:

> Substances found in edible fruits and vegetables that may be ingested by humans daily in gram quantities and that exhibit a potential for modulating human metabolism in a manner favorable for cancer prevention (Jenkins 1993).

Examples: The list is long of the many examples of such plant chemicals. It includes the following:

- Carotenoids—beta-carotene, orange- and yellow-pigmented, and green leafy vegetables
- Flavonoid group of pigments—many fruits and vegetables
- Indoles, isothiocyanates—cruciferous (“cross-shaped blossom,” cabbage family) vegetables
- Isoflavones—soybeans, tofu
- Limonoids—citrus
- Lycopene—tomatoes
- Phenols—many fruits and vegetables
- Polyphenols—grapes, green tea, red wine
- Protease inhibitors—beans
- Saponins—legumes – beans and peas
- Sterols—broccoli, cabbage, cucumbers, egg plant, peppers, soy, whole grains
- Sulfur-containing allyl sulfide and sulforaphane—garlic, leeks, onion
- Terpenes—cherries, citrus peel

These are among the plant chemicals that may be effective in disease prevention.

http://lpi.oregonstate.edu/infocenter/phytochemicals/flavonoids/#disease_prevention

Appendix E

Medical Foods

Medical foods are regulated by the FDA Office of Special Nutritionals on a case-by-case basis. They are used as enteral foods (not administered into a vein parenterally, but not traditional foods) to improve nutritional support of the hospitalized patient. In 1988, Congress provided the first legal definition of "medical food" as food formulated to be consumed or administered enterally under the supervision of a physician and which is intended for the specific dietary management of a disease or condition for which distinctive nutritional requirements based on recognized scientific principles are established by medical evaluation (U.S. Congress 1988).

The medical foods can be ingested via tube feeding or the mouth and are strictly foods designed to meet specific nutritional requirements for people diagnosed with specific illnesses.

Medical foods may either supplement the diet or be the sole source of nutrition and are used based on medical evaluation. Currently, such medical foods are not available "over-the-counter" and may not be subject to NLEA labeling regulations, as they are not considered the same as foods for special dietary use. The fact that both categories of foods often overlap poses new FDA policy/regulatory discussion.

The FDA regulates medical foods and considers such foods to be "formulated to be consumed or administered internally under the supervision of a physician, and which is intended for the specific dietary management of a disease or condition for which distinctive nutritional requirements, on the basis of recognized scientific principles, are established by medical evaluation" (Huffer 2012). Nutraceuticals and dietary supplements do not meet these distinctive nutritional requirements and are not classified as Medical Foods.

The USDA recognizes medical foods as non-prescription nutrition used for dietary management of a disease or condition. It needs to be noted that such foods are not the same as reduced-fat or low-sodium for example. They are not used by the general public and are not available in supermarkets.

Hippocrates once said "Let food be thy medicine, and medicine be thy food."

Appendix F

ChooseMyPlate.gov

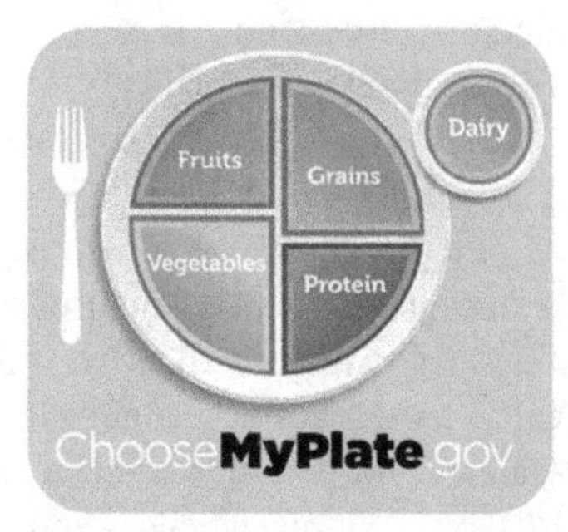

MyPlate illustrates the five food groups that are the building blocks for a healthy diet using a familiar image—a place setting for a meal. Before you eat, think about what goes on your plate or in your cup or bowl. To learn more about building a healthy plate, select a food group below. Vegetables, Fruits, Grains, Dairy, Protein Foods

10 Tips Nutrition Education Series

The Ten Tips Nutrition Education Series provides consumers and professionals with high quality, easy-to-follow tips in a convenient, printable format. These are perfect for posting on a refrigerator.

These tips and ideas are a starting point. You will find a wealth of suggestions here that can help you get started toward a healthy diet. Choose a change that you can make today, and move toward a healthier you. These tips are also available in Spanish.

- ChooseMyPlate [color; b/w]
- Add More Vegetables to Your Day [color; b/w]
- Focus on Fruits [color; b/w]
- Make Half Your Grains Whole [color; b/w]
- Got Your Dairy Today? [color; b/w]
- With Protein Foods, Variety Is Key [color; b/w]
- Build a Healthy Meal [color; b/w]
- Healthy Eating for Vegetarians [color; b/w]
- Smart Shopping for Veggies and Fruits [color; b/w]
- Liven up Your Meals with Vegetables and Fruits [color; b/w]
- Kid-Friendly Veggies and Fruits [color; b/w]
- Be a Healthy Role Model for Children [color; b/w]
- Cut Back on Your Kid's Sweet Treats [color; b/w]
- Salt and Sodium [color; b/w]
- Eat Seafood Twice a Week [color; b/w]
- Eating Better on a Budget [color; b/w]
- Use SuperTracker Your Way [color; b/w]
- Enjoy Your Food, But Eat Less [color; b/w]
- Make Better Beverage Choices [color; b/w]
- Make Celebrations Fun, Healthy & Active [color; b/w]
- The School Day Just Got Healthier [color; b/w]
- Choosing Whole-Grain Foods [color; b/w]
- Be Food Safe [color; b/w]
- MyPlate Snack Tips for Parents [color; b/w]
- Healthy Eating for an Active Lifestyle [color; b/w] **NEW**
- Be Choosey in the Dining Hall [color; b/w] **NEW**
- Mini-Fridge Makeover [color; b/w] **NEW**
- Stay Fit on Campus [color; b/w] **NEW**
- Be an Active Family [color; b/w] **NEW**
- Be Active Adults [color; b/w] **NEW**

Vegetables	Fruits	Grains	Dairy	Protein foods
Eat more red, orange, and dark-green veggies like tomatoes, sweet potatoes, and broccoli in main dishes	Use fruits as snacks, salads, and desserts. At breakfast, top your cereal with bananas or strawberries; add blueberries to pancakes	Substitute whole-grain choices for refined-grain breads, bagels, rolls, breakfast cereals, crackers, rice, and pasta	Choose skim (fat-free) or 1 % (low-fat) milk. They have the same amount of calcium and other essential nutrients as whole milk, but less fat and calories	Eat a variety of foods from the protein food group each week, such as seafood, beans and peas, and nuts as well as lean meats, poultry, and eggs
Add beans or peas to salads (kidney or chickpeas), soups (split peas or lentils), and side dishes (pinto or baked beans), or serve as a main dish	Buy fruits that are dried, frozen, and canned (in water or 100 % juice), as well as fresh fruits	Check the ingredients list on product labels for the words "whole" or "whole grain" before the grain ingredient name	Top fruit salads and baked potatoes with low-fat yogurt	Twice a week, make seafood the protein on your plate
Fresh, frozen, and canned vegetables all count. Choose "reduced sodium" or "no-salt-added" canned veggies	Select 100 % fruit juice when choosing juices.	Choose products that name a whole grain first on the ingredients list	If you are lactose-intolerant, try lactose-free milk or fortified soymilk (soy beverage)	Choose lean meats and ground beef that are at least 90 % lean
				Trim or drain fat from meat and remove skin from poultry to cut fat and calories
For a 2,000-calorie daily food plan, you need the amounts below from each food group. To find amounts personalized for you, go to Choose**MyPlate**.gov				
Eat 2½ cups every day	**Eat 2 cups every day**	**Eat 6 ounces every day**	**Get 3 cups every day**	**Eat 5½ ounces every day**
What counts as a cup? 1 cup of raw or cooked vegetables or vegetable juice; 2 cups of leafy salad greens	What counts as a cup? 1 cup of raw or cooked fruit or 100 % fruit juice; ½ cup dried fruit	What counts as an ounce? 1 slice of bread; ½ cup of cooked rice, cereal, or pasta; 1 ounce of ready-to-eat cereal	What counts as a cup? 1 cup of milk, yogurt, or fortified soymilk; 1½ ounces natural or 2 ounces processed cheese	What counts as an ounce? 1 ounce of lean meat, poultry, or fish; 1 egg; 1 Tbsp peanut butter; ½ ounce nuts or seeds; ¼ cup beans

10 tips
Nutrition Education Series

choose MyPlate

10 tips to a great plate

Making food choices for a healthy lifestyle can be as simple as using these 10 Tips. Use the ideas in this list to *balance your calories*, to choose foods to *eat more often*, and to cut back on foods to *eat less often*.

1 balance calories

Find out how many calories YOU need for a day as a first step in managing your weight. Go to www.ChooseMyPlate.gov to find your calorie level. Being physically active also helps you balance calories.

2 enjoy your food, but eat less

Take the time to fully enjoy your food as you eat it. Eating too fast or when your attention is elsewhere may lead to eating too many calories. Pay attention to hunger and fullness cues before, during, and after meals. Use them to recognize when to eat and when you've had enough.

3 avoid oversized portions

Use a smaller plate, bowl, and glass. Portion out foods before you eat. When eating out, choose a smaller size option, share a dish, or take home part of your meal.

4 foods to eat more often

Eat more vegetables, fruits, whole grains, and fat-free or 1% milk and dairy products. These foods have the nutrients you need for health—including potassium, calcium, vitamin D, and fiber. Make them the basis for meals and snacks.

5 make half your plate fruits and vegetables

Choose red, orange, and dark-green vegetables like tomatoes, sweet potatoes, and broccoli, along with other vegetables for your meals. Add fruit to meals as part of main or side dishes or as dessert.

6 switch to fat-free or low-fat (1%) milk

They have the same amount of calcium and other essential nutrients as whole milk, but fewer calories and less saturated fat.

7 make half your grains whole grains

To eat more whole grains, substitute a whole-grain product for a refined product—such as eating whole-wheat bread instead of white bread or brown rice instead of white rice.

8 foods to eat less often

Cut back on foods high in solid fats, added sugars, and salt. They include cakes, cookies, ice cream, candies, sweetened drinks, pizza, and fatty meats like ribs, sausages, bacon, and hot dogs. Use these foods as occasional treats, not everyday foods.

9 compare sodium in foods

Use the Nutrition Facts label to choose lower sodium versions of foods like soup, bread, and frozen meals. Select canned foods labeled "low sodium," "reduced sodium," or "no salt added."

10 drink water instead of sugary drinks

Cut calories by drinking water or unsweetened beverages. Soda, energy drinks, and sports drinks are a major source of added sugar, and calories, in American diets.

Go to www.ChooseMyPlate.gov for more information.

DG TipSheet No. 1
June 2011
USDA is an equal opportunity provider and employer.

Appendix G

Food Label Health Claims

Health claims describe an association between a nutrient or food substance and disease or health-related condition. Food processors may choose to use any additive, including nutrients or non-nutrient supplements, in the manufacture of food products. Regardless of what is used, they must comply with all Nutrition Labeling and Education Act (NLEA) regulations regarding the contents and stated **health claims** of their products. They must use vitamin and mineral additives judiciously (not just to enhance the values on their food label), and then only make label claims regarding nutritional benefits that are allowed (Chap. 20). A research dietitian with the USDA Agricultural Research Services has said:

> Although compounds in foods that must be concentrated to obtain physiologic effects should be regulated as drugs, foods and purified food constituents in amounts commonly consumed should not be classified as drugs simply because they are being tested for potential health effects or disease prevention. Research should not be discouraged by requiring investigative new drug procedures for substances in amounts available in the diet. (USDA)

Improving nutrient density may be researched more via reading (Berry 2012)

Examples of Approved Health Claims

- Calcium and lower risk of osteoporosis
- Sodium and a greater risk of hypertension (high blood pressure)
- Saturated fat and cholesterol and a greater risk of coronary heart disease (CHD)
- Dietary fat and a greater risk of cancer
- Fiber-containing grain products, fruits, and vegetables and a reduced risk of cancer
- Fruits, vegetables, and grain products that contain fiber (particularly soluble fiber) and a reduced risk of CHD
- Fruits and vegetables and a reduced risk of cancer
- Folate and reduced risk of neural tube defect
- Sugar alcohols and reduced risk of tooth decay
- Soluble fiber from whole oats and psyllium seed husk and reduced risk of CHD
- Soy protein and reduced risk of CHD
- Whole grains and reduced risk of CHD and certain cancers
- Plant sterol and plant stanol esters and reduced risk of CHD
- Potassium and reduced risk of high blood pressure and stroke

The FDA allows health claims to be one of three types:

1. Unqualified health claims (allowed since 1993). The claim must meet the SSA standard
2. Qualified health claim (2003). SSA not met. Use of the term "may."
3. Structure/function claims. The effect that a substance has on the structure or function of the body—not a specific disease. That is, calcium and strong bones

Appendix H

Research Chefs Certification as a Culinary Scientist and More

The Research Chefs Association (RCA) certifies that food science, as well as culinary knowledge, is held by an individual. www.researchchef.org or http://www.culinology.com

The Research Chefs Association Certification Commission (RCACC) was founded in 2003. It was to "promulgate policies, procedures and criteria which will enhance the certification process for Certified Research Chefs (CRCs) and Certified Culinary Scientists (CCSs). To guide its activities, the RCACC strives to meet National Commission for Certifying Agencies (NCCA) standards." (RCA)

According to the RCA, it is "the leading professional community for food research and development. Its members are the pioneers of the discipline of Culinology® - the blending of culinary arts and the science of food."

Certified Research Chef Eligibility (CRC)

In order to be eligible to become a Certified Research Chef, applicants first must meet eligibility criteria in the categories of Education, Food Service Experience, and Research and Development Experience. Having done so, candidates then must pass a certification exam on their knowledge of food science and related subjects. (RCA)

Certified Culinary Scientist Eligibility (CCS)

In order to be eligible to become a Certified Culinary Scientist, applicants must first meet eligibility criteria in the categories of Education, Food Science Experience, and Food Service Experience. Having done so, candidates then must pass a certification exam on their knowledge of culinary arts and related subjects. (RCA)

RCACC Approved Definitions for Eligibility

A *Research Chef* for purposes of certification is defined as one who works in food product development, has expertise in culinary arts, and a baseline knowledge of food science.

A *Culinary Scientist* for purposes of certification is defined as one who works in food product development, has expertise in food science/technology, and a baseline knowledge of culinary arts. *Food Science Related Degrees* include Culinology®, Food Technology, Microbiology, Chemistry, Nutrition, Biochemistry, Meat Science, Dairy Science, Cereal Science, Biology, Fish Science, Poultry Science, and Food Engineering. (RCA)

Appendix I

Human Nutrigenomics

By definition *genomics* refers to the *complete* genetic makeup of an organism. Some human genomic test assessments began in the 1980s, and are now becoming more available. Results may show a patient's individual risk of disease or its recurrence, which can help medical care providers and patients make better informed and more personalized treatment decisions.

Nutrigenomics is the new science of *nutritional genomics*. It is the application of the science of genomics to human nutrition—and it views the relationship between nutrition and health. Research in nutrigenomics covers *cellular and molecular processes* and the relation to many diseases, including degenerative diseases such as atherosclerosis and cancers, as well as aging. Research is focused on the *prevention* of disease and requires understanding of nutrient-related interactions at the level of the gene.

Nutrigenomics can be approached in more than one way. For instance, personalized nutrition may be offered to people based on small differences in their genome (single nucleotide polymorphisms, or SNPs) compared to another person. Nutrigenomics can also be considered when molecules in certain foods have the ability to change the expression of genes in an individual, by increasing or decreasing the expression of a gene into a protein. An example of each of these is given below.

Personalized nutrition may be used as evidence accumulates that certain foods may be harmless to some people yet detrimental to others. For instance, some people have a version of a gene that codes for a protein that makes them susceptible to myocardial infarction (MI) with intake of caffeine because they metabolize caffeine slowly, while others with a different version of the same gene make a protein that metabolizes caffeine fast and their risk for MI is lowered. People with the slow version gene could be counseled about their increased risk for MI with caffeine intake (Cornelis et al. 2006).

Food also comes into play when considering gene expression in general. Not all genes are expressed in all tissue; only those genes necessary for that tissue are expressed into proteins. Researchers are finding that certain molecules in foods can affect the expression of certain genes. For instance, research has shown that vitamin D can increase the expression of genes that code for anti-inflammatory proteins and decrease the expression genes coding for pro-inflammatory proteins, thus helping with chronic inflammatory conditions such as autoimmune disease. Vitamin D is fat-soluble and should not be taken in excess (Mark & Carson 2006).

(This Nutrigenomics Appendix I is written with the assistance of B.L. Mark Ph.D., R.D.)

Appendix J

Product Development: Innovation

"Innovation is alive and well in the food industry, albeit within the strictures of consumer acceptance—along with government regulation and public safety of course" reports the editor of *New Product Design* magazine. (Kuntz, LA. Innovation in the food industry. *New Product Design*. 2012. October: 12.)

Glossary

Biotechnology Biogenetic engineering of animals, microorganisms, and plants to alter or create products that have increased resistance to pests, improved nutritive value, and shelf life.

Culinary Scientist For certification—defined as one who works in food product development, has expertise in food science/technology, and baseline knowledge of culinary arts.

Drugs Intended for use in the diagnosis, cure, mitigation, treatment, or prevention of disease or to affect the structure or function of the body.

Enrichment The addition of nutrients to achieve established concentrations specified by the standards of identity.

Foods Products primarily consumed for their taste, aroma, or nutritive value.

Fortification The addition of nutrients at levels higher than those found in the original or comparable food.

Functional foods Any modified food or food ingredient that may provide a health benefit beyond that obtained by the original food; the term has no legal or general acceptance in the United States, although it is accepted by some as food for specified health use.

Genetically Modified Organisms (GMOs) Genetically modified seeds for crops.

Medical foods Food formulated to be consumed or administered enterally under the supervision of a physician and which is intended for the specific dietary management of a disease or condition for which distinctive nutritional requirements based on recognized scientific principles are established by medical evaluation (U.S. Congress 1988).

Nutraceuticals The name given to a proposed new regulatory category of food components that may be considered a food or part of a food and may supply medical or health benefits including the treatment or prevention of disease; a term not recognized by the FDA.

Nutrigenomics The new science of *nutritional genomics*. It is the application of the science of genomics to human nutrition.

Phytochemicals Plant chemicals; natural compounds other than nutrients in fresh plant material that function in disease prevention; they protect against oxidative cell damage or facilitate carcinogen excretion from the body and exhibit a potential for reducing the risk of cancer.

References

Huffer, L. (2012) California to vote on GMO labels for foods. New Product Design. October:14–15

Jenkins MLY (1993) Research issues in evaluating "functional foods". Food Technol 47(5):76–79

Goldberg I (ed) (1994a) Functional foods: designer foods, pharmafoods, nutraceuticals. Chapman and Hall, New York

Goldberg I (1994b) Functional foods. Chapman and Hall, New York

Hasler CM (1998) Functional foods: their role in disease prevention and health promotion. A publication of the Institute of Food Technologists' expert panel on food safety and nutrition. Food Technol 52 (11):63–70

Sloan AE (2000) The top ten functional food trends. Food Technol 54(4):33–62

ADA Position of the American Dietetic Association (1995) Phytochemicals and functional foods. J Am Diet Assoc 95(4):493–496

Pszczola DE (1998) Addressing functional problems in fortified foods. Food Technol 52(7):38–46

Staff Report (2001) Combining nutrients for health benefits. Food Technol 55(2):42–47

Berry D. (2012) Improving nutrient density. New Product Development: 60–66

Cornelis MC, El-Sohemy A, Kabagambe EK, Campos H (2006) Coffee, CYP1A2 genotype, and risk of myocardial infarction. JAMA 295:1135–1141

Mark BL, Carson JA (2006) Vitamin D and autoimmune disease: implications for practice from the multiple sclerosis literature. J Am Diet Assoc 106:418–424

Index

V.A. Vaclavik and E.W. Christian, *Essentials of Food Science, 4th Edition*, Food Science Text Series,
DOI 10.1007/978-1-4614-9138-5, © Springer Science+Business Media New York 2014

N

CPSIA information can be obtained
at www.ICGtesting.com
Printed in the USA
LVHW011909070719
623369LV00002B/6/P

9 781461 491378